Atle Ottesen Søvik

A Basic Theory of Everything

Atle Ottesen Søvik

A Basic Theory of Everything

A Fundamental Theoretical Framework for Science and Philosophy

DE GRUYTER

ISBN 978-3-11-077092-6
e-ISBN (PDF) 978-3-11-077095-7
e-ISBN (EPUB) 978-3-11-077104-6
DOI https://doi.org/10.1515/9783110770957

Library of Congress Control Number: 2021953515

Bibliographic information published by the Deutsche Nationalbibliothek
The Deutsche Nationalbibliothek lists this publication in the Deutsche Nationalbibliografie; detailed bibliographic data are available on the Internet at http://dnb.dnb.de.

Printing and binding: CPI books GmbH, Leck

www.degruyter.com

Acknowledgments

It has been a great joy to write this book. There are probably many relevant things I do not know that I do not know, but which I hope to learn more about from criticism and discussions of this book. Many thanks to the following people for reading either the whole manuscript or parts of it, and offering many valuable comments: Lorenz Puntel, Sivert Ellingsen, Øystein Elgarøy, Anders Kvellestad, Einar Duenger Bøhn, Hedda Hassel Mørch, Anders Strand, Øystein Linnebo, Andreas Brekke Carlsson, Alan Padgett, Asle Eikrem, Ragnar Misje Bergem, Svein Jåvold, Hans Robin Solberg, Katarina Pajchel, Andreas Hvidsten, Robert Hartman, Michael Mørch, Sorin Bangu, Alan White, John Daniel Andersen, Jarle Frette, Tollef Graff Hugo, Morten Magelssen, Martin Jakobsen, Gunnar Björnsson, Manuel Vargas, Olav Søvik, Martin Søvik, Michael Gisinger, Frode Bjørdal, and some anonymous referees. Thanks to De Gruyter for publishing the book and for excellent service in the process (especially from Mara Weber and Konrad Vonderobermeier).

As always, I am grateful to my family for all the meaning they bring into my life: Andreas, Jenny, Kristian and Elise.

Bærum, Norway, December 2021 Atle Ottesen Søvik

https://doi.org/10.1515/9783110770957-001

Contents

Part Two: **The Mind**

Part Three: **The World**

List of Abbreviations

AI	Artificial intelligence
BWBW	Best way to the best world
CV	Curriculum vitae
DNA	Deoxyribonucleic acid
EEA	European Economic Area
EPR paradox	Einstein-Podolsky-Rosen paradox
GPS	Global Positioning System
GR	General theory of relativity
GRW	Ghirardi-Rimini-Weber theory
IIT	Integrated information theory
NATO	North Atlantic Treaty Organization
PBR theorem	Pusey-Barrett-Rudolph theorem
PSC	Physical substrates of consciousness
RNA	Ribonucleic acid
SFA	Self-forming action
SI	Système international d'unités
SR	Special theory of relativity
UN	United Nations
UNSDSN	United Nations Sustainable Development Solutions Network
WMAP	Wilkinson Microwave Anistropy Probe
WWII	World War II
ZFC	Zermelo-Fraenkel set theory with the Axiom of Choice

https://doi.org/10.1515/9783110770957-002

Part One: **The Theoretical Framework**

1 Introduction

There are many classical and unsolved problems in philosophy. Some examples: What is truth? How does the mind work? Do we have free will? What are the basic constituents of the world? What is time? What is goodness? And so on.

There are many different answers to these questions, but not only are there different answers, there are quite different ways of understanding the questions and their presuppositions. There are different theoretical frameworks, which means that there are different ways to conceptualize and categorize the basic structures and contents of the world. These theoretical frameworks determine how questions are understood and what are considered to be good or bad answers.

Different theoretical frameworks can be used to give different answers to problems, but in the end all theoretical frameworks will come to a point where no further explanation can be given. At these points, something is often said to be either a brute fact or something irreducible, and I will explain later why no theory can explain everything.

An interesting fact to note is that while some theoretical frameworks say that we have come to an irreducible entity that cannot be further explained, other theoretical frameworks may say that the entity in question does not exist at all, or that the whole problem is a misunderstood pseudo-problem.

Here are some examples (and I will get back to details on all of them): Different philosophers have different theories about what a cause is and what an effect is and how they are related. Others will say that there are no such things as causes and effects or an influence between them and that these concepts are superfluous; they are just imprecise ways of expressing something else that is really happening. While some find it important to discuss how to understand the influence from cause to effect or how absences can be causes, others will say it is pointless.

Another example: Different philosophers have different theories of free will. Some will say that being an agent is something irreducible and that agents cause their actions through an irreducible form of agent causation. Other philosophers will deny that there are agents in this sense at all, that there is agent causation at all, or that there is free will or responsibility. Instead, everything is just physical processes in the brain, or something else. While some find it interesting to discuss how the agent can have a will that is free, others find it pointless as they claim there is no such thing as a will that can be free.

A common situation in philosophy is thus that people working within one theoretical framework can present something as a profound problem and sug-

https://doi.org/10.1515/9783110770957-003

gest that something is the deepest answer that can be given, while people working in another theoretical framework can dismiss the whole problem as a pseudo-problem and say that the suggested answer refers to something that does not exist at all. Something that seems like an unanswered question about the world could be nothing but a problem resulting only from an incoherent theoretical framework introducing vague concepts that may not refer to anything at all.

Here are some more examples of problems where there is disagreement on whether there is a problem to solve at all: Should we try to solve the problem of how the present can move forward in time, or is there no present moving forward in time at all? Should we try to understand how the soul or consciousness can move the body, or is there no soul or consciousness at all? Should we search for objective truth, or is there no objective truth? Should we try to find out what the correct reference of a sentence is, or is there no such thing as the correct reference of a sentence? Should we try to find out how the laws of nature make things move, or are there no laws of nature making things move? Should we try to find out what kind of mind-independent existence numbers and geometrical figures have, or do they have no such existence at all? Should we try to find the exact probability of an event or a theory, or are there no such exact probabilities? The list could go on.[1]

In this book I am going to argue that very many problems in both philosophy and science come from theoretical frameworks that are ultimately incoherent, and that they can be solved (sometimes by being dissolved) by being translated into a more coherent theoretical framework. I will be suggesting a theoretical framework to test on several problems and argue that translating the relevant problem into the new framework shows either how the problem can be solved or why there was no real problem to be solved at all, but instead just a question with incoherent presuppositions.

---

1 In addition to the specific questions, the question of whether we have a problem to solve might make us ask when it is reasonable to seek for an explanation of something. The probability of life in the universe is very small, and researchers tries to explain cases of so-called fine-tuning. The probability that I should be born is also very small, but nobody tries to explain why it happened, so what calls for an explanation? My own view is that as a point of departure, it is a goal to explain as much as possible. When it comes to actual resources used on finding answers, we find some questions more interesting than others. In many cases, chance is a perfectly good explanation, for example in the question of why I was born, and we see no reason to look further. In other cases, there are interesting alternatives on the table, for example in the question of fine-tuning, where multiverse hypotheses are discussed against a God hypothesis and others. Interest in the question and interesting alternatives is thus what makes us continue looking for answers, even though ideally it would be nice to know the cause of as much as possible.

Let us look at some examples of what various people think of as irreducible constituents of the world: physical particles, forces, gravity, space, time, fields, energy, mass, charge, spin, the wave function, laws of nature, causality, determinism or indeterminism, probability, modality (contingency, possibility, necessity), truth, logic, language, understanding, thinking, intentionality, subjectivity, consciousness, persons or selves, freedom, responsibility, substances, properties, relations, dispositions, essences, universals, structures, numbers, moral values, beauty – and many more examples could be added to the list. The reader may want to ask him- or herself which of these entities are irreducible and necessary ingredients in a full description of the basic structures of the world and how it works.

While some claim that these are irreducible entities, others argue that they can be ontologically reduced, which means that what the concepts refer to is nothing but something that other concepts also refer to more precisely.[2] A classic example is physical heat, which is ontologically reducible to molecular motion. That is not to say that we should get rid of the concept of physical heat, only that we can understand it even more precisely with other concepts and do not need heat in addition to explain what heat is. All theories of the world will have some basic constituents that cannot be reduced to something else, but these theories disagree on which are actually basic constituents.

Is it good to have as few irreducible entities as possible in your theory? Some are generally skeptical to reductionism, while others (myself included) find it a virtue in metaphysics to try to have as few loose ends as possible, in order to answer as many questions as possible. If you take the reductive approach, the following question is of high interest: What is the lowest number of basic constituents you need to explain everything else?[3]

---

2 Robert Bishop suggests some good distinctions concerning how to understand ontological reduction, where alternatives can be defined in terms of necessary and sufficient conditions. Reduction is often discussed in terms of two alternatives: either reduction understood as an underlying domain providing both necessary and sufficient conditions for what happens or exists at the higher level, or radical emergence, where the underlying level provides neither necessary nor sufficient conditions for what happens or exists at the higher level. Bishop suggests two alternatives in between called contextual emergence and multiple realizability, where the underlying level contains either some necessary, but no sufficient conditions, or only sufficient conditions (Bishop, 2019, p. 3–2). Like Bishop, I defend the positions in between, both because I defend indeterminism and since the result of laws depend on the contexts in which they work.

3 As will become clear below, I do not suggest that a secure foundation for knowledge can be found. Rather, any theory must be justified by its coherence. But since a theory with fewer irreducible and unexplained entities than another theory will be more coherent (other things being equal), this legitimizes a search for entities that are basic in the sense of irreducible and unex-

In this book I argue that the number is approximately four. That's right: 4 (and I will get back to why I added "approximately"). These four constituents (which will be thoroughly defined below) are (1) An area (or a field), (2) values that can be actualized at different places in the field, (3) something that actualizes the values, and (4) some rules that the actualizing follows. Everything else we have a name for (John, horse, jumping, brother, four, red, and absolutely everything else) is a structure we can discern among the values that can be actualized in the field (and even the event of discerning structures is a structure among the values in the field).

I wrote "approximately four" above, since I will return to the question of number of values and number of rules and whether these four entities can be reduced to each other. The rules will turn out to be implied by the structure of the actualizer, so one could well say that the number of basic constituents is three. But as an introductory statement, I will argue that any theory of the world must include values being actualized at a place according to rules, which means that no more basic coherent explanation can be given.[4] I will also be arguing that very many problems in philosophy and natural science get solved or dissolved (with no interesting question left unanswered) when we translate the problems into this theoretical framework. The reader should of course initially be very skeptical towards such a claim, so the book itself must then be the defense of this claim.

1.1 Intended readers and goal of the book

When writing this book, I have imagined a stereotype reader, which is a young person studying first year of philosophy wanting to know how everything in the world works – and I have been this young person myself earlier in life. Presumably, the book will be of greater interest to those who are familiar with some standard philosophical problems. But I have also noticed how many different kinds of people enjoy reading Bill Bryson's book *A Short History of Nearly Every-*

plained entities within the theory, and searching for theories with as few as possible of such entities.

4 Of course you may say that everything that exists are structures or individuals or something like that to make the number of entities one, but then they will be fundamentally different structures or individuals. We are looking for a theory that is informative and not just trivially true. As mentioned, I will discuss possible ways that even the four basic entities can be reduced to each other.

thing or Yuval Noah Harari's books *Sapiens* and *Homo Deus*, books which draws very big lines through all of history pointing out connections.

Reading such books gives a good feeling of understanding the big picture of how things work and how things are related in the world. When writing this book, I wanted to give readers the feeling that they can understand the workings of the world and the mind, and that even with just very simple mathematics, one can have a basic understanding of the relativity of time, the twin paradox, $E = mc^2$, the strange world of quantum mechanics, particle creation and annihilation, consciousness, free will, truth, and much, much more. Even if the theory I suggest in this book is wrong in different matters, at least it offers a good basis for considering other alternatives.

Another type of reader I have had in mind is a typical person working within science and getting in contact with bigger philosophical questions than what are usually dealt with within the particular discipline that he or she works within. The desire to see how things fit in a bigger picture is a desire I hope to contribute to satisfying. The theoretical framework in the book should at least fit well with your standard natural science.

There is not much history, context or background to each topic (which would have made the text extremely long), but instead the text is guided by numerous questions, to let the reader see how all these questions can be answered in a coherent way and get the feeling of understanding. Of course I may be wrong in many of the answers I suggest and still there will be questions that are not answered, but at least the book gives a unified suggestion of how to understand very many different basic things. It is not common for a book to cover such a large amount of topics, but it is a kind of book that I know I have been looking for, and maybe someone else have too.

The text may be longer than preferred by many, but I wanted it to be thorough enough not to be immediately rejected by people knowing the topics. If someone reads about these topics elsewhere after reading this book, I hope to have written enough about the topics here to be a serious discussion partner with other positions.

There are relatively more figures and details in the parts on physics than in those on ethics and politics. This is partly because the book has a greater focus on fundamental building blocks than on macro structures in society. Partly it is also because I assume that the general reader needs more help with understanding mathematics and physics than ethics and politics.

I realize that the outline of the book may be a bit confusing. The reason is that I want to do three things that are connected, in an order that makes it understandable, but with the unavoidable result that the big picture does not get really clear until the end. Some parts of the text are probably easier to understand if they are read again after having read the whole book. The three things I want to do is to answer a lot of different questions, but also to show how the answers are related and support each other, and finally to show how many things are reducible to the theoretical framework presented in Chapter 3. The topics are sorted in a way that let me explain first things I need in order for later explanations to make sense, but since everything is related, everything also gets gradually clearer.

1.2 Outline of the book

Here follows a detailed outline of the book. At the end of this section, there is a figure which summarizes the main points.

In Part One we look at what a "theoretical framework" means at all (in Chapter 2) before I present the specific theoretical framework that the book will be defending (in Chapter 3). Chapter 2 deals with several methodological topics in addition to the concept of theoretical frameworks. What is it that we do when we understand something? How can our mind understand the world outside of the mind? What is truth and how do we know whether something is true?

I argue that we understand something by placing it at a certain place in a certain theoretical framework and relating it to other entities in the framework. A theoretical framework does not have to be words and sentences. Just relating images in our mind to each other without words is, in a broad sense, a theoretical framework and, in the deepest sense, what it means to understand something: to relate it to something else.

Understanding happens in the mind, but the mind wants to understand the rest of the world which is not mind (the world "outside" of the mind). However, we have no access to the world which is not mediated through our mind. How can the mind then discover something true about the world when there seems to be an unbridgeable gap between them?

The required link between mind and world is truth. I argue that the mind discovers what is true about the world by discovering what is the most coherent understanding of our experiences with the world. That a theory is coherent means that it is consistent, but also that it is more coherent the more data it is able to integrate and the more connections it is able to explicate between the data. I explain how coherence is a criterion of truth, which means that the most coherent theoretical framework is also the one best justified as true.

In Chapter 3 I present my own theoretical framework describing the basic ontological constituents of the world. This will be the theoretical framework used to answer all the other questions in the rest of the book. I use the terms "ontology" and "metaphysics" interchangeably simply to mean a theory of the most fundamental structures of the world.

I start by presenting some background knowledge from quantum field theory in physics. In this theory, there is a field for every elementary particle in the world, which includes the forces. Every field has its own field equation, and by using these we can find that certain qualitative values like mass, charge and spin come in different quantitative values (expressed by numbers) at different points in the field in accordance with the rules called field equations.

Similar to this picture from physics, I suggest that everything that exists is a field where qualitative values can be actualized in quantitative values according to rules. This idea is not very revolutionary when it comes to the physics, but I present a new suggestion on how to understand consciousness as a specific kind of qualia values actualized in a qualia field. Later in the book, I analyze different components of conscious experiences and how they could be structures of qualia values. This approach to understanding consciousness has some great advantages when it comes to the interaction between consciousness and the physical world, which I will say more about below.

According to the theory presented here, the basic constituents of the world are values actualized in a field according to rules. I argue that no ontology can be simpler, for any theory of the world must include something corresponding to fields, possible values, rules and actualizations, which again implies that the simplest ontology is the one employing just these basic entities and nothing more. But how should we understand more precisely the relation between these entities? What makes motion happen in the world according to rules?

I try to find the answer that gives as few unanswered questions as possible, but I also argue that whatever the most basic structure is, it has to be something quite incredible. The reason is that anything that has existed or happened in the history of the universe must have been possible from the very start, since otherwise it would not have existed or happened. That means that there must from the start have been something with an extraordinary potential for bringing forth things and events.

I find that the simplest combined understanding of the basic entities is to think that there is an actualizer which has a structure that implies that the values being actualized follow certain rules. This actualizer is again a structure in a field of possible values. This is certainly a complex structure, but it seems unavoidable to believe in the existence of its parts, and their combined existence is the simplest way of envisioning it.

Several other topics are discussed in Chapter 3. One of them is the question of existence. What is existence? What is it that all things that exist have in common? It seems very different to say that numbers exist, that horses exist, and that unicorns exist in our mind, but what is it that they have in common that legitimizes the use of "exist" in each case?

In the widest possible sense of the term "exist", anything that has a structure exists. This then includes possible structures, like mathematical entities nobody has yet thought of. However, it is more common to use existence in a narrower sense than just referring to any possible entity having a structure. In the narrower sense of existence, anything that exists can be described by several features: to exist is to be part of the fundamental structure, to be actualized, to be localized, to be registrable, and to have causal effect in a very wide sense of the term "causal". This power to have causal effect, which could be thought of as the deepest sense of existence in the narrow sense, is the power of the basic structure to actualize values according to rules.

Another question discussed is the topic of modality. The three basic modal terms are "possible", "impossible" and "necessary", but what does it mean that something is possible, impossible or necessary? Is modality an irreducible part of the world, like many philosophers have suggested?

My suggestion is that modality is a theoretical framework where we use the concepts of possible, impossible and necessary to categorize entities in the world based on an initial set of presuppositions. That something is possible means that it expresses a consistent combination given the presuppositions. That something is impossible means that it expresses an inconsistent combination given the presuppositions. That something is necessary means that it is implied in the presuppositions and thus inconsistent to deny.

Here are some examples: Physical possibility, impossibility and necessity have the laws of nature as their presuppositions. Assuming that the laws are correct: If you describe something consistent with these laws, it is physically possible. If you describe something inconsistent with these laws, it is physically impossible. If you describe something implied by these laws, it is physically necessary. For example, in our universe it is physically possible to move slower than light; it is physically impossible to accelerate to a speed faster than light; and it is physically necessary that light travels at light speed in a vacuum.

Logical possibility, impossibility and necessity have the meaning of terms as their presuppositions. If you describe something which is consistent with the meaning of the terms used in the theoretical framework, it is logically possible. If you describe something which is inconsistent with the meaning of the terms used in the theoretical framework, it is logically impossible. If you describe something which is (deductively) implied by the meaning of the terms used in the theoretical framework, it is logically necessary. For example, given that bachelor means unmarried man, it is logically possible that a bachelor is 40 years old, it is logically impossible that a bachelor is married, and it is logically necessary that a bachelor is unmarried.

Other kinds of possibility, impossibility and necessity are also dealt with in the chapter. We do not need modality as an irreducible entity on its own to describe the world. It suffices that there are rules according to which motion occurs, and that there are theoretical frameworks which can be consistent or inconsistent.

Other problems solved or dissolved in Chapter 3 are the following: What is a substance? How can a substance have properties? What is the haecceity or identity of a substance? What are universals, and how are they instantiated in particulars? What is the difference between abstract and concrete? What are dispositions and how do they work? I suggest that the concepts of substance, property, haecceity, universals and dispositions can all be ontologically reduced, and that they give rise to many pseudo-problems.

Having established a theoretical framework in Part One, we can start using it to solve problems, and in Part Two we look at problems in the philosophy of mind. How does the mind work? How can we think rational thoughts? What is consciousness and does it play a causal role in the world? Do we have free will and responsibility?

Some argue that there is something unique and irreducible to being an agent or a self who acts in the world. Persons cannot be reduced to nature; consciousness is non-physical, and we have free will and responsibility for our actions. Others argue that everything is reducible to causal processes in the brain and the body. Consciousness is physical and there are no such things as persons, selves, wills, freedom or responsibility.

In this part, I will be defending a causal theory of the mind. I must therefore start with a discussion of what causality is in Chapter 4. This will be important also in the chapter on free will, since I shall argue that free will is about being the cause of your own actions, and that how we understand the role of contrasts when selecting causes is important for understanding free will.

In Chapter 4, I discuss what causes and effects are and how they are related. What is the connection between a cause and the effect which makes the cause have the effect it has? There are two main positions in this question. According to the first view – called probability-raising – causation is something that makes something more likely to happen. For example, throwing a rock at a window is the cause of the window breaking since throwing a rock at a window makes it more probable that the window will break. According to the second view – called process linkage – causation is a physical connection between cause and effect. This physical connection can be understood for example as transfer of energy or momentum or force or something else.

My own suggestion is somewhat similar to the analysis of modality. Categorizing something as causes and effects is a theoretical framework that can be used to categorize many relations in the world in many different disciplines. For example, a biologist could say that a mutation causes a new skill for an organism, a psychologist could say that shame causes a man to blush, and an anthropologist could say that education for women causes them to have fewer children.

The theoretical framework of cause and effect is a coarse-grained and imprecise framework for analysis. There is nothing unique called causation that happens between these causes and their effects. There is no influence moving from the cause to the effect that is a physical connection or does something to raise the probability of the effect.

What happens is that we often see similar states of affairs followed by similar states of affairs and assume that this cannot be a coincidence, but rather we categorize the first state of affairs as the cause and the second state of affairs as the effect. But nothing called causation happens between the cause as cause and the effect as effect. Rather, calling them causes and effects are coarse and shortcut descriptions of something that happens at a more basic level where values are actualized according to rules. Causation is thus an efficient way of saying that B follows (is caused by) A in virtue of laws of nature interacting with (among others) (the constituents of) A. Causation as an influence from cause to effect is thus ontologically reducible, and many philosophical problems of causation are pseudo-problems, like how can an absence have a causal effect, e.g. how could not watering the flower cause the flower to die?

After having discussed in Chapter 4 what causation is, I continue in Chapter 5 with describing how the mind can be understood causally. There are different philosophers, for example agent causationists or substance dualists, who will reject the idea that we can understand the mind as a normal causal process (similar to other causal processes). In order to argue that it is superfluous to include irreducible agents or souls into one's ontology, I must present a detailed causal understanding of the mind. I present in detail the topics of mind, thinking, consciousness and free will in one chapter each in order to show how it is possible that persons thinking and making free choices can be ontologically reduced to causal processes between values actualized in fields.

In Chapter 5, I start by presenting how mind could evolve as a causal process. We see many reasons to think that causal processes in the brain cause the content of our conscious experiences. When it comes to the experience of being a self, I use Antonio Damasio to distinguish between the autobiographical self, which is a storage of memories in the brain, and the core self, which is a stream of conscious impulses.

It seems strange to think that a choice could be a causal process. Some philosophers will say that a choice means that certain motives or reasons have been considered rationally by an agent who then makes a free choice among them, and that this cannot be reduced to a normal causal process.

However, this description of agents making choices is a quite coarse-grained description of a choice. How does one of several motives get chosen and how does it lead to action? A common answer to questions like these is that irreducible agents employ irreducible agent causation through irreducible rationality. I argue instead that the claims of irreducibility show that the theory is unable to answer the questions because it uses the wrong categories to answer, but that we can do better.

Here is a way to understand a choice as a causal process: a sense impression enters the brain and activates from memory alternatives for action which again activate desires connected to each alternative for action. The alternatives also activate memories from the autobiographical self about previous experiences with the alternatives, and these memories are connected with emotions. The remembered emotions can influence which alternative we desire the strongest. When a desire is strong enough that a threshold has been reached and nothing blocks it, a signal is sent to the motor neurons to activate an action scheme, and this constitutes making a choice.

Even if thinking is part of the mind, the topic of thinking has its own chapter in Chapter 6. Many could probably accept that emotions, memories and desires are causal processes, but how can thinking be rational if it is a causal process? Several philosophers have objected to the idea of thinking as a causal process by arguing that it undermines itself: thinking cannot be rational if it is causal. Then it is just particles being pushed around by laws of nature with no rational goal.

The chapter is based on the grounded cognition theory of mind, and argues that thinking works mainly by organizing parts into wholes. The brain registers features through feature-detecting neurons, and organizes them into wholes that we call objects, consisting of parts, which are their properties, all of which are stored in memory. Events are also stored in memory and can be understood as wholes consisting of parts (subevents).

Reasoning, like thinking in general, is about finding out which parts belong together in which wholes. Deductive reasoning is to determine whether something is part of a whole. For example, does the whole "All humans are mortals and Socrates is a human" include as a part "Socrates is mortal"? Inductive reasoning, on the other hand, is to group something together as a whole. For example, a lot of white ducks is grouped together as the whole "all ducks are white"; or "Jones having this and this motive", "Jones having his fingerprints on the knife", and "White being dead" is put together in the whole "Jones killed White with a knife for this and this motive".

Through evolution, rationality has evolved as a means to reach goals, and finding out what is plausibly true is a useful means to many goals. Brains that operate with causal processes that also give rational results have been selected through evolution. I explain this in further detail in the chapter on the causal role of consciousness.

In Chapter 7, I discuss consciousness. What is consciousness, and how does it relate to the physical world? This topic contains some really big questions, and I will be proposing several new theories.

Conscious experiences are often defined by saying that there is something it is like for a subject to experience them (Nagel, 1974, p. 436). Presumably it is not like anything for a hat to be a hat or for a hat to hear a trumpet sound,[5] but there is something it is like for me to be me, or for me to hear a trumpet sound, and the trumpet sound is different from a flute sound because they are different conscious experiences, the content of which are often called qualia. Qualia defined like this are the same as phenomenal conscious experiences, the qualitative experiences we are aware of, like sense impressions, thoughts, feelings, and desires. I use "consciousness" and "qualia" in a very wide sense. Qualia are all the structures you can be conscious of: a red tomato that you consciously experience is a quale, consisting of the qualia red, round, tasty etc., and so is the conscious thought of a tomato, all of which I argue *it is like something* to have.

What is consciousness? We should distinguish the content of conscious experiences (like a tomato or a chair) from what qualia are made of (their ontological status). We know of physical values that can be actualized in physical fields, and I argue that qualia are values that can be actualized in a qualia field. The specific values give the content of the qualia. For example, conscious experiences of light come in degrees of qualia values of hue, brightness and saturation; conscious experiences of sound come in degree of qualia values of pitch, loudness and timbre;[6] and in this chapter I describe a similar analysis for tastes, odors, emotions, and thoughts.

How can something physical produce conscious experiences, which seem so different from everything physical that they deserve to be called non-physical? The interaction problem is a famous problem concerning how to understand the relation between consciousness and the physical. It is very difficult to under-

5 I comment on the possibility of panpsychism in the chapter on consciousness.

6 The content of these conscious experiences should not be confused with their physical correlates.

stand how two such different things could interact at all, and it is difficult to combine with energy conservation and the fact that there always seems to be sufficient physical causes for every physical event.

I propose the following solution: From physics we know that fields can interact with each other. I hypothesize that physical activity in the brain – either at the neuronal level or at the fundamental level – can activate excitations in qualia fields – either at the fundamental qualia level or at higher levels of configurations of qualia values. This explanation has a great advantage when it comes to the interaction problem, since field interactions in physics happen at the level of their field equations. If there are rules for how physical fields behave and rules for how qualia fields behave, it does not seem like an impossible interaction that these rules could be combined, just like other rules combine at a mathematical level in the interaction between fields in physics. The physical events do the causal job, and so no energy conservation is disturbed. It thus solves several problems, but others are still unanswered: why is there any connection at all between consciousness and the physical world?

How could brains be connected to specific qualia values through evolution? The theory I propose is that in the beginning, brains connected to random qualia values in a chaotic manner from a much larger reservoir of qualia values than we know today, but over time, those brains that connected to qualia that were useful for survival were the ones that survived. But how is it possible for consciousness to have a beneficial effect when it seems that consciousness plays no causal role at all in our world? It seems that non-conscious zombies could have done everything we do with no causal role for consciousness.

Another huge problem to deal with is the problem that there are many reasons to think that consciousness can play no causal role in the world, since there seems to be a closed system of physical causes making anything happen where energy is conserved and where there is no place for consciousness. On the other hand, it seems that evolution has selected conscious beings for a reason, which gives us a reason to think that consciousness must have a causal effect. There are thus good reasons to think that consciousness plays *no* causal role and to think that it *does* play a causal role. How can this problem be solved?

To solve the problem, I argue that we must distinguish between on the one hand how neural patterns in the brain are connected to qualia structures in our conscious mind and on the other hand how neural patterns in the brain are connected to physical structures in the physical world (and this distinction is usually not made in discussions of the topic). Neural patterns have gotten consistently related to qualia structures that are simple and useful representations of the world. While there are many similarities, the qualia structures are also different from the physical world structures, and it has had a beneficial effect for brains to be consistently related to qualia structures as opposed to just being consistently related to structures in the physical world, especially when we make choices. The reason they are useful lies in being simpler than but partly structurally similar to the world, which allows for more efficient choices. What I argue in the chapter is that our brains work more efficiently when – through evolution – they have been related to physical structures in the world via simple conscious structures in our conscious mind than if they had been merely connected to physical structures in the world and working non-consciously only.

The causal role of qualia is then not to be found in particular choices, because neural patterns do the causal work in every particular situation. But these specific neural patterns have been selected through evolution because of the qualia they were connected to. They have been selected because qualia which are structurally similar to – but simpler versions of – world structures have allowed neural patterns to make more efficient and beneficial inferences and to guide action in beneficial ways. This is probably not so easy to understand in a first presentation, and I must refer to the chapter for details. Several other problems on consciousness are also discussed in this chapter, such as why evolutionarily beneficial actions feel good; how to understand subjectivity; where consciousness is located; and how *my* consciousness is connected to *my* body and *your* consciousness to *your* body. Especially the theory of what consciousness is, how to understand the interaction between consciousness and the physical, and the causal role of consciousness are new theories presented here for the first time.

Chapters 4 to 7 on causality, mind, thinking and consciousness present a detailed understanding of the mind, which can be used to discuss free will and responsibility in Chapter 8. Having free will means that a choice is up to you, and that you are the ultimate cause or source of the choice, but how is that possible given that the world is either determined or indetermined? If the world is determined, what happens was determined long before we were born, but if the world is not determined, it seems that there is too much good luck or bad luck that make us who we are and what we do, so that we do not have enough control to have free will and responsibility.

I argue that a choice is a causal process and that the autobiographical self can influence which desire becomes the strongest and leads to action. This does not seem to help secure free will, for even if the autobiographical self can cause choices, there must be other causes that made the autobiographical self what it is.

I describe in this chapter how the autobiographical self can be the cause of its own content over time. It happens when the autobiographical self causes choices which cause experiences which are stored in the memory of that autobiographical self. These experiences change the autobiographical self, which again causes new choices, and in this way the autobiographical self shapes itself over time, which I call becoming a more independent autobiographical self (self-formation).

I argue that the world is indetermined, and that there will be indetermined scenarios where it is right to select the autobiographical self as the ultimate cause of the choice. Free will (and responsibility) come in degrees, and by developing a more and more independent autobiographical self which can be the cause of choices, we become more and more free and responsible.

To have control is fundamentally to be the cause of a choice (nuances to this claim will be added later). We are certainly influenced by luck, but over time choices are less and less dependent on luck to become what they become. Different people have different degrees of free will anyway, so the fact that luck influences us does not show that we do not have free will to different degrees.

I argue that holding people responsible is a general practice of cultivating moral behavior through blame and praise, even if this may not be the motive in many specific cases. This practice presupposes people being responsible in the sense that they can take praise and blame into consideration in a normal process of deliberation.

This chapter on free will and responsibility concludes Part Two on the mind. All central concepts like persons, selves, thinking, deliberation, choices, free will, among others, are understood as configurations of values being actualized either in the physical field or the qualia field, and they are related to each other as elements interacting in a causal process.

Part Three of the book deals with various elements in the world different from the mind, and again an important goal is to show how very many things that could seem to be irreducible entities can be reduced to values actualized in a field.

I start with the concept of time in Chapter 9. There are many different questions that should be answered by a coherent theory of time, some of which are the following: What is time? Does time flow? What is "now"? Is time relative? What is the structure (topology) of time – how does past, present and future relate to each other? Are statements about the future true? What, if anything, is the cause of time? Are there points of time? Is time continuous or discrete? Why does time move forward – why does time have an arrow? Is time travel possible? How does time relate to the mind? Does time pass when there is no motion? Is time infinite in the past and forward directions? Was there time before the Big Bang? When will time end?

A central question in the philosophy of time is the question of whether time flows or not. On the one hand, you find those who defend a block universe where past, present and future all exist at once. This is called eternalism. All motion is an illusion according to this view, and it is the most prominent among philosophers of time. On the other hand, you find those who defend presentism – the view that only the present exists, and that now or the present is a special moment of time moving through history.

If eternalism is true, then we do not have free will in the sense that I described in Chapter 8. I will defend presentism in the chapter on time. An important argument in favor of the block universe is that there is no universal simultaneity in the theory of relativity. The argument is roughly as follows: There is no universal simultaneity, which means that what is future to me is past to someone else, but how can the future be open if it is already in the past of someone else? The lack of universal simultaneity thus implies that everything is fixed and no motion happens.

To deal with this argument I present the theory of relativity and explain why it is a very useful theoretical framework in physics. However, it is not impossible to define a universal simultaneity even if physics does not need it (and sometimes physics does need it, such as when determining the age of the universe).

I suggest a thought experiment to show that we can make sense of universal simultaneity: Imagine the reference frame of the universe as a lattice work of synchronized cameras permeating the whole universe and taking a picture at one point of time, as defined by this frame. The pictures could be combined into one big snapshot of the universe where one could see some objects being contracted. This cannot be done in practice, and black holes would disturb parts of it, and it is a definition of universal simultaneity which physics has no need for, but it is a definition of universal simultaneity that makes sense, contrary to those who say that no such concept can make sense given relativity theory.

For many of the problems in the philosophy of time, things get muddled because central terms can be understood in different ways. I solve many of these problems by specifying definitions of central terms like time, now, simultaneous, past and future. I argue that the basic meaning of "time" is motion, which means that time is not an entity with its own existence beyond motion. The word "time" can also be used for measurement of motion. Such measurement can be understood as an abstract and hypothetical measurement of time for the whole universe, but we also use it for actual measurements of time, either globally or locally, and with different means for measuring time. Finally, the word "time" is used for our conscious experience of motion. We should specify whether we speak of time as motion, time as an abstract or hypothetical measurement of motion, time as different actual measurements of motion, or time as experience of motion. Since time is merely motion which can be measured, it means that time is reducible to values being actualized in an area. The other terms connected to time are defined in Chapter 9.

The main point is that time as measurement is a theoretical framework that is used to categorize motion and events, which are actualizations of values in a field. Time is not needed as an irreducible ingredient in the world beyond the actualization of values according to rules.

Chapter 10 is on mathematical entities. Together with moral values they are typical examples of entities that many think cannot be reduced to something else, but rather they must have their own existence in a kind of platonic world. It is thus an extra challenge to see if it is possible to reduce it to the theoretical framework suggested in this book.

Mathematical truths are special in seemingly being a priori (not depending on natural states of affairs), necessary (they could not have been otherwise), and abstract (not part of space, time or causal chains). How can they be so? What are mathematical entities like numbers and geometrical figures, ontologically speaking?

Mathematical entities are possible structures and patterns among qualia values – structures that are thinkable, and which were possible to think about even before somebody actually thought of them – and many of the ones we have thought about, we may still only have a coarse understanding of, meaning that we do not fully see their implications.

When mathematicians develop theoretical frameworks they have a roughly clear set of entities they think of as mathematical entities that they try to relate to each other coherently. Mathematical theories are developed by starting with some axioms and rules and exploring implications. Sometimes such exploration leads to inconsistency and the theory is thrown away, but often it also happens that one either faces questions that cannot be answered, so that one need more axioms, or there is a need for distinctions because of ambiguities (or to dissolve an inconsistency), or something that seemed meaningless can be made meaningful by introducing new elements to the theory. In other words: problems get solved by expanding the theoretical framework.

Here are some examples: All numbers can be multiplied with themselves – squared – or taken the square root of, and some will have a whole number as a result when you take the square root. Taking the square root of negative numbers was considered impossible until someone invented imaginary numbers, where the imaginary number i is defined as the square root of -1. Some questions do not have definite answers given the theoretical framework. What is 0 divided by 0? It can be 0, since 0 divided by something is 0; or it can be 1, since a number divided by itself is 1; or it can be infinite, since a number divided by 0 is infinite. Some will say that the question of 0 divided by 0 is wrong or illegitimate since it is not definable in the framework, but that is just to support my point. Asking for the square root of minus 1 was considered to be a wrong or illegitimate question until imaginary numbers were invented thus extending the framework. Distinguishing between infinities of different size would be yet an example of expanding the framework.

When discussing the truth of mathematics, we should distinguish between truth in two different senses: truth relative to theoretical framework and truth relative to the most coherent theoretical framework. Truth relative to theoretical framework, means that a statement is true if it is coherently connected with the rest of the framework of which it is part. For example, the statement that the sum of angles in a triangle adds to 180 degrees is always true in Euclidian flat geometry, but not always true in Riemannian curved geometry. It is true that $2 + 3 = 3$ in tropical geometry, since it is that theoretical framework that defines the meaning of the terms. It is the theoretical framework that determined what plus means and it is the theoretical framework that must determine how to answer the question of what 0 divided by 0 is. This is then truth relative to framework, determined by whether the framework is coherent. But we can also make sense of mathematical truth in the sense of what would be true relative to the most coherent mathematical theoretical framework (as opposed to just being true relative to *a* theoretical framework).

How can mathematical statements be true? They can be true relative to a theoretical framework by being coherently integrated in a theoretical framework, but we can also single out certain mathematical statements as the ones that are true in the most coherent theoretical framework. In any case they are made true by the integration in the theoretical framework and the fundamental possibilities of patterns in qualia values.

Why are mathematical truths a priori? They are a priori partly because their truth depends only on the meaning of the basic structures of the theoretical framework and not on our experiences, and they are a priori because they depend only on what are possible patterns among qualia values and not on which patterns have been actualized in the physical world.

Why are mathematical truths necessary? They are necessary in a theoretical framework because their truth is given by the basic axioms of the theoretical framework. That something is necessary means that it is true given the basic assumptions in the theoretical framework, and thus inconsistent to deny given the presuppositions, but the axioms themselves are not necessary (unless they are logically necessary in the sense that they are inconsistent to deny). They are also necessary because they are different descriptions of the same structure. There is a structure such that $2 + 2$ and 4 are both descriptions of its parts.

Why are mathematical truths abstract? The mathematical truths we have discovered are abstract because they are narrow relations abstracted by the mind where we disregard the individuals they can relate. Fundamentally, mathematical entities are possible patterns in qualia values, and thus abstract as opposed to actualized physical values.

Chapter 11 is on probability. This topic is important for understanding in more detail probabilistic causation, the role of probability in quantum mechanics, and the role of estimating probability when making ethical judgments about what to do.

What is probability? Is it an objective feature of the world which give events a certain probability of occurring? Is it a quality of theories depending on how much evidence there is for the theory? Or is it a subjective degree of belief – how probable an individual finds something to be?

In this chapter I argue that we make a mistake if we try to find out what probability really is. There are many different kinds of probability, and what makes them into types of probability is that probability is a theoretical framework with rules for how to calculate probability, which fits (more or less) with different phenomena in the world. Probability is not an entity of its own, but a set of concepts we can use to analyze and describe in more detail different phenomena in the world, such as the relation between causal powers, different pieces of evidence or different beliefs.

In other words, probability is a theoretical framework, the status of which can be understood more precisely in terms of the theoretical framework developed in this book. I distinguish between ontological probability and epistemic probability. *Ontological* probability claims are about states of affairs that will happen, and they are made true by how causal powers work in the world. *Epistemic* probability claims are about the relation between evidence and explanations. As *objective* statements they are made true by the coherence of the explanation. *Subjectively* understood, they are about the belief of persons, and are made true by mental processes in the mind of the persons.

I also distinguish between past and future, because it is useful for understanding epistemic probability claims. When subjective or objective *epistemic* probability claims are made about the past, what happened has *ontological* probability 1, and thus it is only the evidence which determines the suggested probability value. But when subjective or objective epistemic probability claims are made about the future, the ontological probability together with the evidence determines the suggested probability value.

Take for example the claim "the probability of rain tomorrow is 50 %". This may be a person's statement expressing his own belief that he is very uncertain about whether it will rain, but without knowing any evidence. But an expert may also have seen all the evidence of what the weather will be and say that the probability of rain is 50 %, and then the ontological probability of rain may be 50 %, while the epistemic probability of that hypothesis may be almost 100 % (because it is very certain that the world is such that rain may occur with 50 % ontological probability).

These three kinds of probability thus relate to each other in the following way: Ontological probability is about states of affairs in the world; objective epistemic probability is about the evidence for which states of affairs actually occur in the world; and subjective epistemic probability is about how individual persons evaluate that evidence. Each kind depends on the previous to become what it is, while it is not the other way around. In other words, the subjective probability I assign to something is based on the evidence I have access to, which is based on facts about states of affairs in the world.

I have dealt in more detail with the basic qualia values in the chapter on consciousness, but throughout the book I often speak about basic physical values without going more into detail. Chapter 12 will look more into the fundamental entities in physics. In this chapter I deal with physical values like energy and mass. By understanding some simple rules on how mass, energy and momentum relate, we can understand surprisingly much about how particles and objects move in the world, and even how particles are created and annihilated.

But what *are* values like force, mass and energy, and spin? Physicists can give very precise formal definitions relating the terms to each other, but there are very few suggestions at all when it comes to the ontological status of spin, charge, mass or energy (what they are "in themselves" – what their internal structure is as opposed to their relation to other concepts).

My hypothesis is that (most of) them are general relations that hold between many different individuals in the world, which fundamentally arise from structures to be found in the rules that nature follows in guiding motion in the world. This means that concepts like mass and energy, etc., do not refer to something that has its own physical existence at all, but instead express some common structures in the rules that nature follows.

This hypothesis is supported by the fact that there is nothing we can isolate that is a piece of energy or a piece of mass or a piece of momentum or a piece of spin or a piece of charge – nor are these properties that consist of something known. Their ontological structure is unknown, and we can only describe them in terms of their effects. Of course, they may be configurations of some deeper physical and yet unknown structure like superstrings or quantum foam, which physics may discover. I make a suggestion that physical quantities may be reduced to a force having magnitude and direction in points, and I relate it to the reflections previously made on the meaning of existence and actualization, unifying all three: the energy of motion, the essence of existence, and the power of actualization are one and the same fundamental actualizer.

Having looked at the physics of relativity and basic concepts in physics, reflecting on the basic structures of the world would not be complete without a visit to quantum mechanics, which is the topic for Chapter 13. It is a non-technical presentation focusing on how to solve the typical mysterious problems of quantum mechanics.

I start by presenting a list of problems that a good theory of quantum mechanics should solve: The double slit experiment where particles behave like waves, and how the interference effects disappears when you monitor the particles; the effects of scrambling and reversing results depending on what you choose to measure; entanglement where measuring one particle has immediate effects on another particle regardless of how far away it is; understanding the important Born rule and understanding the measurement problem: why do measurements have determinate outcomes, the probability of which is given by the Born rule instead of there just being a determined evolution giving superpositions, which the formalism would make you expect? And finally: what does quantum mechanics tell us about the world? What should we take to exist as irreducible entities and what are merely mathematical representations?

Contrary to what one should think reading much about what is written about quantum mechanics, surprisingly good answers can be given to these questions. In this chapter I lean heavily on how Tim Maudlin argues that the problems should be solved, ending with some suggestions of my own relating quantum mechanics to the rest of the book. The chapter also contains an excursus reflecting on why we have the laws of nature that we do.

The chapter on quantum mechanics ends Part Three, and then Part Four deals with the future, starting with Chapter 14 on ethics. It is placed in the part on the future, because I define the good in light of a future possible world.

The existence of moral values is another typical example of something which many claim cannot be reduced to something natural. Most ethicists are realists about moral values, but have very different views on what makes moral claims true. On the one hand, moral claims seem to have the same form as other sentences that can be true, and it seems perfectly fine to say that it is true that one should not kill for fun. But how should one check to find out that it is true?

This is a central question in metaethics, where the supernaturalists refer to the will of God to explain moral values but non-naturalists refer to non-natural moral values – the good in the platonic world would be a typical example, while naturalists believe that moral values are something natural. One version of such naturalism holds that normative terms can be translated to naturalistic terms, and I defend such a reductionist naturalistic view.

I argue that the good is a concept we make up to describe a particular possible world in the future that we think is the best possible world according to a particular standard, and we hypothesize that there is also a best way to reach the goal. This is enough to give meaning to normative terms like "good" and "should", and there is no normative force or moral values existing in a platonic world beyond the concepts invented by humans to describe a possible world. Ethics is thus a theoretical framework describing a path towards the future that the supporters of the framework suggest that people should follow.

In more detail, I suggest the following definition of The Good: The Good is that possible world which an all-knowing and all-sympathetic being would know would most probably be valuated the most by the most. In simpler words, ethics is about finding the best possible way to the best possible world, based on what would be valuated the most.

The definition of The Good is actually a definition of what is, morally speaking, the best. A lot of possible worlds and actions could still be good without being the best, so there is a sliding scale between good, better and best. On the other side, The Bad is that possible world which an all-knowing and all-sympathetic being would know would most probably be disvaluated the most by the most. The Bad is then actually the morally worst possible world, so there is a sliding scale from bad to worse to the worst possible world. The terms "should" and "ought" can now be descriptively defined as means to a goal. Given that The Good is the goal, one should or ought to do this and this, while one should not or ought not do this and this, given that we want to avoid the bad.

I discuss this view against a series of objections and test cases, especially the trolley dilemma and the question of human value and human rights, and whether one can defend the view that all humans have the same value and also greater value than animals.

Chapter 15 is about the future, and there I try to draw some more concrete implications for three different topics about what has been said so far. The first topic is what to think of the meaning of life of individuals – what is the goal of an individual life and why is it worth trying to reach that goal? The second is what the theory of mind and free will in this book implies for the possibility of superintelligent machines. The third is what the ethics implies for more practical political goals – how to proceed in reaching the best way to the best world.

Finally, Chapter 16 is an epilog reflecting on the deepest possible level of explanation, and even daring a speculation on how to answer the question of why anything exists at all.

While I claim to answer a huge list of problems, I assume that many of the readers of this book will think that I have not really solved any big problems, but instead merely redefined terms and escaped the real problems. Instead of having explained things I can be accused of having explained them away. Instead of explaining causes, time, agents, substances, universals, etc., I have only explained something else and let the causes, agents, etc., remain unexplained.

My standard reply to such challenges is that I have tried to give a more coherent explanation in my theoretical framework of what those concepts are trying to express in their ultimately incoherent theoretical framework, and that the traditional understanding refers to something that does not exist. Whether I am right depends on whether there is something left unexplained that we have good reason to believe exists. Just throwing out a term for something I have not explained is not a good argument if we do not have a good reason to think that it exists.

This short comment is reminiscent of a discussion between P. F. Strawson and Rudolf Carnap where Strawson challenged Carnap by saying that he did not solve problems, but changed them instead. Carnap replied that it is good if the doctor uses a scalpel instead of a pocket knife to solve the problem,[7] meaning that sometimes a vague question needs to be translated into several more precise and different questions. Herman Cappelen thinks Carnap's answer is evasive, while I find it good: sometimes the best answer to a question is to point out that the problems are created by how the question is formulated.

This book is about the interrelations between many large topics, which means that there is not space to discuss each of them in the depth they deserve. The central point of the book is the theoretical framework presented in Chapter 3. I develop a specific theoretical framework, and apply it to many different problems to show its usefulness. This approach allows me to show the significance of the main idea for a broad range of topics. As will be clear, the main issues in metaphysics are interdependent, and thus it is very useful to show how the main idea creates coherence across different areas. My intention is to show how the coherent interrelation between the topics support each other, and I hope that readers can see the potential of the main approach even if they have quarrels with details.

The disadvantage with this broad approach is that while I do respond to main objections, there remain many objections that I do not answer. I have published an appendix online with answers to objections for every subchapter that one can consult, and I appreciate being sent objections not yet answered.[8] The

7 Recapped in Cappelen (2018, pp. 98–106).
8 https://atleottesensovik.mf.no.

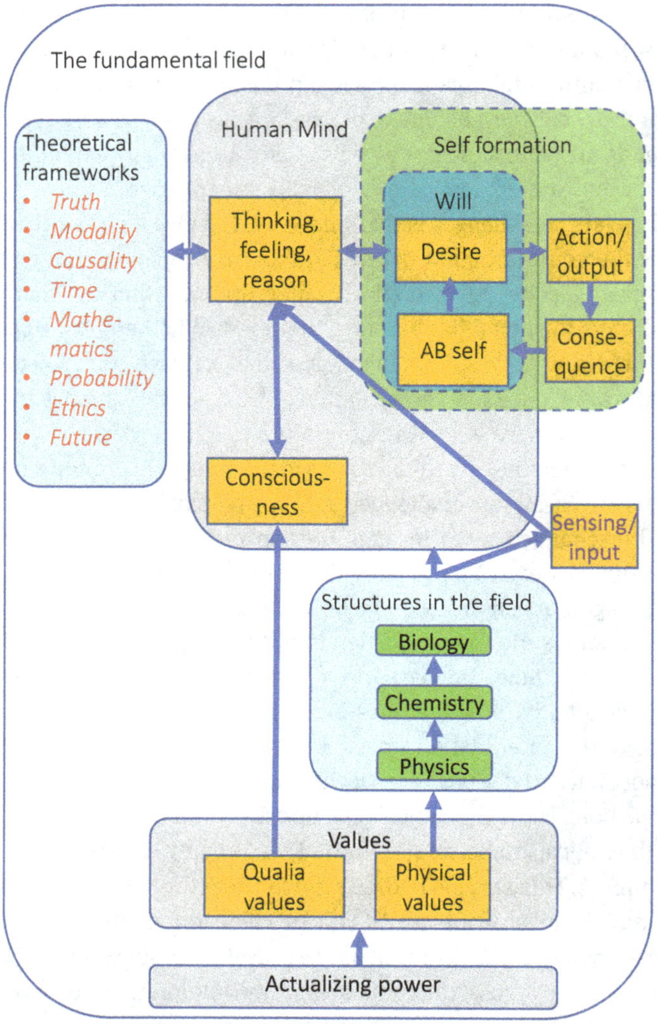

Fig. 1: The world

appendix will be continuously updated, so it will always have a version number on it that one should refer to if citing it.

After this long presentation, I have tried to summarize it in a figure (Figure 1).[9] There are three basic constituents: the actualizer, the possible values

9 Thanks to Svein Jåvold for pushing me to make this figure, and for helping me with content and design on all the figures.

and the field. These three together give us the content of the world with its laws, geometry, objects (including humans) and conscious experiences. When it comes to truth, modality, causality, time, mathematics, probability, ethics and the future, these are all theoretical frameworks that we use to sort the contents of the world (patterns in the values) and describe relations in the world (also patterns in the values).

Part One of this book covers the boxes in the bottom, describing how the fundamental actualizing power actualizes patterns in values. Part Two takes us through chemistry and biology to the two boxes above representing human mind and free will. Part Three unfolds the content of the box in the top left corner, representing different theoretical frameworks that the mind has constructed, describing the world outside of the mind. Part Four is about the last points in that box – ethics and the future. This description of the four parts is only roughly true, since various subtopics have been moved around in order to be presented in the order that makes it easiest to understand.

2 Methodological Topics

This chapter deals with several methodological topics in addition to the concept of theoretical frameworks. What is it that we do when we understand something? How can our mind understand the world outside of the mind? What is truth and how do we know whether something is true?

Since the rest of the book will be about understanding different things, the concept of "understanding" is a good place to start. I explain in Section 2.1 how we understand something by placing it at a certain place in a certain theoretical framework, relating it to other entities in the framework.

Understanding happens in the mind, but the mind wants to understand the world outside of the mind. In Section 2.2 I discuss the relation between the mind and the world outside of the mind, and argue that we can only relate to the world outside of the mind from within the mind.

What is then the relation between the mind and the world outside of the mind, and how can the mind discover something true about the world? In Section 2.3 I argue that the mind discovers what is true about the world by discovering what is the most coherent understanding of our experiences of the world. I explain in Section 2.4 how coherence is a criterion of truth, which means that the most coherent theoretical framework is also the one best justified as true.

A theoretical framework which is to be as coherent as possible also needs to be formulated in concepts which are as understandable as possible. All concepts get their meaning from being related to other concepts, but some concepts are more basic in the sense of being simple and central concepts which many other concepts are understood in light of.

In Chapter 3 I present the concepts that express the basic *ontological* constituents of the world, but in Section 2.5 I describe the basic *semantic* concepts that are used to describe ontological entities. These basic semantic concepts have similar meaning, namely "structure", "individual", "relation", "whole" and "part". All structures are wholes and parts and all structures are individuals and relations. The concepts are basic since in order for anything to be understood it must be a structure, which means that it must be a part integrated in a whole, and it must be a relation between individuals. The basic concepts are again analyzed in terms of similarity, difference and part.

When I have presented the theoretical framework of actualized values at places in Chapter 3, we shall see that the basic concepts of structures among actualized values can solve and dissolve a lot of philosophical problems, like: What is a substance? How can a substance have properties? What is the haecceity or identity of a substance? What are universals, and how are they instantiated in partic-

https://doi.org/10.1515/9783110770957-004

ulars? What is the difference between abstract and concrete? I suggest that the concepts of substance, property, haecceity, and universals can all be ontologically reduced, and that they give rise to many pseudo-problems.

The chapter ends with an excursus on mind, world and language (Section 2.6) discussing how sentences can express the world. Some classical debates on the meaning and reference of sentences and propositions are discussed there.

2.1 Understanding

In this section I define the basic concepts that are needed to understand the rest of the book, and I start with the concept of understanding. What does it mean to understand something? When raising this question I am not talking about what it means to understand something *correctly* as opposed to *mis*understanding; rather I am interested in what it means to understand anything as anything at all. To understand a state of affairs at all means to relate it to something else – either to its own parts or to another state of affairs. Then you understand it in relation to something, which means that you understand it *as* something, which is to understand at all.

If there is something and you do not know what to relate it to, you do not understand it. For example, you may hear about or see a spark plug. If you do not know what to relate it to, you do not understand what it is beyond the basic understanding that it is an entity in the world. If you think it is part of a TV, you understand it in a broad sense because you understand it as something. However, a narrow and daily sense of "understanding" is that to understand something is to understand it correctly, and if you think the spark plug belongs in the TV, you do not understand it correctly. Rather, you misunderstand it – whereas you understand it correctly if you think it is part of a motor (Gravem, 1996; Puntel, 2008, pp. 249, 342).[10]

One may have a conscious experience of something which one does not understand beyond it being something in the world, for example a spark plug. In a minimal sense one could say that you still understand the spark plug as something – as an object existing in the world. This means that the term "understand" is a gradual concept where you can understand something more and more by being able to relate it to more and more other states of affairs. In daily life we use the term "understand" only when one understands more than that some-

10 Later we shall see that explaining is the same as understanding: clarifying the place of an element in a larger theoretical framework.

thing is an object existing in the world, and so if we have no idea what a spark plug is, we say that we do not understand what it is, and say that we understand something only when we know the most important connections. This means that in daily speech, there is a (fuzzy) border between understanding and not understanding, but more precisely speaking, understanding is a gradual concept: understanding comes in degrees, and as long as you are aware of something, you understand it in a minimal sense (for example as something that you are aware of).

Later in this book I will discuss further the concepts of understanding, intentionality, subjectivity and related terms, and say more about the role of consciousness and subjectivity for understanding, but here I just make the basic point that we understand something by relating it to something else. This understanding of the concept of understanding is probably broader than how many people are used to using the concept. However, I find it very useful for understanding a lot of other topics and their close relations, and to explain why many things come in degrees instead of being either/or. Throughout this book we shall see how integrating parts into wholes (as we do in understanding) is fundamental to understanding also meaning, truth, beliefs, experience, and thinking. We often make strong distinctions between the concept of understanding and other concepts, whereas I will argue that they are more closely related, and that the concept of understanding is the unifying factor. This is the case also in the next topic, on the relation between mind and world.

2.2 Mind and world

We have good reason to think that there is something outside of our minds, in the sense that there exists something which is not identical to contents of *mind*, here used in the sense of conscious mental representations of the world. Some common reasons to think so: It seems clear that we can be wrong and that everyone can be wrong, for example if everyone thought the sun orbited the earth. It seems clear that some understandings are better than others, for example that it is better to believe that the earth is round rather than flat or square. And it seems clear that we cannot with our mind determine what the world should be like, but rather the world causes parts of our mind to have certain contents different from other contents. For example, I cannot choose to run happily through a hard wall; rather it is the case that if I try, the world then causes the content of my mind to be that my head hurts from trying to run through a hard wall.

On the other hand, we all experience that we have no access to the world except the access we have to the contents of our minds. When we discover that we were wrong about something, new content in our mind has replaced old content. If I, instead of thinking about a chair, go to experience the chair itself, all I experience is content in my mind – the visual impression of the chair, the feeling of the hardness of the chair, etc. All empirical tests of ideas that we make in order to grasp the world itself, and all data we collect, are experiences that we know as content of our minds. There is no experience and no access to the world that is not given to us as content in our minds. If you say to someone "Do not tell me how you think the world is, but how it actually is", this is an impossible order, since nobody can say other than how they think the world is (Rescher, 2010, p. 5).

Note first the important difference between individual minds and mind at all. We can know and experience something outside the mind of a particular individual, but we cannot experience something outside of mind at all, or know anything about it. Note secondly that there is a distinction to be drawn between mind at all and what is outside of mind at all, but this is a distinction that we draw within mind. When we talk about what is inside of mind and outside of mind, everything we think about and talk about is still inside of mind. Yes, dinosaurs existed before humans and the universe existed before mind evolved on earth, but all of this we can relate to in our mind only.

There is still a difference between mind and something which is not identical to mind. We have reason to believe that something which is not mind makes it the case that the content of our minds is a certain way rather than another. The fact that our experiences are structured a certain way indicates that there also is structure outside of our minds.[11] But we cannot say, think or know anything about it except as we relate to it in our minds. When we say something is true or exists or is independent of mind and language, all we talk about are still contents in our mind.

I find this frustrating. There is good reason to think that there is a difference between an individual's mind content which is that "John sits in the chair" and the state of affairs in the world that John sits in the chair ("states of affairs" will be further defined below). And still there is not one thing that we can say or

---

11 One could object that it only follows from our experiences being structured a certain way that our mind has a certain structure. However, different people will have experiences that are similar to each other when they are at one place and time in the world, and then another kind of experience when they are at another place and time in the world, and the most plausible explanation of this fact is that there is actually a structure outside of minds influencing the content of mind instead of the mind producing its content independently of the world outside.

think or know or experience about the state of affairs that John sits in the chair which is not also the content of some mind. There are and have been many states of affairs in the world that are possible for minds to relate to before any mind has actually related to it, and yet we can only relate to it through our minds. As seen above, there are many reasons to think that there is something which is not mind which makes us have a certain experience of John sitting in the chair, but everything we want to say about this mind-external something is still content of our minds. Similarly, we have good reason to think that there are atoms, but everything we know about atoms we know as contents of our minds. When we want to relate to the world directly, the mind is an impenetrable wall outside of which we cannot come.[12]

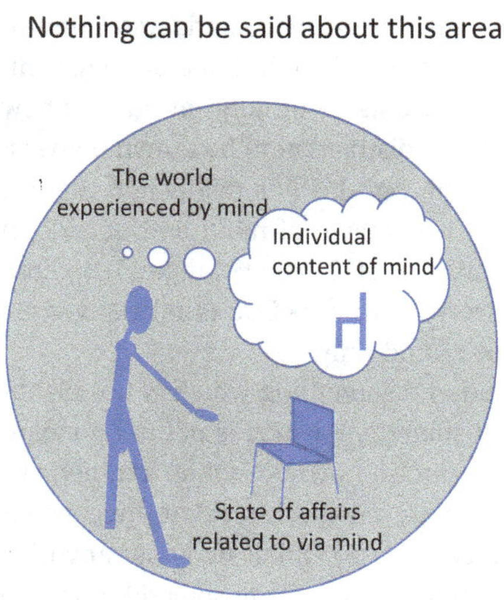

Fig. 2: Mind and world

A bit more detailed comment on the illustration: nothing can be said, thought or known about the area outside the circle because whatever you then say, think or know will automatically be located within the circle. From within the circle you

12 Asle Eikrem refers to a discussion between Hegel and Kant and distinguishes between "Grenze" and "Schranke", where a Grenze is a line between two different areas where we know each side, and a Schranke is a wall we meet where it does not make sense to speak of that which is beyond (Eikrem, 2013, pp. 172–173). Hegel criticized Kant's distinction between things in themselves and things as they appear to us, since everything we say or know about things in themselves are the things as they appear to us. The discussion in this chapter is an attempt to explain how we can say something true about the world on these conditions, and I argue that we can do this by understanding truth in terms of coherence, as explained below.

can know that it is true that nothing can be said about the area outside of the circle because it would be inconsistent to deny this (since if you say something about the area outside of the circle it is automatically located within the circle). Within the circle you may hypothesize about the area outside of the circle, but anything you hypothesize is understood on the conditions within the circle, which means that anything you might discover would be something previously unknown but still within the circle.

In this section I have spoken only of mind as a process where entities are understood by being related to each other. Such a mind process can be understood as a very simple theoretical framework for understanding. In more advanced forms of understanding, theoretical frameworks are "instruments that make possible the articulation, conceptualization, and explanation of theoretical contents or subject matters" (Puntel, 2008, p. 24).[13] A theoretical framework will have a language (semantics and syntax), a logic and conceptuality, and the different components of the theoretical apparatus (Puntel, 2008, p. 9).[14] While usually expressed in normal words in spoken languages, I use the term "language" very broadly here to refer to any system of signs related to each other. This means that images in our mind related to each other also constitute a simple theoretical framework that we use to understand the world as long as there are some systematic relations between them.

Above I said that we understand something by relating it to something else. Another way of saying the same is that everything we understand we understand by placing it at a certain place in a certain theoretical framework. Nothing is understandable outside of a theoretical framework, since placing something at a certain place in a certain theoretical framework is what it means to understand something. You will never find an example of something you understand which is not something you understand in a theoretical framework. This means that anything gets its cognitive meaning from its place in a theoretical framework,

13 I use Puntel's understanding of theoretical framework, which again is a modification of Carnap's concept of linguistic framework (Puntel, 2008, p. 9). "A theory" and "a theoretical framework" are overlapping in content, but the framework is broader, so that you can have different theories within the same framework. Two theories are in the same framework when they use the concepts of the framework and certain relations therein, but they are still different theories if they give different answers to more detailed questions.

14 See also the helpful description of the same understanding of theoretical frameworks in (A. White, 2014, pp. 24–29). The concept of theoretical framework is similar to what Luciano Floridi calls "levels of abstraction" (Floridi, 2014), but in my view much more precise and useful than Floridi's concept.

and that its meaning is relative to the theoretical framework.[15] We will get back to what that means for the question of truth.

The way that natural science tries to relate directly to the world is through predictions and experiments, which gives us good reason to think that certain claims about the world are true or false. But there is no direct access to the world, and it is not the world itself that corrects us. Rather, there are new understandings of the world which replace old understandings of the world because the new understandings are more coherent than the old ones. Understanding the relation between mind and the world is closely connected to how we understand truth. For this reason, I will now say a little more about truth.

2.3 Truth

Truth is often understood as a correspondence between what is in our mind and what is in the world. We have seen above what the problem is with this definition of truth, namely, that we have no access to the world outside of our minds. The correspondence relation is thus in fact only between different mental contents in our mind. But what then makes an idea true? Instead of finding a correspondence with something outside of mind, what actually happens when we search truth is that we try to get a better grasp of the world by getting a more and more coherent understanding of the world. For example, we have good reason to think that the earth is round instead of flat because it makes many more experiences (or "data", to be defined later) much more coherent. Likewise, we have good reason to believe that Jupiter is the biggest planet in our solar system, but not because we have access to Jupiter itself independent of our experiences of Jupiter. Instead, we have access only to experiences of Jupiter and experiences of experiments involving Jupiter.[16]

One could object that I am confusing epistemology and ontology; that I confuse what we know with what is true or that I confuse the act of thinking with what I am thinking about. I reply that I do not, since my point is to show the

15 It is common to describe similar points by saying that meaning and thus truth is relative to "context", "paradigm", "perspective", "tradition", "viewpoint", or something similar. I replace all these vague concepts with the concept of theoretical frameworks (Puntel, 2008, p. 338). Note that one can use "perspective" in a narrow sense to distinguish a location within a theoretical framework from the theoretical framework as a whole. I will add details on relating theoretical frameworks to each other in the section on truth and in the chapter on mathematics.

16 In later chapters I argue that the brain makes representations of the world, and not that our mind directly grasps or contains the world.

close relationship between epistemology and ontology: even the ontological facts about what is true and what we are thinking about are inaccessible to us outside of our minds. There is an important distinction to be made between mind and world, but the distinction is still made within our minds.

Is it then possible to use coherence not only as a criterion of truth, but also to define what truth is? I submit that it is. We can define truth as the description of the world which is as coherent as possible. What makes a particular proposition true is then that it belongs to (i. e., is coherently integrated in) this maximally coherent description of the world. We cannot have a realistic hope of ever finding this maximally coherent description of the world or know that it is the maximally coherent description. Nevertheless, this understanding of truth is an understandable concept which can work as a regulative idea in the sense that it gives a coherent justification of our practice of searching for theories and comparing them with the aim of finding the theory best justified as true (at the present time), which is the most coherent theory (at the present time).

It may seem like this theory disconnects truth from the real world, but it does not. The link to the real world is in the idea of the description of the world which is as coherent as possible. The content of the idea of the maximally coherent description is a hypothetical mind knowing all expressible data and what the most coherent description of them are. It may seem unnecessarily complicated to hypothesize such a mind, but the reason for including it is that it does not make sense to speak of the world as it is independent of mind, and at the same time the world is obviously more than what any humans have thought of (and was there before human minds evolved), so this understanding of truth can integrate all this while avoiding the problematic reference to the world in itself.

This idea of the most coherent description of the world can retain the intuitive understanding of truth in a coherent way. Intuitively, we think that a true proposition must express how the world really is. A proposition is classically said to be true if it says that the world is so and so and the world really is so and so (Tarski, 1944). The previous sentence has two parts that are similar, but the second part says that the world *really* is like the sentence says.[17] What the word "really" here adds is the idea of a most coherent description of the world: A proposition is true if it says that the world is so and so and the world *as described in the most coherent way possible way* is so and so.[18]

---

17 Puntel quotes Quine saying that a sentence is true if the world is *really* like the sentence says, and Tarski saying that a sentence is true if the world is *indeed* like the sentence says (Puntel, 2008, pp. 225–226).

18 This theory of truth is greatly inspired by and very similar to the one presented in Puntel (2008, pp. 222–245).

Before I proceed, I shall answer to an objection to the idea of a maximally coherent description of the world. There seems to be an infinite number of truths and an infinite number of incomparable ways to describe the world, implying that the idea of a maximally coherent description of the world does not make sense. For example, I could say, "It is true that I could name my pencil Mr. Pencil Number One", "It is true that I could name my pencil Mr. Pencil Number Two", and so on for infinity, giving us an infinite number of truths.

I argue in this book that the world is a finite field, which means that it does not have infinite extension, and that it has a finite set of possible qualitative values that it can actualize. This means that the world has a finite basis, on the basis of which an infinite number of truths can be constructed (Puntel, 2008, pp. 426–430).[19] It further means that even if more finely grained connections between entities in the world can always be uttered, there is a maximally coherent description of the finite basis of the world, in the sense that all additional truths can be deduced from the maximally coherent description of the basis (Puntel, 2008, pp. 426–430). This is comparable to infinite sets which have an infinite number of members, but which nevertheless have a finite description of criteria for membership in the set.

There is a famous incompleteness theorem by Kurt Gödel which implies that we cannot have a complete description of the world (for details, see the chapter on mathematics), and of course any explanation will leave something unexplained, but truth is here defined not as a complete description of the world. It is only the most coherent one. One may object that there can be very different descriptions of the world which are equally coherent, but in that case I assume that they will be very structurally similar (in order to actually give equally coherent descriptions of the world), so that they are comparable and that they can be translated to each other. In the history of science, new theories have replaced old ones, but the descriptions that work well are nevertheless structurally very similar. Even if quantum mechanics is better than just understanding the atom as consisting of neutrons, protons and electrons orbiting the core, the main structure of the old model remains in the new model. When theories are detailed, it is hard to imagine that you could have completely different theories nevertheless being equally coherent. This is why I assume that equally good theories will be structurally similar and translatable.

---

19 Note the difference between the existence and articulation of everything in the world. A totality may exist even if it cannot be articulated. On the other hand, language seems to have an infinite potential for articulating anything, since new signs can be combined in infinitely many ways (Puntel, 2008, pp. 429, 374).

One could also object that there could be descriptions of the world so different that it is impossible to compare which is the more coherent, but remember that we are talking about a regulative idea, which means that we have a reason to search for truth and compare theories, but it does not mean that it will always be possible in practice. What is important is that there is no principled reason why two descriptions should not be possible to integrate in a larger theoretical framework within which they could be compared.[20] We then have reason to try as well as we can, even if we do not always succeed. This is very different from a situation where we could see no reason to search for truth at all.

2.4 Coherence as a criterion of truth

Since being true is defined as being part of the most coherent description of the world, coherence can be used as a criterion of truth when discussing which theory is more likely to be true. Other theories of truth can *define* truth without referring to coherence and still use coherence as a *criterion* of truth, but when truth is defined in terms of coherence, the criterion of coherence makes more sense than when definition and criterion are separated.

If one theory is more coherent than another, it gives us reason to believe that it is closer to the truth (the maximally coherent theory) than the other. But it will always be possible that both theories are wrong, and a more coherent theory will appear in the future. Nonetheless, coherence is the best criterion we have for truth, and most other common criteria for truth can be interpreted as falling in under this general criterion, as I will soon show. But first: what is coherence?

The American philosopher Nicholas Rescher has suggested that coherence should be understood as a concept with three aspects (Rescher, 1973, pp. 31–38, 168–175). The first aspect is consistency, which means that two elements in a theory cannot contradict each other. A contradiction is to say both

---

20 Skepticism is a position arguing that there can be no truth, but as Puntel argues this is a self-defeating position. Skeptics presuppose that they can argue understandably and meaningfully for their position, and that presupposes a functioning language and logic in order for them to have a point (Puntel, 2008, p. 64). However, it can seem that relating true sentences to theoretical frameworks is itself a kind of relativism. Relativism is commonly said to have a self-reference problem since if everything is relative, then the claim that everything is relative is also relative, but then everything is not relative. This problem is avoided here, since there is a regulative idea of truth where theories are better justified when they are more coherent, and something may be true in all theoretical frameworks, for example that A = A or that the world is one (Puntel, 2008, pp. 242–245).

p and not-p at the same time and with regard to the same. Consistency is a fundamental part of coherence, as a theory that is not consistent is not coherent.

The second aspect is comprehensiveness, which means that a theory is more coherent the more data it can integrate. Data is understood as truth candidates. This definition of data is especially important because a common critique of coherence theories of truth is that they are not connected to the real world. Some might say, for example, that theory can be coherent or consistent without being true. Yet, what a sophisticated coherence theory of truth (like the present one) says is that the theory that is most coherent (integrates the most data in the most coherent way) at the time of consideration is the theory that is most likely to be true. In other words, it is not the case that according to this coherence theory of truth, any consistent theory is true. Rather, it is the theory that is best able to integrate what we believe to be the facts that is best justified as true.

The third aspect is cohesiveness, which means that the more connections there are between the data, the more coherent a theory is. Connections can differ in strength, in the sense that a deductive connection is stronger than an inductive connection. Lorenz Puntel offers some more detail on how to understand such connections. They do not refer merely to inductive and deductive connections between elements. Any description of any relation between the data (e.g. causal relations, spatial relations, and so on) gives a more finely grained description of the states of affairs that the theory describes and a more precise integration in the theory, which thus makes it more coherent (Puntel, 2008, pp. 439, 464, 476).[21]

Puntel does not answer the following question: How should we understand the relation between *logical* connections such as consistency, deduction and induction on the one hand and on the other hand *any kind of* connection such as spatial relations, causal relations, being taller than, being the uncle of, and so on? In what way is a theory supported by *logical* connections like consistency, deduction and induction, and in what way is a theory supported by being able to explicate *any kind* of connections like spatial or causal, etc.?

I suggest the following answer, and start with the logical connections. Inconsistency in a theory means that you have not clarified whether a certain part be-

21 Puntel is inspired by Rescher and thinks of coherence as having the same three aspects (Puntel, 2008, pp. 475–476). He adds as criteria *depth*, which means the degree to which structures are valid in many different theoretical frameworks, and *grainedness*, which means differentiation, detail and specificity (Puntel, 2008, pp. 408–409). I understand depth as falling under comprehensiveness and grainedness as falling under cohesiveness, but they are worth mentioning to deepen further how the aspects should be understood. The main goal is the universal coherence which describes the whole structure of the world (Puntel, 2008, p. 439).

longs to the theory or not, and thus the theory cannot be evaluated before this is clarified. A deductive connection means to explicate something which is already part of the theory: If A is true, then B must be true, because B is part of A. An inductive connection means that we are able to integrate a part in a consistent way, which gives inductive support because the truth is the theory that is able to integrate the most parts in a consistent way, and so being able to integrate one part extra gives you reason to believe that you are one step closer to the truth (although you cannot know for sure).

An element being integrated gives some inductive support to the theory and if it fits much better in one theory than alternative theories, the support is extra strong. The more connections of any kind (spatial, causal, etc.) that a theory can integrate in a consistent way, the more finely grained it becomes, and thus the more coherent the theory becomes.

What is then the relation between the two kinds of connections – logical and others? When you describe any kind of relation (spatial, causal, etc.) you describe in more detail what your theory is. The more detailed a theory is while still being consistent, the more coherent it is because more parts (in the sense of more details) have been integrated consistently. This means that integrating any kind of relation in a theory makes it likely to be closer to the final truth, thus integrating any kind of relation gives a theory inductive support.

A classical problem when comparing theories is that they can integrate different elements in different ways. One theory can seem to integrate more important data than another theory, but how do we determine what data are the most important or relevant? The aspect of cohesiveness helps to explain which data are most relevant, since data with many connections to the topic at hand are more relevant than data with few connections. However, you could have vague data with many connections to other entities, like yin and yang connected to everything. Data that are fine-grained and can be integrated in a fine-grained way also give us reason to believe that these are important and relevant data to include. For example, we have better reason to think that many of the fine-grained data of physics should be included in a theory of the world instead of vague data of astrological influence from the stars, which should be integrated as fiction instead of as real. The more connections there are between data with many connections or between fine-grained data, the more important they are. Note then that coherence comes in degrees. If a theory is consistent, it can be more or less coherent depending on how comprehensive and cohesive it is.

Commonly, people will refer to the hypothetical-deductive method as a test for truth, or say that one should be able to make testable predictions, or many other such things. But all of these different tests and criteria can be understood as ways of showing the coherence of a theory. No normal test criteria of truth or theoretical virtues in science contradicts the idea of coherence here presented, but rather it falls under them and explains why the criteria suggested are useful. Note however, that there are some criteria often appealed to that do not support a theory being true, but rather are pragmatic criteria. An example is the criterion of simplicity. By "simple" I mean structures with few unique parts. If two theories are equally coherent, we select the simplest one, not because there is a good reason to believe that truth must be simple, but rather because it is easier for humans to relate to a simple theory than a complex one if there is nothing else to gain by using the more complex one.

An objection to this theory of truth could be to say that if you choose one theory of truth, its criteria will be thought of as the best, but another theory of truth will say that other criteria are the best, and so there is no theory-independent way of saying which theory of truth is the best, but then again it will be relative what is true, because it depends on a non-justified theory of truth.

How can we choose among basic criteria for what we should think of as truth at all? Will it necessarily become a viciously circular process? How can you say that a theory of truth is best, if every theory of truth defines what "best" means? Philosophers of truth are guided in their search for the best theory of truth by looking for the theory that solves most of the problems that they have when discussing and understanding what truth is. When you have defined truth, we can continue with a clear discussion on the conditions that we are now talking about truth in this sense. But if we want to discuss what the term "truth" should mean at all, this will be a more pragmatic discussion guided by problem solving. Philosophers of truth start with a coarse-grained understanding of truth and search for the theory of truth which is best at solving problems of understanding truth. No problematic relativism follows from this fact.

2.5 Excursus: Basic semantic entities

So far, our focus has been on various criteria for a good theory, but now we shall turn to looking at the building blocks in the theory to be suggested. I shall look at the basic concepts to be used in the theory and what the basic ontological building blocks of the world are. Note the difference between basic semantic entities and basic ontological entities: When I discuss "structure" as a basic semantic entity (a concept), I discuss what it means for anything to be a structure, but without saying what the fundamental structures of the world are. When I speak of the specific structures that are the basic ontological entities of the world however, I speak of specific entities with specific structures. This distinction is often unclear in discussions of structuralism.

The following discussion might seem a bit lengthy for some readers, but it is important for the following reason: I will be arguing that many of the things in the world can be reduced to a few basic elements, and it is, therefore, incumbent upon me to be clear about how these basic elements should be understood, which again requires that the concepts are as clear as possible. The building blocks of an ontology should be as understandable as possible in order to be as coherent as possible, which indicates a close link between meaning and truth.

You will find many philosophers and scientists saying that some element in their theory is basic, irreducible, primitive, undefinable, spontaneous, or something like this.[22] While it is true that any theory will have some basic elements that cannot be further explained, there are so many and such different candidates around that I find it very plausible that when people say such things it usually means that their theory has run into problems that their theoretical framework does not have resources to handle. The problem is that their theoretical framework is not coherent enough, not that there are all these irreducible and undefinable entities in the world. In any case, it speaks in favor of a theory if it can explain something which another theory calls unexplainable,[23] and since I shall claim that I do so many times, it is important that I spend some time defining my basic concepts. To this I now turn.

What I have said so far about theoretical frameworks and truth still leaves many questions open. What are the fundamental ontological constituents?[24] How does the mind work? How can language express reality? I shall return to all these questions to fill out more of the picture. So far focus has been on understanding what we do when we try to understand something and how we can argue that a particular understanding is true. But there is still more to say about the fundamental building blocks of the theoretical frameworks – the basic semantic or meaningful entities, which are concepts that can be used to express the world.

Defining basic concepts seems tricky. It would seem to require even more basic terms to define them, but how can any terms then be defined as the most basic terms? The answer lies in seeing what it means to define a concept. To clarify the meaning of a concept is to relate it to other concepts in a theoretical framework. The more connections one clarifies between the concept to be defined and other concepts, the more the concept is defined. Some (inspired

---

22 See for example Lowe (2002), where many concepts are suggested to be primitive or irreducible, for example causality, identity, existence, and agent causation (Lowe, 2002, pp. 191, 213).

23 At least if the alternative theory as a whole has fewer unexplained entities. Explaining one unexplained entity by introducing several others may be of no help.

24 When I speak of fundamental constituents, fundamental means the same as irreducible.

by Rudolf Carnap) distinguish between an explication and a full definition, where an explication reveals some of the meaning of a concept by clarifying its relation to other concepts, and a definition gives the full meaning of the concept (Puntel, 1990, chapter 2.1).

I argue below that meaning, like understanding, comes in degrees. I describe how concepts interpret our experiences and our experiences change how we understand the meaning of terms, in a continuous process. However, we can say that some concepts get a full definition in the sense that it gets a definition where there is not more to understand about the concept than what is offered in the definition. Heat can be defined as molecular motion, and there is always more to learn about molecular motion, but heat is still fully defined by molecular motion in the sense that heat is reducible to molecular motion. The concept of existence, on the other hand, is a basic concept which can be explicated by being described by many features, but where there is still something left unsaid about what existence really is.

I will mostly be explicating the basic concepts, relating them to concepts that are often less basic than themselves but are nevertheless clarifying (Puntel, 2008, p. 414). Since the concepts are connected to other concepts that are again connected to other concepts, most of the concepts in this book will be more and more explicated throughout the whole book, which is to say that they can then be understood better and better. This means that by calling concepts basic here, I only mean that they are simple and central terms in the theory that many other concepts are defined in light of, but in practice all concepts are understood in light of each other with the goal of having the most coherent conceptual framework overall.

There is no correct definition of any term. Rather everybody learns a language which gives a coarse understanding of the world, and the challenge then is to make it more coherent. Some argue that there are correct terms, and that the book of the world should be written in the world's own terms, for example that we should use green, but not grue ("Grue is a thing which is green before (let's say) the year 2100, and blue afterwards") (Sider, 2011, p. 61). According to the view presented in this book, it does not make sense to speak of the "objectively correct terms" or "the world's own terms", but it does make sense to speak of the most coherent description of the world. One could then say that the terms of that description were the correct terms (in the sense of being the terms of the most coherent theory), but in any specific discussion there is no point in appealing to the correct definition of a term – the only thing one can do is to demonstrate that a theory as a whole is more coherent than another.

For example, it is most coherent today to use the term "green" for green things, but if they suddenly turn blue in the year 2100, we will discover that it is more coherent to think that they were "grue" after all. I thus reject the idea of natural terms having a natural meaning or a real or essential definition, and argue instead that any definition is always stipulative: we suggest a coherent way of describing the world by giving certain terms certain meanings and relating them to each other. This is not to say that since we define concepts, the world cannot correct us; our experiences help us discover what is the most coherent way to define concepts.

This is how one should think of all the definitions I offer in this book, and not argue against a definition I offer that that is not what the term really means. If you disagree with my definitions, I would like to learn why your definitions give us a more coherent theory. I try to stick as close to everyday use (sometimes called the reportive definition) as I can in order not to create unnecessary confusion.[25] But if the everyday terms do not give us a coherent understanding of the phenomena under investigation (but rather cause confusion), I suggest revised definitions in order to get a more coherent understanding.[26] The whole book should be understood as departing from an everyday and inconsistent framework aiming to make it more coherent which often includes revisionary definitions where old definitions are causing problems. With this note of definitions made, it is time to start defining terms.

In order for anything to be a part of our understanding or thinking or a part of a theoretical framework, it has to have some similarity over time. It has to be so in order for us to be able to think about it, refer to it, give it a name, talk about it, and so on. If something just flashes in our mind for a moment and then disappears from our mind (including memory), it cannot be thought about or be given a name so that it can be used as a part of a theoretical framework.

25 Herman Cappelen argues that we cannot say what the limit is for revising a concept, since it is itself a matter of revision (Cappelen, 2018, p. 116). I agree, but would add that our main guide in concept revision is to create coherence while avoiding misunderstanding. Gradual revision with overlapping meaning can create more coherence while avoiding misunderstanding, while a strong and quick revision will create misunderstanding, implying that one should use another word instead.

26 Herman Cappelen argues that conceptual engineering is an inscrutable process over which we have no control (Cappelen, 2018, p. 72). What he means is that we have no control over how people actually understand concepts, whereas my intention is merely to "engineer" what I mean by a concept and try to convince the relevant scholarly disciplines that this is the best way of understanding it. Cappelen argues that even when we give stipulative definitions, we do not determine the meaning, since we are still using (for example) English to describe the meaning (Cappelen, 2018, p. 76). I reply that I just intend to clarify with the means by hand without presuming that I will be perfectly understood.

Some similarity over time is thus required, and this may be just similarity over time in the mind.

If something has sufficient similarity[27] over time (or even remains identical, in a sense to be explained soon) it can be understood as a *structure*, for something can be a structure both at a time and over time. More precisely, a structure is something which is not simple and not unconnected, rather it is a collection of elements with a certain relation between (Puntel, 2008, p. 27). We are unable to understand something if it is not connected to something else, and thus we would not be able to understand something completely simple not connected to anything else. Every element in a structure is then always a sub-structure in (or part of) another structure. Note that this is a wider sense of structure than understanding structure just as the abstract relation between elements, but disregarding the elements.

Structures can vary from being very loosely connected and very unstable over time to being tightly interconnected and stable over time. Individuals we have given a name to, are typically quite stable and interconnected structures. As will become clearer in the next chapter, the simplest possible structure that can exist is a value actualized at a point for the briefest possible time, and this value will then be part of the field where it is actualized. This simplest possible structure can only be understood by us as part of a larger structure and by keeping it in mind over time.

What connects the elements in a structure? The answer, which will be clearer later, is that laws of nature connects physical structures, including brain structures, which can cause us to think about structures in our mind. As mentioned, these can vary from being loosely connected and unstable to being densely connected and stable.[28]

What it means for structures to be stable or identical over time or not should be further explained. Structures can be identical, similar or different, and parts of structures can also be identical, similar or different. It is not easy to define terms like "identical", "similar" and "same" without giving a circular definition. The term "same" is especially ambiguous, since it is used for things said to be identical *over* time in addition to identical *at* a time. I shall use the terms "similar" and "identical" instead of "same", and by "identical" I mean "identical at a certain point of time", saving the question of identity over time for later. Since the idea here is to find a definition that works for basic semantical entities in a theoretical framework, we can explicate the meaning of *identical* structures

27 By "sufficient similarity" I mean sufficient enough for a mind to perceive it as similarity.

28 Thanks to Alan White for helping me to think clearer about structures and their connections.

as being indistinguishable structures. That structures are *different*, on the other hand, means that they *are* distinguishable. That structures are similar means that they are almost or quite indistinguishable. Structures are distinguishable when it is possible for a mind to keep them apart from each other in mind by noticing differences, which is something not identical.

To say that identical structures (at a point of time) are indistinguishable still leaves open whether we are talking about type or token identity. Two structures are type identical if all their internal parts (as opposed to external relations) are identical, while they are token identical if their external relations are also identical (including location in space and time). What does it mean to say that two structures are token identical? If they are token identical, it can only be one structure, so how can I speak of two structures? More precisely, it means that we can discover that what we thought were different structures – for example the Evening Star and the Morning Star (which are both the planet Venus), or the holocaust and the Shoah – refer to one and the same structure only.

By saying that they are token identical we do not just say that they are two individuals that are type identical in virtue of having identical internal parts, but instead they are token identical, or referring to the same event. The structures in our mind when using different terms like "Atle Ottesen Søvik" and "Kristian Ottesen Søvik's father" are not token identical, since they have different content. But as I will explain in detail below by means of an example with Clark Kent and Superman, these different structures can be understood as coarse-grained structures that are both part of one structure which can be described in a more finely grained way.

Anything that can be understood is a structure, as shown above (Puntel, 2008, pp. 11, 168).[29] By definition, a structure relates individual elements. In a

29 Several people (like John Wheeler, Seth Lloyd, Paul Davies, James Gleick or Vlatko Vedral) have suggested that information is a fundamental entity in the universe (giving us "it from bit"). Einar Duenger Bøhn argues that since information can be coded into different mediums, the information itself is something more basic than any physical medium (Bøhn, 2019, chapter 3). My own view is that there is no such deeper level of information. "Information" is a vague concept used in many different ways, which have in common that there is an understandable structure (which is understandable because it is a structure). That information can be coded into different mediums does not show that there is something deeper called information, it just means that a structure can be similar in different mediums, and related to something, like an idea to a word. For example, the year rings in trees is not information that anyone has coded into the tree, but rather it is a structure with a known cause that we can relate to our calendars and understand how old the tree is (get information about the age of the tree). DNA is information that has come about in the same causal way as year rings. There is no need for something called information in addition to explain what happens.

broad sense of the terms "individual" and "relation", all structures are individuals and all structures are relations. That is because to be one structure is to be one individual in a broad sense of individual. And to be a structure is to be a relation in a broad sense of relation, since all structures contain relations. But there is a narrower definition of the terms "individual" and "relation", where individuals and relations can be different entities.

In the narrow sense, individuals have a quite stable and quite interconnected internal structure (Puntel, 2008, p. 264). For example, a human body is quite stable and interconnected in its internal structure, having an outer border, arms above, legs below and a head on top, a genetic code, etc., and the structure of the body remains quite stable while the body moves around in its environment.

In the narrow sense of relation, there are two main types of relations. The first is relation in the sense of being a sub-structure of an individual. An example would be "having an arm" or "being green", which is often called monadic relations or properties in the philosophical literature. The second is relation in the sense of being structures that contain individuals, but where we disregard the individuals. For example, you can have a structure where one individual is larger than another individual, and then you disregard the individuals, and what is left is the relation "is larger than". This is often called a dyadic or relational property in philosophical literature. Such a relation need not be only a relation between individuals. It can be a relation between relations or between an individual and a relation. For example, you can have the relations "larger than" and "smaller than", which are related by the relation "being opposite of" since "larger" is opposite of "smaller". From now on I use the terms "individuals" and "relations" in their narrow senses.

Speaking of individuals and relations this way is more intelligible than speaking of substances or things having properties. That is because it is difficult to understand what a substance is when you take away the properties since then nothing seems to remain, and it is difficult to understand how a substance can "have" properties. I have substituted "substances" with "individuals" and "properties" with "relations". In the terminology suggested here, to be a substance is to be an individual in the narrow sense. It is a configuration of parts that have a quite stable and quite interconnected internal structure. For a substance to have a monadic property means that whatever is referred to as the monadic property is part of the individual. That a substance has a relational property means that there is a structure that has the individual as sub-structure.[30]

---

30 The idea of understanding a substance as a configuration of parts I have from Lorenz Puntel, first in Puntel (1990), but further developed in Puntel (2008). The definitions of individuals and

There is a classical objection to the picture I have drawn above, for there are many criticisms of substance ontology favoring instead relational ontology, structuralist ontology, trope ontology or something else, and the classical critique that substance ontologists give them is to say that something must carry the properties. There cannot be relations "all the way down"; there must be some basic individuals first, which then can have properties or enter relations or form structure. While substance and relation may both be basic *semantical* terms, the argument is that *ontologically*, you need a substance before you can have properties, relations, and structures (Ladyman, Ross, Spurrett, and Collier, 2007, p. 138). Such is the critique, and my response will be unpacked when we look at ontological entities below, but briefly put I will argue that we do not need substances. Instead we have fields with values at points, and then we have a mind which can experience these values and find structures or patterns among the different field values and make concepts for them.[31] Ontologically speaking, no substance is needed, and we can get rid of many superfluous concepts like essences, this-ness, or haecceities, which are vague concepts meant to explain what makes something what it is or unique.[32]

Concerning the objection that there must be individuals before relations: Recall the distinction between the concept of a structure and the specific structures that exist. When it comes to the concept of a structure, we cannot understand the simplest parts of a structure in complete isolation, since understanding something means to relate it to something else. What we can do is to clarify relations as much as we can in order to understand as much as we can. When it comes to specific structures that exist, I argue that the simplest structures are the values that are actualized at points in fields, to be described further in the next chapter.

Another advantage with this focus on structure is its ability to deal with the following problem. On the one hand there is the problem of theory change: it seems that old theories are constantly replaced with new theories with new terminology, even when the old theories seemed quite good. Can we then trust our

relations are my own, but I would not be surprised if others have defined these the same way without me knowing it (I may have read about it and then forgotten it). What I call a "relation in the narrow sense" is called "pure structure" by Puntel (Puntel, 2008, p. 29).

31 I replace vague concepts like the "essence", "thisness" or "nature" of a thing by the concept of the structure of a thing (Puntel, 2008, p. 476). I use both "structure" and "pattern" to mean the same, but since "structure" gives associations to something static and "pattern" gives associations to something dynamic, I often use "structure" if it is mostly static and "pattern" if it is mostly dynamic or both terms if it is both.

32 Rejecting these ideas means that if you ask me whether it was possible that somebody could have been born by my parents, have my body and live my life, and still not be me, I answer no.

own theories being approximately true? On the other hand, there is the no-miracles argument. Since we have landed on the moon and can make phone calls to Africa, it seems that our theories must be approximately true; otherwise it would be a miracle that they work so well. James Ladyman argues with John Worrall that structuralism is the solution to this dilemma, since through theory changes and new terminology similar structures remain, which indicate that the theories are approximately true (Ladyman et al., 2007, chapter 2). By focusing on structure in my own ontology, I hope it will be better able to integrate changes in other theories. What I mean is that even if I refer to natural science in many places to support my views, I hope that the ontology I present will be able to integrate new theories that might possibly replace the ones we have today.

In the next six paragraphs, I will answer some objections against different understandings of relations to show how my account solves these problems, but they may well be skipped by those who are not especially interested in the analysis of relations. The problems are as follows: What is a relation? Where are relations located? How do relations relate? Some are skeptical to relations because of Bradley's regress: Either a relation is nothing – but then how can it relate? – or a relation is something. Then it seems that a relation that is something between A and B itself needs to be related to A and B, but then you get a regress of relations where each of them needs a new relation to be related to the others. So, what are relations and how do they relate?

I believe these questions can easily be answered within the theoretical framework suggested here, and suggest some brief answers. I will also say more about the existence of individuals and relations when I write about existence later. Now first: What are relations? When we call something a "relation", what are we referring to? As explained above, relations in a wide sense include individuals, so that Socrates is a relation. But in a narrow sense, they are either sub-structures of individuals (e. g. having a body) or structures that have individuals as parts but disregarding the individuals (e. g. Socrates being married to Xantippe – disregard Socrates and Xantippe and "being married to" is the relation that remains). This last type of relation comes in different meta-versions and combinations, since you can have a meta-relation between two relations, meta-relations between meta-relations, etc., or combinations like a relation between a relation and an individual, etc. For example, the relations of a circle's circumference and its diameter are related by the meta-relation pi, and pi squared is a relation between the two meta-relations "pi".

In all these cases, relations are structures or patterns that we find in theoretical frameworks expressing the world, and in almost all cases these structures are found to occur many times and many places in the world as individuals of the world move around. We give them names like "left", "square", "being the uncle of", and "being the square root of", and these are very useful concepts when we want to describe the world with all its individuals moving around. It is much more efficient to say that someone is the uncle of someone than to describe how he either became the brother of someone who had a child or married a woman whose siblings had children, etc. And it is much more efficient to say that something is square than to describe the values in all the points in that area.

Where are relations located? Sometimes it makes perfect sense to say that relations can have a location, if we use the term location in a wide sense to refer to an individual or in a narrow sense to refer to a part of an individual. So for example the royal palace of Oslo is located in Oslo, and the front of the royal palace of Oslo is located in front of this palace. But many relations are just terms used to describe recurrent patterns like "being the uncle of". Becoming "the uncle of" is a long process of brothers or husbands being born and having siblings (in-law) having children, and it happens many times and many places. "Being the uncle of" is not located in the world outside of individual minds – it is a structure in a theoretical framework. Uncles are individuals who are located at places in the world, and we can give useful information about two individuals by saying that one is the uncle of the other. We can use this structure from our theoretical framework to say that this recurrent pattern of being an uncle has happened in the world in this case also: the man in this case has siblings (in-law) that are the parents of the nephew or niece in this case.

How do relations relate? Relating is not an action or activity that an individual performs; rather, relating is an abbreviation for a certain structure in a theoretical framework. When it comes to Bradley's regress, his point was that either relations are nothing – but then they cannot relate – or they are something, but then they need relations to relate them to others in an infinite regress. This is a pseudo-problem since relating is not an activity that you need a causally efficient individual to perform, but rather it is what I just described above: a pattern in a theoretical framework.

Notice how "identity" or "similarity" was important for understanding the stability of a structure. Sometimes a structure can be type identical from moment to moment (in the sense that all internal parts are type identical), like an elementary particle, and an exact definition is possible. But sometimes a structure is just *similar* to itself from moment to moment, like a human being, and what it is to be a human being may not have an exact definition, but rather be a concept with fuzzy borders. The same goes for individuals, who need only be quite stable over time and quite interconnected in order to be an individual, although an individual might also be a precisely defined structure. And the same goes for relations, which include individuals in the definitions I have given here – sometimes structures, individuals and relations and the specific examples of such will be imprecise concepts.

This means that what it is for an individual to be identical to itself over time may not always be clear, since selecting something as an individual means to find a stable pattern over time, but without it being possible to decide exactly where the limits are. The examples are well known, like where is the exact limit for something to be baldness, or exactly when did a boat being built become that boat, or how drastically can the boat be changed and remain the same boat, being identical to itself over time.

Some try to solve such problems with haecceities or other creative solutions, whereas I think it is better to acknowledge that individuals are patterns we describe among values actualized in points (to be described in the next chapter), and thus we can understand why sometimes coarse-grained concepts for individuals become unable to describe exactly what is happening, but we still have full understanding of the situation. When a boat is being built or a person is getting bald, there is no mystical point at which suddenly an ontological universal of baldness descends upon a head, or the idea or haecceity of the boat enters into the materials. "Boat" and "bald" are useful concepts, but with fuzzy edges, and more precise descriptions are possible (of exact locations and numbers of hairs and exact structure of boat in progress).

This question may nevertheless feel more important when it comes to personal identity over time, and to whether it is conceptually possible to survive death, or fission (your body and brain being divided in two and connected with new halves), or teleportation, etc., and I will return to this question in the chapter describing the self.

I argue in this book that very many philosophical problems are pseudo-problems that arise because they are formulated within theoretical frameworks and with concepts that are ultimately imprecise and incoherent. When you start to ask precise questions about agents, free will, causation, time, substances, identity over time, etc., insoluble problems arise because the theoretical framework is useful in many ways, but ultimately incoherent. One could argue that I just ran into problems with the concept of individual, but that was only concerned with individuals at complex ontological levels. These are configurations of more basic individual parts that I will describe more exactly in the next chapter. This theory is not vague in its understanding of individuals, but rather explains why many specific concepts for specific individuals (like boats) are vague.

What I have tried to establish in Chapter 1 is that the primary goal of a theory should be to develop the most coherent theoretical framework. In this book I will try to offer the most coherent theoretical framework for ontology that I can think of, which is a theoretical framework having as its main components a field of points where possible values are actualized according to certain rules. Fields, points, values and rules are structures, individuals and relations. They would not be understandable if not. But while structures, individuals and relations are the most basic *semantic* entities, I shall argue that fields, points, possible values and rules are the most basic *ontological* entities in the sense that they are the basis for the existence of minds with language to understand these ontological entities. I am not saying that structures, parts, individuals, relations, etc. are the basic building blocks of the world, but that they are the basic linguistic/mental concepts to use for best understanding the basic building blocks of the world.

Many questions are still unanswered, but I will turn to ontology in the next chapter, where several parts will support and explicate the methodological part. In any case, the theoretical framework here offered is only meant to be the best theoretical framework currently at the market. Another may prove better in the sense of being more coherent or with a more well-argued criterion than coherence. I end this chapter with an excursus with more details on how mind, world and language are related.

2.6 Excursus: More on mind, world and language

When discussing truth, I argued that a true proposition is one which is a part of the most coherent description of the world, but I should give a more detailed presentation of linguistic entities like sentences, propositions, references, and so on. Questions to be dealt with are questions like how it is possible for a string of symbols to have meaning and to refer to states of affairs in the world. What is the connection between the signs of a sentence and minds, meaning, and the rest of the world?

For readers who are not familiar with discussing such concepts or have a particular interest in the questions, this may feel to be a bit too detailed. That is why it is placed in an excursus. As the topics covered in this section are closely related to other issues discussed in this book, keep in mind that they will be further clarified as we look at physics, consciousness, mind and other topics later in the book. However, a preliminary introduction is given here. I start by offering short definitions and answers to questions before I address some relevant controversies in the academic literature and answer some possible objections.

First, let us consider what a sentence is. Sentences are strings of words which are composed of physical signs or sounds that can be read or heard. I shall only be concerned with sentences that in normal circumstances can be understood as expressing something that can be true or false about the world.[33] This means that I leave out sentences like "yikes!", "go to your room", "Tuesday jumped greenly" and "Gnuk saic hoc", but include sentences like "2 + 2 = 4", "it's raining" and "Es regnet" (which in German means "It's raining"). The same sentence *type* (having the same internal structure of signs) can be uttered (usually written or spoken) or perceived (usually read or heard) at different times and places, which would then be *token* sentences (same type, but at different times or places).

Sentences are said to have meaning. I shall be concerned with three different meanings sentences can have:

1) the meaning of a sentence is the mental content that somebody who utters a sentence wants perceivers of the sentence to have;
2) the meaning of a sentence is the mental content that is activated by the sentence in the mind of an individual who perceives or thinks about a sentence; and
3) the meaning of a sentence is the mental content that is commonly activated by the sentence in the mind of individuals who perceive or think about the sentence (at the period under consideration). In all cases, there could be more than one meaning to a sentence.

When it comes to mind and mental content, I will say a lot more about this in Part Two of this book. However, note the difference between, on the one hand, a non-conscious mind, which is a form of brain activity that can become conscious, and a conscious mind on the other. Conscious mind *consists* of something which makes it conscious mind as opposed to non-conscious mind (this something is qualia values, which I will describe later). But conscious mind will also have a *content*, a structure that one is conscious about, for example an apple or someone sitting in a chair. By "mental content" I refer to such structures in mind (like apples and chairs) that one can be conscious about or not.

The three kinds of meaning presented above are all mental contents in the minds of individuals. Yet we should distinguish between *actualized* and *possible* mental contents. Mental contents can be *actualized* in the mind of somebody now, but there are also many *possible* mental structures that can be actualized in the mind of someone either now or at another time or never.

There are different kinds of mental content. One kind of mental content is an understanding of the world. I here use "understanding of the world" in a wide sense to include a perception or representation or understanding of either how the world actually is or how a world could possibly be. Such a wide under-

33 Such sentences are often referred to as "statements".

standing of the world can be true or false as an understanding or perception or representation of our world. This wide sense of "understanding" is what I mean by "understanding" in this subchapter. Our world is the one we inhabit.[34]

A proposition is a mental content which is an understanding of the world which can be true or false. A sentence expresses a proposition when it activates such a mental content in somebody who utters, perceives or thinks about the sentence. In this subchapter "mental content" refers to understandings of how the world or a world could be. I describe how such mental content is activated in the mind later in the book. A sentence type has the potential for activating mental content also when nobody is uttering it, reading it, or thinking about it. The potential for activating mental content is what it means that a sentence expresses a proposition in general, whereas the actual activation of a specific mental content in a specific situation is what it means that a certain token sentence expresses a certain proposition at a certain time and place. It is not the case that a specific sentence type expresses one proposition or that we can find the correct proposition that a sentence expresses. What is closest to the idea of "the correct proposition" is the mental content that is commonly activated at when people perceive or think about the sentence (at the period under consideration).[35]

As mentioned, a proposition can be true. What it means for a proposition to be true is for it to be part of the most coherent understanding of the world. When an understanding of the world is true, it can be a coarse-grained or a fine-grained truth. The more precision and details and connections it involves, the more finely grained it is. The recognition that true propositions can be coarse-grained truths is going to solve many problems in how to understand the relation between propositions, facts and states of affairs.

A fact is nothing but a true proposition (Frege, 1956, p. 307). I here use the term "true" to mean true about our world and a fact in our world. "Propositions" and "facts" are not the same, since false propositions are not facts, but true propositions and facts are the same. Like propositions, facts can be coarsely or finely grained.

The next concept I will discuss is "states of affairs". Here we need to distinguish between two levels of analysis. The standard level of analysis is when we

34 I define different meanings of "world" in Chapter 3.

35 Let us say that a person intends to convey a message to a large group of people, but that 99% of the listeners misunderstand it. It may seem strange to say that their misunderstanding is the correct proposition, but on the other hand it also makes sense to say that the person chose wrong words to communicate his or her message. In such a situation I find it most coherent that the audience misunderstood the intended meaning, but not the correct meaning of the sentence.

divide between mental contents in the minds of individuals on the one hand and, on the other hand, states of affairs different from the mental contents in the minds of individuals, which in many cases are a matter of how physical objects are related to each other in the world outside of individual minds. On a deeper level of analysis, we take the deepest and broadest possible perspective on the world and recognize that the distinction between mind and states of affairs is still a division made within mind, and thus that we only know states of affairs as they appear to us in mind. This is true even if we hypothesize that there are states of affairs in the world which are not mind and which make certain understandings of the world more coherent than others.

This distinction between levels of analyses has the following implication for an analysis of states of affairs: At the standard level of analysis, states of affairs are parts of the world different from mind and including the mind, which the mind can perceive, understand, represent and think about as mental content. These might be things, events, relations and so on. At the deep level of analysis, everything that we call states of affairs are accessible to us only as mental content. We are subjective conscious selves interacting with conscious experiences only (I write more about subjectivity and the self later in the book).

This means that for us, when concerned with any particular state of affairs and what it is like, at the deep level of analysis there is no fundamental difference between an actualized state of affairs and a true proposition or a fact, since all of these can only be experienced as conscious mental content. A proposition is a possible state of affairs, and a true proposition is an actualized state of affairs, which is also a fact. When we feel that we experience a state of affairs like a chair, we experience our own conscious representation of the chair. When we say that the world offers resistance to our understanding of it, it means that we have a conscious experience which contradicts what we thought before. When we say that we discover what the world is really like, what we actually discover is a more coherent understanding of the world than what we had before.

Still, we have reason to think, even at the deep level of analysis, that states of affairs are something different than mental content, in order to explain why some understandings of the world are more coherent understandings of our experiences than others. There must be something about the world outside of mind which make some understandings true and others false. We cannot say anything about that, which is not itself mental content.[36] If we describe physical properties

36 A common goal of metaphysical theories is that they should "carve nature at its joints" (Sider, 2011). However, "nature's joints" is just a vague metaphor and obviously a content in our mind, but it seems to aim at what I am after here: that outside of our minds is that which makes some theories more coherent than others.

of physical particles, we are describing hypotheses and experiences which are all mental content, so we cannot know what it is like independently of how we experience it.

Does this mean that we cannot say what the world is like outside of our mind? Can we know nothing about what makes it the case that some understandings of the world are more coherent than others? We can hypothesize that the world (here understood as that which is not identical to contents of any mind but making it the case that contents of mind are one way rather than another) is quite close to the way our theories say that the world is – that there are such and such elementary particles, etc. And we have quite good reason to think that it is, for it would be a miracle for our technology to work so well if it was not the case that it is based on theories on electrons, etc., which are quite close to the truth.

I move now to the topic of "reference". The *meaning* of sentences is sometimes contrasted with their *reference*. When it comes to the question of reference, we should distinguish between, on the one hand, what it means for a sentence to refer to something and, on the other hand, what the reference of the sentence is. We want to know what the relation is between the sentence and the reference, and what the referring mechanism is that makes reference possible. Not only sentences, but also words and names can refer, so in the following I will speak about how "words" refer, and by "words" I include sentences, expressions, names and body language, like referring by pointing.

What does it mean for words to refer to something? We should here distinguish between the intended reference of someone who wants to refer to something and the perceived reference of someone who takes a word to refer to something. There is no correct or actual reference that a word has, there are only intended and perceived references. What is closest to the intuition of a correct reference, is the reference that it is most commonly perceived that the words refer to (at the period under consideration among the most regular users of the word), either given a specific type of context or just in the context of the language that the words are a part of.[37]

Before I continue to answer what it means that words refer, I must say something about what the reference of words is. The reference of one or more words is a state of affairs, either the state of affairs that it was intended to refer to (intended reference), the state of affairs that it was perceived that one or more words referred to (perceived reference), or a state of affairs that it is most commonly perceived that one or more words refer to (at the period under consideration

37 As the variables in the definition show, there is no correct definition of a reference beyond what we decide to call the correct reference.

among the most regular users of the word), either given a specific type of context or just in the context of the language that the words are a part of (normal reference).

How do words refer to their reference? In the case of an intended reference, that a word refers to a state of affairs just means that a person thinks about a state of affairs and uses a word for it that he or she associates with that state of affairs. In the case of perceived reference, that a word refers to a state of affairs just means that the word activates a representation[38] of a state of affairs in the mind of the perceiver. There is no direct connection between words and states of affairs in the world; they are only connected via the minds of individuals.

This understanding of reference implies that in many specific situations referencing will fail. For example, a person may intend to refer to a state of affairs in our world which actually is not a state of affairs in our world, and persons perceiving the words may not know what they are meant to refer to, or may think about something other than what was intended. However, a person may succeed in referring to state of affairs X in the world outside of her mind if this person and the perceiver both think about X – even if what they think about X is wrong. A common example of this is if person A refers to "the man drinking champagne in the corner" and person B thinks about the same person in the corner, even if that person is not drinking champagne (Kripke, 1977, p. 256). In this case, the referencing succeeds in the sense that intended reference and perceived reference is the same, even if the proposition that there is a man in the corner drinking champagne is false. There exists no "objectively correct reference of the sentence" beyond intended and perceived reference. What is closest to the idea of an objectively correct reference is the normal reference as it was defined above.

I said that words refer by activating in the mind representations of states of affairs that are associated with the word, but how do words become associated with states of affairs? This can happen in many ways. One can make up a term or a name or a description for something that one is thinking about for oneself, something that small children often do in referring in creative ways to different states of affairs. Or one can write some crosses and circles on a blackboard and say that it refers to football players, war ships, or something else.

More often, one has learned what words are commonly used for different states of affairs, such as "chair" or "Bob". This again requires either that some-

38 I will say much more later in the book about what representation is and how it works. One could worry that it does not help to explain propositions by referring to mental contents if it is equally mysterious how mental contents are able to represent the world, but I suggest a theory of how this works in the chapters on mind and consciousness.

one at one point decided that these states of affairs should be referred to as either "chair" or "Bob", or it has evolved as a common use, and what is common use may change over time.[39] Some words refer to many states of affairs, like "chair" referring to many chairs, and some words refer to unique states of affairs connected to space and time, such as Elvis Presley or the Jurassic age.

In addition, there are words and body language that can be used to refer to many different states of affairs in specific contexts, such as "this" and "that" or pointing at something. I now proceed to discuss some common debates and objections connected to these topics, especially the debate between Frege and Russell on propositions and reference. I will look into some common questions and objections in these debates to explain how my own proposal deals with them.

Gottlob Frege thought that propositions can neither be in the outer world nor in the inner mental world, and thus they must exist in a platonic realm of abstract entities (Frege, 1956, pp. 302, 311), while Bertrand Russell thought that propositions are just the states of affairs in the outer world that sentences refer to (B. Russell, 2010 [1903], p. 47). There are important arguments in favor of propositions not being parts of the inner mental world of individuals. First, it seems that propositions could be true even if no mind existed. Second, it seems that propositions are shareable between individual minds. Since I argue that propositions are mental content I need to answer these objections, which I will do presently.

The first argument is that it seems that there could be true propositions even if no mind existed. For example, it seems that we could say about a possible world containing only gas that the proposition "gas exists" is true and exists in that world even if no minds exist in that world. Since I define propositions as mental content, I reply that the proposition "gas exists" does not exist *in* that world, even if we can say it truthfully *about* that world that gas exists there, but that is exactly because mind exists in our world.[40]

The second argument is that propositions are shareable – it seems that one proposition can be in the mind of several people and expressed by several differ-

39 Saul Kripke is a famous defender of the theory that names get their meaning by being dubbed or baptized, which fixes the reference, and then the reference is transferred or borrowed by others (Kripke, 1980, pp. 96–97, 135–140). I describe this process in a vaguer manner by saying that the commonly perceived reference of terms evolves and changes over time. This is a true description of what happens, which then avoids problems connected to when people make mistakes in dubbing or borrowing references (for example, that people started using Madagascar for the island outside of Africa even if another area was first called Madagascar (Evans, 1973, p. 196).
40 This distinction between "in" and "about" a world is another way of making a distinction between truth *in* and *at* a world (Adams, 1981, pp. 20–24).

ent sentences. My response to this is that there can be mental contents in individual minds that are very similar to each other (or even type identical), but that shareability is no warrant for thinking that propositions must exist outside of mind, since it only means that minds can think of similar things.

Russell thought instead that propositions (the meaning of sentences) are the states of affairs that the sentence refers to (B. Russell, 2010 [1903], p. 47). This seems problematic if the sentence refers to something that does not exist, since the sentence then does not have a reference. One could try to escape this problem by saying that it refers to something in a possible world and not in our world, but the problem gets worse if a sentence refers to something that could not possibly exist, for example the largest possible prime number or a person who traveled back in time and killed his grandparents before he himself was born. A sentence about the largest possible prime number or a time-traveling person killing his own grandparents before his birth seems to have meaning, but presumably they do not have a reference in any possible world. How can meaning of a sentence then be the reference of a sentence?

The most famous challenge to the Russellian view is Frege's puzzle. Frege made a famous distinction between "Sinn" and "Bedeutung" or "meaning" and "reference" (Frege, 1948). His point was that we cannot say with Russell that the proposition is just the reference – they must be distinguished. He then used some examples to show this, which are referred to as Frege's puzzle(s). Consider these sentences: "The Morning Star is the Evening Star" or "John believes that the Morning Star is the Evening Star". They seem like non-trivial, non-necessary, informative sentences. The Morning Star is a bright star one can see in the morning and the Evening Star is a bright star one can see in the evening, but in fact they are both the planet Venus, which means that both the Morning Star and the Evening Star refer to Venus.

If Russell is right that both the meaning of Morning Star and the meaning of Evening Star is Venus, then the sentences above are like saying "The Morning Star is the Morning Star" or "John believes that the Morning Star is the Morning Star". They then become trivial, unnecessary and non-informative, which they did not seem in the other case. Frege makes a good case for a distinction, but how should we understand more precisely the propositions and references of these sentences?

When it comes to "The Morning Star is the Evening Star", the *proposition* that is most likely intended and most commonly perceived is the mental content that what we know as the brightest star in the morning is actually the same star as the one we know as the brightest star in the evening. When it comes to the same sentence, but starting with "John believes ...", the most likely intended and most commonly perceived proposition of the sentence is that John believes the proposition just described (that the two stars are one and the same).

The *reference* (most likely intended and most commonly perceived) of the sentence "John believes that the Morning Star is the Evening Star" just is the state of affairs that John believes the proposition that the Morning Star is the Evening Star, as described above. But what is the reference of "The Morning Star is the Evening Star"? Does it refer to one or two states of affairs?

As argued above, there is no correct reference, only an intended and (most commonly) perceived state of affairs that the sentence refers to. What makes the question tricky is that the state of affairs in this case can be just Venus, or it can be Venus shining in the morning or Venus shining in the evening. The sentence refers to different states of affairs, and what is most likely intended and most commonly perceived as reference is two states of affairs being part of the same state of affairs. The two states of affairs of the star shining in the morning and the star shining in the evening are both part of the same state of affairs, namely Venus shining in both the morning and the evening.

Compare this with "Clark Kent is Superman". The state of affairs of "a man with glasses working at the *Daily Planet*" and the state of affairs of "a flying superhero" are parts of the same state of affairs of "a man sometimes working at the *Daily Planet* and sometimes flying heroically around". Morning Star and Evening Star or Clark Kent and Superman are coarsely grained descriptions of states of affairs that could be described in a more finely grained way. One could make many examples with coarse-grained and fine-grained descriptions of one state of affairs. Imagine a newspaper writing stories about three birds observed in a forest where no birds were before. A family reading about it realize that it must be the three nightingales they have observed in the forest and say to each other, "The three nightingales are the three birds in the forest".

The last question I want to address when speaking of propositions is the question of indexical propositions, for example "I am here now". There are many different nuances to be made in a thorough discussion of indexicals, for while indexicals typically are words that locate a speaker, many other words are similar to indexicals in how dependent their meaning is on their context. I shall only make a brief and overall comment here to address the question of whether indexicals are irreducible to other propositions.

One could argue that indexicals are reducible in the way that one can specify the content of the indexical terms with non-indexical terms. For example, instead of saying "I am here now", that proposition could be translated to "Atle is in Oslo at 9 am 27 March 2021". But one could also argue that indexicals have an irreducible meaning, for example "I am here now" seems infallibly true, while "Atle is in Oslo at 9 am 27 March 2021" could very well be false. Of course, "I" and "here" and "now" can refer to many different things, but when "I am here now" is uttered by a person, it seems that in that context it is infallibly true. How is such a difference to be accounted for?[41]

I believe that the general picture drawn above accounts for the interesting things that seem special about indexicals. Most statements are coarse-grained truths that presuppose a theoretical framework to have the meaning they do, without there being an objectively correct meaning. In this book I will show that "I" can mean many different things, "exist" can mean many different things, and "now" can mean many different things, and of course "here" can be understood in a narrow or broad sense.

We give a sentence – indexical or not – one of several possible meanings by locating it in a theoretical framework. It is only infallible if we give it a tautological meaning, like "the statement being uttered now is being uttered now" or "the one who is speaking now/standing here is speaking now/standing here". In other words, there is no special infallibility to indexical propositions; they are reducible to non-indexical propositions and become infallible when we make presuppositions it would be inconsistent to deny.

With the distinctions and definitions offered here, I have tried to offer explanations of how sentences can have meaning and refer to the world, without the need for introducing any irreducible entities like propositions or a reference relation between language and the world. Further details on how the mind works are offered in Part Two.

41 I am thankful to Florian Fischer for making me aware of this objection.

3 Fundamental Ontological Entities

In this chapter, the goal is to present a theoretical framework describing the most basic constituents of the world that exist and how are they related. This will be the theoretical framework used to answer all the questions in the rest of the book.

I start in Section 3.1 with some background knowledge from quantum field theory in physics. In this theory, there is a field for every elementary particle in the world, which includes the forces. Every field has its own field equation, and by using these we can find that certain qualitative values like mass, charge and spin come in different quantitative values at different points in the field, in accordance with the rules called field equations.

Similar to this picture from physics, I suggest that everything that exists is a field where qualitative values can be actualized in quantitative values according to rules. In the next parts of the chapter (Sections 3.2–3.6), I define each part of the previous sentence in detail, and say what I mean by "field", "values", "actualized", "can" and "rules". When discussing the concept "can" (in Section 3.5), I also discuss the question of modality, since "can" implies "possible", which is a modal term.

Some philosophers argue that dispositions are a fundamental part of the world; namely, that things have dispositions that determine their behavior. I argue that the concept of physical possibility as understood here makes the concept of dispositions superfluous, and offer further objections to dispositionalism in Section 3.7. An objection that could be raised against my understanding of modality is that there are other kinds of necessity than the ones I discuss, for example a posteriori necessity. I discuss this objection in an excursus at the end of the chapter (3.12).

Sections 3.2 to 3.6 present the basic entities of the world, but how are they related? This is the topic for Sections 3.7 to 3.9. I have said that the world consists of values being actualized, but a deep and difficult question is to understand what actualizes the rules: What makes the rules being followed? What makes motion happen the way that it does?

I enter this discussion by presenting a main distinction in how laws of nature are discussed in philosophy: are laws of nature merely formulations of coincidental patterns or is there something making the world behave according to rules? I do not conclude in this section (I conclude in Section 3.10), but bring out the arguments since they are relevant for helping us decide how to understand the relation between the basic entities.

https://doi.org/10.1515/9783110770957-005

Another important question for deciding how to understand the basic entities is to discuss whether it is possible that fields, values, rules and actualization do not exist. This is discussed in Section 3.8, and I argue that any ontology must include something corresponding to fields, possible values, rules and actualizations, which again implies that the simplest ontology is the one employing just these entities and nothing more.

Having concluded that we must include these four entities, how do they relate to each other? In order to have a criterion for choosing an answer, I discuss in Section 3.9 whether the basic ontological entities in a theory should be simple or complex. On the one hand, in any theory of the basic constituents of the world there will be something unexplained and we have no good reason to think it is simple. On the other hand, it is reasonable to seek simplicity in the sense of having the fewest entities, the simplest structures and the fewest connections, because this gives us the fewest unanswered questions. It is a goal for a coherent theory to have as few loose ends as possible.

We shall see that we must have fields, points, possible values, rules and actualizations and actualizers in our ontology. But how can this all be combined in a way that gives as few loose ends as possible? I discuss in Section 3.9 different alternatives in light of philosophy and physics and conclude that the simplest combined understanding of the basic entities is to think that there is an actualizer which has a structure that implies that the values being actualized follow certain rules. This actualizer is again a structure in a field of possible values. This is certainly a complex structure, but it seems unavoidable to believe in the existence of its parts, and their combined existence is the simplest way of envisioning it.

Even if I have already been discussing what is the most basic entities that exist, I have not defined existence. The reason is that I needed to land on how to understand the basic structure of the world before I could define existence, since I will use it in my definition of existence. What does it mean to exist? What is the difference between saying that something exists merely in mind or that it exists outside of mind or that a possibility exists? What does everything that exist have in common which makes it true to say that they all exist? These are the topics for Section 3.10. The chapter ends with a short discussion of the concept of "nothing" in Section 3.11, before an excursus on analytic and synthetic a priori and a posteriori statements in Section 3.12. I end with a comparison of my theory with important alternatives in Section 3.13.

3.1 Background: Fields in physics

The part of natural science that mostly deals with the fundamental entities of the world is physics. Many people think that the world is built up from elementary particles. Such a view has a great intuitive appeal since one can easily imagine that complex things are built from simple parts, as in the case of Lego bricks. In addition, many will have learned in school that everything is built from atoms, which again are made from protons, neutrons and electrons.

What are considered the elementary particles in physics today is a different picture. The standard model of particle physics lists the elementary particles according to various distinctions.[42] The main division is between bosons and fermions. Crudely put, fermions take up space while several bosons can be at the same place.[43] The fermions are then divided into quarks and leptons depending on whether or not they interact with the strong force. Quarks interact with the strong force and come in two types depending on whether their electrical charge is $+2/3$ or $-1/3$. Among quarks we shall only care about the up-quark and the down-quark, since different combinations of these give the proton and the neutron.[44] Leptons are also divided depending on whether their electric charge is 0 or -1, but here we shall only care about one lepton with charge -1, namely the electron.

Protons, neutrons and electrons are in the group called fermions, but there is also another group called bosons. Except the newly discovered Higgs boson, these particles are force carriers. The four known forces are electromagnetism, the strong force, the weak force and gravity. Electromagnetism is carried by photons, the strong force is carried by gluons, and the weak force is carried by W- and Z-bosons, while gravity may be carried by hypothetical gravitons, but no such particle has been discovered. In the standard view on gravity, coming from Einstein's theory of relativity, gravity is a geometric effect of spacetime bending. I will return to that topic when we discuss space and time later.

---

42 In this presentation I stick to conventional particles and leave out anti-particles.

43 More precisely: Fermions have half-integer spin, while bosons have integer spin, and according to the Pauli exclusion principle, particles with half-integer spin cannot occupy the same quantum state within a quantum system simultaneously.

44 More precisely: Since quantum mechanics describes probabilities for what happens in experiments, we should, instead of saying that protons and neutrons consist of up- and down-quarks, say that it is the case that in certain interactions with protons and neutrons (at lower energy levels), up- and down-quarks are the particles we are most likely to find, in a way that makes it roughly right to say that protons and neutrons consist of up- and down-quarks.

There is an alternative to thinking that the world is fundamentally made from elementary particles, and that is the view that the fundamental entities are fields, and that particles are to be understood as field oscillations (S. Carroll, 2013). But what is a field? The field concept has a long tradition in physics, where it had roughly the meaning of an area or space under the influence of or within the range of some agent (McMullin, 2002, p. 14). The modern field concept arose with the work of Michael Faraday and James Clerk Maxwell. Faraday understood electromagnetism as physical lines of force, while Maxwell thought of the field in terms of a medium called the ether, and found equations describing how electromagnetism would propagate through the ether (Hobson, 2013, pp. 5–6). Einstein found field equations describing how gravity works in a gravitational field, but he dismissed the ether as medium, and thought instead of the field as a state or condition of space itself (Hobson, 2013, p. 6).

The electromagnetic and gravitational fields were classical fields in the sense of being continuous fields, and they described the forces of electricity, magnetism and gravity. In quantum field theory, the fields describe the interaction of particles (which includes forces), and they can take discrete values.[45] Richard Feynman defines a field as "any physical quantity which takes on different values at different points in space" (Feynman, 1964). With this definition, we could consider the air around us as a temperature field with a number value in each point. Or the same air could be considered a wind field with a value in each point for strength and direction.[46] And the same area[47] is a gravity field with a value for gravity in each point.[48] In each point of spacetime, numerous different kinds of values can be actualized.

But how can particles be fields? The idea in quantum field theory is that all particles (those carrying force, those who constitute atoms, and all the rest) have one field each and one field equation each. By solving the field equations for the relevant particles we are investigating, we can find what is called a quantum state for the system being studied. When you know the quantum state, you can use the Born rule in order to find the probability for finding specific values actualized at specific times or places. Such specific actualized values could be to find a particle at a location, or to find an interaction happening, or some other

45 The way to quantize a field is to assign an operator for each point in order to get specific values.
46 The value would then be a vector, since vectors have length and direction.
47 By "area" (or "field") I do not mean a two-dimensional area, but a three-dimensional space, but I use "area" (or "field") as a more open term than the term "space" as understood in physics.
48 The gravity value is a tensor, since the field has a complex curvature describable by Riemannian geometry.

property you are interested in measuring. To find a particle at a location would be the same as finding certain values of mass, spin and charge at a location, since the different particles are distinguished by different values of spin, mass and charge.

Finding specific values of spin, mass and charge, and changes in such values, are what we interpret as particles interacting with each other. Another way of describing it is to say that fields interact. We combine field equations (usually into what is called a Lagrangian formalism) to describe the interaction between particles/fields with one equation each. Field interactions can be combined into bigger and bigger configurations to get atoms and tables and chairs, but a guiding idea in quantum field theory is that everything (at least everything physical) can be understood as field interactions governed by field equations.

It may seem strange to think that fields should be what give rise to particles instead of just seeing fields as mathematical devices that can be used to predict the behavior of particles. But considering fields as fundamental and particles as field effects have many advantages, and have convinced renowned physicists and philosophers that fields have their own existence – for example Sean Carroll (S. Carroll, 2013), Steven Weinberg (Hobson, 2013, p. 2), and Marc Lange (Lange, 2002).

The view that particles are reducible to field interactions seems to solve many problems. For example there are certain reactions where one particle turns into another without it making any sense that one particle should be hiding inside the other, but it makes good sense to understand it as a reaction in a field (S. Carroll, 2013). A lot of the paradoxes in quantum physics concerning particles that seem to be at several places at once get solved if elementary particles are understood as values at points in fields.[49]

A quantum field theory hope is that one day every physical entity and every motion can be understood as the result of field interaction according to field equations. Physics of today is not there yet. In 2012 the Higgs field was discovered, which solved some problems in the standard model of elementary particles, like why particles get mass. It is widely believed that there is a lot of unknown dark matter in the universe, which may be unknown particles arising from unknown fields and equations, and there is dark energy equally poorly understood.

But maybe all physical interaction can be understood as field interaction, and maybe the many fields and equations can be reduced to a few or even

49 A long list of reasons for thinking that fields exist instead of particles can be found in Hobson (2013).

just one field and one equation. Physical theories like superstring theory are theories trying to reduce all particles to different vibrations on one fundamental entity like strings or membranes. There are many so-called effective field theories which are limited theories that work well at limited areas and limited energy levels. This makes physicists like Steven Weinberg assume that the theory we have today is probably also an effective field theory, which is an approximation of a more fundamental theory which could have been revealed at higher energies than the ones we are able to explore today (Weinberg, 1997, p. 21).

This was all about physics, but later I shall introduce the idea of qualia fields where conscious experiences are configurations in a qualia space with different values in points, which may interact with physical field activity like brain activity. But this is enough for now as background to the idea of the field. Now I will try to use this idea to suggest a theoretical framework for doing metaphysics.

3.2 Fields (with points)

In the following, I shall suggest a fundamental ontological theoretical framework. While the previous chapter dealt with fundamental semantic concepts for understanding, such as "structure", "individuals" and "relations", this chapter is about the fundamental building blocks of the world, such as fields, values and rules. It is not meant as a detailed interpretation or framework for quantum field theory as it exists today. Rather, it is meant as a deeper framework which hopefully is coherent with whatever physics today and in the future ends up with.[50] Think of it as a theoretical framework for the most fundamental ontological level and then there are probably several ontological levels of increased complexity and decreased generality[51] up to the level where physics of today are operating, which is still at a more fundamental level than most other theoretical frameworks in the other sciences today.

I present first the main concepts in this theoretical framework, and I start with the concept of a field. The concept is not identical to fields in physics, so to distinguish the concept from them, I call it the fundamental field, or the F-

50 This may sound like an unfalsifiable theory and in general unfalsifiability is something to be avoided, but at the deepest metaphysical levels it can be a virtue to discover something which cannot be falsified. For example, Descartes discovered the unfalsifiable truth that it is certain that there must be thinking, which is not a problem but a great discovery. More on good and bad kinds of unfalsifiability can be found below.

51 By ontological level I just mean different levels of complexity and generality, so that higher ontological levels contain more complex structures and are less general in what they describe.

field. From now on, I shall often just call it the field and mean the F-field. If I want to refer to fields in physics, I call them "fields in physics" or the context makes clear that that is what I am talking about.

What is the F-field? The F-field is the area encompassing all areas where any value of any kind can be actualized in accordance with certain rules. To explain this further, I shall now present in detail every part of the definition: area, value, can, actualized, and rules. Since these are fundamental concepts, they are partly defined in light of each other – not in a viciously circular way, but rather in a way that clarifies more and more detailed structure as we move along. All ontologies will have this challenge of circularity with their fundamental concepts. We start now with the concept of "area".

What does "area" mean? An area is anywhere a value is actualized or where a change can happen from one point of time to another.[52] While "area" may sound like a two-dimensional plane, "area" should here be understood to cover all three spatial dimensions (dimensions are further defined later). I start by presenting further the concept of change. What is a change? A change happens when one or more structures are replaced with different structures at the same place, which again means that the same area can be coherently described by different signs before and after the change. One may criticize these definitions by saying that time and place have become relative after the general theory of relativity, but I shall return to the topics of time, place and relativity later, so for the time being, consider what is said as taking place in the reference frame of the area under discussion.

If we include all areas – anywhere a change can happen or a value is actualized – into one big area, we get the fundamental F-field, which is the actual world, as opposed to fictive worlds.[53] This is an area greater than our universe into which our universe is growing, since our universe would not have expanded if it was not possible for it to expand, which means that there is an area outside of our universe where it was possible for our universe to be actualized. One should not confuse the area defined here with the spacetime of physics. It must be the case that there is an area greater than our universe into which our uni-

52 In the reference frame of the area where a change is claimed to have occurred. "Happen" is the same as "be actualized", to be defined below.

53 I define "world" in more detail later. To put it briefly here: The real world is the field and the values that are and have been actualized. These are actualized in the physical field and the qualia field, which corresponds to the physical world and the world of consciousness. The real world is to be distinguished from fictive worlds, which could have been actualized either in the physical field or as imaginations in the qualia field. As a possibility of the real world, fictive worlds are part of the real world.

verse is growing when “area” is defined as I have done here, otherwise it would not have been possible for the universe to expand.

That a change can happen in the F-field should be taken in a very fundamental sense, meaning that the F-field has points where values can be actualized, but it may well be the case (especially at the most fundamental level) that after a value has been actualized at a certain place, it will stay like that forever, so that no change can happen there any longer. Some values may have been actualized from the beginning, remaining forever in their place. This is why the definition says that either a change can happen or a value is actualized, to include the possibility that somewhere a value may be actualized and no more changes can happen because of the rules for actualization.

Is this field one united large area of points where either changes can happen in every point or values are actualized, or could there be gaps in the field where no values can be actualized and no changes happen? I think that there are no such gaps in the encompassing fundamental field since lack of gaps would make the field less complicated, but I cannot rule out the possibility that the field has gaps. I define the F-field as the area that encompasses all areas where changes can happen, so if there are gaps in this area, they are nevertheless included in the area that the fundamental field covers, since these parts are encompassed by the F-field and related to the other areas. Field theories in physics assume that the field can take on values anywhere, and these theories work well, so inductively we have reason to assume that the fundamental field has no gaps.

The field can be divided into points. A problem with the common concept of a point is that it is in theory infinitely small. But how can something infinitely small exist, and how could anything be located there? To solve this problem, I define a metaphysical point (as opposed to a mathematical point) as no smaller than the smallest area where a change can happen that it is metaphysically possible to register (notice or become aware of).[54] Maybe some continuous smaller change could occur, but no theory will ever need a metaphysical concept of a point so small that no change there is metaphysically possible to register. That something is metaphysically possible to register means that a metaphysically possible being could notice it, and thus a change that no possible being will ever notice is of no use in an ontology. The definition allows for metaphysical points that have a finite size, instead of being infinitely small, making it no problem to say that such points exist as parts of the fundamental field. Note that mathematical points as used in equations may still be infinitely small.

54 “Metaphysical possibility” will be defined below.

If there are areas where no change can happen and no values can be actualized, it could still be hypothetically divided into points of the same size as the points defined above. Since no change can happen there anyway, it does not matter how we think of the sizes of the parts of such areas.

That was what I had to say about the term "area". But what does it mean to say that the field and the areas or points in the field *exist?* I shall return to the definition of existence after I have defined the different parts of the concept of the field. The next term in the definition of the field (the area encompassing all areas where any value can be actualized according to certain rules) is "value".

3.3 Values

The term "value" can mean just a number of one parameter or another, for example the number 27 of a measurement of temperature or 50 kilograms of mass. Such a value number is a *quantitative* value, since it is measurable by a number. But here I use the term "value" more broadly to include the parameter that is being numbered, so that temperature or mass are different kinds of values. Such values I call different qualitative values. The qualitative values are simple structures that can change at points. Comparing with common terminology, physical quantities like mass (expressed in units like kg), would correspond to my qualitative values, while their magnitude (for example the quantitative value of 5 kg), would correspond to my quantitative values. From now on, I use the term "value" to refer to qualitative value, unless otherwise specified.

Some values will be fundamental in the sense of being effects of general rules applying to the whole field. Some of the elementary particle fields seem to be such general values, for example electrical charge. Other values are less fundamental in the sense either of being effects from less general rules (rules that have certain conditions (if-then rules) for when they are fulfilled), or in the sense of being configurations of more basic values being actualized in some area. For example, heat is a value that requires that there be molecules in motion, which again presupposes more general values being actualized in order for there to be molecules.

"Value" is thus an extremely broad term expressing anything that can occur in a point or a larger area. Hopefully – since it would be easier to understand – there are quite few fundamental values (or even just one), so that all other values (everything else occurring at places) can be understood as configurations of more basic values. In the course of this book the number of different basic qualitative values will be very reduced. The particles in the standard model of particle physics are distinguished by different quantitative values of three main different qualitative values, namely charge, spin and mass. Maybe everything physical can be understood as configurations of some few values at different points, such as charge, spin and mass, which again may be configurations of even more basic values. I will return to the question of fundamental physical values in Chapter 12 on the fundamental entities in physics.

We do have some inductive reason to believe that everything physical are configurations of more basic values, since the standard model is so successful in explaining almost everything physical as effects of the interaction of some (quite) fundamental fields. What characterizes the fundamental values is that they come in degrees that can be numbered and that they can be actualized in all points in the field. Consider some non-fundamental values that can occur at a relatively large area, like the presence of skin, grass or rock. If there is skin, grass or rock somewhere, it will be in a very limited area, and it will not be at very small points, since at a very small point there will only be, for example, one atom, and it will not make much sense to say that it is one atom of skin or grass or rock. Skin, grass and rock are configurations at larger areas of more fundamental values being actualized in that area.

Why do I have such a broad definition of "value" to include things like skin, grass and rock as values? Would it not be better to use the term just for values that come in degrees and stretch over the whole field? The reason for my broad definition is for the framework to be applicable even if we should find out later that the standard model picture is wrong. You may object that such a framework can become too vague and unfalsifiable: I will answer this objection below.

Another possible objection is that this value approach seems to be interesting in the case of physical things, but what about the world of conscious experiences? I will treat this topic in detail in the chapter on consciousness, but give a short answer now. Our conscious experiences seem also to consist of certain conscious values (called qualia) being actualized in points in our conscious experience (also called qualia field) and they seem to come in degrees. For example, experiences of color seem to come in degrees of hue, brightness and saturation, while experiences of sound seem to come in degrees of pitch, loudness and timbre. The suggestion I will present in detail later is that everything physical and everything conscious (in other words: absolutely everything) is to be understood as structured configurations of a few basic values in points in fields.

3.4 "Actualized" (actualization)

Values are actualized when they occur or happen at a point. That they are actualized, occur or happen means that they are localized at a certain place and that they can have causal effects in a wide sense, which I will further develop later. Here I just explain the choice of definition. If a value in a point cannot have a causal effect in any way, neither it nor its implications can ever be experienced by anyone, thus there is no point in saying that it might possibly exist, since it will never play a role in the ontology.

This may seem unproblematic when it comes to physical values, but quite problematic when it comes to conscious experiences, which we want to say are actualized in the mind, but which do not seem to be located or have causal effects. In the chapter on consciousness I show what the causal role of conscious entities is, and discuss where they are located, which shows that the definition works for such conscious values as well.

Another way describing actualization is as follows: That values are actualized also means that they are the effects of that which actualizes them, and this actualizer will be further described below.[55]

Being actualized also means that the qualitative value does not have the quantitative value zero, but instead a number. I shall say below that the field has values that can be actualized in every point, but that the quantitative value may be zero, and so saying that a value is not actualized or actualized with the quantitative value zero means the same in this book.

What does it then mean to actualize, and who or what actualizes the values? That is a discussion I will save for below, since it is clarifying to talk about possibilities and rules first. I have thus only described the passive form of actualization here – "actualized" – and will describe the active form – "actualize" – below.

3.5 "Can" (modality)

The definition offered of the F-field says that values *can* be actualized in the field. Another way of saying that is to say that it is *possible* that values are actualized in the field. But what is possibility? And what is necessity? Why is some-

55 Sometimes a value can have the quantitative value of zero in a point and make it epistemically the same as if the value was not actualized in the point. Nevertheless, there is an ontological difference between whether the qualitative value could possibly take on another quantitative value than zero in the point or not.

thing possible, impossible or necessary? I shall start this entry by offering some distinctions. They will be between logical, physical, metaphysical and epistemological possibility (and impossibility and necessity); between type and token possibility; and between narrow and wide possibility. After the distinctions I explain what possibility, impossibility and necessity, as such, are.

That something is logically possible means that its description in a theoretical framework is consistent (not contradictory, as explained in the presentation of coherence above). That it is logically impossible means that it is inconsistent. That it is logically necessary means that it is inconsistent to deny it.[56] For example, it is logically possible that John is a bachelor, while it is logically impossible that a bachelor is married and logically necessary that a bachelor is unmarried.

That something is physically possible means that it can be or happen as it does because of laws of nature and states of affairs in our universe. That it is physically impossible means that it cannot be or happen as claimed because of laws of nature and states of affairs in our universe. That it is physically necessary means that it must (cannot not) be or happen as it does because of laws of nature and states of affairs in our universe. For example, it is physically possible for a car to move faster than a horse, while it is physically impossible for a car to move faster than light and physically necessary for the car to move slower than light.

That something is metaphysically possible means that it can be or happen as it does because of the fundamental structures of the world. That it is metaphysically impossible means that it cannot be or happen as claimed because of the fundamental structures of the world. That it is metaphysically necessary means that it must (cannot not) be or happen as it does because of the fundamental structures of the world. When it comes to examples of what is metaphysically possible, impossible and necessary, these will only be hypothetical suggestions depending on what the person suggesting the examples thinks are the fundamental structures of the world. Of course anything that is physically possible or actual is also metaphysically possible since it can or has occurred. But whether other things are metaphysically possible or whether something is metaphysically impossible or necessary is something we do not know, but can only hypothesize. The fundamental metaphysical structures are the basis for

56 In many cases we need to draw implications from what is said to see that it would be inconsistent to deny it, and it may be controversial whether denial is inconsistent if people disagree over the meanings of terms. For example, we may need to agree on what "bachelor" means to determine whether it is inconsistent to deny that the pope is a bachelor, and we may need to learn that both the Morning Star and the Evening Star is Venus to see that it is inconsistent to deny that the Morning Star is the Evening Star.

all other kinds of possibilities. These fundamental metaphysical structures must themselves have been possible in the wide fundamental sense of being not-impossible, but other than that other kinds of possibility presuppose these fundamental metaphysical structures as their basis.

To say that something is epistemically possible just means that the speaker does not know whether it is logically or physically or metaphysically possible, impossible or necessary.[57] So let us say that I do not know whether every even number is the sum of two prime numbers or I do not know whether there exist gravitons (spin-2 particles carrying gravity) – then I can say that it is epistemically possible that such is the case.

Concerning physical possibility, we can distinguish between type and token physical possibility. Something is *token* physically possible if it is possible (for X) here and now, while it is *type* physically possible if it is possible (for X) in general. For example, it is type physically possible for me to open a door, but maybe it is not token physically possible for me to do it right now (even if there is a door in front of me) because of something happening in my brain, or because the world was determined such that I would not open the door right now. There are many other similar distinctions people have suggested and given different names, like opportunity, ability, wide and narrow possibility, etc., but I find that the type-token distinction, to my knowledge, covers it. This distinction will be important when we come to the topic of free will.

However, there is a distinction which is similar to the type-token distinction and relevant for my project here. I call it narrow and wide possibility. A narrow possibility is possible *for the time being* (at the point of time under consideration), while a wide possibility is possible *in the future* (relative to the point of time under consideration). So for example, I can say both that it is possible and not possible for me to play guitar or speak Italian. In the narrow sense it is not possible for me to speak Italian or play the guitar since I do not know how to do it. But in the wide sense it is possible for me to speak Italian or play the guitar since I can learn how to do it. For the time being it is not possible, but in the future it is possible.[58]

57 And, as pointed out to me by Sivert Ellingsen, if a speaker knows that something is necessary in one way or another, then it follows that it is also epistemically possible.

58 The distinction is similar to how Aristotle distinguishes between first- and second-order potentiality and first- and second-order actuality, where the second-order potentiality is the first-order actuality (*De anima*, book 2, part 5). Being able to learn Italian would be first-order potentiality, learning it would be first-order actuality, but then also second-order potentiality, since one can then speak Italian, while actually speaking it would be second-order actuality. While

Now we have the distinctions needed to explain the little word "can" in the definition of field. That values *can* be actualized in the field means that it is metaphysically possible in a wide sense for values to be actualized there. The fundamental field thus encompasses all areas where anything can happen at any time, now or in the future.

While these descriptions explain the difference between different kinds of possibility and necessity, I have not yet explained what the terms "possible", "impossible" and "necessary" mean in general. What makes something possible or necessary in any version of possibility and necessity? Differently put: how should we understand modality?

I suggest that the different kinds of modality express different alternatives for combination of parts in a theoretical framework given certain physical, metaphysical or semantical presuppositions. That something is possible means that it expresses a consistent combination, given the presuppositions. That something is impossible means that it expresses an inconsistent combination, given the presuppositions. That something is necessary means that it is implied in the presuppositions, and thus inconsistent to deny.

This can be exemplified with physical, metaphysical, and logical possibilities, impossibilities and necessities. When it comes to physical modalities, the presuppositions are the laws of nature in our universe. If you describe something consistent with these laws, it is physically possible. If you describe something inconsistent with these laws, it is physically impossible. If you describe something implied by these laws, it is physically necessary. As exemplified above, in our universe it is physically possible to move slower than light; it is physically impossible to accelerate to a speed faster than light; and it is physically necessary that light travels at light speed in a vacuum. If we discuss token physical possibility or narrow physical possibility, it presupposes also states of affairs in the universe – in other words, what exists and has happened before the event we are discussing.

When it comes to metaphysical modalities, the presupposition will be a hypothesis about the fundamental structures of the world. A description which is consistent with this hypothesis will – by hypothesis – be metaphysically possible. A description which is inconsistent with this hypothesis will – by hypothesis – be metaphysically impossible. A description which is deductively implied in this hypothesis will – by hypothesis – be metaphysically necessary. For example, if the presupposition is the hypothesis of reductive physicalism, it is, according

Aristotle distinguishes them by their dependence, I distinguished them by time, but both distinctions clarify their relation.

to this hypothesis, metaphysically possible that there are chairs; metaphysically impossible that there are ghosts; and metaphysically necessary that all things are physical. If the hypothesis is true, these examples are *actually* metaphysically possible, impossible and necessary, which in this case would mean that the rules actualizing values allow for chairs, but not for ghosts, and will always actualize physical things.

When it comes to logical modalities, the presuppositions are the meaning of terms. If you describe something which is consistent with the meaning of the terms used in the theoretical framework, it is logically possible. If you describe something which is inconsistent with the meaning of the terms used in the theoretical framework, it is logically impossible. If you describe something which is (deductively) implied by the meaning of the terms used in the theoretical framework, it is logically necessary. As exemplified above, given that "bachelor" means unmarried man, it is logically possible that a bachelor is 40 years old; it is logically impossible that a bachelor is married; and it is logically necessary that a bachelor is unmarried.

This approach to understanding modality also explains the other ways we use modal terms, such as when we say that something is possible, impossible or necessary because of the law, or ethics or the rules of chess, etc. The law or ethics or the rules of chess are the presuppositions and then we make statements about what is consistent, inconsistent or implied by these. For example, given the rules of chess it is possible for the queen to move sideways, but impossible for the queen to move like the horse, and necessary for the queen to be at D1 or D8 from the start. What I am saying is that we do not need modality as an irreducible entity on its own. It suffices that there are rules according to which motion occurs, and that there are theoretical frameworks which can be consistent or inconsistent.

This approach also explains what dispositions are. When we speak of physical possibility we speak of what is consistent given the laws of nature. When we speak of the disposition of a thing, we speak of what is physically possible for that thing to do (or what can happen to it) given the structure of that thing. Dispositions do not add anything that we do not already have in virtue of the other components of the ontology. One could still argue that dispositions should replace other entities in the ontology, such as laws of nature. I discuss this later in relation to the topic of laws of nature.

This approach to modality is similar to the approach in the philosophy of modality called combinatorialism, but it differs in its focus on theoretical frameworks, consistency and given presuppositions. I believe that it is more coherent than the common forms of combinatorialism,[59] but there is not space to discuss it here.

3.6 Rules and their actualizers

A peculiar fact about our world is that it is so structured. There are many patterns that are regularly followed. While some seem more accidental, others seem to be without exception everywhere and at all times. These patterns that seem to be followed without exception can often be formulated very easily or with short and exact formulations. For example, the behavior of all electrons can be formulated with a short field equation that all electrons seem to follow everywhere and at all times without exception.

I shall use the term "rule" for a structure or pattern that entities behave in accordance with. Note that I am not at this point saying that there is something that exists and makes the world act in accordance with the rules, I am just calling the patterns that everything follows rules. It seems impossible to deny that there are such rules, in the sense of there being patterns that everything follows (at least up till now – nobody knows the future). One could suggest that there are just some very stable patterns that have evolved over time. But then it seems that this evolution follows some deeper rules that make some patterns stable and spread. But maybe these deeper rules are also coincidental, so that everything that happens is a matter of coincidence. But then it would seem to be a rule that everything can happen by coincidence. Maybe that rule can also change over time. Well, then it seems to be a rule that that rule can change over time. At some level or another it seems unavoidable that there are some general rules that are followed everywhere and at all times. In fact, there seems to be many detailed rules discovered by science that are followed without exception and which are poorly explained as mere coincidences.

A very interesting question is how to understand these rules further. Are these rules followed by accident and for no reason, or is it the case that the rules *must* be followed and *cannot* be broken? Are there laws (in the sense of existing truthmakers) that make the world behave in accordance with the rules, or is it something else that happens which has as a result that we can formulate such rules (for example that electrons have certain dispositions which make it the case that we can formulate laws for the behavior of electrons)?

59 See for example Armstrong (1989).

These questions are debated in the philosophy of laws of nature, and there are three common responses to how we should understand laws of nature. At the one side is the view that there are no laws of nature and that the patterns we find in the world are mere coincidences. It is commonly described with the example that you can empty a bucket of colored stones on the floor and find some patterns in the mosaic afterwards that are simply the result of a mere coincidence. The laws of nature are then understood as description of some patterns that happen to be universal.

The most advanced version of this view is the best-system approach associated with Mill, Ramsey and Lewis. A description of everything that is true in the world would be extremely complicated, but some generalizations can simplify a lot while still letting us deduce what is true. The best system would be the simplest system that still lets us deduce what is true to the greatest degree, and the basic generalizations in this system are the laws of nature. So for example, instead of describing how much each object attracts each other, the generalization $F = GMm/R^2$ is a simple generalization that sums it all up and thus has the function of being a law of nature in that system.

Some main arguments against this view are that it seems unbelievable that the world should just happen by coincidence to be so consistent in its lawful behavior (J. W. Carroll, 2008, p. 76). It seems that something must make it the case that there are these seemingly exceptionless regularities in nature, that coincidence is not a sufficient explanation, but that laws of nature can explain it (Armstrong, 2004b). We have advanced field equations which give exact descriptions of how fields evolve and interact, which it seems implausible to believe is just a coincidental pattern in nature.

Further, it seems that science works on the assumption that we should distinguish between mere regularities and laws of nature. For example, scientists distinguish between regularities that are laws of nature and regularities that stem from so-called fine-tuned initial conditions of the universe, and they try to explain the last ones but not the first ones, instead of just treating them all as general regularities (Roberts, 2008, pp. 16–20).

David Lewis, who of course defends the Mill-Ramsey-Lewis understanding of laws, finds that the worst problem with his own theory is that it seems to depend on humans what the laws of nature are since it is our minds that decide what we consider to be simple and strong, but he thinks it would be lunacy if his theory implied that we can change the laws of nature by changing our thinking (D. Lewis, 1994, p. 479). His solution to this problem is to say that it only partly depends on us, assuming that there is something about nature that makes us find certain laws simple. If nature is kind to us, we can hope that some laws are clearly simpler than others, which makes them robust candidates not merely depending on how our minds happen to function (D. Lewis, 1994, p. 479).

On the other side of the debate is the view that there are laws of nature that govern the world – that there are ontological entities that make it the case that the laws of nature must be followed and cannot be broken. That is how I interpret "govern" in this sense: that there is this direct link between some entities making it possible or necessary for other entities in the world to behave in certain ways. This is realism about laws of nature, and the view is defended by different thinkers for different reasons, for example John Carroll (J. W. Carroll, 2008), Marc Lange (Lange, 2000) and Tim Maudlin (Maudlin, 2009).

The main reason to defend such a view is that science seems to work on the assumption that there is a difference between coincidental regularities and lawfully necessary regularities, as exemplified above. So, for example, Tim Maudlin says that we need laws of nature to explain the difference between what seems coincidentally true and (physically) necessarily true (Maudlin, 2009).[60] And John Carroll says that there are some generalizations that are true *because of something in nature* (for example, there are no gold spheres larger than a mile in diameter because of the amount of gold in nature and how it is spread) and some generalizations that are true *because of nature* (for example, there are no enriched solid uranium (U235) spheres larger than a mile in diameter because that would be impossible for uranium) (J. W. Carroll, 2008).[61]

But many also have problems with the view that there should be laws of nature in the sense of entities that govern nature and make it behave in a particular way by necessity. The main problem is that it seems to be a claim about some very special entities that we have no empirical access to, nor do we have any uncontroversial theories about how they should be able to govern the world this way. It would be ontologically more economical if one did not need to include laws of nature in one's ontology.

For these reasons, many seek a middle way. They want to say that there is an explanation for the exceptionless regularities in nature, but that there are no laws of nature making nature behave a special way. Rather, something other than governing entities are what make it the case that the world behaves in such a lawful way. But what could that something else be?

A noteworthy candidate is dispositionalism. Dispositions are the capacity of things to behave a certain way. For example, glass is fragile, which means that glass has the disposition to break. So things have dispositions to act in certain

60 For a similar view, see Lange (2000).

61 The uranium example is from Van Fraassen (1989).

ways, and these dispositions can be summarized as laws of nature. It is not the laws of nature that make the electrons behave a certain way. Rather, electrons behave a certain way because of their dispositions, which can be summarized as laws about electron behavior. This is a kind of Humean realism, and a famous defender of this particular view is Stephen Mumford (Mumford, 2005, pp. 408–409).

An argument against dispositionalism is that there are laws that seem to be about the whole world, such as symmetry laws and conservation laws, and it seems strange that these should be the result of dispositions of individual things. Does everything have the disposition to contribute the right part to the symmetry of the world? These laws are good candidates for being laws of nature (J. W. Carroll, 2008) and are used by Angelo Cei and Steven French to argue against dispositionalism about laws (Cei and French, 2014). Alexander Bird has defended dispositionalism and argued that symmetry laws are pseudo-laws that will disappear from physics, while Cei and French argue that they are so central that they contradict dispositionalism.[62] Bigelow et al. have argued that conservation laws and symmetry laws may be a disposition that the world as a whole has. But Cei and French reply that "being a world" is an extremely coarse property, and an extremely coarse explanation. It is like explaining all of Socrates' features in terms of "being Socrates".[63] Furthermore, if something is a disposition of the world as a whole, what should be the trigger outside of the whole to manifest that disposition (Esfeld, Deckert, and Oldofredi, 2015, p. 25)?

Regardless of whether dispositions are part of individuals or part of the world, dispositions are mysterious entities. What are they, how do they work, and why does a disposition actualize one kind of behavior instead of some other? If we want to say something more than that they are some things that just work, it seems we must say that they are some things that have certain possibilities that can be actualized, and they do so according to some rules. But then it seems that we need some rules that make dispositions work at all – and work one way rather than another – if we are to understand how dispositions work. If I just say that something happens the way it does because of its disposition, that is just a name and no explanation. It is like saying that sleeping pills work because they have the disposition to make you sleepy.

In order to understand dispositions, we need to see that something actualizes some possibilities according to some rules. But then we just have the laws of nature that dispositions were meant to help us get rid of. This shows that dispo-

62 Cei and French (2014), referring to Bird (2007, p. 214).
63 Cei and French (2014), referring to Bigelow, Ellis, and Lierse (1992).

sitions do not explain anything until you add laws of nature. But then dispositions are superfluous entities that we might as well get rid of. One could insist that they should be included to explain what the actualizers of possibilities are, but as seen in the paragraph above, they do a poor job.

Mumford argues that it is ontologically economical to leave laws of nature out of your ontology, and Ted Sider makes a similar point: Physical facts do not explain more by being called laws of nature, and thus they can be discarded (Sider, 2011, pp. 15, 22). Or as put by David Lewis: non-Humean laws of nature do not do the work they are supposed to do if we cannot explain how they make regularities happen (D. Lewis, 1986b, p. ix). This line of reasoning makes many conclude that it is better to leave out laws of nature than to include something mysterious which does not explain anything at all.

Contrary to this argument, I will make the following point: Even if one does not know what the laws of nature are or how they work, one may still have a good reason to think that there must be something *that* makes regularities happen, and for lack of a better name call it laws of nature. Laws of nature can explain *that* there are regularities even we cannot (yet) explain *how*.

The critic can respond that it is then better to leave mysterious laws of nature out and just say that the regularities that happen to exist without further explanation. But the problem with this response is that if there is no reason why something happens as opposed to something else happening, we should expect less regularity and more chaos. We have to assume a brute fact: either regularities without cause or regularities with cause. Regularities without a cause is ontologically more economical than regularities with a cause, but regularities without a cause is inconsistent or at least highly implausible since if nothing makes the regularities happen we should expect chaos instead of regularity (since there are so many more possible states of chaos than order). Regularities caused by laws of nature is thus a brute fact which gives a more coherent theory overall than regularities without a cause.

Here is an additional reason to opt for regularities caused by laws of nature. There is motion in the world, and specific kinds of motion as well. We use little-understood terms such as "energy" and "force" to explain it, but "laws of nature" is a good candidate which can make concepts like energy and force reducible and explain motion. I will return to this point in more detail later in this chapter and in the chapter on fundamental concepts in physics.

This presentation of the discussion of laws of nature in philosophy is obviously not meant as an exhaustive presentation. Rather, I wanted to show some main positions and arguments in order to set the stage for the following discussion on how best to explain fields, points, values, rules and actualization. I shall be concluding the discussion on what actualizes the rules in Section 3.10, and suggest a kind of middle way between the main positions we have been discussing. But before I do so, I need to discuss two questions which we shall see will be relevant in that discussion. The first question is "Could fields, points, values, rules and actualization not exist?", and it will be discussed in Section 3.7. The second question is "Should the basic ontological entities be simple or complex?", and this will be discussed in Section 3.8. The whole discussion in Sections 3.6 to 3.9 is on how to understand the relation between the basic building blocks of the world, and will be concluded in Section 3.9.

3.7 Could fields, points, values, rules and actualization not exist?

I have not yet defined the term "exist", since (as will be clear below) I need to discuss the explanation of fields, points, values, rules and actualizers before I define "existence". On the other hand, I already need to discuss whether it could be possible that they do not exist in order to defend their place as fundamental concepts. A temporary working definition of existence which covers what I find relevant given the later definition of existence will suffice for the following discussion. So far then, I shall say that to exist is to be part of the world of causal interaction. I know I have not yet defined causal interaction either, since I need to define the existence of fields, rules and actualizers before I can explain causation. But the definition should nevertheless be understandable and fit well with how many people think of existence. It is not important to have a clear definition from the start, since I think that all will agree that fields, points, values, rules and actualizers in the definitions I have given above exist given any common definition or understanding of existence. Unfortunately, I cannot explain everything at the same time, so there has to be this piece by piece progress, and we will get to all the central terms as we move along.

This is the question to be discussed here: Could fields, points, values, rules and actualization not exist? I have mentioned some different definitions of these terms, some narrower and others wider, but here I ask given the most minimal (or wide) definitions I have presented. Then the *field* is the area that covers all other areas, and an area is anywhere a value is actualized or a change can happen. The *points* in the field are just very small areas down to the smallest area where a change can happen that it is metaphysically possible to register. When something happens somewhere, a value is *actualized* there, for *value* is just an extremely broad term expressing anything that can occur in a point, small or large. I added that in order for a value to be actualized it must be able to have a causal effect, since if it did not, neither it nor its implications could ever be experienced by any mind, and then there is no point in saying that it exists. A *rule* is a structure or pattern that everything behaves in accordance with, and even if some rules are not really universal it seems that it is unavoidable to include some (meta-)rules in an ontology.

I think that hardly anyone will reject that these entities must exist, and here is why: As long as you agree that something happens somewhere, the field is just all the places where something happens. One can hardly reject that something happens, and thus one can hardly reject that fields exists in this sense, where field is not defined as a medium or substance of any kind. Now, there are some physicists and philosophers who think that time is an illusion, so in one sense nothing *happens*. However, I said that when things happen it means that values are actualized in points, and the same physicists or philosophers who think that time is an illusion will usually be happy to grant that values are actualized in points. Points are just small areas where something happens, and so by the same reasoning, nobody can deny that there are points in the world.

Values being actualized also just means that something happens, so while the field is the location, values being actualized is what happens at the locations. If values being actualized just means that something happens, you cannot deny it without self-contradiction, for something happens when you deny it. And that some values *can* be actualized (that some values are *possible*) is necessarily true when some values have been actualized since it could not have happened if it was not possible that it should happen. Finally, that there are rules in the sense of being patterns that everything behaves in accordance with, also seems impossible to deny, since no matter what happens it seems to fall in under some general description.

Could these entities be reduced to something else that makes it superfluous to include fields, points, possible values, rules and actualization? Could it, for example, be just a great number of simple constituents and mere randomness which, given enough time and space, could evolve into what we later understood as fields, points, values, possible rules and actualization? Well, in order for anything to happen some values must be actualized, which thus requires possible values, actualization and a field of points. Further, evolution into something stable requires some rules for that to happen. Evolution as an explanation – even at a fundamental level – does not help us get rid of the basic entities.

Could the basic ontological entity just be an omnipotent and omnipresent God who just made the world with his or her will and power, so that fields, points, possible values, rules and actualization are not fundamental ontological entities? Well, even "being God" is a value that must be actualized in order for God to exist. And even if God is not an object to be found in spacetime, God must be somewhere in order to exist. God cannot be nowhere, but is usually said to be everywhere (omnipresent). This is typically presented as God being present with God's power everywhere – in other words, the value "God's power" is present at every point – and *everywhere* is the fundamental field. And even being God follows certain rules that God did not create, but which must have been in place before God could make a choice, because it does not make sense to say that God chose God's own being before existing. It is a traditional view to say, for example, that God cannot do anything evil. If there is an omnipotent and omnipresent God there is a field which God covers, there are values actualized in God, and how God is and acts follows certain rules. Introducing God does not simplify or reduce the need for fields, points, possible values, rules and actualizations.

Could we say that the field is a substance which has properties, and thus reduce this ontology to standard substance ontology? Well, since the substance needs to be located somewhere, and since properties must be actualized somewhere according to rules, it still seems that we need the entities I have suggested in this chapter, while we do not need the categories of substances and properties in addition. You could choose to call the four irreducible entities the four substances with special properties, but there is nothing to gain by it, and the language of substances and properties are superfluous at higher ontological levels.

Would it be easier to go with the Aristotelian ontology, where the world consists of substances made of form and matter, and the matter has potential which is actualized as its form? Aristotle believed in a set of fixed eternal forms that teleologically causes individual substances to actualize their form, for example as when the acorn becomes an oak tree because it has the oak tree form. The ontology of this book has no teleological causation or a set of eternal forms, rather individual substances and their form are reduced to patterns among the values.

If you want to translate the ontology of this book into Aristotelian concepts, one could say that the field is a substance which has the values as its matter and the actualizer as its form. The form should then be understood as a process driven forth by efficient causation and not as a teleological cause having an effect. The possible quantitative values that the different qualitative values can actualize would be like formal causation determining the potential of the matter. There are thus similarities in structure between this ontology and an Aristotelian ontology, but this ontology is simpler than the Aristotelian in getting rid of very many forms, substances, and teleology.

As far as I can see, any ontology one can suggest must include something corresponding to fields, points, possible values, rules and actualizations, which again implies that the simplest ontology is the one employing just these entities and nothing more. Does this give us a problem of unfalsifiability, since it seems to give a theory which cannot be falsified? Usually, we want theories to be falsifiable, since otherwise they are just as uninteresting as innumerable unfalsifiable theories like "everything is caused by unobservable spirits". But actually, unfalsifiability is a virtue at the most fundamental metaphysical level if it is otherwise part of a coherent theory.

For example, it is unfalsifiable that something can be experienced or that thoughts or language in a broad sense exist, and that is a good thing giving strong support to the truth of these claims. The same goes for fields, points, values, rules and actualizations. Denying that they exist presupposes their existence. This means if any ontology must include fields, points, possible values, rules and actualizations, that is a good thing. Ontologies do not have to include precisely these terms, of course, but it seems they must include something that can be translated as fields, points, possible values, rules and actualizations.

I should explicitly say that it is a certain kind of unfalsifiability we desire at the deepest metaphysical level, and that is the kind of unfalsifiability where it is a claim that it is either inconsistent to deny or where it is impossible to think that it cannot be true; for example, that there are thoughts. We do not want the kind of unfalsifiability which you cannot reject because it is too vague or it avoids all objections even if it can easily be thought to be untrue. In other words, we do not want theories talking about undiscoverable teapots or theories about a secret organization which has hidden all evidence of its existence or a theory saying that everything good comes from Zing and everything bad comes from Zang.

So far I have argued that we must think that these basic entities in one sense or another exist, but there is more to say about the relation between the basic entities and their ontological statuses. Are fields individuals with structures while rules are just abstractions of what the fields do? Or are the rules entities with power to actualize themselves, while the field is nothing but the place where the actualization happens? Can some entities combine several of the aspects together? What actualizes the possible values at the places they are actualized in accordance with rules? I shall soon discuss the relation between these entities, but in order to evaluate the quality of the different answers, I need to make a small discussion first on whether the basic ontological entities are more likely to be simple or complex.

3.8 Should the basic ontological entities be simple or complex?

It is a common desire in physics and metaphysics that the basic ontological entities should be simple. Daniel Dennett argues against God as an explanation by saying that an explanation of everything should be like a crane built from simple parts and not a skyhook hanging in nothing (Dennett, 1995, chapter 3).

This claim is reasonable in many contexts of explanation, like for example evolution, where species are well explained when we see the small steps that brought them forth. One could understand the intuition also when it comes to an ultimate explanation that one wants the basic building blocks to be as close to nothing as possible, so that it is a gradual way from nothing to almost nothing to everything. But there is no way from absolute nothing to something. Absolute nothing means no potential and no possibility for anything, and then it is impossible for anything to come from nothing, neither something simple, nor something complex.

We have to start with something, and it may in fact seem better to start with something complex than something simple since it will be easier to explain something complex by something complex. But physicists and metaphysicians hope to find something simple as an explanation of everything. Note that "simple" can mean many things. In one sense, the basic ontological entities must necessarily be complex. Why? Because they must necessarily have in them the potential for everything that exists and happens and has ever existed and ever happened. Since everything must have been possible from the start (in the wide sense of metaphysical possibility), the start must have been complex enough to fathom all that, and so in that sense it is necessarily complex.

A field of values which can be actualized according to rules may seem like an implausible starting point for an ontology since it sounds so fantastically complex, but it seems that any ontology must include these parts and that any starting point must be something with an incredible potential, thus starting with the smallest amount of necessary ingredients seems like the best start. Basic physics today does involve some amazing equations according to which everything behaves, so this starting point does have support in its explanatory capacity, even if it is complex.

But complexity, like simplicity, can also mean different things. Is it simple in the sense of being few entities and simple structures with few parts and connections? Or is it rather complex entities in the sense of entities with complex structures, but simple in the sense that they leave few questions unanswered? Of course, fewer entities and simpler structures with fewer connections will usually also leave fewer questions unanswered, although not necessarily, if they are too simple to explain what they are meant to explain.

We do not have an ontological reason to assume that the basic ontological entities are simple structures, and by "simple structure" I mean structures with few unique parts. There will necessarily be something about the fundamental ontological entities which is not explained. That is because of the way that explanation works. We explain something and understand something by integrating it in a larger framework. But the fundamental entities that are the outer frame of the fundamental framework cannot be explained by putting it all in an even greater framework, for then it would not have been the fundamental entities and the fundamental framework. This means that fundamental ontological entities may well be complex and they will not be fully explained.

On the other hand, it is reasonable to seek simplicity in the sense of the fewest entities, simplest structures and fewest connections, because this gives us the fewest unanswered questions (more structure means more to ask about). That is reasonable because the theory with the fewest unanswered questions is the theory that is most coherent and thus best argued as closest to the final truth. The theory may still be wrong, and maybe the most coherent possible theory has some very complex fundamental ontological entities in it. But methodologically, the best we can do is to strive for as much coherence as we can and for that reason look for the simplest basic ontological entities we can, while still granting that not everything can be explained and that the basic ontological entities may very well be quite complex. They are after all meant to explain particles, fields, forces, galaxies, conscious experiences, thoughts, time, energy, the internet, and all the rest. The goal is to have as few loose ends (unanswered questions) as possible.

3.9 What fundamental ontological entities give as few loose ends as possible?

We have seen that we must have fields, points, possible values, rules and actualizations in our ontology. And actualizations seem to require actualizers. How can this all be combined in a way that gives as few loose ends as possible? My goal now is to consider whether some of the suggested basic entities can be understood as following from some of the others, or being reducible to some of the others, to see what can be the simplest solution.

It may from the outset seem that it is not possible to reduce the number of fundamental entities, since they all seem to presuppose each other in order to exist. It seems that any entity needs to be at an area in order to exist, for how can it exist if it is nowhere? And it seems that any entity must consist of some actualized value of some kind in order to be something at all. It also seems that anything must behave according to rules in order to be understandable as an entity which has a certain structure, otherwise it could turn into something completely different the next second.

This could be a limitation on our understanding – that we cannot understand anything as an entity existing in the world unless we locate it and make it consist of some values and behave according to some rule. This means that our theory of what are the most basic entities in the world is limited by what it is possible for us to understand as the most basic, but that is an unavoidable fact, so we just have to continue on those conditions. We should strive for the most coherent theory, which also means the most understandable theory. If someone offers a theory with incomprehensible building blocks, we have no good reason to believe it is true, and saying it was true would not have understandable content.

Even if we cannot imagine a world without an area, values, rules and actualization, it is possible to understand these entities in different ways and relating to each other in different ways, so we can still ask for the simplest possible way of understanding them and their relation. Are they separate entities, for example, or parts of one big structure? Maybe you have one actualizer which can actualize values according to patterns, and then the field and the rules follow from that? On the other hand, this actualizer seems to presuppose an area and values in order to be what it is in the first place. Maybe then it is better to think that there is a field which has a structure that actualizes certain values at times and places in accordance with rules. The rules may just follow from the way the field is structured, so that if you have such a structured field, you do not need some special generators of possible values and actualizers of values at times and places. John Carroll and Stephen French are examples of persons opting for this approach (J. W. Carroll, 2008; Cei and French, 2014).[64]

It seems to me that this last approach is also closest to the most typical attempts in physics of finding a fundamental theory. In quantum field theory there are fields for each particle which behave according to a field equation. The theory suggested here is that there is a fundamental field which can take on many different qualitative values in all points, which would be similar to the idea of there being a field for each particle. There can then be different quantitative values for each qualitative value in each point, and this quantitative value may well be zero for many values in many points. There could also be deeper rules that the field equations come from – maybe even just one. In physics today, theories like string theory will typically try to find the underlying structure for all values at higher levels, while the idea of one basic rule is a hope for physicists working with supersymmetry. Wave function realism would be another example of how

64 At least this is my interpretation of them.

different theories consider it to be one rule governing all motion in the universe (Ney and Albert, 2013).

There are in fact an amazing amount of symmetries in the laws and among the particles, where for example you have similar particles but with some opposite property, for example their charge or spin. In quantum field theory every particle has an antiparticle with the opposite charge, which makes the total charge of the whole universe sum to zero. An even more fascinating fact is that it also seems that the whole universe has positive and negative energy summing up to zero (S. Carroll, 2016, pp. 182, 200 – 201).[65] Yet another fascinating fact is that there are many very similar or identical mathematical relations in the laws of nature guiding seemingly unrelated things. For example, massive bodies gravitate very similarly to how charged particles attract or repel; and waves are very similar even if they are gravitational waves, water waves or light waves. Many more examples could be given (Siegel, 2018).

These are hints that the universe might follow one or more deeper rules which have to do with symmetry, balance or harmony in the universe. According to Tim Maudlin, one of the deepest insights from entanglement and wave function collapses in quantum physics (which imply action at a distance where measuring one particle at one end of the universe can give the opposite result in another particle at the other side of the universe) is the holism it implies for the whole universe (Maudlin, 2003, pp. 482 – 486).

Are there other ideas in physics which could shed light on how a rule-following universe could come to be? Physicist Lee Smolin has suggested that universes are born from black holes, and that every time a universe is born from a black hole in another universe, the laws and initial values are slightly changed (Smolin, 1997). This is a theory that has some implications. Universes with many black holes will give rise to more universes than universes with few black holes, and so a typical universe will have many black holes. Our universe is estimated to have at least 10^{18} black holes. A typical universe would then also be fine-tuned for star formation, which our universe is, and one specific kind of fine-tuning for black holes would be that no neutron stars have a mass more than twice that of the sun. Smolin predicted that in 1992, and while many neutron stars have been shown to have almost twice the mass of the sun, none has had more. There

65 Thanks to physicist Anders Kvellestad for making the following two points: While physicists believe that the total charge of the universe was 0 from the beginning, it has accumulated a slight surplus of charge over anticharge, observed today. We should distinguish between, on the one hand, the symmetries found in the equations or laws of nature and, on the other hand, the symmetries found in the universe corresponding to the solving the equations.

are also other predictions he made, like inflation having to be single field inflations, etc., which has also stood up to scrutiny so far.

This theory still needs a field, points, possible values and actualization according to rules. But maybe the black holes are the actualizers of possible values according to rules in limited areas. And maybe the rules are varied a little for each start. That could be a sort of explanation for why the rules are what they are in our universe, although it would not explain why there are rules at all. Nor would it explain the rules that black hole generation follows or what the generator of black holes in the first place is. And it does not explain consciousness. At the basic level it seems we just have to presuppose a structure at a place which can actualize possible values in different points at different times.

There is a possible way to simplify the basic structure. There seems to be a problem that on the one hand any motion follows rules and on the other hand something must make it true that everything follows rules, for how can something which makes other things follow rules itself follow rules? It seems easiest, then, to say that instead of rules there just happened to be a pattern without a cause, but above we saw the problem with this response – namely, that it makes us expect fewer regularities than what we actually see.

It seems to me that the best way to think of the basic structure is that the actualizer itself must have a structure which makes that which is actualized follow certain rules. We have to have an actualizer, and the actualizer has to have a structure, and this structure could then be of a kind which makes that which is actualized behave according to certain rules – the rules then being the implications of how the actualizer is structured.

We should thus distinguish between the rules that the actualizer follows and the rules that actualization of values in our world follows. The actualizer itself seems to follow rules, but in the case of the actualizer and the rules it follows it is probably best to say that these rules are just identical to the structure the actualizer and its motion has – otherwise an infinite regress will follow. But when it comes to the values being actualized in our world, these are actualized according to rules, and these rules follow from how the actualizer actualizes the values because of the structure of the actualizer. Humeanism is thus rejected with regard to the values actualized in this world, but the rules that the fundamental actualizer follows are just the brute fact that the fundamental structure of the world has the particular structure it has (which could be seen as a kind of Humean insight).

This actualizer is then a kind of fundamental power to actualize different quantitative values for the qualitative values, or in other terms: the power to make the qualitative values take certain quantitative values when and where they do. This fundamental power has a certain structure which is the reason why the values are actualized in the patterns they are.[66] The field contains all the possible qualitative values in all points and the actualizer determines what the quantitative value at all places are (and very often the quantitative value in a given point is zero). This is the simplest way I can see for how to relate the actualizer, the field, the possible and actualized values, and the rules.

It is probably the case that the fundamental power actualizes values partly in response to what values are already actualized at a place at a time. This would explain why some laws seem to operate at more complex levels only, or why states of affairs can seem to work as constraints on how the laws of nature work, or why we seem to have cases of so-called top-down-causation. Robert Bishop argues well that the specific results of laws of nature are context-dependent (Bishop, 2019).[67]

If the actualizer itself is part of the structure which is the all-encompassing field of possible values, we end up with one field which has a structure which implies rules, and this seems to me to be the best starting point we can have in the sense that it is the simplest starting point giving the most coherent theory as a result.[68] The rest of the book will have to support this claim.

To those who now shake their head from the idea of such an implausible value actualizing structure, I want to remark the following: We already know that our universe follows some amazing rules, usually formulated in short and beautiful equations. According to special and general relativity the universe follows the same rules independent of reference frame, and light moves at the same speed independent of the speed of the light source. Because of these rules, objects contract and motion slows down so that the rules are followed. This is already quite mind-blowing. Since there are values being actualized, we know that there must be possible values that are being actualized. To suppose that there is a field of possible values and an actualizing structure working by maybe just one fundamental rule seems to be as simple as it can get in our fantastic world. It is

66 The structure of the fundamental actualizing power could be set for ever or be of a more exploratory and changing kind. I return to this question in the excursus on why the scientifically formulated laws of nature are as they are.

67 See for example pages 6–1 and 6–10.

68 As far as I can see, this basic view should be shared regardless of whether one believes in God or not. If God exists, God also has a fundamental structure (and it is a brute fact that God happens to have the fundamental structure God happens to have even if some properties follow from others) and then the world is fundamentally such that the basic structure of the fundamental field is one of an omnipotent conscious being. If God does not exist, then the world is fundamentally such that the basic structure is something impersonal, like a process of symmetry exploration and symmetry breaking. Combining is also possible, that there is a God creating a universe with symmetries, but the question is both what exists and what is fundamental.

complicated and advanced, but to repeat a point made above: anything that has existed or happened in the history of the universe must have been possible from the start, since otherwise it would not have existed or happened. Any starting point must have the capacity to produce everything that has ever existed and happened, and thus seems to be a quite remarkable and complex entity (Puntel, 2008, p. 456). This means that there must be something incredible existing from the start, and this analysis has made a suggestion as to how this could be as simple as possible, even if it is still very complex and fantastic.

This discussion completes my presentation of the basic entities of the world. However, the whole discussion has presupposed that these are entities that exist, which again requires a discussion of what it means that something exists and what existence, as such, is. This discussion now follows.

3.10 Existence

What is existence? A question about existence which has been discussed for a long time in the history of philosophy is whether or not existence should be thought of as a property of things. For example, while Anselm argued that God being perfect must have the property of existence, Kant criticized the idea of existence as a property you could add to a thing.

Bertrand Russell gave some influential reasons for us to think that existence is not a property of things. If existence is a property of things, it seems that there can be things which do not have this property. In other words, there are things that do not exist. This seems contradictory – that there *is* a thing which does not *exist*. To say that a unicorn does not exist seems to say that there is a thing called unicorn, but it does not exist, which again sounds contradictory – there is a thing which does not exist. Alexius Meinong held the view that there are things which do not exist and was criticized by Russell for it (B. Russell, 1905).

Russell's alternative to saying that existence is a property of things was to say that things are properties that are either instantiated or not: being a horse is a property which is instantiated in our world (there is an *x* which has the property of being a horse), but being a unicorn is a property which is not instantiated in our world (there is not an *x* which has the property of being a unicorn) (B. Russell, 1905).

Another alternative to both Meinong and Russell is to distinguish between existing as fiction or not. For example, Peter van Inwagen quotes Meinong saying, "There are objects of which it is true to say that there are no such objects" (Inwagen, 2003, p. 131). He then argues that there cannot *be* something which *is* not – unless "to be" has two different meanings (Inwagen, 2003, p. 133). Van Inwagen's solution is then to distinguish between a narrow sense of existence, which is to participate in being, and a wide sense of existence, which includes existence in fiction (Inwagen, 2003, p. 139). It is then not contradictory to say that "there are unicorns, but they do not exist," since "there are" is used in the wide sense first, and then their existence in a narrow sense is rejected.

Many debates on existence remain at this level, discussing the difference between existing in fiction or in the world, or existing now versus existing in the past or the future. But one can also try to dig deeper and ask what it is that all things that exist have in common and what it means to exist at all. I have been speaking about the fundamental entities of field, values, actualization, and rules, and how most objects we are familiar with are structures in the actualized values in the area. I shall now discuss first the existence of structures actualized in the values before moving to the deeper question of the existence of the fundamental entities and what it is that all existing entities have in common.

We are now considering structures actualized in the field commonly said to exist, but which seem to have different forms of existence. How are we to understand the differences in existence when it comes to mammoths, horses, unicorns, numbers, ideas, relations, holes and shadows? I shall now present different distinctions between different forms of existence, which will also allow me to analyze the concepts of "concrete"/"abstract" and "universal"/"particular". I start with a common distinction which is often called the distinction between existence in the mind and outside of the mind.

Everything we can think of, experience or have a word for exists as a structure which consists of parts which are either just possible values or also actualized values, either qualia values or physical values. I now use the term "the qualia field" as a term for all the qualia values and "physical field" for all the physical values. I shall explain how the qualia field works in much more detail later, but so far it can be thought of as consciousness. We are very often interested in whether something exists only in the mind of someone or also outside of the mind. For example: do unicorns exist merely in the mind or also outside of the mind, like horses do?

The difference between the existence of horses and unicorns is that horses are individuals that have as parts values that are actualized in physical fields instead of merely in the qualia field. Unicorns are structures that only have as their parts actualized qualia values when someone thinks of them (then they exist in their minds). When nobody thinks of unicorns they exist as possible patterns that values in either the qualia field or the physical field can actualize. The same goes for uniflatworm, which presumably I was the first ever to think of right now, a flatworm with a single horn in its forehead.

The different forms of existence here mentioned are firstly existence as a pattern in actualized values in the qualia field, and I shall call that "existence in consciousness". Then it is existence as a pattern in actualized values in the physical field, and I shall call that "existence in the physical world". That is a bit misleading, since consciousness and possibilities also exist in our universe, but in many ways it gives the right associations to people interested in the distinction between existence in the mind and outside of the mind.

It is possible to include a third form of existence, but that would be broader than what is typically meant by "existence". This is existence as possible patterns in values that are not actualized, either in the qualia field or the physical field. Regarding this third form, all the values exist as possibilities to be actualized in the field at all times, so here I speak of the patterns that can be actualized when specific values are actualized at specific times and places. If someone says that the geometrical figure of a dodecahedron obviously existed before anyone thought of it, I suggest that the only coherent interpretation of this claim is that it existed as a possible pattern we could think of later (more on this in the chapter on mathematical entities). But before it was imagined it did not exist in the narrower sense of existing either in the physical field or the qualia field.

When it comes to existence in the physical world, we need a distinction between patterns that have actualized values as their parts and patterns that do not. This is the same distinction as that between individuals and relations that do not have individuals as parts. For example, a horse or a chair will cover an area in the physical field with actualized values in the same area at the same time. In other words, they have actualized physical values as their parts. Relations like "higher than" or "being the uncle of" are patterns in actualized physical values, but they do not have actualized physical values as their parts. They do not cover an area of actualized physical values at a specific point of time. Rather, they are patterns we have noticed and named with a word and which we can think of in the sense of actualizing the pattern in the qualia field (in our consciousness). If we say that a hole exists, we should mean that holes exist in the sense that they are coarse-grained descriptions of actualized values where a hole-shaped structure can be discerned.[69] Some relations are relations that nobody has thought about, but they are possible patterns that can be discovered.

69 Discussing the existence of holes is a serious topic in metaphysics, see for example D. Lewis (1983, chapter 1).

This distinction between individuals and relations that do not have individuals as parts is how I suggest that the distinction between concrete and abstract should be understood. Individuals are concrete, while relations that do not have individuals as parts are abstract. In this last category we also find mathematical entities and logical structures. The number four is a pattern that we find many times where four individuals are involved, and then the pattern is noticed and abstracted while the concrete individuals are disregarded. I write more about this in the chapter on mathematical entities.

When it comes to individuals existing in the actual world, we also need to distinguish between existence at a specific point of time and existence at some point of time. This distinction lets us understand the difference between how horses and mammoths exist, since horses exist now and mammoths existed before in the physical field (whereas both exist now in the qualia field when I think of them). I will define "time" and "now" in the chapter on time, and discuss there in more detail the distinction between existence now and existence before.

Now I have introduced enough distinctions to make a suggestion as to how the relation between universals and particulars should be understood. Universals, as they are usually referred to in philosophy, are just any pattern we can find. I suggest that there is no platonic world where all the things that are usually called universals exist. Particulars are individuals that have actualized values as their parts at a time and a place. Human is a pattern, and thus a universal in the traditional sense, but my colleague John covers an area at a time of actualized values, and is thus a particular. What does it then mean for a particular to instantiate a universal? If we refer to the typical examples of universals, it just means that an individual at a time and place has a structure and relations found elsewhere as well.

However, we could use the term "universal" to refer to the list of fundamental values, and say that particulars instantiate these universals when they have actualized values as parts. Then most individuals would instantiate values like mass and charge. If we include less fundamental values, they could instantiate values like hair and skin. But relations like "being an uncle", etc., are not actualized values that individuals have as parts, and so they are not universals instantiated in the sense of being values individuals have as parts. Thinking of the world in terms of particulars instantiating universals is thus in most cases more confusing and leads to pseudo-problems rather than being of use.

This is a good place to define also the concept of "world". In the broadest sense, the world is any possible or actualized world, but when I use the term "world", it refers to the values that are or have been actualized. These can be actualized in the physical field or in the qualia field, which would refer to the physical and the non-physical world (or the world of consciousness). The context shows whether I refer to just the physical world or the whole actual world (including consciousness). By "nature" or "physical", I refer to the values actualized in the physical field. The real world is to be distinguished from fictive worlds, which could have been actualized, either in the physical field or as imaginations in the qualia field. As a possibility of the real world, fictive worlds are part of the real world.

So far, existence has been defined as actualization of values in a field, but not much has been said generally about what it is to exist or to be actualized. Now I am ready to dig deeper into the concept of existence. Recall that definitions of basic concepts are explications of some connections between those concepts and others, and not full descriptions of their internal structure. Even if existence is basic, and cannot be contrasted with anything, it can be explicated without using terms that are even more basic (Puntel, 2008, pp. 414, 436). The terms get more and more explicated as they get related to other terms as the book develops.

When discussing existence, Lorenz Puntel distinguishes between beings which exist and Being (with capital B) as that which all beings have in common. Concerning Being, he distinguishes between "Being as such" and "Being as a whole", where Being as a whole includes beings (said to exist) and Being as such is Being considered without beings (Puntel, 2008, pp. 417–418). Since the term "being/Being" can be quite confusing in English, I shall use the term "existence as such" when I speak of that which all existing things have in common (= Being considered without beings).

We now move on to the question of understanding existence in itself. As mentioned earlier, it might be that it is only possible to explicate existence by describing certain features without giving a full definition or explanation. In the following I shall try to explicate certain features that characterize everything that exists to see how far we can come in understanding existence itself. The features we shall start considering are membership in (or being part of) the basic structure, structurality, actualized value, localization, registrability, and causal influence.

I have described a basic structure, which is a field where values can be actualized according to rules. A feature common to everything that exists is membership in (or being part of) this basic structure. Everything that exists is contained in the basic structure as a part of it: The area, actualizer, values and rules are all part of it. Every other entity we can think of (such as mammoths, horses, unicorns, numbers, ideas, relations, holes and shadows) also exists as part of the field, more specifically as a structure or pattern in either the possible values or actualized values, either qualia values that one can be conscious of (as

I shall argue below) or physical values that can become physical entities. I use the term "physical" here in a very broad sense as something which is non-conscious and quantitatively measurable. The fundamental field can take on qualia values and physical values, and maybe other values, but these two kinds of values seem to suffice for describing the content of our universe. Yet maybe the fundamental field itself is of another kind of value than conscious or physical values.

Another feature common to everything that exists is structurality. Something that we can understand as existing must have a structure; otherwise we could not have an understanding of it at all. Not even the concept of existence makes any sense unless we can give it some structure. If we think of existence in the broadest possible sense, which means to include possible entities as existing before they were thought of (like a dodecahedron), then existence seems to just mean structurality. In the widest possible sense, anything that has a structure exists, and it having a structure is what it means that it exists.

However, we usually want to make a distinction between merely possible entities and others that are said to exist as opposed to being merely possible. I shall now proceed to considering the features that are common to entities that exist in a narrower sense than just having structure. The first description is to say that everything that exists has actualization in common. This is not very helpful, since "actualized" is just a synonym for "existing" in this narrower sense. One could add that "actualized" means "caused by the actualizer", which again does not say much, but if combined with a theory of the actualizer (as above), it does provide some more determination of the concept.

The next feature that everything that exists has in common is localization. That is to say that in addition to actualization it includes actualization at a certain place. This is not so controversial when it comes to localization in the physical field, but quite controversial when it comes to mental entities, since our thoughts do not seem to be located at a particular place. I defend this view in further detail in the chapter on consciousness.

The next feature is registrability. Here I add to actualization that if something is actualized at a place it should be registrable by some kind of mind, since if no possible mind can register it, there is no point in including it in an ontology. What I suggest is that all things that we think exist – physical and conscious – can be registered directly or indirectly by a mind. Most likely many things exist which have not *actually been registered* by a mind, but they must be registr*able* in order to exist and be part of an ontology. If you want to insist that something can exist without being registrable or understandable, you are unable to make sense of the claim that it exists (for if you are able to make sense of it, then it is registrable and understandable after all).

Another feature closely linked to registrability is the ability to have causal effect, both on minds and other entities. If something cannot have causal effect on anything, then it cannot have causal effects on minds and be registered either (I here take "being registered" as a direct or indirect causal effect of that which is registered), and then it cannot have any role to play in an ontology either.

I now have a list of features common to everything which exists. In the broadest possible sense, anything with a structure exists, but in the narrower sense which I shall mean by "existence" in this book, existence as such is characterized by membership in the basic structure, structurality, actualized value, localization, registrability and causal influence.

These are still just descriptions which mean something other than "existence" itself, so what is the internal structure of existence itself? I have described it as being part of the basic structure, which in one sense means that the description of the basic structure is a description of what existence is. But on the other hand, we can imagine that the basic structure did not exist. It seems to be something that all its parts have in common, which makes it exist as opposed to not existing, and what is that?

I think that it is the ability to have a causal influence – the energy, the motion, the "fire in the equation", the fundamental force or power, God, or whatever you want to call it. It is this ability to influence and be registered, but what is that then? It is the power of this basic structure to actualize values according to rules, and we do not have a deeper understanding of what it is, but we do know that there must be this mind-blowing capacity of the fundamental structure which was there from the start since it was necessary for it to be there from the start for all things that have happened later to happen. This power of actualization, which we have now described with many characteristics, but which we do not know in detail, is what existence is to the best of our knowledge. We know that something spectacular like this makes everything exist, but we do not know at bottom what it is. It is the same basic actualizing power that was a stopping point beyond explanation above.

3.11 Excursus: Nothing

Since we have now talked about being, let's talk about nothing for a while. We saw above that Lorenz Puntel argues that we should distinguish between, on the one hand, beings that exist and, other the hand, Being as a whole and Being as such (where "Being as a whole" is Being as such and beings that exist, while "Being as such" is Being considered without beings – that which all beings have in common). Corresponding to this distinction, Puntel argues that we should distinguish between two different senses of the concept of nothing, which Puntel calls *nihilum relativum* and *nihilum absolutum*, but which I here will call "nothing in the weak sense" and "nothing in the strong sense", respectively. A world without beings would be an empty world, thus nothing in the weak sense. It would be a field where no values were actualized. But it would not be nothing in the strong sense, which is the absence of Being as a whole, including Being as such (Puntel, 2014).

Puntel argues that it is only possible that there could have been nothing in the weak sense, but it is not possible that there could have been nothing in the strong sense of the term, and according to Puntel, the concept of nothing in the strong sense is a pseudo-concept with no coherent meaning. He offers several arguments for this, which I have discussed elsewhere (Søvik, 2018). Here, I shall only consider one line of reasoning which continues insights discussed above.

If we are going to understand the content or meaning of a concept at all, the content has to have some structure. We cannot have a thought which has as its content no structure since there has to be some structure in a thought that is to be understandable or even be a thought. If we try to think about the meaning of the term "nothing" in an absolute sense, we will only be thinking about nothing in the weak sense. We try to imagine no structure, but in reality we are thinking about an empty area, a world where all things have been removed, but still a place which is an empty world, and not nothing in an absolute sense.

One could argue that while the attempt of having a conscious image of nothing has structure, one can have a non-conscious, abstract understanding of the concept with no structure, but even that requires a non-conscious state of mind, which has structure (as I argue further in the chapters on mind and thinking). To have an understandable concept requires structure, and structure is existence in the widest possible sense of the word, so one cannot think the concept of nothing in its absolute sense.

This is not much of a problem, since one rarely needs the concept of nothing in an absolute sense. We can imagine the removal of everything, which would still be an idea of an empty world, and most often the term "nothing" is needed only in its weak sense. But for a few discussions on the relation between mind and world, and existence and nothing, it is interesting to note this fact about our understanding of nothing: when first there is mind, "nothing in the strong sense" is unthinkable.

I agree with Puntel that it is impossible that there should at one time have been absolutely nothing, and here I will only refer one traditional line of thinking, similar to a point I have already made: Since something exists, it must always have been possible (in the widest sense) that something could exist, otherwise something could not have existed now. In other words: There must always have been a world where it was possible that something could exist, which is to say that something (namely this world with this possibility) has always existed. To deny that would be to say that it is impossible that something exists now, which is self-contradictory. It is thus logically necessary that something must always have existed, but it does not follow in more detail what it is that have always existed. I will return to that question in the end of the book.

I have now described the fundamental ontological entities and said what it means that they exist. This concludes Part One, which describes the theoretical framework to be used in the rest of the book. The next big topic to be discovered is the topic of mind, which is what Part Two is about. Part Two will come after an excursus on analytic/synthetic and a priori/a posteriori statements, and an excursus comparing the theory in this chapter with similar theories.

3.12 Excursus: On analytic/synthetic and a priori/a posteriori statements

It might seem that my analysis of modality in Section 3.5 is not sufficient, for example that it lacks an understanding of a posteriori necessity. While the essence of this objection is briefly considered in a footnote above (footnote 56), I include an excursus here to deal with the objection in more detail.

There is a famous distinction in philosophy from Kant between analytic and synthetic statements, and between a priori and a posteriori statements. How to understand these distinctions is a matter of debate, but as an introductory rough definition, we can state the following: Analytic statements are true by virtue of the definition of the terms of the statement, while synthetic statements are not. A priori statements can be known to be true without experience of that which the statement is about, while a posteriori statements must be verified by experience. "Bachelors are unmarried" would thus be analytic and a priori, while "John is a bachelor" would be synthetic and a posteriori.

A big issue of debate is then how to understand the distinctions more precisely, and what to think of cases that seem to fall in between. For example, Kripke has argued that while usually it is a priori analytic statements that are considered to be logically necessary (like "all bachelors are unmarried"), some statements are a posteriori necessary, like "water is H_2O" or "The Morning Star is the Evening Star". These are statements that we discover the necessity of through experience.

What makes this tricky is that propositions get their meaning in the theoretical framework within which they are placed. It is always possible to deepen our understanding of concepts. Often we have a coarse-grained understanding of concepts, which can change as we get a different understanding of the world. We can then discover that the concept meant something other than we thought. Let us look at some examples.

We may have an idea of bachelor as an unmarried man, meaning that it would be self-contradictory to say of an unmarried man that he is not a bachelor. And yet, if someone says that the pope is not a bachelor, we might discover that our idea about what it means to be a bachelor does not include the pope. Next, we may have an idea of water as a familiar liquid. We can discover that water is H_2O, and call it an a posteriori necessity. And yet, there is much in the world that everyone calls water which is not the normal H_2O. For example, there is $H_2{}^{17}O$, $H_2{}^{18}O$, $HD^{16}O$, $D_2{}^{17}O$, $T_2{}^{18}O$, etc. (Weisberg, 2006).[70] Glass cannot be made of cardboard. But if someone discovers how to manipulate cardboard to make a hard see-through material that can break like glass, could glass then be made of cardboard? A week is seven days, but what if earth's rotation had changed dramatically? A bachelor is an unmarried man, but what if gender activists make us include women in the concept of bachelor?

The examples show that we can discover that the world was different from what we thought, or the world can change, or our use of words can change. It thus seems also that what is a priori, a posteriori, analytic, synthetic, contradictory or necessary can change, but two objections to this should be answered.

First, there is a distinction between a sentence which has a necessary or contradictory form, like A = A or A = not-A, and those that do not. "A bachelor is unmarried" does not have a contradictory form (it says F is G) while "a bachelor is not a bachelor" does have a contradictory form (it says F is not F). In the first sentence we must deduce the contradiction by defining the terms, but the second sentence seems obviously contradictory. However, we can come to discover that the terms were unclear, so that they do not necessarily mean the same even if the same word is used (F and F did not mean the same, and were instead F and G). Some bachelors may not be bachelors, if the first "bachelor" means any unmarried man and the second "bachelor" qualifies it to adult men eligible for marriage, or something like that. An unmarried man who is 16 years old is a bachelor in one sense, but not a bachelor in another sense.

Secondly, what if we assume that the words mean the same and that they are related to the same theoretical framework at the same point of time? Then it seems clear that something can be analytic or contradictory or a priori, etc., at that point of time, even if we later come to understand the world or the words

---

70 What we call H_2O is more precisely $H_2{}^{16}O$.

differently. It seems that we can say that given a certain meaning, sentences can be contradictory, a priori, etc.

I agree with this argument, and will thus say that given a specific meaning, something can either be consistent, inconsistent, or inconsistent to deny. But as shown above, this can change over time. I will discard the vague concepts of a priori, a posteriori, analytic and synthetic since these seem to presuppose a clear border where there is actually a continuous negotiation of how to define terms and interpret experiences. It will suffice with the distinction between "consistent", "inconsistent" and "inconsistent to deny", as long as the meanings of terms are determined.

3.13 Excursus: Comparing the ontology of this book with important alternatives

In this excursus I compare the ontology of this book with some important alternatives. This kind of comparison is done throughout the book with many alternatives, so the focus here is on comparing my view with others who have made a similar main move in trying to reduce most entities to a set of values in points. The natural place to start is then a comparison with David Lewis, who has been working on a program he called Humean supernevience, where he has argued that everything in the world supervenes on local qualities in points.

Humean supervenience is the view that "all there is to the world is a vast mosaic of local matters of particular fact" (D. Lewis, 1986b, p. ix). The idea is that all that exists is a geometry, which is a system of external relations of difference in space and time between points, and at those points there are local qualities, which are natural, intrinsic properties. Everything else supervenes on that (D. Lewis, 1986b, pp. ix – x). There is no room here to present in detail how Lewis reduces all sorts of metaphysical concepts to local facts, but here is a very brief overview of some important concepts, to give an impression of his philosophy:

Laws of nature are just facts about the world that are both simple and very informative in the sense that a lot of truths can be deduced from them (D. Lewis, 1986b, p. xi; D. Lewis, 1973b).[71] Causation is counterfactual dependence, in the

71 In addition to simplicity and strength, Lewis later added a third criterion of fit to include indeterminism and chance in the laws, see (D. Lewis, 1994, p. 480). Lewis thought that he could reduce everything else to the basis of natural facts, but saw one severe problem, which he called the big, bad bug. This problem was how to understand chance, in the sense of the objective chance at a point of time for one event to happen (D. Lewis, 1986b, p. xiv). Lewis thought

that laws of nature supervened on the actual facts of the world, but these laws should then state what the objective chance at a point of time for an event to happen is. The problem is then as follows: first there is a particular set of facts that is the world history up until today, and based on these facts we formulate laws of nature as the best general descriptions of the regularities. But the facts that have happened until today are compatible with different futures, which means that different laws with different chances will be equally good today and only the future can tell which was best. Laws do then not explain chances, but supervene on chances instead (D. Lewis, 1986b, p. 127).

If chances supervene on the history of particular facts, this contradicts the principle Lewis calls the Principal Principle, which says that our credence at a point of time that an event should happen should be conformed to the actual chance that the event should happen. If there is an objective chance at a point of time of 50 % for the coin landing heads, then our credence for the coin landing heads should also be 50 %. Let us say that at a point of time there is a 20 % chance for event A, which means that our credence for believing A to occur should be 20 %, but then the future arrives and A does not happen. If chances supervene on the facts, then this means that the chance for A was in fact zero, which means that we should have believed that the chance was 0 %. The Principal Principle seems to imply both that we should believe the chance to be zero and not to be zero, and this is then a contradiction (D. Lewis, 1986b, p. 129; D. Lewis, 1994, p. 485). The problem arises since facts about the future can undermine what we take to be chances in the present if chances are made true by the facts only (D. Lewis, 1994, p. 485).

Having struggled with the big, bad bug for years, Lewis finally thought he had solved the problem in 1994, and presented the solution in an article called "Humean Supervenience Debugged" (D. Lewis, 1994). Lewis here starts by arguing that if the Principal Principle tells us to conform our credence to the objective chance of an event, and the objective chance of that event depends on the future, then this is a fallacious use of the principle, since it depends on information about the future, which is inadmissible information (D. Lewis, 1994, p. 485). This does not seem like much of a solution since it seems to imply that the Principal Principle never applies to establish the credence of a future event. But we often want to know what the chance of a future event is, and it seems we are often correcting in estimating it. Again there seems to be a contradiction in saying that on the one hand we cannot estimate the chance of a future event since it depends on inadmissible information about the future, and on the other hand we seem to be able to do it well all the time (D. Lewis, 1994, pp. 485–486)?

Lewis finishes the solution by saying that admissibility comes in degrees. If some evidence only depends minimally on future events it is almost admissible, and the Principal Principle can be approximately correct. For example, information about the future may in some cases be irrelevant to establishing the present chance, and then the Principal Principle gives the right answer, but in other cases the future may undermine the present chance, and then the Principal Principle does not work (D. Lewis, 1994, p. 486).

Lewis then corrects the Principal Principle and makes a new version where our credence of event A should not be conformed with the objective chance of event A, but instead of the objective chance of A given a complete theory of chances. On the one hand, then, we have objective chance which is made true by the patterns of facts in the world, including past, present and future. On the other hand, we have our credence about chances, which is made true by a true theory of chances. Since the future can undermine our present theories, we may be wrong in estimating chances, but if nature is kind to us, we will often be able to estimate chances reliably. Lewis agrees that this understanding of chance is not exactly as the intuition of an objective chance at

sense that A causes B means that if A had not happened B would not have happened (D. Lewis, 1986b, p. xii; D. Lewis, 1973a).[72] How can we know that a counterfactual statement is true when it describes something that has not happened in our world? It is by comparing with worlds that are similar to ours – the truth of counterfactuals supervene on worlds that are similar to each other (D. Lewis, 1986b, p. xii; D. Lewis, 1973b). Lewis is famous for explaining counterfactuals and reducing modality by arguing that what we call possible worlds actually exist – they are of the same kind as our world (D. Lewis, 1986a). This allows him to explain modality without using modal terms, even though it presupposes the existence of very many worlds.

Lewis is a materialist (D. Lewis, 1986b, pp. x–xi), but what does that mean? When stating what the fundamental facts are, he says that they are natural, but he never gives his consent to a specific definition of what it means to be natural. However, he does assume that the world is more or the less the way physics describes nowadays (D. Lewis, 1986b, pp. x–xi). When it comes to the topic of mind, he is an identity theorist and a functionalist about mind, which means that there are no non-physical qualia (D. Lewis, 1966). Concerning free will, he is a compatibilist, which means that free will is compatible with the events of the world being determined (D. Lewis, 1986b, p. 291).

Compared with my own views, I share the basic intuition of Lewis to think that many concepts are reducible to a basis with some fundamental values being actualized at points. However, we think differently about the basis, which means that almost all other concepts are reduced in different ways. Again, there is only space for a brief overview here, but the rest of the book will unpack my views.

The two most important differences between me and Lewis is that I include laws of nature and non-physical qualia, and this leads to many overall differences. Lewis includes natural properties only (understood as physical properties), while I include values (physical and non-physical) being actualized according to rules. I find that this allows me to have more coherent reductions in the second round than Lewis, but he could have argued that his basis is simpler than mine. However, I think contemporary physics already contradicts his understanding of the basis that there are only local properties at the fundamental level. Instead

---

the present seems to imply, since it depends on the future, but he thinks that it is close enough and a good approximation (D. Lewis, 1994, pp. 486–489).

72 Lewis modified his view several times, in the direction that if A had not occurred a certain way, B would not have occurred a certain way, see D. Lewis (2000).

values are being actualized according to holistic rules, and Lewis' incompatibility with physics is a common critique of him (Maudlin, 2009).

What are then the advantages of starting with my basis? Chance is not a problem since it is part of the rules. This is not a kind of solution that Lewis would like, but I have given my arguments above for why I think we cannot dismiss the idea that nature is guided by rules, and thus why laws of nature cannot be reduced to patterns in particular facts. Lewis leans to heavily on nature being kind to us, but more chaos would be expected if nothing makes the regularities happen, and thus we should believe that there is such an explanation.

Since I have laws of nature, causation can be reduced to the work they do, whereas I reject using counterfactuals as a basis, since I do not believe that counterfactuals have truth value – they can only be very probable (more on that below). Rejecting true counterfactuals, I do not need to posit innumerable real worlds the way Lewis does. I reduce instead modality to consistent and inconsistent descriptions based on presuppositions.

I find it incoherent to reject non-physical qualia as long as their properties seem inconsistent with the properties of physical things, without us being able to explain how that can be. While it is possible to affirm free will in a weak sense compatible with determinism, I argue that we can defend a stronger and incompatibilist form of free will.

Moving on from David Lewis, there is another philosopher who has an alternative to Lewis, which is similar to my own position in many ways. Michael Esfeld criticizes Lewis for being a supersubstantivalist (since he includes spacetime as fundamental in his ontology) (Esfeld, Deckert, Lazarovici, Oldofredi, and Vassallo, 2018, p. 47), and has called his own approach Superhumeanism (Esfeld et al., 2018, p. 47).[73] The name is chosen because Esfeld reduces spacetime with its structure to laws of nature and physical properties also to laws of nature, leaving only fundamental particles in motion. This is similar to the way I reduce spacetime and many physical values (in later chapters of this book), and for this reason I will compare our ontologies. I start by presenting Esfeld's ontology here, and then I will discuss the differences, arguing in favor of my position at the points where we differ.

Esfeld has a very reductive approach arguing that we can have a sufficient ontology for the whole natural world with only two fundamental entities. The first fundamental ingredient is point particles, which have no other structure

---

73 Esfeld has written with various articles together with different people, but for simplicity I refer to Esfeld only. The book relied on much in this presentation is co-written with Dirk-André Deckert.

than being at a certain distance from other point particles (Esfeld et al., 2018, p. 4). These particles exist eternally, since Esfeld does not think it makes sense that something can come out of nothing or turn into nothing (Esfeld et al., 2018, p. 28). The second fundamental ingredient in Esfeld's ontology is change; that the distance between point particles is changing (Esfeld et al., 2018, p. 4). He argues that everything else can be reduced to this. I shall now present in more detail how that happens.

Esfeld argues that all physical quantities can be reduced to laws describing how particles move. For example, mass, charge, energy, etc. are all concepts introduced to physics through their dynamic role, said to influence how particles move (Esfeld et al., 2018, p. 6). He argues that quantum mechanics shows that these concepts cannot be intrinsic properties of the particles since they apply to systems of particles (Esfeld et al., 2018, p. 7). To put it shortly, he says that particles do not move like electrons because they are electrons, but we call those particles electrons that move electronwise (Esfeld et al., 2018, p. 7). The idea is then that the particles we call electrons are composed of smaller, fundamental particles identical to all other fundamental particles, but some groups of them move according to laws that make it convenient to refer to them as electrons.

When it comes to space and time, Esfeld finds these to be reducible. Time is reduced to change, and how we measure that change is relative (Esfeld et al., 2018, p. 32). Esfeld is a presentist, since this provides the simplest ontology (Esfeld et al., 2018, p. 152). There is no space with a metric or geometry, according to Esfeld. While general relativity may seem to imply that there is such a space with its own unique geometry, Esfeld argues that general relativity should be thought of as describing patterns in how particles move (Esfeld et al., 2018, pp. 27–28, 152–154).

This way of reducing physical properties to laws and reducing spacetime to laws is similar to how I reduce them in this book, except that I reduce them to structures in values in fields whereas Esfeld reduces them to patterns in motions of point particles, and we have a very different understanding of the laws. I will come back to these differences.

Esfeld rejects that fields should be an irreducible part of one's ontology, since he finds them extravagant and unnecessary adding only mathematical problems. Fields are only mathematical devices used to describe the motion of particles (Esfeld et al., 2018, pp. 9, 135). While most things are reduced to laws, Esfeld also reduces laws to the best-system account of Lewis et al. No laws exist guiding motion in the world. Change is a brute fact, but patterns in the changes can be described in ways that are simpler and more informative than others, and the best descriptions are the ones we call laws (Esfeld et al.,

2018, p. 48). I do not reduce fields and laws this way, and will discuss this move below.

There are several issues where Esfeld does not go into the matter, but remain neutral or agnostic. His ontology is an ontology of the *natural* world, which means that he does not state his opinion on matters like mind, consciousness and normativity (Esfeld et al., 2018, p. 8). He is neutral on the topic of modality (Esfeld et al., 2018, p. 14). While arguing against a realist view of laws of nature, he also says that he is agnostic on whether there is a logos that drives the evolution of the cosmos (Esfeld et al., 2018, p. 56). I take that to mean that he is open to the possibility that laws of nature exist, even if he thinks that it does not add coherence to the theory to add them in your ontology.

I will now discuss topics where I differ from Esfeld. The two most important ones are whether there are irreducible laws of nature and whether everything can be reduced to point particles, and I start with discussing laws of nature with Esfeld. My view is that there are patterns in nature which are so clearly systematic that they must be caused instead of uncaused, non-random instead of random. While Esfeld says that gravity is obviously just a mathematical device for describing motion (Esfeld et al., 2018, p. 136), I find the systematic patterns of gravity clearly pointing to something real causing the systematicity.

Esfeld has a reply to the argument that something must cause the motion, which is an argument also made by Lewis (D. Lewis, 1986b, p. xii), namely that adding laws does not add explanatory value, since the laws are defined in terms of their role (Esfeld et al., 2018, p. 52). The idea is that if you explain B by saying that A did it, and then all you can say about A is that "it is what causes B", then no explanatory value has been added. Esfeld thinks that adding something to an ontology is only justified if it increases the coherence of the theory (Esfeld et al., 2018, p. 43), but he argues that adding laws instead creates new problems since we do not understand how the laws guide the particles (Esfeld, 2017, pp. 12–13).

I disagree since I find the laws of nature to increase the coherence of the theory, since an ontology with laws says that the patterns are caused and non-random which is more coherent than a theory without laws where the patterns are instead uncaused and random. Patterns that seem caused and non-random are thus data that my ontology integrates better than Esfeld's ontology, and then it is more coherent (by the definition of coherence in terms of integrating data, which was offered earlier) even if I do not have a deeper explanation of the laws or how they work.

Someone could criticize my response and say: what if we add a superlaw to explain the laws, would that then be more coherent since it integrated one more piece of data? Or a super-superlaw to explain superlaws, etc.? My response is that adding superlaws to explain laws does not increase coherence since it only integrates one piece of data by adding one piece of unintegrated data, in an ad-hoc way since there is no other support for superlaws. The situation is different with laws since there are so many things together indicating both that they are not random and that they are caused, which is then explained if there is in fact something causing it non-randomly. Even if this deeper cause is defined in terms of its role, the whole theory will now call patterns caused and non-random that clearly seems to be caused and non-random, which makes the whole theory more coherent in the sense of having better integrated data.

I now move on to discussing whether everything can be reduced to point particles, with no other structure than distance to other point particles. I struggle to make sense of what it means that such particles exist, since Esfeld says that they do not have extension (Esfeld, Deckert, et al., 2015, p. 9). They are only defined as being at a certain distance from another point, but then what it is *to be a certain distance* is to be a distance between two point particles, and then particles and distances are defined in terms of each other only, which is viciously circular.

To put the point differently: what distinguishes a matter point from nothing? How can you add points without extension to objects with extension? What if I speak of a point two meters in front of me, thereby defining a distance of two meters in front of me: what is the difference between there being nothing two meters in front of me and there being a particle two meters in front of me, if particles are defined only by being at a certain distance from something else? I can define all sorts of distances in all angles from myself, does it mean that there are particles at all these points, or why not?

Esfeld claims that to say that there is a particle somewhere is to say that the point is occupied instead of empty (Esfeld, Lazarovici, Lam, and Hubert, 2015, p. 143), but what is the difference between "occupied" and "empty"? Esfeld discusses the difference between an empty and an occupied point, and says that the difference is whether it has a distance relation (Esfeld, Deckert, et al., 2015, p. 9), but I do not understand that answer, since as mentioned above it seems that both an empty and an occupied point can be two meters in front of me.

Esfeld says that he defined matter in terms of extension just like Descartes (Esfeld et al., 2018, p. 47), but Descartes defined matter as an extended substance (Descartes, 2008, pp. 209–210), while Esfeld says that matter points do not have extension. The extension is between the substances (particles) in Esfeld's account, and that is what causes the problem.

I argue in this book that not all physical values can be reduced to rules, but that we instead need to keep forces with a certain strength in a certain direction, and that we need non-physical qualia values, and the arguments can be found at the relevant places in the book.

These were the main differences between Esfeld and myself, and it leads to several other differences as well. Esfeld is a determinist, but does not think that it matters for free will, since the laws are not determining anything (Esfeld et al., 2018, p. 51). I am a non-determinist, arguing that what free will means is that sometimes our autobiographical self is the fundamental cause of our choices. Esfeld's point is that the laws do not determine anything, they are just convenient summaries of patterns, and so they do not compete with our determining things. I find it very implausible to believe that a random pattern of natural events matched perfectly with it really being us determining events. Either there is real determination in the world, and then determinism undermines free will the way I define it, or everything is just random patterns, but then our choices are also random patterns, which again undermines free will the way I define it.

To sum up, my main problem with Humeanism and Superhumeanism is the rejection of laws of nature and the reduction of everything to physical values or particles. I agree with David Chalmers when he says of Humeanism á la Lewis that its biggest problem is that it cannot account for phenomenal and nomic truths (Chalmers, 2012, p. 428). Chalmers' own alternative is to have what he calls a scrutability base – a base of truths from which all other truths can be derived – that contains physical truths, phenomenal truths, indexical truths, and a "that's all"-truth saying what the totality of truths is (Chalmers, 2012, pp. 5–7, 22, 428). He thinks that laws are necessary since otherwise all the regularities in the world would be a gigantic coincidence (Chalmers, 2012, p. 428), and that some phenomenological truths about qualia are fundamental (Chalmers, 2012, p. 338). Contrary to Chalmers, I argue in this book that indexical truths are reducible. I am not concerned with "that's all"-truths, since I only care about which theory is the most coherent compared with an ideally most coherent theory.

Chalmers has some interesting reflections on Russellian monism and panprotopsychism with some overlaps and differences compared with my own position (to be presented later in this book) (Chalmers, 2016). Russellian monism is similar to my own view (and that of Esfeld) when it comes to how several physical values (like mass and charge) can be understood as describing relations only, being reducible to something else. Panprotopsychism is similar to my own view when it comes to how parts that are not themselves conscious can combine to constitute subjects.

Chalmers argues that variants of panpsychism can include the virtues and avoid the vices of materialism and dualism in the way it finds a place and a causal role for consciousness (Chalmers, 2016). My own way of understanding the causal role of consciousness is different, as will be shown in the chapter on consciousness. Chalmers discusses how different variants of panpsychism have problems with the combination problem of how conscious parts can combine to form subjects. He presents a specific solution called panqualityism, where unexperienced proto-conscious parts form subjects, but rejects it with a zombie argument.

Panqualityism, especially as defended by Sam Coleman, is similar to my account on several points, which gives reason to compare the views here. The most important similarity is that I share with Coleman the idea that unexperienced phenomenal qualities can combine to form conscious subjects (Coleman, 2012). However, I think differently from Coleman when it comes to how phenomenal qualities combine to unities, where I offer an evolutionary account, described in the chapter on consciousness. While Coleman thinks that the self is an arrangement of intrinsically unconscious mental qualities (Coleman, 2018), I think of the self as a structure in the brain that can actualize unified conscious episodes.

We also differ at several other points. Concerning causation, Coleman thinks that consciousness is causally efficacious as an intrinsic property with causal power (Coleman, Unpublished), whereas I think the physical properties are the causally efficacious in concrete episodes, while consciousness has only had a specific causal role through evolution (described in the chapter on consciousness). He also seems to think of qualia as grounding physical structures, different from how I think (Coleman, 2015, section 81).

When it comes to non-conscious thinking, Coleman thinks that there can be non-conscious phenomenal qualities that give intentional content to our mind, and which can become conscious (Coleman, Unpublished). Contrary to this, I think of non-conscious thinking as a mere causal process in the brain, but which can become conscious (further described in the chapters on mind, thinking and consciousness).

In addition to these comparisons, I compare my views with other scholars that are similar to me at various topics throughout the book, like Tim Maudlin, Lorenz Puntel, Nicholas Rescher, Antonio Damasio, and Alfred Mele.

Part Two: **The Mind**

4 Causality

Part Two of this book is on the mind, including free will. How does the mind work? How can we think rational thoughts? What is consciousness, and does it play a causal role in the world? And can we make free choices?

Some argue that there is something unique and irreducible to being an agent or a self who acts in the world. Persons cannot be reduced to nature; consciousness is non-physical, and we have free will and responsibility for our actions. Others argue that everything is reducible to causal processes in the brain and the body. Consciousness is physical and there are no such things as persons, selves, wills, freedom or responsibility.

In this part, I will be defending a causal theory of the mind. I must therefore start with a discussion of what causality is in Chapter 4. The topic of causation has an important role in many chapters, and for that reason it is important to have a clear understanding of causality. Chapter 5 is about the human mind as a *causal* process; Chapter 7 is about the *causal* role of consciousness; Chapter 8 is about human free will, which implies sometimes being the *cause* of your own actions; and Chapter 11 argues that causation is important for understanding the concept of objective probability. To understand all these different topics, a quite detailed understanding of causality is important, including the role of contrastivity in causality and how causes are selected.[74]

In Chapter 4, I discuss what causes and effects are and how they are related. What is the connection between a cause and its effect which makes the cause have the effect it has? I will be arguing that cause, causation and effect are the central concepts in a theoretical framework that we use to categorize the world into causes, causations and effects. I will also argue that causation is reducible. Causation is not an irreducible entity in the world. Rather, it is a coarse-grained and shortcut description of regularities at a more fundamental level than that between what we normally call causes and effects.

Before I start explaining causality, we need to make some preliminary definitions. There is a *causal relation* where we have the *cause* on the one side and the *effect* on the other side. The cause bringing about the effect is *causation*, while this whole relation is *causality*. When it comes to understanding causality and causation, there are different questions to be answered. First there are two questions about the relata in the causal relation of cause and effect: I discuss in Section 4.1 what kind of entities they are, then in Section 4.2 how many there are. Then there are questions about how they relate to each other: I discuss in Sec-

74 For a longer defense of my understanding of causation, see Søvik (2016, chapter 2).

https://doi.org/10.1515/9783110770957-006

tion 4.3 how they are connected and in Section 4.4 how they are selected (Schaffer, 2007, section 1).

4.1 What kind of entities are causes and effects?

Question number one is what kind of entities causes and effects are. The standard view is that they are events, although they can also be understood as objects, facts, or states of affairs, among other things (Schaffer, 2007, section 1). Note that those who use the term "event" will usually include static states of affairs. For example, a rock can be considered an event since there is something being a rock from moment to moment. Both the terms "event" and "states of affairs" are thus in line with the majority view, as long as one is considering entities in the world, and not some abstract facts outside of time and space. I agree with the majority view in thinking of causes as entities in the world which may be events or static states of affairs. In the terminology introduced earlier, events and states of affairs are structures or patterns in the world at a time or over time that have actualized values as parts.

Below, we shall see that some argue that there is a special kind of causation called agent causation which is not reducible to event causation, whereas I argue that so-called agent causation is reducible to event causation. In order to show that I need to say quite much about minds and choices first, so I will return to this topic later.

4.2 How many relata are there in the causal relation?

Question number two is how many relata there are in the causal relation. It may seem like a strange question, for is it not obvious that there are just the two, cause and effect? However, some argue that causation requires that we set up contrasts on the cause side or the effect side or both. The point is that we can give many examples where what the cause is depends on the contrasts we set up. And this is quite important, for example in discussions on free will, where it can determine whether it makes sense to say that a free person is the cause of an action. It is also important for the question of determining reference class in probability estimates, a topic to which we will return in Chapter 12.

I think it is clear that sometimes setting up contrasts determines what the cause is. Consider the following example: John is on his way to a party. He comes to a crossroads where he has planned to meet Neil, who knows the way. But Neil is late, and John wants to get to the party, where he will get food, because he is hungry. He is fairly sure that he should go to the right as he has a relatively good memory of being given directions and recollects being told that he should go right at this crossroads. However, it could also be left, as the sign pointing left shows the name of the street where Sarah lives and John knows that she lives close to the party. So, although not absolutely sure, he is fairly certain that he should go right and so decides to takes the chance since it is quite likely that it is the correct way.

In this example, we can ask the cause of why John went right without setting up a contrast on the effect side. But we can also choose two different contrasts on the effect side:

1) Why did John go right as opposed to going left?
2) Why did John go right as opposed to waiting for Neil, who knew the way?

In question 1, the cause is that he had a memory of being given the direction to go right. In question 2, the cause is that he was hungry and wanted to get to the party. However, as it comes to the question of whether it is an exact number of relata on each side or whether this is context-dependent, I must postpone answering until we come to question number four on selection, as I will then have resources to explain my answer.

4.3 How are causes and effects connected?

Question number three is a big question which needs some space to be answered: How are we to understand the connection between cause and event? This is the big question of what causation is, and what it is that makes a cause a cause and an effect an effect. There are two main answers: causation is probability-raising or it is process linkage (Schaffer, 2001, p. 75).[75]

According to the first view (probability-raising), causation is something that makes something more likely to happen. For example, throwing a rock at a window is the cause of the window breaking since throwing a rock at a window makes it more probable that the window will break. According to the second view (process linkage), causation is a physical connection between cause and effect. This physical connection can be understood, for example, as transfer of energy or momentum (Fair, 1979, p. 220) or force (Bigelow, Ellis, and Pargetter,

---

75 Counterfactual causation is sometimes treated as a category in itself. Its most famous defender is David Lewis, see D. Lewis (1973a). Lewis introduced probability-raising into his theory in 1986 (Hitchcock, 2010, section 4.1), so Schaffer treats it as a subcategory under probability views.

1988; Bigelow and Pargetter, 1990) or some conserved quantity (Dowe, 2000, p. 90), as an interaction in quantum field theory (Heathcote, 1989, pp. 102–105) or as an instantiation of a law of nature (Heathcote and Armstrong, 1991).

Jonathan Schaffer argues that neither of these two main understandings of causation (probability-raising and process linkage) is sufficient in itself, no matter how they are refined. Against probability-raising views, Schaffer offers the example of Fred throwing a brick against a window and missing, while Pam throws and hits. As long as both have a chance of hitting or missing, Fred's throw raises the probability of the window breaking, but his miss is not the cause of the window breaking (Schaffer, 2001, pp. 79–81).

Against process linkage views, Schaffer argues that there can be causes without process linkage. Imagine Pam standing in front of a window, this time with a catapult. There is a rock in the catapult and the catapult is ready to launch. Imagine further that Pam releases the lever so that the catapult throws the rock through the window. No relevant physical quantity such as energy or momentum has been transferred from Pam to the catapult or the rock, nor does any persisting entity connect them, so she is not process-linked to the breaking window, but we still want to say that Pam is the cause of the window breaking (Schaffer, 2007).

Schaffer argues that probability-raising and process linkage should be combined, so that one should choose among processes that are process-linked to an effect to see if they raise the probability of the effect. But he still finds that there are cases that cannot be explained this way either, and concludes that the causal relation remains mysterious.[76]

Other people have also argued against the process linkage view. James Woodward argues that causes are sometimes found at a higher level than the more basic level of energy exchange in the world. For example, if there is freezing weather and the price of oranges rises, we think of the freeze as the cause of the increased prices (Woodward, 2009). Another argument against both probability-raising and process linkage views is that it seems natural to include laws of nature as causes, since often these are the only cause we can give for a phenomenon. But it is impossible to decide whether or not the laws raise probability compared with a situation with different laws, and it does not seem to be

76 See Schaffer (2001) and Schaffer (2007), where Schaffer gives the following example against his own theory: Pam throws a brick at an aquarium so that it breaks and spills water, while Bob (the more reliable vandal) holds his throw to see what happens. We want to say that Pam is the cause of why the carpet gets wet, but her throw lowers the probability of the effect and she is not process-linked to the spilling water, only to the breaking glass.

any known physical quantity that is transferred from laws of nature to their effects.

I will now present my own theory of causation. I argue in this book that there is a theoretical framework where fundamental entities are values actualized at points according to rules. These values being actualized according to rules are what creates motion in the world. But all sorts of structures and patterns can be discovered among these actualized values and expressed through different theoretical frameworks. Sometimes we see that similar states of affairs are followed by similar states of affairs, which makes us assume that this is not just a coincidence. It seems that the first state of affairs in some way produces, creates, leads to, or at least influences the other state of affairs. Since the same state of affairs often leads to the same state of affairs, we assume that all states of affairs are created or produced or are the result of other states of affairs. "Causation" is the word for this assumed influence between cause and effect, as well as all the other words used when we say that a cause "produced", "created", "influenced", or "led to" the effect. All these words – "causation" included – tend to be used without clear meaning, trying to express that which goes on between cause and effect; namely, that which makes it the case that after *this* state of affairs *that* state of affairs follows.

Many people think that something happens between cause and effect. Some think it is a link of some kind, for instance, the transfer of energy. Some think that the cause does something that raises the probability of the effect. Often, however, there is nothing happening between cause and effect at the level we are investigating; rather, something is happening at a more basic level of values being actualized according to rules. Some examples: When a billiard ball causes another billiard ball to move, it is because of the interaction between electron clouds at their surfaces due to the Pauli Exclusion Principle and the momentum of the particles, all of which are excitations in the relevant fields according to the relevant rules. The stretched rubber band causing something because it is stretched and then released can also be ontologically reduced to fundamental forces between the constituents of the rubber band. Muscle activity can be ontologically reduced to motor neurons triggering muscles, and this neural activity is again reduced to cell parts, ions and their charges and the rules guiding electrical charges.[77]

The schema of causes producing effects is then to be understood as a theoretical framework that we try to use at all ontological levels and as part of other

77 Of course, this does not answer how neurons and muscles came to be the way they are, which was by means of evolution.

theoretical frameworks. In different frameworks, some states of affairs are identified as causes of other states of affairs and are confirmed as such if there is a regular relation between them.[78] For example, in biology a mutation causes a new skill in an animal, in psychology shame causes a man to blush, in the social sciences a woman being educated causes her to have fewer children, in economics increased demand causes increased prices, and so on. However, all these causes only have effects in virtue of being complex states of affairs in interaction with the combined result of different rules being actualized.[79]

As a side note I want to add here the important point that different theoretical frameworks allow us to see different patterns in the world that are not expressible in the concepts used by physics, and for this reason we need different scientific disciplines for a greater understanding of the structures of the world. While this book shows how things that exist are often ontologically reducible to more basic values, this does not imply that scientific disciplines not working with fundamental values are superfluous.

While there are some fundamental rules, some rules are "if-then" rules that only apply to higher ontological levels (levels where complex structures are actualized), or regularities that arise as the combined effect of more basic rules and interactions. But I believe all sorts of suggested kinds of causation can be reduced to this, for example the famous top-down causation also known as downward causation. An example of this is the atoms in a wheel rolling downhill where it seems that the wheel is exerting top-down causation on the atoms in the wheel (Steward, 2012, p. 233). But this is fully explained as the forces holding the atoms together in the wheel and the road and gravity combined with the rules that they actualize. While I have seen many examples of top-down causation, I have not seen any that could not be reduced to ordinary causation as here understood (mental causation is often presented as the prime example of top-down causation and will be treated in its own chapter).

---

78 But it will not be finally verified as the cause, since the regular relation may just be a correlation with another cause.

79 After having sent this book manuscript to proofreading, I learned that there is a theory of causation which makes a similar main point that I do when saying that causation is a theoretical framework. It is the epistemic theory of causality by Jon Williamson, where he argues that "causality is a feature of the way we represent the world rather than a feature of the agent-independent world itself" (Williamson, 2009, p. 205). Causality is a map of inferences according to Williamson, made true by the physical world (Williamson, 2009, p. 205), and I agree with this main point. However, as this chapter will make clear, we differ at many points when developing the further details. I say specifically that the laws of nature make the regularities happen while relating causes with effects is a theoretical framework we use at many ontological levels. In my account, the role of contrastivity in our selection of causes is also very important.

It is often much more efficient and useful to pick out causes and effects at high ontological levels in different theoretical frameworks because we can then generalize and simplify. As long as one state of affairs is regularly followed by another, this is useful knowledge; by calling the first "cause" and the second "effect", we hypothesize that at a deeper level there is a lawful connection between them. That does not mean that there must be a law governing the high ontological level, for there may be laws governing lower levels pulling in different directions, but still the interaction is such that there is a general regularity.

For the sake of understanding, we take all sorts of shortcuts to establish some general rules. In theoretical frameworks describing high ontological levels, concepts often have imprecise boundaries. Many concepts have fuzzy edges (like "bald"), but even imprecise concepts describing high ontological levels may be much more explanatory than precise concepts in theoretical frameworks describing low ontological levels. We understand the causes of World War II better in terms of imprecise concepts such as racism, imperialism, and so on, than with terminology from physics describing how many atoms moved around in Europe in 1940. This means that many descriptions of high ontological levels, even when quite imprecise compared with the terminology of physics, are still efficient in terms of generating understanding, and so we often describe causes this way. Other shortcuts are to cite one cause which is elliptical for a series of causes or to describe several processes as one. For example, "Pam caused the window to break" can be short for "Bob loaded a catapult with a rock and Pam released the lever, which made the catapult project the rock through the window". In this instance, a lot of information is omitted because the speaker expects that the enquirer only wants to know the person responsible.

What, then, is the connection between cause and effect? What is causation, or what does it mean that the effect "causes", "produces", "creates", or "influences" the effect? Concerning "causation" and all the other related words used to describe the relation between cause and effect, such as "produce", "create", "influence", "contribute", and "lead to", I repeat that, in most cases, there is nothing emanating from the selected cause that influences the effect or creates motion in any way just in virtue of the cause being the state of affairs that it is. It is the rules being actualized at the basic ontological level that make motion happen. Saying that a cause caused an effect is most often an efficient way of describing some regularity where there is a chain of events, and this can be ontologically reduced to a combination of laws of nature acting on complex states of affairs. For example, if I say that the helium gas in a balloon caused the balloon to ascend, there is actually nothing in helium that propels the balloon upwards. Rather, it is a shortcut description for saying that gravity pulls the heavier air molecules around the balloon downwards, and because of the rule expressed in the Pauli Exclusion Principle, the air molecules then push the balloon upwards.

What about the actualizer of rules that makes states of affairs behave as they do? Is that genuine causation? One could always say there is only this one process which is the real causation and nothing else is. However, that would imply that almost everything we call causation is not really causation, and so it would be quite confusing to suggest that this should be the real meaning of causation. I suggest instead that we continue to speak of causation as we normally do, but that we realize that it is a coarse-grained but efficient way of saying that B follows A in virtue of rules being actualized involving A and B (or a little less precise: laws of nature interacting with states of affairs in the world). I shall soon say more about how cause A is selected.[80]

Before I discuss selection of causes, I will answer one objection to the account above. The objection is that it seems that absences can also be causes. For example, me not watering the flowers seems to be the cause why the flowers died. But if a non-event is a cause, there is no deeper actualizing of values occurring which makes the flowers die, so something seems wrong in my account.

Jonathan Schaffer argues that absences are causes since they figure in many paradigm examples of causes. For example, decapitation causes death by preventing oxygenated blood from reaching the brain (Schaffer, 2005, pp. 329–331). Phil Dowe argues against absences as causes. He says that even though we might say something like, "The father's lack of attention to his child running into the street caused the child to be hit by the car", we intuit that there is a difference between such causation and other causation, for we do not think that the father's inattentiveness made the child run into the street. Schaffer agrees that there is a difference between what he calls the intrinsic relation between events and citing absences as causes, but he argues that it is a hopeless procedure to establish a concept of causation which leaves out many paradigm examples of causation (Schaffer, 2000, pp. 291–293). So, should absences be included or not?

It is only values being actualized that make things move in the world. Absences do not. However, learning about absences also makes us understand what is

80 Ladyman and Ross raise the objection that fundamental physics might come to show us that the laws of nature are time symmetrical, and then the relation between cause coming first and effect afterwards may break down (Ladyman et al., 2007, p. 277). I shall argue later that even if the laws of nature turn out to be time symmetrical, that will not destroy the causal relation. The reason is that time symmetry has to do with processes being reversible (ice cubes may melt or freeze in coffee or hair may grow in or out), but I shall argue that time would nevertheless flow from past to future and laws of nature would be the cause of motion even if they make hair grow in or out.

connected and what is not, and this is useful for prediction and manipulation in a similar way as discovering other causes is. For example, it is useful to learn how to kill an unwanted plant, and if we expected a certain plant to be watered and live, it is clarifying to learn that the person we expected to water the plant did not do so. We often have expectations about what happens and how things are connected and so we are informed when we learn about absent connections – like the fact that no one watered the plant. When we enquire about a cause, we often do not know whether it is an absence or an obtaining state of affairs we are seeking. Absences and disconnections are intertwined in our descriptions of many normal processes; thus, in daily speech we do not distinguish between causation by absence and causation by interaction between laws and states of affairs.

Nonetheless, there are important differences. Laws of nature and states of affairs exist in the world, and many descriptions of causes supervene on these. Absences are not ontological constituents of the world and they are not ontological additions to the world. Rather, absences should be understood as statements about the world (Armstrong, 2004a, p. 448). Sometimes, they simply point out that something is absent in the world at a particular time. Such a statement can be a true statement made true by the totality of the world or the relevant area (Armstrong, 2004b, chapter 5). For example, if I say that there is no unicorn in my office, that statement is true in virtue of all the contents of my office not including a unicorn.

However, there are other counterfactual statements that are counterfactual conditionals of the form "If (counterfactually) A had (or had not) happened, then B would have happened". For example, "If the father had paid attention, then the child would not have been hit by the car". Some argue that such counterfactual statements can be true and their truth is commonly understood as depending on what would have happened in the closest possible world (D. Lewis, 1979). I believe that many such statements cannot be true because of problems with determining what the closest possible world is and what happens there (especially if our world is undetermined, as I believe it is).[81] One may certainly give good inductive reasons for believing that if I do not water the plant, then it will die. But it is *not true* that if I had not watered the plant last week it would have died, for it is impossible to know for sure what would have happened to that plant if I had not watered it. That something is true means that it is part of the world, and counterfactual states of affairs are not.

81 Good arguments against such counterfactual claims having truth value can be found in Lowe (2002) and Bowie (1979).

Should absences then be included in the concept of causation when there are so many similarities even if there are differences? It depends on what you want from the concept of causality. Is the main goal of the concept of causality to understand why states of affairs are as they are or change or move as they do? This speaks in favor of leaving absences out, since basic motion is due to existing states of affairs only. Or is the main goal of the concept of causality to clarify connections and/or a lack of connections in the world? In that case, absences are very useful tools.

In most theoretical frameworks, high ontological levels are described, and absences and causes play roughly the same role of showing which events are regularly followed by other events. But if we want a detailed theoretical framework for understanding motion in the world at the deepest level, it is more precise to leave absences out of the concept. Also for the detailed theoretical framework I will present soon in understanding free will, it will be clearer to leave absences out of the concept. From now on I will use the term "cause" without including absences in the concept. This does not exclude efficient and less precise descriptions of cause, but my goal is that the shortcuts I make may be understood as shortcut versions of a fuller description which can be ontologically reduced to the interaction between laws and states of affairs. In such shortcuts absences may be a part, but this should again be understood as an efficient way of describing the actual states of affairs involved.

For example, in the case of Pam and the catapult, saying that Pam caused the rock to be thrown includes an absence because the rock is released by means of disconnection; this is an efficient way of saying that Pam caused the blocking device to be moved and the catapult caused the stone to be thrown. We give such shortcut descriptions because we anticipate that if someone asked what caused the rock to be thrown and we answer "the catapult", they will continue by asking who released the lever on the catapult. Citing an absence in this case is simply a quick way of describing two states of affairs in the world: Pam moving the blocking device by releasing the lever and the catapult propelling the rock. This is quite different from saying that not killing Hitler in 1942 caused the war to last until 1945, which is speculation about what counterfactually would have happened in a possible world.

"Causally relevant" then refers to all those states of affairs that are involved in the interaction of laws of nature and states of affairs which made the selected effect happen. This "involvement" includes both states of affairs being pushed or pulled by forces and states of affairs resisting forces (by means of other forces). So where do we set the limits for what it means to be involved? If the limits are too wide, we end up including everything from the Big Bang and after as causally relevant for everything. I will argue in the next section that causal relevance can be limited by means of contrasts, and that we can search for the most precise state of affairs within the contrasts which lawfully leads to the effect.

In the next section, I shall raise the question of whether something can be the "most important cause", and as that is also a question of selection, it is now time to turn to the topic of selection. In many cases, there seem to be many causally relevant states of affairs. Take, for example, a car crash where the road, the tires, the cars, the drivers, the passengers, what the drivers had been doing previously, the weather, and many other states of affairs seem causally relevant to what happened. Does that mean that all these states of affairs are causes? To understand this, we need to turn to the question of selection, which was question number four on our list.

4.4 How are causes selected?

When we seek the cause of an effect, there are many events that can be related somehow to the effect. But how do we select the cause or causes among these? This is not the question of how we select causally relevant events from the set of all events; rather, it is a question of how we, as we often do, select one cause among many causally relevant events. There is wide agreement that the selection of causes is not objective.[82] On the other hand, most people will select the same causes in very many circumstances. If you ask people why a house burnt down, they will refer to the short circuit or the pyromaniac as causes but not the presence of oxygen, even if the presence of oxygen was also necessary for the house to burn.[83] So how do people select causes from mere (background) conditions and can the selection of causes be understood as objective or is it merely subjective?

C. J. Ducasse suggested long ago (1926) that a distinction should be made between *conditions*, which are *necessary* for the effect to happen, and *causes*, which are *sufficient* for the effect to happen. However, this clearly does not work to explain the difference between causes and non-causes. Both oxygen and ignition are necessary for the house to burn. Neither is sufficient alone, yet both can be selected as causes. Hart and Honoré suggest that when we select causes we select the abnormal conditions as opposed to normal conditions and agents as opposed to non-agents. For this reason we select the (abnormal) ignition of the house rather than the (normal) presence of oxygen – but if it had been a lab-

82 Schaffer cites Lewis and Mackie as examples, see Schaffer (2007, section 2.3).
83 Schaffer has this example from Hart and Honoré, see Schaffer (2007, section 2.3).

oratory setting that was supposed to be oxygen free, the unexpected presence of oxygen could have been considered the cause of the fire rather than the ignition (Schaffer, 2007, section 2.3).

Other philosophers have made other suggestions.[84] E. J. Lowe suggests that the only viable metaphysical distinction is between *contributing* causes and *complete* causes, where the complete cause is the sum of all the contributing causes (Lowe, 2002, p. 167). Similarly, Bernard d'Espagnat suggests that all we can speak of is contributing causes, which he prefers to call "conditions which influence". Among those influencing conditions, we select the most abnormal or unexpected one as the cause, according to D'Espagnat. He does not speak of the complete cause, his point being that there are infinitely many conditions that influence any state of affairs (D'Espagnat, 2006, pp. 313–314 and 330–332).

Schaffer argues that the question of contrastivity solves the problem of selection of causes. His point is that we select contrasts subjectively depending on what we want to understand, but when the contrasts are selected, the cause is objective. So those in the laboratory expected there to be no oxygen and wonder why there was a fire as opposed to no fire. When "no oxygen" is the selected contrast, we can agree that the presence of oxygen as opposed to no oxygen is the cause of the fire. On the other hand, when we want to understand why the house burnt down, we expect there to be oxygen present but we do not expect any ignition. "No ignition" is the contrast, thus we select the ignition as the cause.

So what is the best way of understanding the selection of causes? Armstrong and Heathcote argue that we find causes by experiment because experiments indicate that a connection between events follows laws of nature. If you vary the conditions but find one state of affairs which constantly gives the same effect, it seems that this is a process that can be reduced to interaction between laws and the selected state of affairs. For example, iron expands when heated, regardless of whether lumps of iron are of different shapes, mixed with other materials, heated at various temperatures, and so on; thus, we say that heat causes iron to expand. By such experiments, we can try to establish as precisely as possible the state of affairs with which the laws of nature interact, for example, that the laws of nature in this case interact with the molecular dynamics of an object and not its volume or shape. This connection between the laws of nature and the state of affairs can be called a nomic connection.[85]

---

84 For example, John Stuart Mill suggested that we should distinguish between standing conditions (like the presence of oxygen) and causes that are changes (like lighting a match) (Mill, according to Lipton, in (Beebee, Hitchcock, and Menzies, 2009, p. 623).

85 Judea Pearl is especially known for his work on systematically manipulating variables to select causes, see Pearl and Mackenzie (2019).

Sometimes many states of affairs seem to be connected nomically to an effect and yet we still select the same cause among these. For example, both ignition and oxygen are connected nomically to a burning object, yet almost everyone selects ignition as the cause of the object burning – but why? This depends on a combination of interest and background expectation. If you are a scientist who just wants to learn as much as possible about the world, you will probably select both ignition and the presence of oxygen as causes of the fire since both are connected nomically to the fire. If you are like most people, who expect oxygen to be present but not the object being ignited, you select the ignition as the cause of the fire since your general interest lies in learning how to predict and manipulate. If you are a scientist in a laboratory generating ignitions in a setting in which you expect no oxygen to be present, you will probably select the presence of oxygen as the cause of the object burning.

We set contrasts depending on what we already expect and what interests we have. Even if very many states of affairs are involved in every event, we can simplify by describing states of affairs in broad terms that include many other states of affairs. For example, speaking of Jack today may refer, in a wide sense, to everything that has made Jack the way he is today, which then includes numerous events. In addition, we often disregard very many states of affairs as uninteresting because we expect their presence not to be relevant to the contrast we want to understand. Most often we will disregard the burning object being dry, surrounded by oxygen, held together by the strong force of its atoms and so on, and will only be interested in the ignition as the cause of it burning as opposed to not burning.

If we do not know of any cause for a concrete effect, we seek any cause we can find. If we know of several causally relevant states of affairs, we select among them based on expectations and interests. Setting up contrasts allows us to specify more precisely what we are after – to find the relevant nomic connection. But what if we have several candidates for causes which all seem necessary for the effect? Is it then possible to say which cause is the most important? This question is important for a later discussion of the role of agents and indeterminism in causing choices.

It is extremely difficult to develop a general criterion for finding the most important cause because importance is relative. In many cases, we may think of a typical cause of why something is the way it is, but of course it is crucial that the subatomic elements are held together or that God keeps it in existence, and so on. Sometimes we know about strong forces involved, but we are looking for one specific and unexpected small extra force or resistance. The reason we often agree about the cause is that people have much the same interests and expectations and take the same things for granted when they look for causes. Thus,

I believe that the best one can do in order to limit the number of causes is to try to specify as clearly as possible the causal contrasts in order to clarify what one wants to understand. These contrasts serve to exclude many causally relevant states of affairs that do not concern us.

In the case of Pam and the catapult, if someone asks for the cause of the window breaking, we answer, "Pam". We could answer with the rock, the catapult, the spring and all the other states of affairs that were involved in the breaking of the window. But we anticipate that if we answer "the rock" or "the catapult" and so on, the inquirer will keep on asking until we say that Pam released the lever. This is because we assume that the person asking is interested either in finding a person to hold responsible or establishing if it was an accident, i.e., what happened, so that a repetition can be prevented later. However, if a person expected the window to be made of bulletproof glass and asks why the window broke, the cause he might be searching for is that there was a mix-up in the delivery process so that a normal window was fitted instead of a bulletproof one. As the examples show, the strongest force, the highest probability and so on are not what determine how people select the cause; rather, it is a matter of what the inquirer wants to understand against the background of expectations and contrasts.[86]

Although I do not have space to discuss it here, I would like to suggest briefly that the understanding of causation I have presented here also solves many other problems connected to causality and that this supports the plausibility of the theory. Concerning how fragile or how fine- or coarse-grained causes should be understood to be, the answer is that it depends on what we want to understand (the selected contrasts) and on the ontological level and theoretical framework. The level of fragility or grainedness is not the same for all causes but depends on what we want to understand. If I want to understand why Mary turned as opposed to not turning, the cause was that John said hello to her. If I want to understand why Mary turned so quickly as opposed to turning at a regular pace, the cause was that John said hello loudly.[87]

Concerning the question of transitivity,[88] the answer is that some causes are transitive and others are not. At high levels of description giving efficient descriptions of causes, there can be many examples of non-transitive causes. At the level of forces acting directly on states of affairs there is more transitivity, since one thing being pushed or pulled can lead to another thing being pushed

86 The same reasoning as that made here solves the problem case that Schaffer said he could not solve, as mentioned in footnote 76.

87 I have the example from Schaffer (2007, section 1.2).

88 The question of transitivity is: If A causes B and B causes C, does A then cause C?

or pulled. Billiard ball A pushes billiard ball B, which pushes billiard ball C, and the causes are transitive. However, as we have seen, there are many causes that do not involve the transfer of energy or momentum and there need not be transitivity. The hardness of the wall causes the ball to bounce back and the ball causes the flower vase to break, but the wall does not cause the flower vase to break, thus there is no transitivity. We speak of causation to describe regularities at high levels in different theoretical frameworks, but often the regularity depends on the interaction of laws and complex states of affairs at lower ontological levels, and often the descriptions are efficient shortcut descriptions. Naturally, there will therefore be many examples of causation without transitivity.

To conclude, the theory of causality presented here is able to solve many problems and integrate the answers into a coherent framework, thereby supporting this understanding of causality. This will be useful in discussions about the mind, consciousness, free will and probability, especially the role of contrastivity in selecting causes. We turn now to the first of these topics, which is an (event-) causal understanding of mind.

5 Mind

Having discussed in Chapter 4 what causation is, I can continue in this chapter describing how the mind can be understood causally. There are different philosophers, for example agent causationists or substance dualists, who will reject that we can understand the mind as a normal causal process. In order to argue that it is superfluous to include irreducible agents or souls in one's ontology, I must present a detailed causal understanding of the mind. I must present in detail the topics of mind, thinking, consciousness and free will (each in an independent chapter) in order to show how it is possible that persons thinking and making free choices can be ontologically reduced to causal processes between values actualized in fields.

In this chapter, I will argue that our mind functions like a causal process in the same way as other causal processes work in nature. It may sound very strange to believe that the mind functions as a causal process, and many will argue that when we are acting for reasons we are doing something very different from a causal process (McDowell, 1996). But as I shall show, many philosophers also believe that our mind is a causal process, and it is very difficult to understand how the mind works if it is *not* a causal process. When a person considers alternatives and end up choosing one, which she then acts upon, how does this happen if it is *not* a causal process?

I start in Section 5.1 with some biological background to describe how mind has evolved. Such an evolutionary explanation of mind supports a causal understanding of the mind, since it is then a product of natural causal processes.

Sections 5.2 and 5.3 deal with the relation between brain, mind and consciousness. Mind is understood as brain activity that can become conscious. In Section 5.2, we look at different arguments supporting the belief that the brain causes the content of consciousness. In Section 5.3, we look at many different examples of how almost anything we do consciously can also be done non-consciously by the brain. That mind can occur non-consciously in the brain also supports the view that mind is a causal process.

Sections 5.4 to 5.7 look at different components that are part of a choice in order to start understanding how a choice can be a causal process (and later how it can be said to be free). The components are emotion (5.4), memory (5.5.), the self (5.6), and desire (5.7). Especially important is the self, where I use neuroscientist Antonio Damasio to distinguish between the autobiographical self – which is a storage of memories in the brain – and the core self – which is a stream of conscious impulses.

https://doi.org/10.1515/9783110770957-007

In this chapter, I will be leaning quite much on the book *Self Comes to Mind* by the neuroscientist Antonio Damasio. It is a good book and allows for a concise presentation with much support from neuroscience. I shall start with some biological background which will show itself relevant later. I believe that showing how the mind could arise through evolution also helps to support the view that the mind is a causal process.

5.1 Biological background

Although many questions remain unanswered, scientists have quite detailed theories which give a good explanation of how biological life could arise from chemical components. I have in mind especially the RNA world hypothesis developed by, among others, Jack Szostak, Leslie Orgel and Gerald Joyce. The general idea is that simple reactions can cause stable copies of itself.[89] Gradually they could form into RNA molecules, and scientists have been able to replicate RNA from the same ingredients and conditions assumed to have been present on earth 3.5 billion years ago (Lincoln and Joyce, 2009). The exciting potential in RNA is how it can store information, make copies of itself and catalyze certain chemical reactions to make, for example, proteins. RNA, DNA and ribosomes do crucial work in cells, and they are all made of RNA, which suggests a gradual natural process from chemistry to biology.[90] One possible way to go from strings of RNA molecules to cells is that they could have merged with fat bubbles to become simple cells, which can also divide under certain conditions (Budin and Szostak, 2011). This would be some of the important steps towards cells dividing and containing DNA.[91]

Already at the level of a cell, one can see how it resembles a simple version of a larger organism: the cytoskeleton is like the skeleton, the nucleus is like the brain, the cytoplasm is like the tissue and organs, the cilia are like limbs. Like organisms, cells need to get nutrition in and waste out, and turn the nutrition into energy that can be used for reproduction and getting more nutrition in and waste out (Damasio, 2010, pp. 33, 41).

89 As suggested for example in Dawkins (1976, pp. 13–20), and illustrated well by Terrence Deacon, in Deacon (2006).

90 That ribosomes are ribozymes made of RNA is a recent discovery supporting the RNA world hypothesis (Cech, 2000).

91 There is a nice series of videos on YouTube from ibiology.org where Jack Szostak explains all this in detail.

The evolution of the senses is also well understood. I mention this also briefly to support the general evolutionary approach to our mental life as a causal process. Cells sensitive to light evolved into cells in a cavity which could then register the direction of the light. A lens sharpened the signal, and cells reacting to different wavelengths made further discrimination possible. Eyes have evolved independently many times in evolution. The nose started as cells sensitive to chemical stimuli and pheromones and, since it is sensible not to eat everything that comes into the mouth, taste developed in a similar way to smell. Hearing started as reactions to vibrations in the jaw, which evolved into the middle ear, and of course nerve cells in the skin are sensitive to touch.[92]

When cells start to cooperate, everybody can enjoy the benefits of specialization. In organisms, different cells do different kinds of work and get nutrition via the blood system (Damasio, 2010, pp. 33–34). Neurons are a kind of cells which have the specialized ability to change other cells by sending an electrochemical signal (called "firing"), by which they can move the muscles of a whole organism and help the organism survive by moving around (Damasio, 2010, pp. 37–38, 50).[93]

Two more very important things that neurons started doing were to make simple representations and simple dispositions. For example, some neurons could react to something poisonous by firing and activating some other neurons, making the organism move away from the poison or spit it out if it was an organism with a mouth. If this happens regularly, we could say that the first neurons firing in response to the poison represent the poison, and the second set of neurons firing could be said to actualize a disposition to move. "Represent" should here be taken in a minimal sense to just mean a consistent relation between the poison and the neurons that fire, and "disposition" means an "If presence of A, then act by doing B" mechanism, which is triggered by a specific stimulus and gives a particular response (Damasio, 1999, p. 320; Damasio, 2010, p. 134).

Through evolution, representations and dispositions would become more and more advanced. This would then help the organism to maintain body functions and to reproduce. We shall now look more into how the brain could make such representations, and see that we have good reason to think that it is causal processes in the brain that gives us the content of our conscious experiences. The description of the evolution of mind in this and the following sections fits very

---

92 For details, see Joseph (1996, pp. 7–16).

93 This is the classical description of neuronal interaction, but such interaction might also happen in other ways, such as through coherent oscillation (Fries, 2005).

well with the evolution of more and more advanced computers and robots (Nilsson, 1998).[94]

5.2 Does the brain cause conscious experiences?

It would be very helpful to survival and reproduction if neural patterns could represent states of affairs in the world, like food, drink, attacking lion, or potential partner interested in sex. Damasio argues that neural patterns represent the body and the world, and we shall look at how he argues that there is a lot of thinking, feeling, remembering, and desiring that happens non-consciously. We shall look at some examples of this, since it supports the idea of mind as a causal process. In the next section, I shall also discuss whether terms like thinking and feeling should be reserved for conscious thinking and feeling only, as some argue.

Now follow some examples to support the view that the brain causes the content of our conscious experiences. Every time you have a conscious experience of seeing red, the same area of your brain is active. If that area is destroyed, you will not experience red anymore, and if that area is stimulated, you will experience seeing red even if there are no red objects in front of you (Hadjikhani, Liu, Dale, Cavanagh, and Tootell, 1998; Wandell, 2008). If it is destroyed you will even have problems imagining something red (Damasio, 1994, p. 101). Damage to an area of the brain called the fusiform gyrus of the temporal lobe causes face blindness, and stimulation of this same area causes people to see faces spontaneously (Shermer, 2012).

In an experiment carried out by Damasio and colleagues, they found a pattern in the brain which consistently correlated with the conscious experience of a certain sound. What is interesting about this experiment is that the same pattern was active even when the person was just imagining hearing the sound in his mind, even if no sound was actually made in the world outside his mind (Damasio, 2010, p. 134).

The correlations are quite exact, indicating a close connection between the brain activity and the conscious experience. However, correlation does not necessarily imply causation, as two things can be correlated without one being the cause of the other, like the correlation between night and day. But correlation can also indicate causation. Do we have any reason to believe that it is the neural pattern which causes the conscious experience? Yes, because we can stimulate

94 See especially page xix.

neural patterns in the brain and achieve an effect in consciousness from it, which indicates that the effect in consciousness is causally dependent on what happens in the neural patterns.

Many problems are thus solved if there is a neural pattern underlying conscious experiences. Hallucinations are explained: they occur since a neural pattern is activated even if what it represents is not present. For example, you can think that you see a red tomato even if no such tomato is present because you have a neural pattern representing a red tomato in your brain, making it conscious to you. Phantom pains are also explained, since neural patterns representing (now lost) parts of your body can become conscious to you even if you have lost the actual limb. Certain phantom pain phenomena are very well explained by this model. For instance, there was a person who lost his arm but kept feeling it. Then he lost the feeling in his arm, but retained the feeling of his hand, now just feeling that the hand was sticking out from his shoulder. Finally, he lost that feeling as well. This is explained by the fact that the area representing our hand in the brain is much larger than the area representing the arm, so the neuron connections representing the arm faded away before the neurons representing the hand (Joseph, 2006, 18:13 – 19:34).

The process of how the brain can represent the world is well understood, especially vision. Humans have feature-detecting neurons, which fire in response to a certain feature being seen in the world. These neurons fire for at least thirty different types of features, like angle, size, movement, contour, color, distance from observer, etc. (Imbert, 2004, p. 39; Pinker, 1997, p. 20). Wolf Singer has shown that when neurons that fire in response to certain objects fire synchronously, the object is consciously experienced. If two completely different visual inputs are given to each eye, only one of them becomes conscious to the observer at a time, and it switches back and forth which image is conscious to the observer. When the first object is consciously seen, then the neurons detecting the features of that object fire in synchrony. When the other object is consciously seen, then the neurons detecting the features of that object fire in synchrony (W. Singer, 2004a, p. 25).

The brain puts together information from the feature-detecting neurons to make a unified picture. We know this because sometimes some types of neurons do not function, yet the brain creates a unified picture of the rest of the information. One case is color blindness. In another example, a person (known as D. F.) was almost blind, but she did receive information from the neurons detecting color and texture. Everything she saw was blurry, but she could see a banana and guess that it was a banana because of its distinctive color and texture, but could not say what position the banana was in or what shape it had.[95]

Note that when there is a lack of visual input it does not create black holes in the visual field, but rather the rest is turned into a unified picture since the brain creates a unified impression based on the input it gets. Some people with brain damage in one hemisphere do not lose half of the visual field, but rather create a whole visual field from the input sent to the one hemisphere. The point is that the input is spread out to create a unified impression. If the brain is given contradictory sensory input it will merge it together in a kind of compromise (Gazzaniga, 2009b, 50:08 – 51:55 and 51:00:40 – 51:01.40). Various tests have been performed whereby people are given one visual input but feel something else. For example, they sit on a chair rotating 120 degrees while being shown a film indicating that they rotate less; or a big object is placed before their eyes while they feel a similar, but smaller, object with their hand. Their brains then merge the information together in a compromise. When they do not receive a disturbing visual input, they guess quite well how much they have rotated or how big the object is. But when they have disturbing visual input, the brain mixes the information together to form one impression so that they estimate the size differently (Viaud-Delmon and Jouvent, 2004, p. 73).

None of these examples show that the physical side of reality is ontologically more basic than the non-physical conscious side of reality, which idealists hold to be ontologically basic. For example, it could be that a non-physical conscious mind requires a very complex physical structure to interact with before it can become active. But consciousness does not seem to be an independent or soul-like entity as envisioned by most substance dualists or idealists. Rather, it seems very dependent on the brain, thus at least slightly indicating an ontological priority to the physical. Many experiences are better explained as made by the brain for the sake of survival than as an accurate depiction of the world by the mind. Things do not have colors independent of light setting or of someone watching them. But adding colors to objects makes them easier to see and distinguish, so it makes evolutionary sense that the brain should add colors. The spectrum of light that we can see is the spectrum that most of the

95 This phenomenon is further complicated by the fact that we seem to have two distinct visual systems in the brain, where one allows conscious seeing and the other is concentrated on adjusting body movement. Therefore, although D. F. was unable to see a letter box, she had no problem sticking a letter into the slot. The converse is Balint Holmes syndrome, where a patient can see the letter box clearly but would be unable to stick the letter in, see Carruthers (2006, pp. 88 – 89).

radiation from the sun and stars come in (Fernald, 2001),[96] so again it makes sense that evolution used this spectrum of light as a basis to make conscious experiences of color.

Sometimes the process goes wrong. For example, there are reports that some people see colors when they hear sounds (Gray, 2004). Neurologist V. S. Ramachandran and philosopher W. Hirstein suspect that such stories may be more metaphorical than real color experiences, but they cite an even better example. A person lost his sight; he became totally blind. But after a while he could start to see clearly the objects he was feeling in his hand – not just imagining them in his inner eye, but having an experience which was like seeing the ruler he was holding in his hand. This happened with all kinds of objects (Ramachandran and Hirstein, 1999, pp. 96–97).

Another fascinating experiment shows something similar with blind and blindfolded people: cameras sent output stimuli on either their back or their tongue, and these stimuli followed patterns consistent with what the camera was filming. After amazingly little training, both blindfolded and blind people reported that they started seeing images, and they were able to recognize faces, describe objects, read, manipulate objects and much more. Several times when the camera suddenly zoomed in on something, the blind(folded) people ducked because they felt that something was being thrown at them (Sampaio, Maris, and Bach-Y-Rita, 2001; B. W. White, Saunders, Scadden, Bach-Y-Rita, and Collins, 1970).

There are also many things we see that are not a correct picture of what we see, and it is useful for us to see things like this. Many optical illusions are based on the fact that the brain distorts what we see to make it fit what it should look like.[97] One example is the moon, which looks a lot bigger if it is close to buildings or mountains than when up in the sky, but there is nothing physical that makes it look bigger. The illusion happens because we know that the moon is much bigger than houses, so the brain makes it look relatively bigger.[98] The illusions are generally useful clues for survival but not correct depictions of the world. If our conscious minds were ontologically unique non-physical entities with capacities for grasping the world, one would not expect these distortions caused by what is accessible to the brain and its survival value.

---

96 More specifically, it is what the first animals in the sea would have been most exposed to in the water.

97 Michael Gazzaniga shows some very convincing examples in Gazzaniga (2009c, 29:30–30:17).

98 At least that seems to me to be the best theory (Kaufman and Rock, 1962). See also the similar evolutionary explanation of the Müller-Lyer illusion, in Sternberg and Mio (2006, p. 117).

To sum up this first point, it seems clear that the brain creates patterns that can become conscious, and that it is the physical side of it that determines how the conscious experience comes to be. The next question is whether or not there can be mind without phenomenal consciousness – that is, non-conscious thinking, remembering, feeling, desiring, etc.[99]

5.3 Non-conscious mind

Damasio argues that the brain constantly makes many neural patterns that represent states of affairs in the world or in the body, and these neural patterns can be consciously experienced as images, but he also says that most of them are never experienced consciously. But even if they are not conscious, neural patterns and dispositions seem able to perform their work as a causal chain and do the same work as a conscious person does (sense, think, remember, feel, desire) – often even better. Numerous experiments support this, and I do think that much philosophical confusion could be avoided if more philosophers accepted that there can be conscious and non-conscious thinking.[100]

Let us take an example from daily life. Many people have found that they can sometimes drive a car "on automatic pilot" while thinking about something other than driving. But even if they are not conscious of seeing signs, red lights and so on, their driving indicates that the signs and lights have been registered as such and acted upon (Armstrong, 1981, p. 59). A more astounding example is people who are blindsighted or deafhearing. They have no conscious experience of ever seeing or hearing anything, and yet they can move well through a labyrinth, catch what is thrown to them, move towards where a sound comes from and so on (Carruthers, 2006, pp. 87–88).

Priming is another good example. People can be shown words or pictures on a screen so quickly that they have no conscious idea what the picture or the word was, but testing afterwards shows that they must have registered and understood

---

99 When I use the term "non-conscious" I do not mean "not awake" but "not phenomenally conscious". I give a fuller description of what this means in the chapter on consciousness.

100 For example, Mark Rowlands argues against thinking as consisting of images in the mind since one can say that the glasses are in the drawer without thinking about the glasses in the drawer, but rather thinking about the game on TV (Rowlands, 2003, pp. 78–79). In this case I think it is obvious that the person is thinking at least non-consciously about the glasses in the drawer, and cannot see any other plausible explanation of how he is able to answer the question of where the glasses are.

the words or pictures (Sternberg and Mio, 2006, pp. 64–65).[101] Damasio uses the example of a cocktail party: You are listening to your conversation, but the brain registers other conversations as well. Suddenly you hear your name or something else in another conversation which is marked as important so that you become conscious of it, and you start listening to that other conversation (Damasio, 2010, p. 173).

These examples are mostly concerned with non-conscious *sensing*, although some interpretation is also involved. Below I look at emotions and the self, thinking and desiring, and give examples of non-conscious feeling, thinking and desiring, but here is an example which shows complex non-conscious reasoning. Damasio and colleagues performed an experiment where people were asked to draw cards from various decks: some decks were good, i. e. leading to a reward, and some were bad, i. e. leading to a punishment. There was also a system determining which decks were good and which were bad, so that if you cracked the code you could just draw good cards. The subjects played the game while their skin conductance was measured. The interesting thing was that it seemed that the code was cracked non-consciously several minutes before the players understood it consciously and before they started drawing only winning cards. After a while they would get one type of skin response just before drawing from every bad deck, and another type of skin response just before drawing from every good deck. This was so consistent that somehow some part of the brain must have cracked the code, but the person could not consciously tell this and would keep drawing bad cards (Damasio, 2010, p. 276; Bechara, Damasio, Damasio, and Anderson, 1993).

Is it right to use words like "see" and "think" in contexts other than conscious seeing or thinking? This raises the question of first-person and third-person descriptions of events. Although some kind of identity theory about mind and body might be right, the most common view is that images in the mind and patterns in the brain are not identical (Kim, 2006, pp. 112–113). They seem to have many different properties, so the prima facie view should be that they are not identical. Even if they are identical, it is in any case helpful to distinguish between the first- and the third-person perspective. For this reason, I shall specify whether I mean conscious images or not when I write about images.

What about "seeing" and "thinking" and words like that, which usually presuppose a first-person perspective? There has been much philosophical critique

101 This is a well-established fact, and the reason why many commercials use subliminal stimulation.

of neurologists who use first-person language to describe what happens in the brain. Many neurologists speak of the brain as a person and say that the brain does things we normally just say that people do, like "seeing", "remembering", "interpreting", and "mapping".[102] It is important to be clear about the distinction between first-person and third-person perspectives when one is writing about the brain and mind. But it can also be very difficult. The reason is that humans can do so many things non-consciously in the same way as when they do them consciously. We can see, hear, smell and so on without ever being conscious about what we see, hear and smell, yet we act as if we have seen, heard or smelled it.

Many examples were given above, like blindsight or driving without paying attention. A telling example is that of split-brain patients, where the connection between the two hemispheres of the brain has been cut so that there is no interaction. If you flash the word "spoon" to the left eye only, so that the visual impression is sent only to the right hemisphere, then ask the person, "What did you see?", she will answer "nothing", since the left hemisphere is where most people have their language modules, and obviously nothing was registered in the left hemisphere. But if the person is allowed to put her left hand (which is controlled by the right hemisphere) in a box, she will feel around and pick up the spoon, indicating that the right hemisphere did see the word spoon and understood it.[103] In examples like this it is very tempting so say that one hemisphere saw the word and the other did not and that the left hemisphere said that it did not see it, whereas the right hemisphere cannot speak. But usually we just use words like "see" and "speak" about people, not about cerebral hemispheres.

John Searle criticizes Damasio for saying that these non-conscious representations are part of the mind. "What is the fact that makes them mental?", Searle asks, suggesting instead that they could be understood as a non-mental step on the way to consciousness (Searle, 2011a). I would answer that the fact that makes them mental is that they are causally related to the objects in the world that they represent and internally related to other representations in a structurally similar way to how objects in the world relate to each other. Several reasons have been suggested for viewing these non-conscious representations as parts of the mind, largely because they perform functions we usually call mental. When a brain process can go on non-consciously and have all the same effects as a conscious process, that makes it reasonable to think of it as part of the mind.

---

102 A list of examples can be seen in M. R. Bennett (2007, pp. 154–156).

103 A film of this experiment can be seen in Gazzaniga (2009a, 25:30–28:00).

People may act as if they were consciously sensing and thinking and yet they are not doing it consciously. The reason they can is that the brain creates neural patterns that represent what happens in the world, and this triggers dispositions that the brain works according to. This happens all the time to everyone, and we are conscious only of a few of the things that go on in the brain. It is possible to describe it in uncontroversial third-person language, but it is much more efficient to say simply that a person "sees" the object instead of saying that the object "activates a representation of the object".[104] For instance, in the card game test described above I said that the brain cracked the code before the players did, since the skin conductance always matched the good and bad decks even if the players drew the wrong cards. The expression "crack a code" is usually used for something a conscious mind does, but in this case it seems a very appropriate and efficient description of what has happened.

It is also difficult to distinguish clearly between first-person and third-person language since there are so many words that have been used as metaphors so frequently that they acquire a literal meaning. For example, it is common practice to write that "an argument shows", but one could complain that only people can show something, depending on how the word is defined. Above I wrote that "the brain creates neural patterns", and again one could complain that only people can create something in one definition of "create". The term can also be used, however, to mean something like "to cause".

So, on the one hand it is difficult to separate first-person and third-person language, and of course I want also to show how similar conscious and non-conscious brain processes are, since I think that what we usually call sensing, thinking, etc., while implying a conscious first-person perspective, may well happen non-consciously and be correctly described in a third-person perspective. It is my aim to show the close connection between a phenomenological description from a first-person perspective and a neurological description from a third-person perspective. On the other hand, I do want to keep the distinction since I believe that consciousness and the subjective perspective make a difference, and we shall come to a discussion of what difference consciousness makes. My solution to the problem of first-person and third-person language is to use terms like thinking, feeling and so on in third-person perspective description but specify whether I refer to conscious or non-conscious activity in the brain. Hence I shall distinguish between non-conscious thinking and conscious thinking,

---

104 A neural pattern being *activated* means that the neurons that actualize the pattern are firing, and then the representation they constitute *occurs* in the mind. Activation need not mean that it is conscious, however, as I argue with different examples of conscious and non-conscious mental events.

non-conscious feeling and conscious feeling, and so on. To those who think that these terms are meaningless when referring to non-conscious events I would say that the examples show that the descriptions make sense after all.

Some brain activity never becomes conscious, like for example autonomous processes in the body, and as far as we know they are *not* consciously experience*able*. Some brain activity is not conscious, but can become conscious (*is* consciously experience*able*, but non-conscious), and some activity becomes conscious (*is* consciously experienc*ed*), and we know quite a lot about where the different things happen. It is quite specific, so that certain areas of the brain can create conscious experiences and activity in other parts of the brain never becomes conscious. Common to the areas that create conscious experiences is that they are complex clusters with massive interconnectivity organized around a gate of input from the world outside or the body, and this fits well with the view that the brain at a certain level of complexity creates conscious experiences out of its input (Damasio, 2010, pp. 86–87).

5.4 Emotion

In the previous section, we considered arguments in favor of thinking that there can be non-conscious sensing and thinking, which again supports the view that they can be understood as a causal process. In the next four sections (5.4–5.7), we shall take a closer look at emotion, memory, the self and desire, and see how they can work non-consciously and consider reasons to think of them as causal processes. An extra reason why it is important to understand how these work is that they all play an important role in the process of deliberation. Understanding how they work will help us understand how free will can be a causal process. We start here with emotion.

We have already seen how small organisms can have life-preserving reaction patterns, like spitting out something poisonous. Another example is how baby birds react to a large shadow flying over them, which makes them huddle and sit still (Damasio, 1999, p. 69). There are many examples where animals seem to act without conscious thinking or feeling and just react with simple dispositions resulting from causal stimulus and response.

The automatic response to stimuli has evolved into the advanced responses that we call emotions. In advanced organisms the brain contains a representation of the body in homeostasis (functional balance) which can be compared with a representation of the body as it is now. Those brains which could detect a difference and make something happen to restore the body to homeostasis would be selected by evolution (Damasio, 2010, pp. 48–49). An example would be to detect low blood sugar and create hunger to make an animal eat. This could happen without consciousness, since all that is required are some "If presence of A, then act by doing B" dispositions in the brain (Damasio, 2010, p. 52).[105]

Another advance made by organisms was the evolved ability to detect likely threats (e.g. animal with sharp teeth approaching) or likely delivery of goods (e.g. partner preparing to have sex). It is important for the organism to have rules on when to move and some motivation that actually makes the organism move. This is achieved by the brain sending molecules through the blood vessels and signals through the nerves to warn the organism and prepare the right response. This is an important part of what emotions are about, and fear is a good example: something is registered as threatening, and the brain sends out molecules that prepare the body for fight or flight. In this way, emotions allow for more differentiated and optimized responses to stimuli than automatic action-responses (Damasio, 2010, p. 54).

Damasio distinguishes between emotions and feelings. An emotion is a series of events happening in the body. A neural pattern representing something in the body or the world outside activates trigger regions of the brain, which sends out various chemical molecules in the blood and different signals through the nerves. This prepares the body for certain actions and usually also triggers certain kinds of thoughts. As Damasio says, running from a gunman, you do not think about what to make for dinner tonight (Damasio, 2010, p. 144). An *emotion* is therefore a series of events which places the body in a certain state. A *feeling*, on the other hand, is a term Damasio uses to denote a neural pattern in the brain representing the body's being in that state of emotion. It is a neural pattern representing the body's being, for example, in a state of fear or happiness or anger. This neural pattern may become conscious or not, which means that we need to distinguish between emotions (a series of events in the body), non-conscious feelings (a neural pattern in the brain) and conscious feelings (a consciously experienced feeling/neural pattern) (Damasio, 2010, pp. 109–110, 114).

Only conscious feelings are experienced from a first-person perspective. It may seem strange to speak about non-conscious feelings, but again there are good reasons to distinguish between them. Non-conscious feelings can become activated and change our body state before we become consciously aware of what we are feeling. Men can be shown pictures of naked women so quickly that they are unable to consciously experience it, and yet their bodily reactions

---

105 For examples, see how Peter Carruthers explains very many workings of our mind by employing such "If A, then B" dispositions in Carruthers (2006).

are as if they had seen them consciously (Koch and Tononi, 2014, 10:40 – 11:32). Several other examples could be given.

The distinction made here is that between emotions as body states and feelings as neural representations of that body state, which may or may not become conscious. There are some universally recognized emotions, namely happiness, sadness, anger, fear, surprise and disgust (Damasio, 1999, p. 50). Several of these emotions have their own distinct physical patterns in the body, meaning that a scientist with the right apparatus can to some degree know what you feel without your telling her (Damasio, 1999, p. 61). Again, there are many aspects of emotions and feelings which support the idea that they have evolved as survival-enhancing mechanisms rather than something ontologically unique in another dimension of mind. The basic emotions have distinct physical patterns, and areas of the brain can be stimulated electrically to make people feel extreme anger or fear.[106] The fact that the intensity of an emotion can be determined with electricity supports the understanding of it as a causal process. The fact that we have the basic emotions we do, with their clear survival value, also suggests their origin in evolution. Even the fact that feelings show in the face can be explained with evolutionary reasons (Pinker, 1997, pp. 414–415). These facts all fit well with an evolutionary account of our mental life.

A survey of feelings becomes much more complex as feelings combine with different thoughts, as then we can speak of many different feelings, although what is happening in the body may be very similar.[107] Damasio himself distinguishes between universal emotions, social emotions, background emotions, moods, drives and motivational states. These distinctions are not so important here, although we shall return to drives and motivations when we look at desire (Damasio, 2010, pp. 22–26). Pain and pleasure are important for our topic, however, so where do they fit into this picture? Many feelings have incentive and disincentive functions. Some are negative and can be experienced as punishment; they are meant to make the organism withdraw from something negative. Others

106 Electrical stimulation of the hypothalamus can cause extreme rage, and electrical stimulation of the amygdala can cause both fear and rage (Joseph, 1996, pp. 173–174, 182–183).

107 Damasio speaks of secondary emotions, which are combinations of cognitive states and basic emotions, and these can evoke numerous feelings with subtle variations, like the differences between euphoria and ecstasy based on happiness, or melancholy or wistfulness based on sadness, or panic and shyness based on fear, and so on (Damasio, 1994, pp. 134, 149–150). Feelings like jealousy, envy, *schadenfreude*, etc., may feel quite similar. In an experiment, people were given a drug which puts the body in a certain state. When different groups were in the same room with an actor behaving a certain way, the people interpreted their own feelings in the same way as the actor behaved (Schachter and Singer, 1962).

are positive and can be experienced as rewards; these are meant to make the organism approach something positive. Pain is clearly negative, but it is almost as basic as an automatic response to a stimulus.[108] It has the function of making the organism withdraw from that which creates the pain. Pleasure is a common name for different good feelings, and motivates the organism for doing things that are good for survival and a good life but also for sexual reproduction (Damasio, 2010, pp. 52–53).

So far I have said that feelings arise when the organism senses something in the outside world or its own body, and I shall argue that this is often an important influence when people make choices. An emotional influence of the body which is important for understanding free will is the fact that we can remember earlier emotional experiences we have had. When we are about to make an important choice we can remember earlier relevant events which evoke feelings that influence the choice we are about to make. Feelings are stored in memory and influence choices, and this is another reason to take a closer look at memory.

5.5 Memory

The brain can reproduce events in our mind regardless of our choosing. Memories are stored together with the feelings that we reacted to the situation with, and more emotional memories are better remembered and more easily recalled. This is generally an advantage for humans, since it often makes us recall important earlier events when we are in similar situations and can take advantage of what we learned the last time. But it can be a disadvantage for particular individuals, for example, people with post-traumatic stress disorder who constantly recall horrible events. It is not the event alone that is stored in the memory, but our relation and reaction to the event and our feeling at the time (Damasio, 1999, pp. 130–132).

---

108 The area which is hurt sends signals to the brain through special nerve cells called C-fibers and A-δ-fibers. The destroyed cells in the area release chemicals and these also send a signal to the brain. The input from these different fibers and nerve cells creates representations in the brain of the body being in pain in a certain area, and as these patterns become conscious we have a conscious experience of being in pain (Damasio, 1999, pp. 71–73). The same C-fibers also mediate itching, but they cannot be used for both itching and pain, so if you are itching in an area which is then hurt the itch will disappear and only the pain will be experienced (Gieler and Walter, 2008, pp. 54–56).

Several different functions of feelings have been mentioned, but feelings have another function as well, which has to do with memory and choice. Damasio has put forward a hypothesis known as the somatic marker hypothesis. The main point is that neural patterns representing events in the world are connected to feelings which give the different neural patterns different levels of importance. This explains what we become conscious of both in sensing and in remembering, for at any one time there are numerous neural patterns that could have become conscious: the ones that are more important because of their connected feeling are selected by the brain, which has a disposition for selecting the ones with the strongest feelings attached to them (Damasio, 2010, pp. 174–175).

Furthermore, the somatic marker hypothesis explains how people make efficient choices despite the so-called "frame problem", which means that in any situation there are numerous things that could be considered before one makes a choice; this would delay the choice for a long time. But when feelings are connected to the different images, the brain sorts them according to importance and generally makes the choice much easier. Damasio supports this hypothesis with findings from patients whose brain injuries left all knowledge and logical capabilities intact but whose feelings of emotions were lost. They then also lost their ability to make rational decisions (Damasio, 1999, pp. 40–42).[109]

All of this is important for the understanding of deliberation and free will since, when we are deciding what to do in important situations, memories and feelings will often be activated (without us choosing that such should happen) and influence what we think and feel about the different alternatives for action, thereby influencing or changing our initial desires. Which alternatives pop into mind may seem undetermined, but may nevertheless be a regular causal process with no indeterminism in it. Much more will be said about this later.

There are different kinds of memories which must be stored somewhat differently since it is possible to lose all memories of one kind but not those of another. Short-term and long-term memory is a distinction most people know of. It is also usual to distinguish between procedural memory (remembering how to play the guitar or ride a bicycle) and fact memory (remembering facts or events). There are some who cannot consciously remember what a thing is or does, but are still able to use it correctly because of their procedural memory. A more interesting distinction for the topic of choices and free will is that between general facts and episodic memories. One patient, E. D., could remember many facts

109 For example, Damasio gave such a person a choice of two dates for their next meeting, and he wavered between the two dates for almost half an hour before Damasio stopped him and decided the date for him (Damasio, 1994, p. 193).

about Kilimanjaro without remembering having been on top of it (Markowitsch, 2004, p. 52). Damasio mentions a person who looked at pictures and correctly identified them as portraying a wedding, but he did not remember that it was his own wedding (Damasio, 2010, pp. 138–140). Most important for the question of free will are fact memories and memories of past events. The reason is that fact memories activate images of alternative actions that can be chosen, which again activate autobiographical memories and feelings connected to each alternative possibility for action (Damasio, 1994, p. 196).

The important thing to note from this presentation of memory is that we have stored facts and memories that can be activated as alternatives for action in the future, and we have stored memories with feelings connected to them which can be activated and influence what we desire most strongly to do. Our memories and our experiences can be the cause of our choices, and some of these memories constitute our autobiographical self, which means that our autobiographical self can sometimes be the cause of our choices. The self is the next topic to be considered.

5.6 The self

Damasio distinguishes between wakefulness, mind and self. The mind is neural patterns that can function non-consciously but which require a self to become conscious (Damasio, 2010, pp. 159–166). Damasio has an influential theory about the self. He thinks that the brain can cause a conscious experience by adding a self process to wakefulness and mind. But how did the self develop? I will spend some time on this since it helps us understand who the agent is who has free will. Damasio argues that the self was built in three different stages. I will present these stages in more detail after a quick overview first to make it easier to follow. The first stage is the proto-self, which is a neural pattern representing the whole organism. This proto-self produces a primordial feeling, which is the feeling of my own body existing, but without any further connection to the world. In addition to the proto-self, the brain creates neural patterns representing objects and events in the world, but it also creates neural patterns representing the relationship between the organism and the outside world. From moment to moment there is a series of neural patterns representing how the organism changes in relation to the outside world. This creates changes in the primordial feeling which are consciously felt as an experience of changes in the world. The representations of change create pulses of core consciousness that together constitute the core self, which is the second stage. The core self is this series of consciously experienced changes that arise because of the neural patterns representing changes in the body in relation to the world outside. Finally, these conscious experiences can be held together in extended consciousness to create the autobiographical self, which is a neural pattern representing the life story of a person, created by memories and continuously reconstructed.

In more detail: Damasio defines the proto-self as "an integrated collection of separate neural patterns that map, moment by moment, the most stable aspects of the organism's physical structure" (Damasio, 2010, p. 190). The proto-self is constituted by three different kinds of neural patterns (which Damasio calls "maps"). The first kind represents the body – not the whole body, but the most stable parts of the body. Damasio argues that that is important, since it can explain the stability of the self process, and this is also the reason for the last part of the definition of the proto-self. It is this representation of the body which gives rise to the primordial feeling, which is the basis for all other feelings. The second kind of neural pattern which constitutes the proto-self is a general representation of how the main parts of the body relate to each other when the body is not moving, and it is then constantly compared with patterns representing how the body is moving right now. The third kind of neural pattern is representations of the sensory portals of the body (eyes, ears, nose, tongue, skin), and these have the function of locating where the body is relative to the sense impressions. Not only do we see and hear, but we also feel that we see with our eyes and hear with our ears. This creates an experience of having a certain perspective and particular location in the world (Damasio, 2010, pp. 190–198).

Damasio does not think that the proto-self and primordial feeling are enough to account for the phenomenon of self which we humans experience today. Had it only been a proto-self its experience would have consisted only of a primordial feeling of being a self from moment to moment with no connection to anything. What is needed is a clear experience of being a protagonist connected to events in the world, and Damasio thinks that this happens in the following way: When the organism encounters something, this changes the organism and makes the brain construct a new pattern representing the change. The change in the proto-self leads to a change in the primordial feeling, and this change is experienced as a conscious experience of the object. But it is also experienced as something happening to a protagonist, to a self. The narrative of objects encountering a body and giving rise to different feelings in the body also suggests that there is a protagonist to whom things happen, who feels and acts and has a sense of ownership of the experiences. The self is, so to speak, deduced from the narrative and experienced as such a self (Damasio, 2010, pp. 201–204). Damasio has also described it by saying that our conscious mind is like a movie, and when we ask who is watching the film the answer is that the watcher is a part of the movie (Damasio, 2004, p. 11).

I find it useful to distinguish between a minimal self or basic subjectivity and the more complete sense of self which the core self is. By "basic subjectivity" I mean the phenomenon that something is like something *for* something or someone, and I shall argue that this is presumably what Damasio means by his "primordial feeling". Something cannot be *like* something unless it is like something *for* something or someone. What is expressed in this "*for* something or someone" is what I mean by basic subjectivity. Even if the core self can be deduced from changes in primordial feeling, it requires a basic subjective element in the primordial feelings in the first place. What I want to do here is to add some support for the idea that the core self can be deduced from changes in primordial feelings.

A sense of self, by which I mean a sense of being a single conscious subject, may not need to imply a physical or non-physical continuous self which has this sense of self. Pete Mandik has argued that a self can be deduced from experiences in the same way that we can see a picture and deduce that there must have been a camera at such and such an angle and distance from what is seen in the picture. The perspective and distance suggest where the camera must have been to take the picture, but the picture may have been computer-generated and given a certain perspective. Mandik's point is that there are a lot of subjectively conscious experiences going on in the brain which may create a sense or thought that it must have been a self having these experiences, but in fact there is just a series of experiences *including* the experience that there is a self having the experiences. We understand temperature as hot or cold and we experience things as good or bad relative to ourselves. But there could be something in the nature of the experience that makes us assume a continuous self is having the experiences even if there is no such continuous self. Perhaps there is just a series of experiences, pulses of consciousness, which give the illusion of a self owning the experiences – and even the sense of a self as the agent of certain events even if they just happen as a causal chain of micro events (Mandik, 2001).

But must it not be *someone* who deduces that there is a self? No, the suggestion is that there are ontologically subjective experiences that can be configured in a unified way which then constitutes a self-experience. Ontological subjectivity is a feature of the world, as argued by John Searle (Searle, 2007, p. 327), which can create a self. This does not explain what basic subjectivity is, or how basic subjectivity is possible at all, but it is an explanation of how a full-blown self can arise from experiences that are subjective in a basic sense. Translated to the theoretical framework of this book, it means that the qualia field consists of values that have basic ontological subjectivity to them, which again means that a unified configuration of these can produce a conscious core-self-experience like the one humans have.

How are then all the different qualia produced by different parts of the brain combined into one coherent picture? What "glues" the parts together so that one subject can be the part of many pieces, sometimes referred to as the subject summing problem? Here I think similarly to Sam Coleman, who says that qualia are arranged by being located at different places on a phenomenal screen into a coherent whole. Being this coherent system is to be a conscious subject (Coleman, 2012, pp. 159–160). To this I would add that the phenomenal screen is a part of the qualia field with a unified set of actualized qualia values, and that the brain has evolved to create such unified parts. Above, we saw examples of how the brain merges contradictory input into a unified whole, and below I shall describe how and why this evolution took place. It is the unifying work of the brain that explains how the pieces combine to have a subject as their sum.

The third stage in the development of the self is the autobiographical self. The autobiographical self consists of our memories, including memories of our thoughts about the future and who we are. Whereas the core self is always present when a person is conscious, either in focus as self-awareness or in the background since attention is on something else, the autobiographical self is either dormant or active. It is important to understand that every time memories are recalled, they are modified a little, and the feelings they are connected with may change a little. This means that the autobiographical self is reconstructed all the time (Damasio, 2010, pp. 210–211).[110]

The autobiographical self can be constructed only because the consciousness is able to hold several elements present over time. Different memories can become conscious to a person and grouped together or seen in the light of each other, thus giving a coherent picture of who that person is. I have already mentioned Damasio's somatic marker hypothesis. It says that every time an image is recalled, it is automatically marked with a certain value in the way that a certain feeling expresses this value it is connected with, and this valuation is constantly revised. The image of who we are – our autobiographical self – is

110 Interestingly, this explains how (parts of) psychoanalysis work(s): non-conscious memories from childhood might be activated in the mind or in dreams, but without becoming conscious to the awake person. If there are bad feelings connected with the memories, these may create more negative feelings or problems like anxiety or depression, while the memory itself is still non-conscious. When the memory is recollected in consciousness, new feelings are connected to it in the same way as all recalled memories. But now one may see the incident in a whole new light, or relate to it in a safe context, and connect new feelings with the memory, which might stop or reduce the negative influence on the person.

also marked by such a feeling and constantly revised. Damasio argues that this marking depends partly on preset dispositions acquired through evolution so that things which are important for survival are considered important. But the marking also depends on values acquired through life – things we have come to see as important for us in the light of our individual experiences and reflections (Damasio, 2010; Damasio, 1994, pp. 177–180). This is important for the topic of free will, because it means that things we have come to see as important through our experiences and thoughts are stored in our autobiographical self and influence the later choices we make.[111]

It is useful at this point to sum up the terminology concerning the self that will be used in the rest of this book. The core self is the stream-like consciousness of what happens here and now together with a feeling of what that is like. It is a series of conscious events – the stream of consciousness consisting of experienced qualia. The core self is a process of conscious experiences, and the physical basis for the core self is patterns representing changes in the interaction between the proto-self and the environment. The autobiographical self is a person's understanding of herself based on memories and their connected feelings. The autobiographical self is a physical neural pattern which can become conscious. This image of oneself also influences how it feels to be that person in any moment of core self-consciousness.

Memories of what has happened – consciously or non-consciously – in the mind (sensing, thinking, feeling, desiring) constantly return to the mind and influence what happens in the mind the next time. A person has experiences and these lead to feelings and thoughts that are stored in the memory. The more memories of thoughts and feelings a person has connected to her experiences, the more these memories will influence what the person desires later, since desires also depend on what we feel about different alternatives (as I argue in the next section on desire). The autobiographical self is a collection of memories of thoughts and feelings which influences the desires and choices of every person. Our choices are not only influenced by the autobiographical self, since we are born with many dispositions and other non-chosen influences (e.g., a non-chosen acquired mental disorder) that can also come into play. But the autobiographical self becomes a larger and larger influence on most people's choices during

111 Let us say that future research in neuroscience rejects Damasio's distinction between the core self and the autobiographical self. There will nevertheless remain something structurally similar, where something corresponds to our conscious experience of here and now, while something corresponds to the important memories that shape how we experience ourselves as persons and what we desire. These parts of a better theory of the self in the future must then replace what I here call the core self and the autobiographical self.

their lives, since a larger and larger collection of thoughts and experiences can be remembered, and then habits and character traits are developed which influence future choices.

I use the terms "person" and "agent" interchangeably for a living human body with a mind and a core self. What happens in the core self is stored as memories and added to the autobiographical self, which in turn influences what happens in the core self. The term "autobiographical self" can be understood in several different ways. Every person has many experiences and these are stored as memories. A neural pattern is constructed in the brain representing the person who has had these experiences, and this is a coherent pattern of important memories. This pattern changes over time and can become conscious. This neural pattern is what I mean by "the autobiographical self", and although it can be seen as a process over time, I use it largely to mean the neural pattern as it is today or at the time of discussion.

The autobiographical self is not all my memories, but a selection, and although it can become conscious, not all details of this neural pattern can be conscious at the same time. For some people there are probably parts of their autobiographical self that never become conscious. This means that they may have a conscious understanding of themselves that does not match their autobiographical self as neural pattern at all points. For example, they may have experienced something disgraceful which they non-consciously deny ever happened as a survival technique. They may think that they handle certain situations well, but in these situations the non-conscious and denied memory may still be non-consciously activated so that they behave strangely in these situations – which they may also fail to realize.

I will end this subchapter by noting some meanings of the terms "self" and "sense of self" on which I do not focus in the rest of the book. I have already mentioned the primordial feeling based on the proto-self, which is the subjective experience of being present, whereas the core self is the experience of being an agent owning the experiences, and this experience is based on representations of changes happening to the proto-self. This experience explains why agents feel that they own their actions even if there is no homunculus inside our body but rather a series of causal processes happening in the mind. The sense of being someone who acts is based on the fact that actions follow intentions.[112]

The term "sense of self" can include many different experiences. There is the sense of *presence*, which is the primordial feeling. There is the sense of being *one*, which is based on our unified conscious experience. There is the sense of

112 Similar distinctions are made in Gallagher (2000).

being *one with your body*, which is based on the core self experience of being a protagonist to whom changes happen, and this is the same as the sense of *ownership* of your own experiences. The sense of *where your body is* and *how it is positioned* is based on a constant comparison between a master map of the body in repose and the position of the limbs relative to this.[113] The sense of where you are located in the world is based on the perspective of your conscious images.

What about the sense of identity over time? Since a person can consciously remember what has happened to this body with this core self, the person will have a sense of persisting identity over time, and identify with that same body and its core self of several years ago. The memories are what give the *sense* of persistent identity, but what *constitutes* personal identity over time, metaphysically speaking? I can only deal very briefly with this here, but I think that at this level of detail, it is possible to describe quite accurately what happens in different difficult cases.

For example, we can ask: if each half of your body was united with another half or each particle in your body was removed and replaced one at a time while you are anesthetized, so that two identical replicas are created, or a teleportation going wrong suddenly caused two replicas, which of them is you? "You" is ambiguous, since it can refer to the person, the autobiographical self or the core self. What happens is that the physical structure of the body and the autobiographical self are made into two versions of the original. These two bodies and autobiographical selves now give rise to one core self each, which in their first waking moment have exactly the same memories, but from the next moment on have distinct experiences, thoughts and feelings and start to change their autobiographical selves into two different autobiographical selves.

One person will then have turned into two, just as it sometimes happens that an embryo splits into two and gives rise to twins. Jack in 2020 becomes Jack 1 and Jack 2 in 2021. Jack 1 and Jack 2 were Jack in 2020, but neither of them is Jack of 2020 anymore. Against this, one can then object that two different persons (Jack 1 and Jack 2 in 2021) cannot be identical to one (Jack in 2020). But that presupposes the classical understanding of identity where all properties are undistinguishable, and while this applies to synchronic identity, it is not given that identity over time or persistence should be defined the same way.

---

113 This sense can be disturbed, and scientists have managed to stimulate an area of the brain which allows them to create out-of-body experiences for the person so stimulated (Blanke, Ortigue, Landis, and Seeck, 2002). There are also many simple experiments that one can perform on oneself to disturb this sense; see for example Ramachandran and Hirstein (1999, pp. 105–107).

Derek Parfit argues that what should matter to people when it comes to such cases is not identity in the sense of all properties being undistinguishable, but rather survival in the sense of psychological connectedness and/or continuity for any causal reason. What that means is that the mental states of a person must be causally related through overlapping series even though there may be gaps in the series (Parfit, 1984, pp. 205–207).

Parfit's argument that this is what matters is as follows: Imagine that half your brain had survived an operation in a new body, and you were still conscious and had many memories intact. Then you would and should clearly think that you had survived. If the doctors then came and said that the other half of the brain had also been saved, and now one more body was conscious with many of your memories intact, then you should not think that you have now died, but instead celebrate it as a double success (Parfit, 1984, p. 254). Asking which person is the original is like asking which party is the original if a political party splits up: It is an empty question – we know all there is to know, and there is nothing more to say or know which makes one of them the original (Parfit, 1984, pp. 260–261).

I agree with Parfit's reasoning and think that persistence or personal identity over time is just that a person has a quite stable internal structure that only gradually changes over time, without there being an exact border for when something persists. This is the same kind of reasoning that I presented with identity over time for any object in Chapter 2. You can raise a version of the Theseus' ship paradox against this view, since Theseus' ship would remain the same ship if it was gradually rebuilt, but not if everything was suddenly replaced. Likewise, we say that a human being is the same when cells are gradually replaced, but we would not say so if all parts of human body were suddenly replaced except the feet.[114] Somewhere between these extremes there is a diffuse border, and the coarse concept of personal identity over time is useful even if not exact.

When we use the concepts of autobiographical self and core self we can explain everything that happens in an imagined fission-case where Jack turns into Jack 1 and Jack 2 and there is nothing more to know and no reason to insist that one must be the original. Jack's wife may have a pragmatic reason to see it differently, but will have to find a pragmatic solution, like making a copy of herself as well, and the two new couples must share house, money and responsibility for their children.

114 This example and argument is from Rescher (2001, p. 86). I recommend this book by Rescher for an overview of how to think of some typical classical paradoxes.

Returning to the self, most important for the question of free will is the autobiographical self, the neural pattern which represents a person's life right now and is a product of earlier thoughts, feelings and actions influencing future choices. Whereas the core self is a series of conscious experiences, the protoself and the autobiographical self are physical neural patterns, and the person is the combination of a physical substrate including the physical mind, representations in the brain, and the conscious experience of the core self.

What about the concept of "the self"? (I will not use this term in isolation after this paragraph, but specify just which self I am referring to.) It is normal to think of the self/agent/person as a continuous entity which is the cause of actions. But the core self is not one entity with a continuous existence; rather it is a continuous series of experiences only interrupted by certain forms of sleep and anesthesia or coma. As I shall argue in the chapter on free will, actions are caused by desires triggering motor neurons, and these desires can be caused by the autobiographical self or other causes. It is the physical substrates which give the experience of continuity, the sense of identity with the body and memories, even if both the memories and the body change gradually. Much of what happens in our conscious mind depends on non-conscious physical activity in the brain and body. So those who wish to speak of the self as continuous or a cause of actions should include a physical aspect of the self. I find it better to leave out the concept of "the self" altogether in favor of more precise terms.

As regards these concepts – agent, person, core self, autobiographical self – what does the term "I" refer to? What do I identify with when referring to myself? When I want to be the cause of my choices, what is the cause that is *I?* I certainly identify with a body with a conscious mind which can act now. But I also identify with the same body with a conscious mind at earlier stages which made choices that shaped the autobiographical self I have today. When the term "I" is used in daily life, it can refer to several different things – a body, a core self, an autobiographical self – now and over time, but even if the term is ambiguous in daily speech, it should be precise here. When I, the author of this book, use the terms "I" or "my" about myself, it refers to a particular body with a mind, a core self, and an autobiographical self from its beginning and development until today. In other words, when people say "I", in my definition they refer to that which I define as a person, and the autobiographical self that persons usually develop over time.

There are so many philosophical problems that it is difficult to give a coherent definition of what a person is and what the conditions are for (personal) identity over time, but the most important question is what actually exists in the world, and that is bodies with minds and core selves which usually develop autobiographical selves over time. I suggest that the term "I" should be thought of as referring to this whole configuration of structures (a body with a mind and core self and usually an autobiographical self), but in daily speech it is often ambiguous what "I" refers to since it refers to different parts of this configuration. Thus, if I say that I am six feet tall, "I" refers to the body which is so tall, but if I say that I am thinking, "I" refers to the conscious brain activity that occurs in the same body. Or I can say that I was 30 years old ten years ago and refer to the body, the memories of which are stored in my autobiographical self now.

So far we have looked at how the memories, emotion and the self can influence what we desire when making choices, but time has now come to take a closer look at desire.

5.7 Desire

The concepts of desire and of desire strength are important for my causal understanding of the mind and free will. I will unpack this later, but start with untangling different distinctions when it comes to desire. Damasio does not have much to say about desires, but he does distinguish between emotions on the one hand and drives and motivations on the other, which he says are simpler constituents of emotion (Damasio, 2010, pp. 109, 111). I interpret Damasio's terms "drives" and "motivations" as having roughly the same meaning as "desires", since he exemplifies them with hunger and thirst, which seem like obvious examples of desires (Damasio, 1999, p. 77; Damasio, 1994, p. 116). But just what are desires? Damasio seems to think of them as dispositions in the brain that are meant to help the organism achieve homeostasis (Damasio, 2010, p. 55). That is a third-person-perspective description of the physical realizer of desires and their function. I shall focus mostly on the first-person-perspective experience of desire and return to the physical side later.

From a first-person perspective, "desire" is a concept that can be defined quite widely to include all sorts of wishes and preferences or even judgments that something is good. On the other hand, it can be defined more narrowly to include a feeling often related to pleasure or displeasure. I may judge that something is good (e.g. that I give money to aid projects in Africa), without having a desire that it should happen that I give money to Africa. Since Damasio uses the terms "drives" and "motivational states", he seems to focus on this *felt* desire which is supposed to lead to action. Hunger is a feeling meant to make the hungry act on their desire and eat, and the same goes for thirst or sexual desire. I thus understand desire as including both a thought that something is good and a feeling that makes the person want the desired state of affairs to happen.

Alfred Mele offers several helpful distinctions when it comes to desires. He defines a desire for A as an A-focused attitude which constitutes motivation for A (Mele, 2003, p. 170). Here one should distinguish between a desire to perform an action (go for a walk) and a desire for a state of affairs to be true (that my team will win the cup) (Mele, 2003, p. 16). Most desires are desires for a state of affairs to come true which include doing something in order for it to come true (I mow the lawn since I desire a nice garden), but not all (I may desire to be someone else). In order to avoid constantly interrupting the text by making reservations about counterexamples, I focus on action desires, which include thoughts about some states of affairs one wants to become true. This way desires are connected to alternatives for action, but the alternative for action is understood as including both the goal and the means, and thus desires are related to both goals and means. I also focus on proximal desires, which include the desire to do something now, although there are also distal desires, which are desires to do something later (Mele, 2003, p. 167).

Mele further distinguishes between occurring and standing desires: I may always desire that there should be peace on earth but that does not mean that that desire is always activated or felt (Mele, 2003, p. 30). Some desires are intrinsic, meaning that one desires something for its own sake (e. g. if one likes to whistle), whereas other desires are instrumental, which is when you desire to do something to achieve something else (go to the dentist in order to have good teeth). Desires can also be a mix of the intrinsic and instrumental; for example, you enjoy swimming for its own sake but also do it in order to get into better shape (Mele, 2003, pp. 33–34).[115]

One might think that there are just two desires – a desire for pleasure and a deterrent desire for pain – and that all other desires are sub-desires. But some of the most common desires seem to have physical realizers located at different places in the brain, and it is good if the vague concept of desire can be related to something empirical. There are different areas of the brain that can be stimulated electrically and the people stimulated will report that they feel a certain desire. Animal tests confirm the same. The lateral hypothalamus is active when we are hungry or thirsty and also when we see food and drink. If it is stimulated electrically, we will feel compelled to eat and drink, whereas when it has been destroyed in animals they stop eating and drinking and must be force-fed if they

115 Note the potential confusion that I follow Mele here, who uses the term "intrinsic" to describe a goal in itself, while in the chapter on ethics I describe how Christine Korsgaard says that "intrinsic" should not be conflated with "goal in itself", and so I avoid using "intrinsic" in the sense of "goal in itself" in the ethics chapter. Hopefully, confusion is avoided with this notification.

are not to die. Parts of the amygdala and hypothalamus can be stimulated and people will suddenly feel a strong sexual desire and perform explicit sexual acts even if these are inappropriate in the situation. There are also famous experiments that have discovered what seems to be a pleasure center in the brain. If they are stimulated electrically in humans, the latter will report that they feel great. When animals are allowed to press a button that stimulates them they will continue to do so ceaselessly until they are exhausted (Joseph, 1996, pp. 171–172, 187).[116]

Obviously, humans have many desires that are not innate. People can desire all sorts of things, like desiring their football team to win, and it is impossible that there should be an innate area in the brain responsible for all kinds of concrete desires. Timothy Schroeder, Adina Roskies and Shaun Nichols give an overview over standard neuroscience on desires. They say that standard neuroscience supports that actions are caused by a physical desire in the brain (Schroeder, Roskies, and Nichols, 2010, pp. 84–87). While we have innate desires, they are changed by beliefs and by emotions, which are parts of beliefs (Schroeder et al., 2010, pp. 87–89, 101). In the brain, different desires point in different directions, all wanting the body to move a certain way. Schroeder et al. compare it with a class of screaming children all wanting to take the others with them in a certain direction, and then the teacher picks out one pupil, who gets to decide. In the same way, the brain selects one desire to cause action (Schroeder et al , 2010, pp. 81–83).

This process of acquiring new desires explains how desires relate to reasons for action. Reasons for action are often divided into three:

1) normative reasons, which are objective reasons for acting a certain way,
2) motivating reasons, which are the reasons a particular individual actually had for acting, and
3) explanatory reasons, which are the reasons why an individual actually acted, which may be something different than the normative or motivating reasons (Alvarez, 2018).

It may be confusing to talk about reasons, since the term "reason" is ambiguous. It can mean either an epistemic reason (as in 1 and 2), which is a proposition that makes another proposition more likely to be true, or it can mean cause (as in 3). Furthermore, epistemic reasons can be moral reasons that everyone have for acting certain ways, or individual reasons why something is good for an individual.

116 These examples strongly suggest that there is a physical side to desire strength and that desire strength can cause action – a topic to which I shall return.

Epistemic reasons connect to both goals and means and thus to alternatives for actions as including both goals and means. For example, I can have reasons to desire a goal, and therefore also reason to desire the means to that goal, which means I have reason to desire the alternative for action that includes both the goal and the desire.

I may seem overly focused on desires as causal explanations for actions – how should we understand the fact that people can act for moral reasons? I believe it happens in accordance with the paragraph before the previous one. We start life with our innate desires, but we can acquire moral goals that we adopt as our own goals and that we desire as good. This happens through different processes as described by psychologists: through parental guidance, development of empathy, processes of recognition, experiencing things as good, developing a moral identity, acquiring habits through practice, imagining scenarios about a good and safe world, etc.[117] This all fits well with a process where the autobiographical self can change itself and acquire new desires that can cause actions.

Desires shaped by beliefs about how they are best fulfilled can make desires compete. A person may desire to live well and desire to avoid pain, and believe that if she abstains from sex God will let her live forever, but if she does not there is eternal pain waiting, so these desires and beliefs together make her act on the desires to live and avoid pain and not on the sexual desires. Or a sexual desire can make a person eat less and work out more to impress someone. Thoughts and feelings influence the strength of the different desires. This is important for the question of free will, and before I discuss it further I must discuss what it means that desires have different *strength*.

Desires as consciously experienced seem to have a variable degree of strength. For example, one can be more or less hungry or thirsty or experience stronger or weaker sexual desire. The strength of desire allows a person to have a preference when they have contradictory desires; the strongest desire is preferred the most. But what is the variable in virtue of which the desire is stronger or weaker? Here I suggest that it is the feeling of pleasure and displeasure which is the main variable, although it comes in different shades, in the same way as different feelings can give different kinds of pleasures or displeasures. The variable is in the amount (how much it occupies the conscious mind) and the felt intensity of the pleasure or displeasure. Pleasure and displeasure are felt in different ways by the one who desires something: she may desire something and feel displeasure since she does not have what she desires, or feel

117 For a detailed theory combined with neuroscience, see Narvaez (2013).

pleasure when she thinks about achieving what she desires and feel pleasure when she actually does achieve what she desires. Note that this description was concerned with different degrees of *consciously felt* strength of desire. I shall complicate the picture later by adding a non-conscious and a physical aspect to the strength of desires to explain why we do not always act on the desire that consciously feels the strongest.

So far I have argued that there are competing desires, and that these are influenced by thoughts and feelings. How does such influence happen? We have already seen briefly how Schroeder et al. describe standard neuroscience by saying that we have innate desires that can be changed by thoughts and emotions, but this will now be combined with more details from Damasio and others.

Different desires are triggered by stimuli either from the body or from the world outside. The situation triggers fact memories, both facts about the situation and possibilities in the situation, and autobiographical memories from similar situations and the feelings that the person had in those situations. Autobiographical memories are especially activated if there are strong feelings connected with them. The thoughts about possibilities for action in the situation may also activate autobiographical memories.

Consider this example: A tired man may see an empty chair and desire rest, but he may at the same time see a woman approaching the chair, desire to get to know her, and remember that offering a chair is a way to make positive contact with someone, while swiping the chair in front of them can be considered rude. Maybe he remembers earlier episodes of either offering a chair or swiping a chair, and more generally of getting to know women or being criticized for his behavior. The sight of the woman and the chair can activate thoughts about many possibilities for action and the possibilities that these actions can further lead to, and memories and feelings connected with all of these.

The feelings activated by earlier memories blend in with the feelings that the first desires give rise to. Maybe the man in the example above first had a strong desire for rest and so wanted to sit on the chair, but as he thought about the possibilities of getting to know a woman by offering her the chair and the possibilities of what would happen if he did not offer the chair, chair-offering was connected with good feelings and chair-swiping was connected with bad feelings, and so the felt desire to offer the chair was strengthened and the felt desire to take the chair was weakened.

My wife is a good example of how recalled feelings blend with desires. In the first stage of her pregnancy, the black coffee she usually drank and the pizza I usually cooked made her sick. When her appetite returned, she still did not desire black coffee or my pizza at all, but she did desire different variants of coffee and different variants of pizza. These different kinds of coffee and pizza tasted quite similar to regular black coffee and the pizza I made, but whereas she would desire some kinds of coffee and pizza very much, she would not at all desire something which tasted almost the same, namely regular black coffee and my pizza. The most likely reason for this is that the memory of that special coffee and that special pizza had a bad feeling of sickness connected with it, which reduced her desire for it strongly.

The process I have here described is in essence the following: a situation triggers alternatives for action that a person desires with different degrees of strength, and activated memories mark the alternatives with feelings which change the strength of the desire for each alternative. In this process, every event was causal, because it was all about neural patterns firing and thereby activating other neural patterns with which they were connected.

Damasio describes this process similarly, although quite briefly. In a given situation different alternatives for action are activated, and Damasio's somatic marker hypothesis holds that the different images are connected with feelings. This somatic marking influences which alternatives become conscious at all, and it influences how long we consider something and what we end up choosing, since feelings influence the strength of desire. Parts of this process are conscious and parts of it are non-conscious, according to Damasio (Damasio, 1994, pp. 174, 184–187, 196).

Peter Carruthers also understands the process of practical reasoning like this, and he refers to Damasio when he describes it: we envision different alternatives and register our emotional response to the different alternatives (Carruthers, 2006, pp. 137–138). Also Chandra Sripada thinks similarly when he describes different ways that what he calls the deep self can influence choices (Sripada, 2016). This is important for the question of free will, since it means that the experiences a person has had in her life (including thoughts and feelings) and stored in her autobiographical memory will influence the future choices she makes. Consequently, her autobiographical self will play a causal role in some of the choices she makes.

What happens between the process where, after some deliberation, one desire becomes the strongest and the point when a person acts on her strongest desire? How does the strongest desire lead to action? Damasio does not have much to say about that process. He says that the somatic markings provide incentives for a person to act and that we have an innate preference system for what we like and do not like and dispositions to act in order to achieve pleasure and avoid pain (Damasio, 1994, pp. 174, 179).

Peter Carruthers has developed in much more detail what happens between when a person considers her desires and when she acts. Since his theory is good and fits well with what Damasio says elsewhere, I add it here. Carruthers argues that we have different desire modules, which correspond to the different innate desires that I mentioned above (Carruthers, 2006, pp. 113–114). There is also a different module which Carruthers calls the practical reasoning module. This module selects as its input the desire it registers as the strongest, then starts to follow a set of heuristic rules. A basic sketch of these rules is as follows: The input is a desire for P. The practical reasoning module scans autobiographical and fact memories[118] to find a memory of the "If I do Q, then P will occur" sort. If it finds such a memory, it will search through a set of action schemata that are stored in memory and check if such an action scheme is doable here and now.[119] If not, then the module will search for memories of the "If I do R, then Q will occur" type in order to make the action scheme doable. If this process goes on for a while without success, the module will turn to the second strongest desire and start the process over again (Carruthers, 2006, pp. 57–58, 131).[120]

The details may be different in real life, but this is a scenario which describes the reasoning process as a causal process. The description above was serial, but the brain probably processes different desires in parallel and then the practical reasoning module selects according to certain rules; for instance, that it is worth considering a very strong desire for a certain period of time before

118 Carruthers uses the term "belief", but I understand beliefs as fact memories and their implications.

119 The theory that there are motor schemata guiding our actions has been developed by Richard Schmidt, among others (R. A. Schmidt, 1975; R. A. Schmidt, 2003). The theory explains why many of our actions are not caused directly by desires. Rather, we desire some overarching goal and then motor schemata actualize the desire. For example, when we write something, we do not make a decision to press every letter; rather there are motor schemata for the different words. There is constant feedback from the world interacting with our desires to find out whether a program should continue or be substituted with another. This happens on small scales and larger scales. An example of a small-scale motor scheme is the fact that when I want to type the word "Damasio" I always write "Damasion", which may well be because I have stored a small motor program for the ending "-sion" on English words. An example on a larger scale is driving familiar routes. When driving from my house I almost always take the same route into a roundabout and further into a tunnel. Sometimes I need to go another way and have to change lanes before the tunnel, but I have often ended up in the tunnel. That is probably because the normal route is a stored program and then I have to think consciously about changing lanes when it is time to do so; otherwise I end in the tunnel. The existence of such motor programs is my answer to the charge that event-causal theories of the mind cannot explain how we do many actions that are not directly caused by desires (Steward, 2012, pp. 164–165).

120 Whereas Carruthers uses this module to describe the whole deliberation process, I employ it to describe the final part of the process: from something becoming the strongest desire to the desire being executed in action. I write more about the physical side of this in the chapter on weakness of will (especially how the basal ganglia get a function from being formed by habits).

choosing to act on a less strong desire instead (Carruthers, 2006, pp. 131–132). The main point is that it is a mechanism which decides the strength of desire and follows a set of heuristic rules to turn the desire into action by selecting an action plan which is sent to the motor neurons for execution (Carruthers, 2006, pp. 131, 138).

Although Carruthers thinks that much of our practical reasoning happens like this, he follows Daniel Kahneman in distinguishing between two systems of reasoning, known as system 1 and system 2. System 1 works fast and quite reliably according to the general rules described above. System 2 is a conscious and slow-working reasoning process, but it is more reliable than system 1 and can trump system 1 decisions (Carruthers, 2006, p. 254; Kahneman, 2011, pp. 20–22). System 2's reasoning processes are typically concerned with important and difficult choices, and since I argued that consciousness plays a causal role this allows for consciousness to have a causal role in important choices.

I have so far argued that the strongest desire leads to action via the workings of a practical reasoning module that translates desires into actions. But there are two objections against this that should be considered, and both have been raised by Alfred Mele. He argues that intentions play an important role in reasoning which cannot be reduced to a combination of desires and beliefs. It is not good enough to say that the strongest desire leads to action, for there are many examples where the strongest desire is not the same as a person's intention. For example, the strongest desire of both Alan and Bob may be to insult Carl at a party, and they do, but still it may be correct that only Bob had the intention of insulting Carl at the party. What the concept of intention adds to the reasoning process, according to Mele, is that intentions settle which desire a person wants to act upon. A person has thoughts and can assess different desires in order to identify the best one and intend to act accordingly. What a person judges to be best can be different from his strongest desire. So in Mele's understanding of the deliberation process there are desires and one of the desires is the strongest, but it is a process of assessment producing an intention which settles which desire will lead to action, and that need not be the strongest desire (Mele, 1992, chapters 8 and 9).

Carruthers also discusses how we should understand the relation between what we judge to be best and what we desire the most. How can the judgment of something as best lead to action? Carruthers answers that we probably have an innate desire to do that which we judge to be best. How then is weakness of the will possible? Weakness of the will is not to do that which you judge to be best. It is easy for Carruthers to answer: even if we have a desire to do that which we judge best, we also have other desires, and sometimes these are stronger. So, even if Bob judged it best not to insult Carl at the party, he did it nevertheless, since his desire to insult Carl was stronger than his desire to do what he judged best (to not insult Carl) (Carruthers, 2006, pp. 391–392). I shall say much more about weakness of the will in the chapter on free will.

What about intentions? How can a person intend something which is not her strongest desire? Both Carruthers and Damasio think that desires can be at work non-consciously, and it seems that Mele does so as well (Carruthers, 2006, p. 400; Damasio, 1994, p. 185; Mele, 2003, pp. 30, 164). The answer is then close at hand: even if a desire is not felt consciously as the strongest, it is still the strongest in virtue of the non-conscious feelings connected with it, and selected by the practical reasoning module as the strongest. The module selects the strongest desire, and this is what settles the reasoning process. One could use the term "intention", but as long as there are thoughts, desires and a disposition in the brain for transforming the strongest desires into action there is no need for intention as referring to a separate entity in the brain.[121] I shall explain later in more detail what desire strength amounts to.

Using non-conscious desires in this way may seem like a non-falsifiable theory, but I shall defend it later when discussing weakness of the will, and answer more critiques from Alfred Mele. As mentioned above, however, both Damasio and Mele also believe that non-conscious desires influence our choices. Carruthers defends this belief by arguing that it works that way in animals, and that an account of the human mind must be evolutionary in the sense that it can show how the mind has been gradually built from animals to humans (Carruthers, 2006, p. 403). Much more could be said about desire, of course, but here I have said what I need to say about the self and the mind in order to lay the foundation for my solution to the problem of free will, and how to understand thinking and consciousness.

---

121 Mele has many suggestions echoing Carruthers, since Mele thinks that there are mechanisms in the brain which by default select what is assessed as best as the intention of the person and activate motor schemata to execute the intention (Mele, 1992, pp. 167, 221, 130–137).

6 Thinking

Even if thinking is part of the mind, the topic of thinking has gotten its own chapter. Many would probably accept that emotions, memories and desires are causal processes, but how can thinking be rational if it is a causal process? Several philosophers have objected to the idea of thinking as a causal process by arguing that the theory undermines itself: thinking cannot be rational if it is causal. In this chapter I try to explain how thinking can be both rational and a causal process. I present the theory first, in Section 6.1, and then answer objections in Section 6.2.

6.1 A causal theory of thinking

Thinking is probably the topic where most people will have problems accepting that the mind can be a causal process, since thinking seems to occur in its own dimension of rationality, and rationality seems impossible if thinking is a causal process. For example, John McDowell argues that thinking and reasoning happens in a normative space, and that this space of reasons is different in kind from the space of nature. The space of reasons is *sui generis*, according to McDowell (McDowell, 1996, pp. xv, xix, 72).

I will defend the view that the workings of the mind, including thinking, are causal processes, ultimately reducible to the actualization of fundamental values according to rules. I am not alone in thinking that the mind is a causal process. Alfred Mele defends an event-causal (as opposed to an irreducibly agent-causal) understanding of the mind against many common critiques. He says that a causal theory of the mind was defended by Aristotle, Thomas Aquinas, Hobbes, Spinoza, Locke, Kant and William James, and that even though Wittgenstein and Ryle were against it, it now seems to be the orthodox view (Mele, 1987, p. 31). Mele thinks that no non-causal theory of the mind has been able to answer Donald Davidson's challenge in 1963, which was to specify what it means to decide for a reason if the reason is not understood as the cause of the choice. In virtue of what is it true that a person acts upon a particular goal if it is not that the intention is the cause (Mele, 2003, pp. 39, 58)? Mele uses many arguments to defend causalism and attack non-causalism.[122]

122 See Mele (1992, chapter 13) and Mele (2003, pp. 39–50). Another famous defense of a causal understanding of mind is offered in Chalmers (2011).

https://doi.org/10.1515/9783110770957-008

Even though I do not know in detail how thinking occurs, I do want to spend some time in drawing a coarse but consistent picture of how the common processes of thinking can occur as a causal process. This view I will then defend against objections and use to answer some questions about thinking. Another understanding may be more up-to-date with recent neuroscience, but it is in any case interesting to see just that there is a way of understanding thinking as a causal process that can refute the most common objections. I expect that there are better causal theories of thinking on the market, but offer this as a general defense of the view that rational thinking can be a causal process.

I start by summing up some of the things already said by Antonio Damasio in the previous chapter. I proceed by adding some thoughts from Lawrence Barsalou which fit well with Damasio's thoughts, but extend quite much on several topics. Both thinkers represent the grounded cognition model of conceptual knowledge, which I also lean on.[123] I discuss the explanatory potential in Barsalou's thinking before I move to objections. I start now with summing up Damasio's understanding of thinking.

In the previous chapter, we saw how Damasio described the evolution of brains that made representations of the world, which he called images, and reacted to them with dispositions. Images are neural patterns representing something in the body or the world (or a possible world in the case of imagined or false images, meaning images of states of affairs that are not part of the actual world), and these can become conscious. Dispositions, on the other hand, do not become conscious, but determine how such images are processed in the brain. For example, the dispositions allow us to process images in different ways, like manipulating the images or applying reasoning to them (Damasio, 2010, pp. 63, 143).

Although the word "image" suggests something visual, Damasio underscores that it is images of everything – not just sights, but all kinds of sense impressions and feelings – and they can be concrete and abstract, they can be conscious or non-conscious. The images either come from sense impressions or from memory or combinations of these, and the process of mind is a continuous flow of images (Damasio, 2010, pp. 71–72). In addition to images, Damasio thinks that we have dispositions which allow us to store and recall memories, and they are also what make imagination and reasoning possible (Damasio, 2010, p. 143).

123 From 1970 to 1990 the leading theory of conceptual knowledge was the amodal symbolic model, but since that the grounded cognition model has become more and more popular (Kemmerer, 2014, pp. 274–275).

Damasio seems to use the term "image" for all kinds of conscious experiences we can have, but he does not offer much detail about how thinking occurs in the brain. Lawrence Barsalou has a theory of cognition which fills in many gaps in Damasio's account, which I will present here.[124] Barsalou argues in favor of understanding images as representations in the mind, and that these are based on perception. Barsalou's theory is that when something is perceived, attention is paid to different parts of the percept, and simpler, analogical structures of what has been perceived get stored in memory. These are called perceptual symbols, and are not like physical pictures or conscious experiences, but rather they are records of the neural patterns underlying the perception. Similar perceptual symbols are stored together and also categorized in larger frames. For example, different perceptions of doors are stored together, as are different wheels, and the frames for doors and wheels are both found in the frame for "car", but also in other frames. For example, doors are also found in the frame for house.

As described in the previous chapter, the brain has certain neurons that register different features in the world and organize them into concepts and categories. This process is well understood, and in more detail than many are aware of. As people learn new concepts, neuroscientists can spot the neural patterns in the brain representing them (so they can know which concept they are thinking about), and they also see that similar concepts are more similar at the neuron level than more different concepts (Bauer and Just, 2015). Not only are objects stored in memory, but also events. When such events are stored in memory, the individual learns which events have which results, which is useful for making choices (Barsalou, 2008, p. 623).

But we do not only register the world, we can also think about things we do not sense. Barsalou explains that this happens by simulators in the brain. A simulator is an ability of the brain to simulate an event or object in our mind, even when we do not perceive it, by employing perceptual symbols from different frames. So for instance, I have watched many chairs, and stored common elements of chairs in the frame for chair, which allows me to imagine a chair anytime I want. A simulator is then a disposition that allows us to imagine something based on information we have stored in our memory. The elements of frames can be combined in new ways to form new objects, new concepts or new propositions, even if these have never been thought of or experienced before. This is a core type of computation that the brain performs (Barsalou, 2008, p. 619).

---

124 This presentation is mainly based on Barsalou (1999). Supporting footnotes are also added.

A simulation based on a frame is the equivalent of a concept, and when people make roughly the same simulations they have a common understanding of the concept. Words are stored together with the perceptual symbols that go with their use, so that the word "car" is stored in the frame for car with everything else there that is associated with cars. Hearing the word activates the rest of the content of the frame. This is what it means that the content of the word is understood, namely that one is aware of its connections to other things.

This is a main idea in the grounded cognition model that understanding words and things means that the brain activates other things that they are related to. Brain scanning shows that when you perceive or think of an object, the brain activates areas that represent the different things you associate with the concept: its shape and color (in the fusiform gyrus), the motion connected with such objects (in the middle and superior temporal lobe), and the actions that agents perform with such objects (premotor and parietal areas) (Barsalou, 2008, p. 626; Barsalou, 2016, pp. 84–85).[125]

What about abstract concepts? Barsalou argues that they are all based on a threefold process where we perceive several similar events, then we select specific common traits for these situations (including what persons feel), and these common traits together constitute the abstract concept. So for example, the content of the abstract concept of "anger" is based on several situations where a situation triggers something that makes persons feel and act in certain ways. The abstract concept of "truth" gets one of its common meanings, truth as correspondence, from comparing what a proposition expresses and what is the case in the world in order to check if it is the same or not. And the abstract concept of disjunction (as expressed in the word "or") gets a basic meaning from an imagined state of affairs with an empty slot where two different candidates can be put – one *or* the other. One might easily imagine how the same recipe could be used with other abstract concepts (Barsalou, 2008, pp. 630, 634). This way of thinking about concepts is how I believe concepts get their intended meaning, and this is how I suggest we should understand what it is to be a concept.

I believe that this is the right explanation of how we humans understand abstract objects: they are understood in light of their connections to different concrete images, most of which do not become conscious to us. If I think of a mathematical "set", I may have a vague conscious image of a kind of container. But I may know the rules for how sets should be applied in mathematics and logic, and there can be some ambiguity, for example, the set of all sets that are not member of themselves, and this is enough to understand what an abstract object

125 For many other examples, see Kemmerer (2014, pp. 274–285).

is or what an abstract word means (Kemmerer, 2014, pp. 335–338; Barsalou, 2008, p. 634). There is no rational dimension of abstract objects into which my mind enters. There is merely the brain activity described above with the grounded cognition model, and parts of this process is conscious to us.

So far, the focus has been on how concepts of objects and events are constructed, stored, understood and used to think new thoughts. But what about rational reasoning? Barsalou argues that frames allow for categorical inferences. If I have a frame for "lion" which includes "eats humans", I can infer from seeing a particular lion in front of me that it also might well eat humans. The inference is this combination of elements. Such inferences can make me make better choices in concrete situations, like move away from the lion. I can also use simulations to plan for the future, simulating different events which will activate relevant associations indicating possible consequences for me to evaluate.

When it comes to logical reasoning, Barsalou used the example of how the term "or" gets its meaning. The same kind of reasoning could be used to explain other logical arguments. The basic logical structures of arguments can be reduced to sentences connected with the terms "and", "not", "or" and "if-then". These terms are also reducible to combinations of "not" and one of the three others.[126] These logical terms can also easily be translated into neural firings in simple processes of a neuron being activated or not.[127]

The meaning of the logical terms could be visual images in the same way as Barsalou described "or". His image of "or" was a box with an empty slot where one of two candidates could be put; one *or* the other. "And" could be a box with an empty slot where both of two candidates was to be put; the one *and* the other. "Not" could be a box with an empty slot where a candidate was absent; it was *not* there. And "if-then" could be a box including a candidate; *if* the candidate is there, *then* the box is there. This is typically described in logic with Venn diagrams, which are visual illustrations of the logical relations.

One may object to this that it has not been shown exactly how rational thinking occurs as a causal process or how inferences occur, but I think Barsalous's point with inclusion of elements in frames is on spot. It seems to me that our

---

126 IF-THEN reduces to NOT A OR B (but also: NOT (A AND NOT B). OR reduces to NOT (NOT A AND NOT B). AND reduces to NOT (NOT A OR NOT B). XOR (exclusive or) reduces to IF NOT A THEN B.

127 Here is one way: Let two inputs, 1 or 0, produce an output, 1 or 0, like this: AND: If both inputs are 1, the output is 1. If one of the outputs is 0, the output is 0. OR: If one of the inputs is 1 or both are 1, the output is 1. XOR – exclusive OR: If only one of the inputs is 1, the output is 1, otherwise the output is 0. NOT: The output is the opposite of the input: input 1 gives output 0, input 0 gives output 1. For another kind of representation, see Pinker (1997, pp. 101–103).

thinking, by and large, happens by parts being integrated into wholes. This is a way of understanding thinking which fits very well with what else I have said about theoretical frameworks, understanding and truth, so let me unpack it a little.

We have already seen how the brain organizes features into objects and stores them in memory. Objects can be understood as wholes consisting of parts (their properties). Events are also stored in memory, and can be understood as wholes consisting of parts (subevents). Does it also make sense to speak of wholes and parts at the physical level of the brain? That a neural pattern is part of a whole in the brain means that the parts are connected into a whole by synapses. The more often a neural pattern is confirmed as being part of the whole, the stronger the connection gets. Connection strength then functions in a non-perfect way as an indicator of how probable it is that a part belongs to a whole, which is a typical way of how learning occurs in neural networks in research on artificial intelligence (Russell, Norvig, and Davis, 2010, pp. 738–748).

I have already argued that understanding something means to integrate it as a part in a whole. If we integrate it in the whole where it belongs, we understand it correctly; if we integrate it in a whole where it does not belong, we misunderstand it; and if we do not know where to integrate it, we do not understand it at all (Gravem, 1996). The same goes for explaining: To explain something is to integrate it as a part in a whole.[128] Below, I will argue that meaning also is about integrating something into a whole.

Reasoning is also about finding out which parts belong together in which wholes. Deductive reasoning is to find out whether something is part of a whole. For example, does the whole, "All humans are mortals and Socrates is a human" include as a part, "Socrates is mortal"? When we make a deductive argument, the conclusion is always implied in the premises. This means that the premises are the whole, and the conclusion is a part of the whole that we explicate. Evolution can have selected brains with a disposition for drawing out parts of wholes, and emotion helps us to focus on interesting conclusions

---

128 While explaining is the same as understanding, the term "explanation" is typically used for clarification of certain kinds of connection. The most common are explanation by cause, explanation by intention or explanation by function (I believe that the last two can be reduced to causal explanations). If we add explaining what something is (the meaning of things, events and words) and how something is possible, explanations are subtypes of understanding which overlap with them totally. Typically, explanations are formulated in language for others with the goal of finding the best explanation, while understanding can be private in mind and not necessarily searching for the best understanding only. However, I see no basis for making a clear distinction between understanding and explanation, as for example in Wright (2004).

(e.g., tribe X now wants to kill all members of tribe Y, and I am part of tribe Y, so now members of tribe X want to kill me).

Inductive reasoning, on the other hand, is to group parts together to a whole, adding pieces that seem to fit into the whole. For example, the part "Jones had a motive for killing his wife" and the part "Jones' fingerprints were on the knife" are put together into a whole, and the piece "Jones killed his wife with a knife" is added. Or: A did X and liked it, B did X and liked it, so I can also do X and like it. Again, evolution has selected brains with a disposition for combining parts into coherent wholes. This is something brains do constantly. The process is a causal process, but the kinds of causal process that have produced offspring are processes that can help the organism achieve goals (more on the goals below).

Having a question corresponds to lacking a connection in a whole, and when we try to find the answer to a question, we test out different candidates by filling them into our gap to see how it makes the other parts connect. For example, my question could be what I could give to my dad for his birthday to make him happy. I want to connect "dad" and "happy", so I test different birthday present candidates to see how they fit with what my dad likes and needs. Hopefully I will find something connecting the parts and then I have an answer to my question. Again, evolution has selected brains that try to fill gaps in wholes or make connections between parts in wholes.

We will get back to free will, motivation and choices in a later chapter, but I mention briefly how this fits into the general picture drawn here. In our minds we have stored wholes describing actions and outcomes. Perceptions from the world and our body can activate such wholes, which again activate desires. For example, seeing the fridge activates the idea of opening the fridge and finding something good, which then activates a desire to open the fridge, which again causes the body to move towards the fridge. Another example would be that you could go to pick up object A, which activates a memory of you often forgetting to pick up B at the same time and having to return, which makes you desire picking up object B at the same time, which makes you do so.

This is of course all extremely brief, but it seems that we have brains that organize parts into wholes and that this can explain how much of our thinking occurs as a causal process. In addition, we need a desire for certain goals in order to get thinking going and in a direction. One of the goals should be a desire for knowing truth, which certainly seems to be something we are born with. I enjoyed hearing my three-year-old asking all the time "why" and asking how things are connected, often with some hilarious suggestions of his own, such as different ways that water might be transported up to the clouds.

Over a very long period of time, humans have learned that some argument structures are efficient means to reach the truth, but we also have a lot of generally useful thinking strategies which sometimes go wrong, as shown in abundance by Daniel Kahneman (Kahneman, 2011). There are many standard ways our minds operate which are not very rational. With this first introduction to how thinking occurs in a causal way, I now turn to some questions and objections which will clarify the position further.

6.2 Objections to a causal understanding of thinking

There are different objections to a causal understanding of thinking, which have been raised by serious philosophers and which deserve to be answered. Not all of them are necessarily raised against the kind of theory of thinking I have presented here, but they are typical objections that could be made and questions which I would like to answer to show the coherence of my approach compared with other approaches. Unfortunately, the answers will, as usual, be briefer than what the topic deserves, but again they indicate the direction of my thinking.

There are several problems that might seem unsolved by what I have said so far. Here are some critical questions which I will now discuss: How can representations in the brain be intentional? How can they give understanding? How can they give meaning? How can thoughts and words express the world? How can representations in the brain make thinking into something normative, happening in a space of reasons? How can such an understanding of thinking be a thinking that guides us to truth? Some other questions and problems like the disjunction problem and the misrepresentation problem will be dealt with as subproblems along the way, but these were the main issues. I now discuss these questions in the order just mentioned.

What makes thinking intentional? Intentionality here means aboutness. Thoughts are *about* something, but how can a neural pattern in the brain be *about* something, like a horse, when they seem to be two different states of affairs? Approaching this problem, it is common to distinguish between similarity theories and causal theories of representation (Stampe, 1977). The similarity theory says that something represents something else by being similar to that which it represents. A problem for this approach is that it seems that something may represent something else without being similar (a dot in a picture may represent a horse far away without looking like one), or not represent something which is very similar (a picture of Mary may be very similar to a picture of her twin Susie without representing Susie).

The alternative theory is the causal theory, which says that a representation is caused by that which it represents. So if the dot in the picture was caused by a horse, it represents a horse, and if the picture was taken of Mary it represents

Mary and not Susie. However, there are problems with the causal theory as well. For if you are hit in the head and that makes you see a dog, we still just want to say that the conscious image of a dog represents a dog and not a hit in the head, even if the image of a dog was caused by a hit in the head.

An additional problem for a representation theory of mind is the disjunction problem: why is a representation of a horse a representation of a horse instead of a representation of a horse *or* a zebra in disguise *or* a hit in the head *or* ... etc. (Fodor, 1984)? A third problem is the problem of misrepresentation. How can a representation be said to *mis*represent the world instead of the representation meaning just whatever it represents? What makes something a *mis*representation as opposed to a representation? If you look at a zebra and just have a conscious experience of a horse in your mind, why is that a misrepresentation of a zebra instead of just being a representation of a horse (Lowe, 2000, p. 92)?

In line with the presentation of mind and thinking above, I suggest the following answers to these problems. The first step is how seeing a horse creates a conscious image of a horse in a person: feature-detecting neurons respond to the different features of the horse. When you have seen a horse (or something else) several times, the brain stores a horse-like image in memory. Seeing something later that looks like a horse makes the person recognize it as a horse, because the neurons that fire in response to horse features activate the stored horse image associatively.[129]

We see the horse as a horse when a stored image of a horse has been activated in our fact memory, but so far no problems have been solved. The next important step is to note that the stored image of a horse in our fact memory is also connected to other facts about horses. The person has beliefs about horses and how horses relate to other things. So on the one hand, there is a horse in the world having such and such features and relating to other entities in the world. On the other hand, there is an image of a horse in a person's mind related to a larger framework in the person's mind of what horses are and do.

Now the questions about representation can be answered. The horse in the world causes an activation in the mind of a horse by light waves through the eyes triggering feature-detecting neurons, triggering a stored image of a horse and its surroundings and other facts about horses and their place in the world, which gives the horse image its horse-meaning. The horse image was triggered because of feature-detecting neurons detecting horse features, and so one could say that both the similarity theory and the causal theory are partially right.

---

129 As argued by Barsalou above, and also in Koch (2011).

Now we can define the term "representation" in a wide sense, and not just in the narrow sense of "consistent relation", as it was defined in Chapter 5. The narrow sense runs into problems of disjunction and misrepresentation, but the wide sense does not. Representation in a wide sense means that an image of a horse in your mind, and the other concepts that it relates to in your mind, is structurally similar to the states of affairs that the horse in the world being represented relates to. A representation in this wide sense is a correct representation as opposed to a misrepresentation.

If a person sees a horse standing on the grass in front of him, the horse can be misrepresented in many ways. If it activates the fact memory of a zebra, making the person see it as a zebra, it is a misrepresentation. If it activates a fact memory of a horse, but horses are understood as falling into the category of cats, it is another kind of misrepresentation. If it activates a memory of a horse, but places the horse in the tree above the person instead of on the grass in front of him, it is yet another misrepresentation. But if it represents the horse as a horse placed at the right place in the world and in connection with true facts about horses, it is a correct representation of an experience of a horse. The reason it is a misrepresentation in the first cases and not the last is that only in the last case the image of the horse in your mind and the other concepts that it relates to in your mind are structurally similar to the states of affairs that the horse in the world being represented relates to.

This relating of a horse to a horse framework in the mind also solves the disjunction problem. The representation of a horse is not a representation of a horse or a zebra since it is tightly integrated in mind with horse facts and not zebra facts. The activated framework determines that the representation is a representation of a horse and not a zebra, since such a representation is what a representation is, in a wide sense of the term.

This relating of a horse to a horse framework in mind is also what it means that a representation is intentional, or *about* a horse. However, there is an important qualification to be made. A concept or a representation in mind can be understood, have meaning, and be intentional by being integrated a larger framework, and this can happen non-consciously. However, these neural patterns in the brain must be such that they could become conscious, and are sometimes conscious, in order for there to be intentionality, meaning and understanding. If they had never been conscious, there would only have been a set of physical structures interacting, and there would not be any content in the terms "intentionality", "meaning" and "understanding" since they require a conscious integration of a concept that is subjectively experienced.

We use these three terms – intentionality, meaning and understanding – to say that something is *about* something *for* someone (it has meaning that someone understands), and when such a conscious subject exists, things can be represented also non-consciously and function in the same way as when conscious. But if there was never a conscious subject, if there was only a pile of cells and neurons interacting with physical objects without consciousness being involved, there would be no experiences about something for someone, which is what these terms mean and should mean. A neural pattern representing a horse cannot be *about* a horse unless it can be conscious, for if no consciousness existed, it would be neurons firing only and no aboutness. It is the image of the horse *in* consciousness that makes the conscious experience *about* the horse, whereas it is the relating of the horse to other horse facts that makes it about a *horse*.

One could then argue that the terms should be used only for conscious thoughts and not also for non-conscious processes. But it seems to me that we should use the terms also for non-conscious processes, since these are so similar to the conscious processes. We have seen examples of how people could solve card games non-consciously before consciously, and so it seems reasonable to say that people can understand meaning also non-consciously, but we should not say this if no conscious subject is ever involved.

This was the answer to how neural patterns in the brain relate to structures in the world outside of the brain in a way that makes thoughts represent worldly entities. By adding consciousness, thoughts can be about objects in the world and have meaning for someone who understands them. But is it right to use the terms "understanding" and "meaning" if it is just a causal process, or does that require more? I shall consider *understanding* first, and *meaning* afterwards.

Can there be *understanding* if what happens in the mind is caused by brain states relating causally to each other? John Searle is famous for his Chinese Room thought experiment, where a man who does not know Chinese is standing in a room getting messages in Chinese. He consults a book and finds a recipe which says that when a string with such and such symbols comes in, send out a string with such and such symbols. To those receiving the output from the room, it seems that someone in the room understands Chinese, but actually there is nobody in the room who understands Chinese (Searle, 1980).[130]

What is understanding? As mentioned before: To understand something is to put it in a larger framework. If you *cannot* put it in a larger framework, you do *not* understand it; if you put it at the *right* place in the larger framework, you *do*

---

130 In this section, by "understand" I mean "understand correctly", and not "understand at all". The person in the Chinese Room does understand that there are Chinese symbols, but he does not have a correct understanding of what they mean.

understand it; if you put it in the *wrong* place in the larger framework, you *mis*-understand it (Gravem, 1996, p. 242). So what determines what is the right or the wrong place? That depends. If we are talking about understanding another person, it must be to place what the other says at roughly the same place as her in a similar framework as she has.

For example, if another person says: "I got lucky with my hunting last night", and by that means she met someone at a bar, she is understood if her utterance is placed in a nightlife framework and misunderstood if it is placed in a hunting-in-the-forest framework. If we are talking about understanding something in the world, then the relation between the object and the world must be roughly similar to the relation between what we are understanding and a framework representing a larger part of the world. So if I see a spark plug, I understand it if I place it at the right place in a motor framework, I misunderstand it if I place it in a TV framework or at the wrong place in a motor framework, and I do not understand it if I have no idea what framework to put it in.

Searle's point with the Chinese Room experiment is to argue that understanding is more than manipulating symbols. Differently put, semantics is more than syntax. But what is the relation more precisely? What extra is required for understanding, or why does the person in the room not understand Chinese? I said that to understand something is to place it at the right place in a framework. In order to do that you need to recognize a pattern so that you can place your piece at the right place in the pattern. The man in the Chinese Room does not recognize patterns he can place correctly, and thus he does not understand them – he just follows instructions blindly. But let us say that the person in the Chinese Room started seeing some patterns after a while, so that he could correctly guess how some sentences should be formed in Chinese. He would still not understand them, but why not?

It is because he is not able to connect them with patterns in the outside world. Since he cannot look out, it is not possible for him to connect the strings of symbols with either the behavior of persons or events in the world. Had he learned which strings of symbols were connected with which events in the world, he could gradually learn and understand Chinese.[131] However, in order

131 There are two common responses to the argument, which Searle already discusses in his original 1980 article. The first is the system response and the second is the robot response. The system response says that it is not the man in the room who understands Chinese, it is the room as a whole that understands Chinese. To this argument Searle answers that we can imagine that man becomes the whole system. He learns the whole instruction book by heart,

for it to be understanding in the full sense, which includes intentionality and aboutness, consciousness is required, as I argued above (and will say more about in the next chapter, especially in Chapter 7.4 when discussing subjectivity).

From the Chinese Room experiment, it could seem like a little man inside our head – a homunculus – is needed for there to be understanding. While in reality, it seems like our neurons are like little blind persons in a Chinese Room without the opportunity to look outside. It takes some space to explain how conscious understanding can evolve in such a scenario, and much space will be given to it in the next chapter. I will argue that processes inside the Chinese Room (our skull) has started producing qualia matching structures in the world quite well, and out of this a conscious subject has emerged. This conscious subject, integrating concepts in theoretical frameworks, can understand the meaning of concepts and what they are about, but more will be said about this in the next chapter.

From *understanding* I now move to *meaning*. The problem of meaning can be understood as the same as the problem of understanding: how can there be meaning (in the sense of something we understand) if what happens in the mind is brain states relating causally to each other? If the problem is thus formulated, I reply as above, since I understand meaning as constituted by how something relates to other things in a larger framework (Puntel, 2008, p. 342).[132] But the problem of meaning can be phrased in another way to criticize the account here given of thinking: is meaning dependent only on internal events in the mind, or does meaning depend on external factors in the world?

---

i.e. what to write if he gets such and such symbols in, and he does all the parts of the process, but still he does not understand a word of Chinese.

The second answer to Searle's argument is the robot answer. It says that a lonely brain does not understand, but when it is in a body and interacts with the world, then we have a human being who understands. Searle answers that the man in the room can learn to control the entire robot room and how to control in response to different characters, but he still does not understand Chinese.

What I have suggested is a combination of the system response and the robot response if you add that the robot has sensors. I said above that to understand something is to put it in context. The man in the Chinese Room does not understand anything, because he does not have the opportunity to connect the Chinese characters to things in the world. He is unable to connect the symbols that come in to people and events in the world outside space. But if he had had a window out and saw what people were doing when they sent and received Chinese characters and how they reacted to the movements of the robot room, then he could start to understand Chinese.

132 This is the sense of meaning as "understandability", and not the sense of meaning as intention/goal or meaning as something positive and valued.

Many accept that meaning depends on external factors in the world, due to some thought experiments developed by Hilary Putnam and Tyler Burges. Most famous is the twin earth example of Putnam, where water on earth is H_2O, while on twin earth everything is identical, except that water is XYZ. So when Oscar on earth thinks of water he thinks of H_2O, while when Toscar on twin earth thinks of water he thinks of XYZ. But Oscar and Toscar are molecule-to-molecule identical. Putnam thinks that this shows that "meanings just ain't in the head" (Putnam, 1975, p. 227).

The problem is that on the one hand, it seems that what is in the minds of Oscar and Toscar are exactly the same. On the other hand, they are thinking about two different things. Should we then think of their mental content as the same or as different? How should the term "meaning" be used? In Chapter 2, we saw how Frege distinguished between meaning and reference, and said that *meaning* is the same as a proposition, which is a mental content, while *reference* is the state of affairs in the world that a person either intends to refer to or which it is perceived that a sentence refers to. With this distinction we can then say that the mental contents of Oscar and Toscar are identical in their internal structure, but they *refer* to two different things in the world. By saying that their mental images are identical, I mean that the mental images are *type* identical, since they have the same structure, but they are not *token* identical, for they occur inside two different heads.[133]

In everyday language it is easiest to use the word "meaning" in the sense that Putnam does, including its reference, for we are generally interested in what people refer to. But in order to have a precise ontology, we should distinguish between what exists in the mind of an individual and what exists in the world outside. So everything in the minds of Oscar and Toscar is type identical, all their conscious images are identical in structure, but one refers to H_2O and the other to XYZ when they think about water.

This solution requires that they refer to different things if everything in their minds is identical, so what makes it true that identical contents of mind can refer to different things? The reason that they refer to different things is the causal relation between H_2O and the word "water" in Oscar's world, and the causal relation between XYZ and the word "water" in Toscar's world. The causal relation is important, since any linguistic expression can express any state of affairs in the world. If I put a series of Xs and Ys on a blackboard, they can express soccer

133 Much criticism against the kind of internalism I sketch here can be rejected by use of this token-type distinction (as employed in Kim (2006, pp. 265–266)). Different truth conditions or different indexical references show that Oscar's and Toscar's thoughts are not token identical, but they may still be type identical.

players, battle ships, or whatever states of affairs there are in the world, and so I must say what I mean by them. In Chapter 2 we saw how a reference relation is established between words and what they refer to.

The meaning of words can then change over time, and people can learn meanings of words without any causal influence from states of affairs in the world, but there must be a first causal link between a state of affairs and a linguistic expression via a mind. This means that if Oscar were to visit Toscar and thought to himself, *I would like a glass of water*, he would be referring to H_2O, which is the object named water that he learned the name of. But if he were to see XYZ, point at it and say, "That water looks nice", he would be referring to XYZ. As long as we do not insist that there must be a correct reference, but merely intended and perceived references, there is no metaphysical problem here.

Many externalists will agree that mental entities are *located* in the head, and only insist on external factors sometimes being necessary to *individuate* thoughts. But some externalists are more radical and insist that the vehicles for thoughts are located outside of the body. Some even argue that consciousness is located outside of a person's body. Those who argue that vehicles for thoughts are located outside the body typically argue that when we use paper and pen to calculate or remember addresses, etc., the text on paper is a vehicle for thought taking part in your cognitive process, which should not be rejected as a vehicle for thought even if it is not located in the head. As Andy Clark and David Chalmers argue, if a data chip in your brain helped you think, it should be considered a part of your cognitive system, even if it were then to be removed from your head but kept doing the same job (Rowlands, 2003, chapter 9).

This is of course a matter of definition, for what is to count as "a cognitive process"? But so far I have only been interested in understanding ontologically the structures that give rise to qualia, and I have argued that these are all in the brain, but maybe someday someone will invent something which can have or cause qualia. Thus I cannot see that vehicle externalism presents an argument against the understanding of thinking that I have presented in this chapter. But some also argue that *consciousness* is located outside of the head, and that may be an argument against my understanding of thinking.[134] I will return to the question of the location of consciousness and qualia in the next chapter.

Having now seen how thoughts can be about something and have meaning, even when they are internal to the mind, I shall consider a final kind of critique.

---

134 See for example Puntel (2008, pp. 354–355) and Pettit and McDowell (1986, pp. 137–168), but I am not certain about this interpretation of any of them.

The charge is that thinking must be more than a causal process in the brain if thinking is to be normative, attending to reasons and even guiding to truth. I shall treat these as two separate critiques, the first about the *normativity* of thinking and the second about the *reliability* of thinking, and I start with the question of normativity.

I have already dealt with this challenge briefly, and mentioned John McDowell as an example of the critique. I will here say a little more about his view that thinking and reasoning happens in a normative space, and that this space of reasons is different in kind than the space of nature – the space of reasons is sui generis (McDowell, 1996, pp. xv, xix, 72).

Empiricists like to think of the empirical reality as a standard for whether our thoughts are correct: do they fit reality or not? This is a concern for John McDowell as well: How can the world be a measurement for whether our thoughts about the world are correct (McDowell, 1996, p. xii)?

But this reality which is a standard for whether our thinking is correct is only accessible to us as another thought that we can relate our other thoughts to. So, normative thinking is a matter of coherence among thoughts, even if we think that there exists something prior to language that language can express. The way that *the world* is a measurement for our thoughts (normative) is that it actually is *the most coherent theory of the world* that does the job as the measurement for our thoughts, as described in Chapter 2 of this book.

But what about the charge that agents act for *reasons*, and reasons cannot be understood as causes?[135] To act for a reason can be understood causally in the following way, using the chair example from before: A person sees a chair and a woman approaching. The visual impression of the chair and the woman activates in him two desires; a desire to sit and a desire to get to know her. The same visual impression also activates fact memories like the fact that he can offer the chair in order to get to know the woman, and that it will be considered rude to grab the chair in front of the woman. These fact memories further activate autobiographical memories concerning being rude before and bad feelings connected to that and of being polite to women before and good feelings connected to that. The good feelings remembered get connected to the desire to offer the chair and the bad feelings remembered get connected to the desire to sit. The desire to offer the chair becomes the strongest and activates the motor neurons that makes the man offer the chair. In this process, every event was causal, for it was all about neural patterns firing and thereby activating other neural patterns that they were connected to. But the process could also be described as a man acting for a rea-

135 As argued in McDowell (1996, pp. 73–75) and Bok (1998, pp. 203–204).

son. He offered the chair; the reason was that he wanted to get to know the woman, which was his strongest desire. Acting for reasons does not exclude acting for causes. Many more examples of how reasonable thinking can happen as a causal chain were given in the beginning of this chapter in the presentation of Lawrence Barsalou.

The final critique to be considered here is whether or not we can trust our thinking if the capacity to think is a result of evolution. Alvin Plantinga is most famous for his evolutionary argument against naturalism, and Victor Reppert has made a similar case with his argument from reason. Both refer to a similar argument made by C. S. Lewis (Plantinga, 1993, chapter 12; Reppert, 2003; C. S. Lewis, 1947). I shall focus on Plantinga's argument here.

Plantinga's argument is that we have little reason to trust our own ability to think if it has evolved naturally, since a naturalistic evolution does not care about truth, only about survival. So if a false belief makes you better suited for survival, evolution will favor the false belief (Plantinga, 1993, chapter 12). The obvious response to this is to suggest that understanding how the world is and works and being able to make valid inferences increases survival fitness. But Plantinga argues that there are several scenarios where false beliefs and invalid inferences produce the same effect as true beliefs and valid inferences. Maybe a person runs away from a tiger because he wants to be eaten by a tiger, but thinks that this tiger will not eat him; or he thinks that the tiger is a cat he wants to cuddle, and thinks that the best way to cuddle a tiger is to run away from it (Plantinga, 1993, pp. 225–226).

Even though such fanciful alternatives can be invented for specific situations, it would be extremely difficult to make a detailed sketch of how a gradual increase of false beliefs and invalid inferences gave more and more survival fitness compared to true beliefs and valid inferences. However, Donald Hoffman has received quite much attention with his "evolutionary argument against reality", arguing that we have no reason to believe that our conscious experiences are anything like the real world since they are evolved for fitness only and not for truth. His point is that our understanding of the world, while still useful, could be an extremely distorted version of the real world. As an example he says that it is useful to just perceive the icons on our PC screen and manipulate them to do things, even if the reality behind the screen is very different from these icons (Hoffman, 2015).

Can we trust our thinking if it is a result of evolution? Several distinctions are useful to answer this question. First of all, this discussion takes place using the reason that we have, and it remains a possibility that that we are deceived without knowing it. This would have been the case even if our thinking was not a product of evolution. Had it been created directly by God, it could still be deceived by sin or Satan or whatever. But even if it is possible that we are deceived without knowing it, it is also possible that we are not so deceived, and we shall consider reasons for believing that we are not. What we are discussing is whether we should be skeptical about the ability of reason, *given that it is a result of evolution.* Does evolution imply that we should be skeptical towards reason?

We should then distinguish between what is a reasonable argument considered as such, and how individuals reason. We know that individuals often reason in poor ways, and that there are some general ways of reasoning which evolution has provided us with, which are often efficient, but quite often also very misleading.[136]

The question of how individuals reason is different from the question of whether we can trust what is commonly taken to be a reasonable way of arguing. Can logically valid arguments be taken as a reasonable means for discovering truths, and what about inductive arguments? Logically valid arguments have proven useful and trustworthy since the days of Aristotle. Inductive reasoning is much more contested, although I think that a good case can be made for inductive reasoning understood as reasoning with the criterion of coherence.

In any case, one can question both deduction and induction and ask whether we can trust in reason understood this way, or whether instead this is just a product of evolution useful for survival, but not actually giving a true understanding of the world. Could the concepts we use in logical arguments be totally misleading when it comes to how the world is structured?

We get a good indication that we must at least approach the truth from the fact that we have well-functioning technology. This is formulated as the no-miracles argument already mentioned by scholars like J. Smart and H. Putnam, who say that we have good reason to believe that our understanding of the world is close to truth, since otherwise it would be a miracle how we are able to fly to the moon, etc. (Ladyman et al., 2007, p. 69). James Ladyman makes the point that many theories have been wrong even if they gave us correct predictions and allowed for development of technology, such as the caloric theory of heat or the ether theory of light. Against this charge, Ladyman replies that even though the theories changed, they retained a structural similarity, meaning that we have good reason to think that the main structures of many of our theories are right even if new concepts may prove better in the future (Ladyman et al., 2007, p. 93).

The no-miracles argument gives us reason to trust in our reason, at least that our reason is on the right track and that scientific methods and criteria are good

136 For numerous examples, see Kahneman (2011).

guides towards truth. We do not know how far away we are from the final truth, and thus we do not know how efficient our reason is. But in any case, it seems clear that a causal understanding of thinking can explain how our reasonable thinking actually occurs in this world.

You could object that we may think the no-miracles argument is good because our reasoning is so flawed, but that is a kind of skeptical possibility which one can always raise, but which we have no reason to believe and which well-working technology gives us reason not to believe. One could still object that given evolution we should distrust reason, but since reason is in fact good, we should reject evolution. Against this, I have described how evolution could give rational thinking, and it is a description which also explains many flaws in our thinking. Overall, I thus find there to be best reasons to believe that a reasonable reason has resulted from evolution. That does not secure me from the objection that I may fool myself, or that we live in a computer simulation, etc., but this is possible no matter what theory of reason you have.

This was the last of the objections on the list, so I conclude that one can have a coherent theory of thinking as rational while still being a causal process. This implies that rationality, as well, is reducible to values actualized according to rules in line with the main picture from Chapter 3. I expect the presentation of thinking in this chapter to be unsatisfactory for many, but in the next chapter I will present in much further detail a theory of how our conscious mind could gradually evolve.

7 Consciousness

In this chapter, we will dig into some really big questions, and I will be proposing several new theories. I start in Section 7.1 by presenting the concept of consciousness, often also called qualia. In Section 7.2 I continue with presenting some of the main problems and theories of consciousness. At the one end of the spectrum of theories you have substance dualism, where consciousness and the physical are two different substances. At the other end, you find reductive physicalism, which rejects that consciousness exists as anything other than something physical. In between, there are different kinds of non-reductive physicalism, property dualism, emergentism, and panpsychism. Since panpsychism has become increasingly popular, I spend some extra time discussing it.

I will be defending a specific kind of non-reductive physicalism, where consciousness is not physical, even if the physical is primary in causing the content of what we are conscious of. A major problem for such a view, but also a problem for all views, is how to understand the causal role of consciousness. Section 7.2 suggests a new theory for how to understand this, and since it is a big and important topic, it covers several pages. This part includes a theory of how the contents of our consciousness could gradually develop through evolution and become what they are.

There are several other difficult questions connected to the topic of consciousness, which I discuss in Section 7.3. The most important one is to look deeper into what consciousness is, ontologically speaking. Here I go into detail on how to think of consciousness as values actualized in a qualia field, which is also a new theory of consciousness – and I present a new theory on the interaction between consciousness and the physical, namely as interaction on the level of field equations.

Section 7.4 deals with other problems of consciousness: how something physical can cause conscious experiences; where consciousness is located; how my consciousness is connected to my body and your consciousness to your body; how subjectivity is possible; and why only the subject has access to consciousness. I start now with defining "consciousness" and "qualia".

7.1 The concept of consciousness

The common understanding of humans and their minds in evolutionary biology is that they are organisms made out of cells. They receive input like light waves and sound waves through their sense organs, and these give rise to neural pat-

https://doi.org/10.1515/9783110770957-009

terns in the brain. In addition, there are consciously experienced images, like the sight of a red tomato, the sound of a trumpet, or the feeling of hunger, and there are thoughts, such as the thought that Hitler died in 1945. But what are these images and thoughts? Where are they? Who is seeing them or thinking them? If you close your eyes and imagine a car (or dream of one), you are not actually seeing a car with your eyes and there is no car in your brain, so where is this conscious experience of a car? A trumpet can make the air vibrate, which again makes little hairs in your ears vibrate, and this gives rise to a consciously heard sound. If some of those hairs are damaged, they can start moving on their own and the brain interprets it as an incoming sound, resulting in tinnitus and constant sound (Matthews, 2000, p. 101).[137] But out there in the world there are only air and hair vibrating, and there is no sound in the brain, so where is this consciously heard sound? And where is the thought that Hitler died in 1945? We might find the neurons that give rise to the thought that Hitler died in 1945, but they are not the conscious thought. Let us assume that the neurons giving rise to the thought that Hitler died in 1945 occupy two millimeters of space in the brain: would anyone say that this thought about Hitler is two millimeters long? And *who* is seeing the car, hearing the sound, and thinking the thought if there is just an organism consisting of a lot of cells? None of the single cells or neurons can see or hear or think, so what is this self that can see, hear and think?

Conscious images are often referred to as qualia, but the term is defined in many different ways. For example, Jaegwom Kim defines it quite narrowly, whereas I follow John Searle and define it more widely to include all mental states with a subjective character, including thoughts (Searle, 2002, p. 40; Kim, 2006, p. 15). I use this wide definition since the key problems seem to be the same: they are subjective experiences requiring a self, they are private and possible to introspect, and they are difficult to locate in space.

A famous definition of qualia is that there is something it is like for a subject to experience them (Nagel, 1974, p. 436). Presumably it is not like anything for a hat to be a hat or for a hat to hear a trumpet sound,[138] but there is something it is like for me to be me or for me to hear a trumpet sound, and the trumpet sound is different from a flute sound because they are different qualia. Qualia defined like this are the same as phenomenal consciousness, the qualitative experiences we are aware of, like sense impressions, thoughts, feelings, and desires.

---

137 The tinnitus example also strongly supports the idea that the brain creates conscious experiences.

138 I comment on the possibility of panpsychism below.

Note the two different aspects that qualia are *like* something *for* someone – there is both the fundamental subjectivity that it is *for* someone and the variety that such experiences can be different, like seeing red or blue or smelling or thinking etc. I return to these nuances later, but the fundamental issue at interest is that there are these different experiences that can be subjectively experienced at all (that they can be *for* someone).

In the following, I will use the term qualia in a very wide sense with the same meaning as mental properties, including both phenomenal perception and intentional states. Qualia here means all the structures you can be conscious of: a red tomato that you consciously experience is a quale, consisting of the qualia red, round, tasty, etc., and so is the conscious thought of a tomato, all of which I think it is like something to have.[139]

7.2 Theories and problems of consciousness

I will now discuss some common theories about consciousness and some of the most common problems they are considered to have. Cartesian substance dualism is the view that there are two different kinds of substances in the world – physical substances and conscious substances. Physical substances have extension; conscious substances do not. Substance dualism is riddled with problems. I will not spend time discussing them here, but will instead list several problems to show why very few philosophers of mind support substance dualism (Kim, 2006, p. 51). How are consciousness and bodies connected? Why does it hurt my non-physical soul when a hammer hits my physical thumb? If souls are non-spatial, how are they connected with the body they control, and why can they only control that specific body and not others? How does consciousness make a physical body move? We know quite well which substances in the brain make the motor neurons in the brain make the body move (for example, acetylcholine), but does a non-physical soul produce acetylcholine or how does it relate to it?

139 While it is common not to include intentional states in the concept of qualia, some do, like for example John Searle (Searle, 2002, p. 40). The view seems to be shared by various neuroscientists, who say that the difference between perceptions and thoughts lie in degree of vividness (Ramachandran and Hirstein, 1999, p. 103; Damasio, 1994, pp. 96–97, 101, 108). Giulio Tononi uses the term consciousness to refer to any experience (Koch and Tononi, 2014, 20:58). By terms such as “conscious mental states” or “consciousness”, I mean the same as qualia in this wide sense.

Descartes thought that consciousness unites with the body in the pineal gland, but nothing is explained by saying *where* it happens. Is the soul a spiritual entity that flies after the body when we walk (or fly in an airplane), or is it stuck in the body? It seems strange that it should fly after my body, but if it is not physical, how can it be connected to the body, or transfer energy to make the body move? When I die, how is the soul released? That a non-physical entity causes physical motion breaks with the principle of conservation of energy, and is not supposed to be possible (Kim, 2006, pp. 46–49). Does substance dualism imply that all insects with minimal mental life have immortal souls? Where does our non-conscious mental life fit in – is it conscious or physical? What happens to the soul when we have dreamless sleep, when our consciousness is turned completely off (Searle, 2011b)?

As shown above, substance dualism has a number of inherent difficulties. We have seen many reasons to believe that the brain causes qualia. These and other reasons have made most philosophers of mind opt for physicalism as an alternative to substance dualism. Many were led to physicalism in order to have a theory that could bridge the physical and the mental, since the mental in itself was then just understood as something physical and mental causation then just like any other event causation: The world influences the brain and the brain makes the body move, and the gap between consciousness and the body has been closed.

However, there are two major variants of physicalism. Reductive physicalism says that everything which exists is physical. This became a popular view at the time of behaviorism, and was adopted by identity theorists who believed that consciousness is identical to brain activity and functionalists who believed that consciousness is a function of the brain. Reductive physicalism has lost popularity in the last 50 years. The reason is that it does not explain the seemingly non-physical properties of qualia, and in many cases rather explains them away by saying they do not exist. Reductive physicalists who are eliminativists, like Daniel Dennett, claim that qualia do not exist, although they obviously do (although it might be that the qualia Dennett rejects are defined differently than how I define them) (Dennett, 1991).[140]

The most popular view in philosophy of mind now is non-reductive physicalism (Kim, 2006, p. 275). There is a narrow and a wide definition of non-reductive physicalism. The narrow view is that consciousness is physical but not reducible

140 The book is often called *Consciousness explained away.* In lectures, John Searle liked to quote David Armstrong, who replied with the following when Searle said he believed in the existence of qualia: "You believe in spooks!"

to brain activity. The wide view is that consciousness is non-physical, and yet the physical has ontological priority. Both views hold that consciousness supervenes on the physical, but supervenience can be understood in a wide or a narrow sense, where the wide sense allows consciousness to be non-physical even if there is no change in consciousness without a change in its physical basis (Kim, 2006, p. 12).

In the case of non-reductive physicalism, the relation between the mental and the physical again becomes difficult to understand, for how can the physical produce consciousness? These are so-called hard problems of consciousness: What is qualia? Why and how does the physical bring forth qualia, and why do qualia become the way they are (Chalmers, 1995)? There are no agreed upon answers to these questions. Non-reductive physicalists who defend non-physical qualia tend to be either property dualists or emergentists. Property dualists hold that physical things can have non-physical properties, like consciousness. Emergentists hold that when physical things relate to each other in certain ways, new properties emerge. For example, wetness emerges when water molecules interact in specific ways, so also consciousness can emerge when the brain behaves in certain ways.[141]

Both property dualists and emergentists still face difficulties, especially in relation to the principle of causal closure. The principle of causal closure says that every physical effect has a sufficient physical cause if it has a cause at all (Kim, 2006, pp. 195–197, 290–291; Papineau, 2009, p. 59). The principle is equivalent to the principle of conservation of energy insofar as there are no other kinds of energy than what physics today recognizes (Papineau, 2009, pp. 55–57). The principle of causal closure is not proved; it is a presupposition with much inductive evidence in its support. The inductive support is that physical causes are usually found for all physical events. Also when it comes to human actions, brain events seem to cause them. As Jaegwon Kim notes, the principle of causal closure is a presupposition of the belief in a complete physics, and so a principle few physicalists are willing to reject (Kim, 1998, p. 40).

141 A distinction is often made between weak and strong emergence. The distinction is made in different ways, but according to Timothy O'Connor the main distinction is whether or not the emergent properties are compatible with physicalism in the sense that they are determined by physical properties (O'Connor, 2020). My own understanding does not fit easily into the established categories because of my understanding of laws of nature taking states of affairs into account, my distinction between content and quality of consciousness, and consciousness having a special causal role to be explained soon. In my understanding, all causal power is in the base, but the base includes non-physical qualia values and laws of nature. It can thus be called either weak or strong emergence, depending on how the terms are defined.

The problem that the principle of causal closure leads to for property dualists and emergentists is the following: Either they allow that consciousness have causal effects, but then they violate the principle of causal closure, or they accept the principle of causal closure, but then consciousness is without causal effect, and that is epiphenomenalism.

Epiphenomenalism is the view that consciousness does not have causal power, it only follows physical processes like a shadow. Most philosophers want to avoid epiphenomenalism since experience seems to support the causal efficacy of conscious beliefs and desires. Further, it seems very unlikely that evolution would have selected conscious processes if they have no effect. Therefore, if you want non-physical consciousness to have a causal effect but do not want to break the principle of causal closure, you end up with qualia being physical, but then the problem is to explain why qualia seem to have completely different properties than physical things. Qualia do not seem to be located, have extension, be composed by particles, be observable from a third-person perspective, be quantitatively measurable, or have other properties we usually associate with being physical, so calling them physical needs justification.

Rejecting the principle of causal closure and opting for emergentism may seem to be the best alternative, but it has other problems too. For how can something physical give rise to consciousness, which is so different from everything physical? When it comes to wetness, solidity and other typical emergence phenomena, we can understand how water molecules moving around each other together constitute wetness or how atoms in certain relations can constitute hardness. Yet there is no analogy for how consciousness emerges from something physical which is so different from itself. For that reason, it seems like cheating to say that it is emergence in all the cases, since the case with consciousness is so different from all other emergence phenomena.[142] Saying that consciousness emerges from the physical is more like giving a name to a mystery than an explanation.

Galen Strawson argues well against property dualism or emergentism as solutions to the problem of consciousness. He concludes that there must be a solution between substance dualism and physicalism. On the one hand, the principle of conservation of energy should not be broken, but then consciousness cannot be a cause outside the physical nexus, and that contradicts substance du-

---

142 Sam Coleman offers the following thought experiment to make a similar point against emergence (and to argue that qualia are basic): Imagine God shuffling colorless tiles in more and more complex combinations. It does not seem possible that such shuffling should ever give the tiles color regardless of how complex it is (Coleman, 2015, section 22).

alism. On the other hand, consciousness should keep its causal power, and then epiphenomenalism should be avoided, which contradicts physicalism.

Strawson thinks that the only possible solution is panpsychism (Strawson, 2006). That is the view that consciousness is a fundamental part of reality, and that all physical things have a mental "inside" or "intrinsic perspective", so that all things are both physical and conscious. Then the mental is part of the physical world so that the principle of conservation of energy is not broken, and at the same time consciousness has causal power because it constitutes an aspect of the physical. Following Strawson's espousal of panpsychism a number of young philosophers have seen it as the right approach to the problem of consciousness (Seager and Allen-Hermanson, 2012). David Chalmers is also amenable to the idea (Chalmers, 1996, chapter 8) as are neuroscientists Giulio Tononi and Christoph Koch (see below).

Panpsychism is a very strange idea, however. It claims that everything is conscious, even elementary particles like electrons. Yet such consciousness must be very different from human consciousness. Does it then deserve the name of consciousness? Nothing suggests that atoms and the like have anything like our consciousness. The problem then is that panpsychism wants to argue that all things must be conscious in order to avoid a big gap between physical things and consciousness. But what remains is still a big gap between some mysterious form of consciousness totally different from ours (which atoms, etc. have) and the regular consciousness that we know, although it may be a smaller gap.

Further, what does the "inside" of these things refer to? Does it make sense to speak of the inside of elementary particles when particles seemingly should rather be understood as excitations in a field than as small containers with an inside? The "inside" is a metaphor, and often other terms are used like how particles are "in themselves" or "intrinsic" or "essential". Tononi and Koch prefer to use the term "intrinsic perspective" (Oizumi, Albantakis, and Tononi, 2014), but these terms are not very clear either. It is often defined as that which would characterize the particle if it were alone in the world, but this does not seem to be a possible picture of a particle, since particles are parts of fields following rules and cannot be what they are without them. If the inside/outside relation or intrinsic/extrinsic relation is not clear, it is not clear how the interaction problem is solved either – it has just been moved down to the micro level. The problematic connection between consciousness and the physical seems just as difficult even if you say that it occurs everywhere and at the micro level.

Yet another problem for panpsychism is the combination problem(s). How do all these small pieces of consciousness combine to give us the kind of consciousness we experience having? And if everything has consciousness, it seems that not only individuals, but also every part of the individual should have consciousness. We know quite well which areas in the brain are active when we are conscious, and they are similar to each other in the sense that they are large collections of intimately connected neurons organized around a gate of informational input. This suggests that a complex structure is necessary before consciousness can arise, so that simple structures like atoms cannot be conscious (Damasio, 2010, pp. 86–87). Further, neuroscience shows that stimulating neural patterns creates effect in conscious experience, which indicates that the content of consciousness is causally dependent on what happens in the brain. These are some of the reasons why famous philosophers of mind like John Searle and Colin McGinn call panpsychism an absurd and ludicrous view.[143]

However, this is still a very general list of problems. To be fair, more specific critique should be pointed against a well-articulated view. Many recognize the theory of consciousness developed by Giulio Tononi and others to be the best theory of consciousness, and it also implies a kind of panpsychism. Since this theory has such a central place in the debate, I will spend a few pages criticizing this theory now, which may be skipped by readers who do not have a special interest in the topic.

The theory of consciousness developed by Tononi and others is called integrated information theory (IIT). It is a theory under development, and it has been presented in new versions with numbers, so the current version as of early 2021 is IIT 3.0. I will now give a brief presentation of the theory, before I offer some criticisms. The theory has five axioms, which are taken to be self-evident descriptions of consciousness. Consciousness is defined as any subjective experience, such as seeing something, feeling pain or thinking a thought (Tononi, Boly, Massimini, and Koch, 2016, p. 450). The five axioms are 1) consciousness exists; 2) consciousness is composed of different parts; 3) consciousness is informative – a conscious experience rules our other possible experiences; 4) consciousness is integrated – one has a unified experience that cannot be reduced to the sum of its parts; and 5) consciousness is exclusive – we only have one complete experience at a time (Oizumi et al., 2014, p. 3).

The five axioms can be abbreviated to:

1) existence,
2) composition,
3) information,
4) integration, and
5) exclusion.

---

143 Searle and McGinn, cited in Seager and Allen-Hermanson (2012).

With these five axioms in place, IIT has five corresponding postulates about the properties that are required for the physical substrate of consciousness. Concerning number 1, existence, IIT postulates that there are mechanisms, which can contribute causally to the system they are part of, such as a neuron in a brain or a logic gate in a computer. They exist in virtue of having causal effects. Concerning number 2, composition, IIT postulates that elementary mechanisms can be combined into systems of mechanisms (Oizumi et al., 2014, p. 3).

For number 3, information, the idea is that the state of system of mechanics can constrain what the possible pasts and futures of the system are. The probability of each possible past and future can be thought of as values on different axes representing the possible pasts and futures. These axes can be thought of as a concept space, and a set of specific probability values for a system is defined as a conceptual structure (Oizumi et al., 2014, pp. 3, 10).

Concerning number 4, integration, IIT postulates that mechanisms only contribute to consciousness if they make an irreducible contribution to the conscious experience. The Greek letter phi is a measure of irreducibility/integration of the system, and it is irreducible/integrated in the sense that you cannot partition it (divide it) without destroying the information in the whole. One can measure the difference that partition makes to the system, and the more strongly integrated it is, the smaller changes one can make without disturbing it. Finally, for number 5, exclusion, IIT postulates that it is the system with the highest phi – the highest level of integrated information, which becomes conscious, and all other parts of the system are non-conscious (Oizumi et al., 2014, p. 3).

With these postulates specified, IIT claims that there is an identity between the informational properties of the system and the phenomenological properties of an experience. More precisely, it is the maximally irreducible conceptual structure which is identical to an experience (Oizumi et al., 2014, p. 3). The term "identity" can be used imprecisely, but Tononi is quite clear: "An experience is identical to a conceptual structure, meaning that every property of the experience must correspond to a property of the conceptual structure and vice versa" (Tononi et al., 2016, p. 452). Unfortunately, the term "correspond" makes it a bit vague whether it means that every property of the experience must be identical to every property of the conceptual structure or whether there is merely a correlation between non-identical properties. But the most straightforward interpretation is that the properties must be identical, since that is what the term "identity" usually means.

This theory is very precise on how to understand panpsychism in detail. It answers many objections by being precise and explaining, for example, why a conscious whole does not have conscious parts (since only the structure with the biggest phi is conscious). However, I still think the theory faces serious objections, which I will present soon.[144] First I have a comment to the arguments that seem to support IIT.

There are many arguments that could be used to support the theory, since it is able to explain many facts and make several correct predictions. But I believe that the reason for this is that they have an advanced theory of information which they use to analyze consciousness, and since information is an important part of conscious experiences, a lot of correct things can be said and predicted. However, I shall argue below that there is no good reason to identify the qualitative side of consciousness with integrated information. Integrated information explains the informative function of the brain, but not the qualitative aspect of experience. Integrated information may even be a necessary condition for a conscious experience to be unified enough to be subjectively experienced, but I see no good reason to argue that integrated information and consciousness are identical. Why not?

IIT claims that a conscious experience is identical to a conceptual structure. Qualia is what it is like to be a conceptual structure (Tononi et al., 2016, p. 451). Does this mean that according to IIT a conceptual structure is an abstract entity existing in an abstract space of possible futures and pasts? It could seem so, for example when Tononi and others say about physical substrates of consciousness (PSC), "Note that the postulated identity is between an experience and the conceptual structure specified by the PSC, not between an experience and the set of elements in a state constituting the PSC" (Tononi et al., 2016, pp. 452–453). Kelvin McQueen is also uncertain about how to understand conceptual structures, and asks what their ontology is since they are located in a high-dimensional space (McQueen, 2019, section 4.1).[145]

However, the point seems to be that a conceptual structure should not be identified with the *elements* that constitute the system, but the specific *state* it is in when the internal relations are a certain way. Tononi specifies that the conceptual structure is physical (Tononi, 2015).[146] We should thus think of a conceptual structure in a brain as a state which parts of the brain are in at a particular

144 In this chapter I have selected some objections to IIT. For more objections, see Bayne, 2018; McQueen, 2019; and Søvik, 2020.

145 McQueen also makes the point that IIT gives no justification for including the possible pasts and futures in their postulates (McQueen, 2019, section 3.1).

146 But it is not entirely clear, since in the footnote after this quote (footnote 28) Tononi writes: "It is intriguing to consider to what extent the physical world has *intrinsic existence* (cause-effect power from its own perspective – in and of itself) in addition to *extrinsic existence* (cause-effect power from the perspective of an observer who can perform interventions on it and sample the results)".

point of time. I understand "being in a state" as meaning that the elements relate to each other in a certain way. Being in that state means that that part of the brain makes certain possible futures and pasts of the same part have certain probability values.

If qualia and conceptual structures are identical in the strict sense, every property should be identical and they should have every part of their structure in common. Instead they seem to have very many different properties and structure parts.[147] On the one hand you have the conscious experiences and their properties. The structure or properties of conscious experiences of color is that they come in degrees of hue, brightness and saturation. The structure or properties of conscious experiences of sound is that they come in degrees of pitch, loudness and timbre. The structure or properties of conscious experiences of taste is that they come in degrees and combinations of salt, sweet, bitter, sour and umami, but this is also very influenced by odor. And so it continues, and odors, feelings and thoughts are far more complex than the examples already given.

On the other hand, you have conceptual structures. In the case of human consciousness, the relevant conceptual structures consist of neurons related in such a way that their action potentials and firing patterns at the given point of time make different possible pasts and different possible futures of that system have certain probabilities. The system being in a state with these probability values are the constellations that are said to be identical with qualia. Maybe they are just a way to express other properties in the relations between neurons that are better thought of as identical with qualia. In any case, these properties seem completely different from the properties of qualia. There is no reason to think that they are the same – and thus identical.

Instead of being identical, the properties of qualia and conceptual structures seem very different and inconsistent with each other. The conceptual structure is composed of neurons which again are composed of elementary particles related to each other in a certain way at a certain location in spacetime (thus it is physical), while many qualia show no sign of being located in spacetime or made of elementary particles at all. Everything about a conceptual structure can be seen and measured from a third-person perspective, while qualia are only accessible from a first-person perspective.

To say that qualia and a physical structure are identical when their properties seem so different is of little value if it is merely a claim without support showing why qualia are nevertheless physical when they seem to lack common

147 "Properties" and "structure parts" refer to the same here.

physical properties. If there is much more evidence of differences than of identity, one should conclude that they are different and not identical. This objection applies regardless of whether a conceptual structure is interpreted physically or non-physically, since the properties of the conceptual structure as it is described in IIT is in both cases different from the properties of qualia.

Another argument against IIT is that it seems that there can be integrated information in the brain without consciousness. I have previously mentioned the card experiment. To recall: Damasio and colleagues performed an experiment where people were asked to draw cards from various decks: some decks were good, i. e. leading to a reward, and some were bad, i. e. leading to a punishment. There was also a system determining which decks were good and which were bad, so that if you cracked the code you could just draw good cards. The subjects played the game while their skin conductance was measured. The interesting thing was that it seemed that the code was cracked non-consciously several minutes before the players understood it consciously and before they started drawing only winning cards; after a while they would get one type of skin response just before drawing from every bad deck, and another type of skin response just before drawing from every good deck. This was so consistent that somehow some part of the brain must have cracked the code, but the person could not consciously tell this and would keep drawing bad cards (Damasio, 2010, p. 276).

This process seems very similar to the kinds of processes that can be conscious, and we have considered several cases of non-conscious thinking, feeling, remembering, etc. (Damasio, 2010). It is hard to see how this could come about with no integration of information, since much simpler systems (like thermostats) are said to integrate information.

Given that there is integrated information in the brain in examples like the one above, this can be used to formulate a dilemma for IIT. According to IIT, if there is integrated information, there is consciousness present. This consciousness would then either have to be part of the conscious experience of the participants in the card experiment, or it would not. If it is not part of their conscious experience but nevertheless is conscious (as IIT implies), then the participants in the card experiment must have a parallel stream of consciousness that they are not aware of.

The dilemma is then the following: Given the presence of integrated information in experiments like the card experiment, and given that integrated information implies consciousness, there must either be a conscious experience of it had by the participants or a parallel conscious experience that the participants are not aware of. If the first option is chosen, the problem is that the theory becomes falsified, because the participants have no conscious experience of cracking the code. If the second option is chosen, the problem is that the theory becomes unfalsifiable because it predicts outcomes that are seemingly falsified, but the theory is rescued by appealing to streams of consciousness that are impossible to verify.

I guess it would be possible to reject the dilemma by rejecting that there is integrated information present in the first place. However, the dilemma can be strengthened. Kelvin McQueen has argued that while activity in the cerebellum is considered non-conscious, there are pockets within the cerebellum with maximal (big) phi spikes that should be conscious according to IIT, but which are never reported by anyone as conscious experiences (McQueen, 2019, p. 150).

The dilemma can then be restated: Should we think of these cerebellum pocket activities as conscious or not? Again: If the first option is chosen, the problem is that the theory becomes falsified, because we have no conscious experience of conscious activity in the cerebellum. If the second option is chosen, the problem is that the theory becomes unfalsifiable. It predicts outcomes that are seemingly falsified, but is rescued by appealing to impossible to verify streams of consciousness.

When it comes to the axioms of integration and exclusion, it does not seem right to say that our conscious experiences are that integrated and exclusive. For example, I can have a unified perception of the scenery in front of me at the same time as I consciously calculate 18 + 18 in my head. These seem to be two very different conscious experiences that are not excluding each other. Should one of them be the most integrated conscious experience and therefore knock the other out?[148]

The whole idea that the most integrated system becomes conscious, also have several problems connected to it. Why is it the case that the most integrated system becomes conscious? IIT explains this with the exclusion postulate which says that only one cause exists (Oizumi et al., 2014, p. 9). However, as argued in the chapter on causality, there are always many causally relevant states of affairs, but we select one cause by setting contrasts depending on our interest. There does not seem to be warrant for the claim that there should be only one cause. The requirement of one cause is said to be "a causal version of Occam's razor", saying in essence that "causes should not be multiplied beyond necessity", i. e. that causal superposition is not allowed (Oizumi et al., 2014, p. 9). This could maybe make sense if there had to be one cause for one conscious experience, but it does nothing to explain why there cannot be many causes producing many different streams of consciousness (Oizumi et al., 2014, p. 9).

148 A similar objection is made by Tim Bayne, see Bayne (2018, pp. 5–6).

After this discussion of IIT, we now we have an overview of some problems a theory of consciousness should try to avoid and some difficult questions a theory of consciousness should try to answer. A huge question is how to solve the problem that qualia seem to have a causal role, but on the other hand this seems impossible. This will be the topic for the next subchapter, and then I deal with other problems afterwards.

7.3 On the causal role of qualia

The question of the causal role of consciousness is a big question with important consequences for how one thinks of persons and minds, and so I spend some time on it. I start by recapping some of the main points from the introduction, expanding some of them, to get the problem at hand into focus.

Many philosophers of mind defend the following six claims:

A) spacetime is physical;
B) causal closure – if a physical event has a cause at t, it has a sufficient physical cause at t;
C) mental properties supervene on physical properties;
D) mental properties are not reducible to physical properties;
E) mental properties are causally efficacious; and
F) there is no systematic overdetermination in the world.

These claims together can be said to constitute the position of non-reductive physicalism (Kim, 2006, pp. 195–197, 290–291).

There is a narrow and a wide definition of non-reductive physicalism, depending on how one understands the terms "supervenience" in C and "irreducibility" in D. A standard definition of supervenience is to say that if A supervenes on B, there can be no change in A without a change in B, so if consciousness supervenes on the brain, there can be no change in consciousness without a change in the brain. However, some will then argue that the supervenient properties must be of the same kind as the base properties, and typically a non-reductive physicalist will say that consciousness is physical. But one could also deny this and think of conscious properties as non-physical, but still hold that they supervene on a physical basis. A good parallel would be that aesthetic properties can supervene on non-aesthetic properties. One could thus be a property dualist, rejecting that consciousness in physical but accepting supervenience, and so be a non-reductive physicalist in a wide sense of the term (Kim, 2006, p. 14).

I will now defend non-reductive physicalism in the wide sense. The supervenience in claim C is standardly defined so that supervenience means that there can be no change in that which supervenes without a change in its basis. But by "irreducibility" in D, I mean that consciousness should be understood as non-physical. Even if we do not know the deepest nature of either the physical or consciousness, the terms have such different and inconsistent properties that it is reasonable to think of consciousness as non-physical until someone can explain how something which seems to have properties incompatible with being physical is nevertheless physical. What it means that something is physical is notoriously difficult to define, but physical things will typically have several of the following characteristics: they are quantitatively measurable, have a location in space, consist of elementary particles, and are observable from a third-person perspective. Consciousness may be quantitatively measurable and located somewhere, but it does not seem to be constituted by physical parts and is unique in how it is only accessed from a first-person perspective. This makes it so different from physical things that it is reasonable to call it non-physical, at least until we learn more about the nature of physical and conscious entities and are able to translate their properties into each other.[149]

The problem is that the six claims seem to be an inconsistent set. There is a conflict between mental non-physical properties with causal effects on the physical world at the same time as it is being rejected that such causal effects are possible. Many philosophers have discussed this, for example David Chalmers, who argues that it is problematic to deny a causal role for consciousness (epiphenomenalism), and to accept it (interactionism), and to reject that consciousness exists (materialism). This leads him to conclude that panpsychism can solve the problem by letting the mental and the physical be constitutively connected at the microlevel (Chalmers, 2016). I reject panpsychism, but will argue that A through F is a coherent set and will focus on how precisely to understand the causal efficacy of mental properties.

There are good reasons to think that qualia have causal effects. At the same time, there are good reasons to think that they cannot. What are the good reasons to think that they have causal effect? The first answer many would give is that they experience it: a conscious desire to raise your arm seems to make your arm go up. But this may be an illusion, and the conscious part of it a mere epiphenomenon.

A stronger argument is that there must be a reason why evolution has selected brains that can create conscious mental states. Consciousness seems to play an important role, and neuroscientists argue that consciousness seems necessa-

149 This means that even if many non-reductive physicalists say that they believe qualia are physical, I find this to be an unjustified claim.

ry in certain cases. According to Dehaene and Naccache, there are three kinds of mental operations that require consciousness: durable and explicit information maintenance, new combinations of operations, and intentional behavior (Dehaene and Naccache, 2001, pp. 8 – 12). In line with this, Crick and Koch suggest that we have consciousness to "produce a single but complex representation and make it available for a sufficient time to the parts of the brain that make a choice among many different but possible plans for action" (Crick and Koch, 1998, p. 98).

This indicates that qualia are not merely an unnecessary epiphenomenon, but rather that qualia, in virtue of being non-physical, have causal effects that have been selected by evolution. Epiphenomenalists will often reply that qualia do not have to have a causal effect, but only be nomologically connected to the brain states they accompany, but the difficulty is then to say what this nomological connection is (Robinson, 2007, p. 29). I will propose such a nomological connection here, but also argue that non-physical qualia have a causal effect.

While Popper and Eccles are often cited as giving the argument from evolution, William James made a similar but slightly different argument long ago (Popper and Eccles, 1977; James, 1890, vol. 1, p. 143).[150] If qualia are mere epiphenomena, it would seem that they could have been like anything. Beneficial behavior like eating or having sex could just as well have been painful if they are merely epiphenomena with no causal role to play. And so the fact that life-threatening events are painful and beneficial events give pleasure seems to be too good a fit for a mere accidental epiphenomenon. It would be a great bonus if a theory of the causal role of qualia also could explain why just these brain states are accompanied by just these qualia, and I will suggest such an explanation below.

As seen, there are good reasons to think that qualia have causal effects, but what are the reasons to think that qualia do *not* have causal effects? There are several reasons. It seems possible to imagine zombies who are like us in all respects, but without qualia. Indeed, neuroscientists are showing that almost any mental process can occur non-consciously, as exemplified in the previous chapters. While there are still some mental processes that are always conscious, this does not show that qualia are causally effective. It could still be the brain states that do all the causal work, even if they are always accompanied by qualia.

Why not just assume that qualia are doing causal work? The principle of causal closure has a very strong hold in science. There is strong inductive support for the belief that if you look for physical causes for events, they will be

150 References found in Robinson (2007).

found. Further, it seems to lead to a whole host of problems to accept that qualia have causal effects. It would be an unknown kind of causal influence which seems to break the principle of conservation of energy. How does it work? What it the causal nexus that combines non-physical qualia with the physical world? How do my qualia influence my body and your qualia your body? I do not go into these problems here since I will suggest a solution that avoids them altogether, but I mention them briefly to motivate a solution that does avoid them. In sum, there are good arguments both for and against believing that qualia have causal effects, and the problem is how to think coherently about the causal role of qualia.

In the following, I will suggest a solution to this problem. In brief, the suggestion is that it is important to distinguish between neural patterns representing qualia structures and neural patterns representing structures in the physical world. Note the important point that I here use the term "represent" again in the minimal sense, where it only means that there is a consistent relation.[151] I use this minimal sense because common and wider senses typically mix together the important differences I want to point out. In a narrow sense of representation, a neural pattern represents a qualia structure in the sense that when this neural pattern in a person's brain is active, the qualia structure will be conscious to the person who has that brain. Further, in the same narrow sense of representation, a neural pattern represents a structure in the physical world if there is a consistent relation between them. This consistent relation can take different forms, but a typical example would be that every time the person sees a tree in the physical world, a certain neural pattern representing the tree is activated.

Since the term "representation" can mean many things and is easily misunderstood, I shall write "representation$_{CR}$" when I mean representation in the narrow sense of a consistent relation. This is a narrower sense than how the term is usually used, and I will return to such a fuller sense of "representation" later. To say that a neural pattern represents$_{CR}$ a qualia structure means that when the neural pattern is activated the qualia is conscious to the owner of the brain. To say that a neural pattern represents$_{CR}$ a structure in the physical world means that there is a consistent relation between the presence of the structure in the physical world and the activation of the neural pattern.[152]

151 This narrow definition is how, for example, Antonio Damasio defines representation (Damasio, 1999, p. 320).

152 Why do I use the term "representation" at all in the narrow and uncommon sense if it is so easily misunderstood? It is to make the reader aware of how something that started as a consistent relation could develop into a representation in a full sense. Since the consistent relation is a connection, I will several times use the term "connected to" in the same sense as "being consis-

Neural patterns have gotten consistently related to qualia structures that are simple and useful representations of the world. While there are many similarities between qualia structures and world structures, the qualia structures are also different from the world structures, and it has had a beneficial effect for brains to be consistently related to qualia structures as opposed to just world structures, especially in choices. For in many choices, we employ in a thought process neural patterns not consistently related to anything in the world yet related to qualia, and they help us make useful choices. Evolution has then selected brains that work well because of how they relate representations$_{CR}$ of qualia structures as opposed to physical structures. I will give different examples of this to show the relevance of distinguishing between neural patterns representing qualia or representing the physical world, and how evolution could select one over the other.

The argument I shall develop will conclude that the causal role of qualia is that the concrete qualia we have (as opposed to qualia in general) have had a causal effect on the evolution of brains (as opposed to being the cause in particular choices), and their causal effect is in virtue of the qualia *content* as opposed to qualia *being qualia* (as opposed to being physical), and the causal form is that qualia are possibility makers – they are part of the structural conditions determining the space of possibilities within which the causal interplay between laws and elementary particles happen. To defend this conclusion, I will first tell an evolutionary story about the evolution of qualia as a solution to the problem of mental causation. After that, I discuss the causal role of qualia in particular. But first, a theory of the evolution of qualia.

How were the neural patterns connected to the qualia they have? We do not know how the brain produces qualia, but there are good reasons to think that it does, since one can in such detailed ways influence conscious experience by manipulating the brain. I will not deal with the problem of how the brain produces qualia here, but save this for the next subchapter. Here I will say something about why neural patterns are connected to the qualia that they are connected to.

Here are some suggestions that I will combine into an answer:

First of all, it does not seem implausible that there are many different qualia which we have not experienced. Qualia have many very different qualities at the same time as they seem to come in different quantities, and it seems we have only gradually through evolution tapped into a much larger potential field of

---

tently related to". When it comes to the causal relations between the connected parts, structures in the world cause neural patterns to be activated and neural patterns cause qualia structures to become conscious to the owner of the brain, but not the other way around.

qualia. If the qualia we experience are the only qualia there are, they seem to have a very unsystematic relationship. Why should there be exactly sound, color and smell in these varieties? This is easier to understand if they are just random samples from a much greater collection, which is more systematic.

Secondly, it seems that brains can turn many different physical patterns into qualia. Some examples: Blindfolded people have had cameras on their head which translated the input to physical patterns touching their back or tongue, and after a while they started having conscious experiences of what the cameras were recording (Sampaio et al., 2001; B. W. White et al., 1970). People have put magnets in their fingertips and gotten conscious experiences of magnetic fields in their vicinity (Berg, 2012). People with synesthesia have different types of qualia mixed up, like seeing colors when seeing numbers or hearing tones (Ramachandran, 2007). Some animals can see ultraviolet light, others can see infrared light, and although we do not know what it's like to be a bat, it seems likely that electromagnetic waves at other frequencies than normal visible light could be consciously experienced by some animals. Some experiments have produced new color experiences in people (Churchland, 2007, chapter 9).

Thirdly, the first two suggestions can be used to form a hypothesis to explain the evolution of qualia. The hypothesis is that early in evolution, neural patterns representing$_{CR}$ physical structures in the world (including the body) were able to activate all sorts of qualia in a very chaotic and un-systematic manner. This may sound implausible since we do not experience that all sorts of neurons can activate all sorts of qualia now, but this has both to do with which qualia have been selected by evolution, as I will come to soon, and also with the fact (that I believe) that there needs to be a high grade of unified integration of qualia before there is an experience of being a subject who can experience and remember qualia (more on that below).

Fourthly, I hypothesize that early in evolution, there were only very simple qualia that could be subjectively experienced by animals, like dark/light or painful/pleasing, but then it became more nuanced like shades and colors, different sounds, different feelings, outlines and features, etc., so that there would be qualia for "big and scary", "big, but less scary", "small, but still scary", etc., making it possible to distinguish more precisely between different events. One support for this idea is that it seems that conscious color perception evolved gradually, with the color blue coming very late (Loria, 2015; Roberson, Davidoff, Davies, and Shapiro, 2006).[153]

---

153 In the article here referred to by Kevin Loria, you can test for yourself whether you can see some specific colors.

Fifthly, we know that some neurons evolved to detect certain features in the environment, like angle, size, movement, contour, color, distance from observer, etc. (Imbert, 2004, p. 39; Pinker, 1997, p. 20). When specific features occurred together in a similar way repeatedly, some neural patterns could represent this as an object to be stored in memory (Barsalou, 1999, pp. 582–583). Sometimes some of the neurons do not work, so people may perceive objects without color or without being able to see them move, what shape they have, or which way they are oriented, but usually this is all bound together into a conscious experience of normal objects (Treisman, 1998, p. 1295). That something is a neural pattern means that they are connected in virtue of synapses which can have different strength according to how easily they activate each other.[154]

Combining these assumptions, the idea is that the brain evolves by having more and more neural patterns that can represent$_{CR}$ the world in a more and more detailed way, and that it evolves into being able to make detailed choices where several alternatives are being reasoned about. Early in this process, neural patterns start to produce qualia that get connected to the brain patterns. While the relation between neural patterns and qualia are chaotic in the beginning, it evolves into a parallel process where both neural patterns are able to give more detailed representations$_{CR}$ of the world and the experienced qualia give a more detailed representation$_{CR}$ of the world.

Why would this happen – that brain patterns connect to qualia which represent$_{CR}$ the world in an increasingly more detailed way – if the brain patterns are doing all the causal work and there is no influence from qualia to the physical world? Having finished the theory of the evolution of qualia, I now return to the question of the causal role of qualia.

I will start by developing an important distinction between a neural pattern representing$_{CR}$ *the physical world* and a neural pattern representing$_{CR}$ *qualia*. While I have read many texts on how the brain represents the world, I have nowhere read anyone systematically reflect on the difference and relation between how neural patterns represent$_{CR}$ physical events in the world and how they represent$_{CR}$ qualia. This is what I will do in the following.

The first point to make is that there is a difference between neural patterns being consistently related to the physical world and being consistently related to qualia. When describing feature-detecting neurons, I mentioned how some important feature-detecting neurons are not representing$_{CR}$ physical entities, but instead qualia which do not exist in the physical world, such as contour and

154 This is known as Hebb's rule, and is often summarized in the sentence "Neurons that fire together, wire together" (Baars and Gage, 2010, pp. 83–85).

color, which are very important in registering objects in the world. Physically, when a person sees an object like a brown horse, there are many different particles reaching the eyes from the area where the horse is standing, but the eyes and brain register only photons of a certain frequency from where the horse is and photons of a different frequency from around the horse (unless the background has the exact same color). Thus, a neural pattern causes the qualia of brown for the color of the horse and a contour of a horse around it.

When a neural pattern is said to represent$_{CR}$ a horse, we should then distinguish between, on the one hand, horse qualia (like a conscious perception of a horse or a conscious thought of a horse) that the neural pattern is consistently related to, and on the other hand the physical events that the neural pattern is related to, such as particles in the world constituting a horse, patterns of photons from a horse reaching the eye, sound waves in the air sounding like the word "horse", or the written text of the word "horse".

There clearly is a distinction to be made between neural patterns representing$_{CR}$ qualia and neural patterns representing$_{CR}$ events in the physical world. Many different physical events can activate the same neural pattern which produces one certain quale. And the exact same physical event can activate different neural patterns representing different qualia, as shown in many optical illusions, where you can either see different things in the same figure (duck/rabbit) or where the same thing looks different in different contexts.

At the same time, the distinction is difficult to present clearly, since all the arguments and examples of things that are different from qualia are themselves experienced as qualia. If I talk about physical particles as opposed to the qualia of blue, our conscious thoughts about physical particles are themselves qualia. And yet we have good reasons to believe that something exists outside of our mind, even if our access to it is as qualia. The distinction I make between qualia and non-qualia is thus a distinction within qualia since "non-qualia" is a concept I am conscious about, but there is nevertheless good reason to believe that there exists something constituted by qualia and something not constituted by qualia, as argued in Chapter 2.

Since there is a difference between neural patterns representing$_{CR}$ qualia and neural patterns representing$_{CR}$ the physical world, we should be more precise than just saying that the brain represents$_{CR}$ the world. Instead we should say that there are neural patterns representing$_{CR}$ (or being consistently connected to) the physical world and neural patterns representing$_{CR}$ (or being consistently connected to) qualia, and while there are many important structural similarities, there are also some important differences.

This is vital, since I will argue that evolution to a large degree has selected neural patterns representing$_{CR}$ qualia instead of neural patterns representing$_{CR}$

the physical world. Again, it is difficult to be clear about the difference, since the neural patterns that have been selected by evolution represent$_{CR}$ qualia that do have structural similarities to events in the physical world, and so it may seem that we should just say that neural patterns represent$_{CR}$ structures in the physical world, with no role for qualia to play. But there are important differences as well, and these are the ones we must consider so as to see the causal role of qualia. In order to be precise about this, I will now take a closer look at the concept of a structure.

As presented in Chapter 2 of this book, a structure is an ordered pattern of parts made into a whole. We find structures in the brain, in qualia and in the world. Neural patterns at one point of time are structures, where neurons connect to each other. Groups of neurons can connect with other groups to form bigger constellations of connected neurons. Neural patterns are also structures when they form a process of neurons activating each other, relating causally as events being part of a larger process. Qualia at one point of time are structures, such as the conscious experience of a blue ball having the qualia of round, blue and soft as parts, which can again be part of a larger qualia structure like the conscious experience of a boy with a ball. Qualia processes over time are also structures, where different events happen in a stream of consciousness as parts of the whole event.

Physical structures in the world seem at the basic level to be extremely complex in consisting of very many parts. Qualia structures are massive simplifications of what is going on in detail at the causal level. There are numerous particles interacting in what we consciously perceive as a boy kicking a ball or what we think of as the beginning of WWII. Qualia structures simplify enormously by sorting the world into mainly colored objects with certain shapes interacting with each other.

Over time qualia structures have become more and more complex. In the evolutionary beginning, there were maybe just experiences of light or dark, pain and pleasure, mom and not-mom, etc., but then qualia perceptions have become more nuanced and qualia thoughts are ever expanding. And while thoughts expand, we still grapple with understanding the basic parts of the physical world, like particles, forces, laws, spacetime, superposition – all of which we understand in quite vague metaphors even if they can be described in precise mathematics.

Summing up this brief overview of structures, we see that there are causally connected neuronal structures in the brain; there are many kinds of qualia structures connected in many different ways in consciousness (including some unique structures like subjectivity and intentionality); and in the world outside of mind and qualia the basic structures are poorly understood (we do not know things as they are in themselves independently from our experience). What I am arguing is that from quite early on, evolution selected brain patterns consistently related to qualia which were structurally similar to the physical world, although very simplified and with many structural differences from the structures in the world.

This is a difficult argument to make, for if neural patterns got consistently related to qualia which were structurally similar to events in the physical world, is it not easier just to think that neural patterns got consistently related to these events in the physical world, with no role for qualia to play? Why should we think that evolution selected neural patterns consistently related to qualia which were structurally similar to events in the physical world instead of just selecting neural patterns consistently related to events in the physical world? It seems very clear that evolution has selected neural patterns consistently related to qualia, but the question is why.

We have seen that physical structures are different from qualia structures, and that they are much more complex. Qualia structures are simplifications of physical structures, but so would a neural pattern have been if it were consistently related to a much more complex structure in the physical world. And of course, it seems that many neural patterns can be consistently related to complex events in the physical world without being conscious. Qualia structures being simpler than physical structures can then not be the whole answer to why evolution selected neural patterns producing qualia.

In addition to being simplifications, qualia structures also have different parts and different relations than what can be found in the brain and the rest of the physical world. Early in evolution, animals would experience simple qualia relations, like "something big approaching is scary", or "something small and moving can be eaten". More advanced animals would divide the world into objects pushing and pulling each other. A lot of causal inferences can be made out of such qualia relations without being the real story of why motion occurs like it does. For example, we tend to think that two things cannot occupy the same space, that hard objects cannot move through each other, that heavier objects fall faster, that vacuum can suck something in, etc., and in everyday life these are useful assumptions to make even if they are not accurate descriptions according to physics, and one can make useful inferences or predictions based on them.

As qualia move from concrete objects to more abstract relations, they can simplify the world enormously and be very useful for survival, even if they refer to non-existing entities and relations. Some examples: We can make inferences about other people's behavior based on reasoning about love or the value of money, or even as complex as reasoning after WWII that dehumanization can lead to genocide and that we should prevent this by insisting that all humans have the same dignity and certain human rights. When learning to drive, I learned that tires have *road grip*, which can be used for three things – breaking, accelerating or turning – but if you break and turn at the same time, there may not be enough road grip for the tire to use. And of course, I can make inferences from something falling under the category of intentionality or qualia, which seem to be clear examples of neural patterns being consistently related to a qualia structure where no corresponding structure exists in the physical world. Qualia structures do not have to give survival value individually, but in general, and in their interrelations they should have enough structural similarity with the world to be useful in guiding behavior, although these examples show that there are also great differences.

There is no point in qualia having a structural similarity with the world unless the neural patterns consistently related to the qualia can be used in reasoning and guiding behavior. In order for that to work, there has to be a structural similarity also between neural patterns and qualia structures. As a crude hypothesis based on what was written about thinking above, I believe that the structural similarity between neural patterns and qualia structures is the following: in a complex qualia structure, the neural patterns representing$_{CR}$ the qualia parts are parts of a neural pattern whole representing$_{CR}$ the whole complex qualia. For example, neural patterns for red, round and tasty are connected in a neural pattern for tomato; while neural patterns for tomato, strawberry and cherry are connected with a neural pattern for red, etc. Such neural patterns can be very complex, for example different typical events connected with a neural pattern for being angry.

While there is this similarity between qualia structures and neural structures, there are also important differences. A conscious perception or thought can express a lot of relations immediately in virtue of the qualia being what they are, for example a girl being taller than a boy holding a blue ball. Presumably the brain must have neural patterns for each part – girl, boy, next to, taller than, blue, round, etc. – and these neural patterns have generally one kind of connection, namely a synaptic connection, and one kind of relation, namely a causal relation. Although there is probably more variation in brain pattern relations, qualia seem to be much more varied in the kinds of connections they express between the parts in their structures, especially as we move upwards to complex qualia thoughts like “confederal”, “religious”, etc.

The brain can make inferences from one neural pattern to another and translate it to useful actions because the neural patterns are consistently related to qualia which are structurally similar to the world. I have tried to argue that it is mainly neural patterns consistently related to qualia structures that are being employed in brain inferences, as opposed neural patterns being consistently related directly to world structures, even if there is a great overlap.

This relation of neural patterns to qualia structures as opposed to physical world structures gets more and more important as brains become more advanced, and especially as it starts making choices. In simple animals with simple brain mechanisms, a neural pattern can be consistently related to a physical pattern with no need for qualia, and it can be selected by evolution in virtue of being a small change generally useful for all. An example would be for a worm to eat or not eat things which have certain molecules in their odor.

As brains get more advanced, neural patterns can relate to very complex qualia structures like the thoughts of "politics" or "just war" to be used in reasoning processes or deliberation. The ability to deliberate what to do is very useful, since it is very context-dependent what is wise to do in a particular situation. Evolution can select some general mechanisms that are generally useful, but what is best for an individual in a concrete situation differ greatly. Is it wise to run or hide or try something new? That depends on, among other things, whether there is anywhere to run or anywhere to hide in that particular situation.

A person deliberating whether to run or not is juggling in her mind different brain patterns that are consistently related to certain qualia, and these qualia are related in a way that is structurally similar to how physical structures in the world are related, so that inferences made based on neural patterns representing qualia give conclusions relevant for action in the physical world. This allows the person to run through different scenarios in her mind before acting in the way that presumably is most beneficial to the person. Note then that in many choices, we employ in a thought process neural patterns not consistently related to anything in the physical world, but related to qualia, and they help us make useful choices. This shows the relevance of distinguishing between neural patterns representing$_{CR}$ qualia and those representing$_{CR}$ the physical world, and how evolution could select one over the other. It is easy to think that the neural patterns are just representing$_{CR}$ the physical world and not qualia, but that is because we think of the world by thinking about concepts that are qualia, while there is good reason to think that the non-qualia-constituted world is very different from our experience of it.

Such advanced deliberation processes are much more efficient than how evolution usually works. Normally, evolution will produce beneficial results one step at a time, which get selected because they are useful for many: A process where A leads to B is selected because it is generally useful. But in deliberation there can be a thought process in the brain where neural pattern A leads to B, which leads to C, which leads to D, before action E, and action E is the action that has an effect which is relevant for survival and can be preferred by evolution.

This process can then happen with brain patterns representing$_{CR}$ qualia which are massive simplifications of structures in the physical world, or even qualia not representing$_{CR}$ anything in the physical world, as exemplified above. Even in complex choices, blind neurons make the inference and execute the action, but they need the qualia structures matching the world in order to be useful and lead to survival. The reason is that the neural patterns in essence work by connecting parts and wholes with desires. For example, neural pattern A is part of neural pattern B, where A is followed by C, which is connected with neural pattern D representing desire, and since the current situation is like neural pattern A, action C should be activated in order to achieve desirable goal D. In order for this process to be beneficial, the neural patterns must be connected to something, and I have suggested that the neural patterns are connected with qualia which are similar to world structures as opposed being directly connected to the world structures themselves.

Since this is easy to misunderstand, I will use an example to clarify. Imagine that we put into the world a huge number of robots that can evolve in a way similar to humans: they have neural networks that can be stimulated by the world and move them around, and they must find limited power sources in order to continue functioning and making copies of themselves. In addition, they have a very special thick screen on the back of their heads with many layers of points that can actualize different and unique kinds of values in various degrees. Somehow, different activation patterns in the neural network of the computer activate different values in the points on the screen, but what is shown on the screen does not interact with how the robot works.

As robots evolve from random trials to processes that actually make them survive, some neural patterns stabilize in how they work. Then some patterns also stabilize on the screen. Since an important part of robot survival is that their neural networks represent what is going on in the world, we can expect there to be some structural similarities between what happens in the world and what happens in the neural networks, and consequently also some structural similarity with what happens on the screen. As robots get better and better at representing the world, we can expect that the screen patterns match patterns in the world more and more as well.

So far, whatever happens on the screen seems merely a shadow of what happens in the robot, with no function or role to play whatever. If the robots had just been filming what they registered in the world and sent the output to the screen, that would have been the case. But there is an important point to notice. As the robots evolve at a quite complex level where they are able to think and choose, they are not just making neural patterns that directly mirror an external event. They combine old neural patterns and make new neural patterns which are not directly related to anything in the outside world, but are directly related to the patterns they produce on the screen.

Some of these screen patterns are structurally interrelated in ways that have some useful structural similarities with the world. Sometimes when the robots combine neural patterns which have these usefully structured screen patterns and act in ways that are useful according to these screen patterns, this is beneficial as there is a useful similarity between the screen patterns and the world. And over time they are selected, and the screen patterns become more and more structurally similar to the world and more and more efficient.

The screen may seem superfluous since the neural patterns inside the robot guide the robot actions. But the neural patterns have been selected because the screen patterns they represent have structural entailment relations that are useful for guiding actions in the world, even if the neural patterns themselves are only blind causal processes. The neural patterns are just different combinations of physical parts and wholes used in causal inferences, so the neural patterns interacting in a deliberation process and finally leading to action could not have been so efficient and useful had they not represented screen patterns with entailment relations matching the world in a useful way. The neural patterns just make the robot move in a certain way or send sound waves through its mouth, but this fits into screen pattern representations of the world.

A short addition in the end, with some new topics that we will return to: When these neural patterns have evolved, they work also when the screen goes black, which is comparable to non-conscious actions in humans, but they had to evolve first and would not have evolved (or at least not as efficiently) if these screen patterns with their relation to the world did not exist at all. And maybe some neural patterns are consistently related to – and only functioning when – the screen is turned on. For there may be structures which are constituted by the screen being on, or constituted by how on-screen structures are integrated into larger wholes. The relevant comparison is, of course, how unified and integrated qualia may well be constitutively necessary for the structures of intentionality and subjectivity.

One more toy example to drive the point through: Imagine a world with objects and beings all built from Lego bricks. In this world, there exist no Duplo bricks, which are big bricks that look like Lego bricks but combine a bit differently. You can combine eight 2 × 4 Lego bricks into the shape of a 2 × 4 Duplo brick. Imagine further that in order to survive and reproduce, these Lego beings need to be able to build certain Lego structures, which look like structures that could also have been built by Duplo bricks. Maybe, for example, there were holes in a wall with certain shapes where you had to put in a matching Lego structure in order to survive. Being able to build advanced structures at the right time (to fit into complex holes in the wall at certain times) is more beneficial than just being able to make simple structures (to fit into simple holes permanently in the wall).

Imagine then Lego beings with random Lego structures in their Lego brains which, through evolution, had gotten consistently related to conscious images of Duplo bricks in different orientations. When needing to build a certain complex Lego structure, these Lego brain structures would represent quite simple Duplo brick operations, like one horizontal Duplo and one more on top facing me, which could be translated into signals making the arms move so as to build Lego structures with the same shape as the Duplo structure.

The Lego brain structures representing$_{CR}$ Duplo bricks look nothing like Duplo bricks, but Duplo bricks are structurally similar – although also different – to more complex Lego structures. Because these Duplo bricks are structurally similar to complex Lego structures in the world, it is easier to use representations$_{CR}$ of them in a mental process testing out different structures to find one matching the one needed at a certain time. The Lego brain structures representing$_{CR}$ Lego bricks would not have evolved unless they actually represented$_{CR}$ something (Duplo bricks) which gave a simplified version of how the world worked.

Now we should be ready to answer in detail some questions about qualia. Many qualia structures are different from world structures, although many have also become more similar as we have understood the world better. Long ago, for most humans the quale for "earth" entailed "being flat", and inferences were made accordingly. Now, the quale for "earth" entails "being round as a ball" and has structurally become more similar to that which it represents in the world. From "earth" and "roundness" ancient people could infer "no edge to fall off of" and dare sail against the horizon.

The fact that different kinds of neuronal activity activates different kinds of qualia at all is a happy accident like other basic interactions in nature. In other words, it is the kind of basic interaction that happens between laws and what they regulate. But when it first happens, evolution has selected neural patterns consistently related to qualia. Why did evolution select neural patterns consistently related to qualia? Because they have useful structural relations that allow for efficient navigation in the world. Was the existence of qualia necessary for humans to evolve? Human bodies could have evolved to interact with the world in quite complex ways, but it has happened much more efficiently with the existence of qualia and their useful structural interrelations. Especially advanced and flexible choices, which require a process with several steps of inference before action, made good use of neural patterns connected with qualia which matched the world well (and increasingly better).

Why were exactly these qualia selected? It is because they have structures that are useful in being simpler than, but partly structurally similar to, the world. They could have been a little different, so for example the neural pattern connected to the qualia of blue could probably have been connected to the qualia of brown, but not to the qualia of itching. It is useful to represent$_{CR}$ something as a brown bear or a blue bear, but not to represent$_{CR}$ something as an itch every time photons of the brown or blue frequency hit your eye. Since qualia enter as parts in different wholes, they will be more efficient in a variety of cases the more they resemble world structures. We can thus assume that qualia will match the world increasingly better over evolutionary time, and probably also that it started evolving quite early since it (presumably) fits so well today.

How should the causal role of qualia be precisely stated? It is neural patterns which do the causal work in every particular situation. But it is neural patterns which happen to be connected to certain qualia which have been selected. They have been selected because qualia which are structurally similar to, but are simpler versions of, world structures have allowed neural patterns to make more efficient and beneficial inferences and to guide action in beneficial ways.

This means that qualia are not the causes of particular choices, but they have a structure which explains why the neural patterns that connected to these qualia were selected by evolution. The existence of qualia structures that neural patterns have connected with has had a causal influence on which neural patterns were selected in evolution, not in virtue of qualia being qualia (as opposed to being physical), but in virtue of the qualia structure (i. e., content). Even if other, partly different, qualia could have done the same job, it happened to be these that did the job.

How exactly should we understand the form of causation that qualia structures have on which neural patterns were selected by evolution? Compare this with the question of how we should understand the causal role of mathematics when our brains have evolved to use mathematics in interacting with the world, or how we should understand the causal role of mathematics if a calculator was to evolve into a calculator that we humans would like to use. If we selected calculators that got better and better at calculating, what would be the causal role of mathematics to explain why we chose those calculators? Mathematics would be what the internal processes in the calculator was matching and explain their increased survival value. It would be a part of the surroundings defining what gives increased survival value. The numbers used by the calculator could have been slightly different – and divided into bases of perhaps 8 instead of 10 – but it had to be structurally similar to what they are today in order to be efficient.

Pi is not a physical structure in the world that brain can get consistently connected to, but pi is a useful quale for the brain to be consistently connected to (or, if you think that pi is a physical structure, substitute the example with the square root of minus 1).

What is then the causal role of the surroundings that give an evolved trait an evolutionary advantage? It is the structure within which evolution happens. Terrence Deacon uses the term "teleological causation" for this; he argues that even if you can explain everything about the parts of hemoglobin and how it works in terms of physics, it does not explain its *shape*, which has become the way it is because of the space of possibilities that evolution has explored (Deacon, 2006, pp. 138–146).

Against Deacon, one could argue that these surroundings are also physical structures interacting with other physical structures and so do not need to be thought of as a special kind of causation. And it is true that the structures within which evolution happens are most often physical structures, like water, playing a part in the causal interplay between laws of nature and elementary particles. Yet some of the structures within which evolution happens also shape and influence how motion occurs in the world, but they are not part of the energy balance in the universe. These are the more fundamental structures of the world determining what is possible, like the truthmakers for the geometrical structures of space and – I suggest – qualia.

The causal role of qualia is that they are possibility makers. They are not part of the interaction between laws of nature and elementary particles, but they are part of the structural conditions determining the space of possibilities within which the causal interplay between laws of nature and elementary particles happen. Whether they then deserve to be called "causes" depends on whether one has a wide or narrow definition of causes. If causes are anything that make motion happen as it does, qualia are causes. If causes must fall under a physical law of nature, qualia are not causes. If we think of possibility-making structures as a deeper set of metaphysical laws (the fundamental rules for actualization of values), the roles that qualia have do fall in under a law. I argued in the chapter on causation that causation expresses lawful connections, and it is not unreasonable to call the laws themselves causes (as in "the law of gravity causes the apple to fall faster over time") since we ask for causes to understand why things move as they do, so qualia could well be called causes in this wide sense of the term.[155]

---

155 Fred Dretske makes a distinction between triggering and structuring causes, and the examples he offers fit well with the distinction I am making here (Dretske, 2010). However, he does not offer any exact definitions of the terms, whereas I have tried in the chapter on causation and in this chapter to unpack in detail what causation is.

Qualia do not have an effect on the energy balance in the universe, and they do not threaten the search for causes in physics. In this sense we can keep the causal closure principle, which says that if a physical event had a cause at *t*, it had a sufficient physical cause at *t*. But we must then specify what is meant by the term "sufficient" and emphasize that it is sufficient when we ask about a cause at a certain time, *t*. When we ask for a cause of something at time *t*, we ask why something happened, as opposed to not happening, given the context and conditions at time *t*. As argued in the chapter on causality, such contrasts and presuppositions are crucial to understanding causation. When we ask why something happened as opposed to not happening at time *t*, there is a sufficient physical cause in the sense that we only need to refer to physical events to explain why something happened as opposed to not happening at that time.

In cases involving qualia, we do not need qualia to explain why something happened as opposed to not happening at a certain point of time, since this can be sufficiently explained by the action of neural patterns. But if we want to know why there are *these* neural patterns doing the causing as opposed to *other* neural patterns, we need qualia to explain this. Evolution selected neural patterns connected with certain qualia as opposed to other qualia because these neural patterns were evolutionarily more efficient because they were connected to qualia that represented the physical world in an evolutionarily beneficial way. The qualia did not cause physical changes in the physical world at a point of time by moving something physical, but they were part of the structural conditions that made some physical processes more fit for survival than others.

The principle of causal closure can thus be kept, but the term "sufficient" should not be taken to mean that non-physical qualia have no role to play at all. It only means that they do not play a role in explaining why something happened as opposed to not happening at a certain time and place. I am not using the term "causation" differently when speaking of mental causation from when I speak of causation in the causal closure principle. The point is rather that causes must always be selected given certain contrasts and presuppositions, and when we ask for the causal role of qualia, they are not necessary to explain why a certain physical event involving qualia happened as opposed to not happening in a particular brain context, but they are necessary to explain how such brain contexts came to be the way they are in the first place.

To conclude on the causal role of qualia: The causal role of qualia is more precisely stated that the specific qualia we have, like colors and sounds (as opposed to qualia in the sense of anything that could possibly be consciously experienced), have had a causal effect on the evolution of brains (as opposed to being the cause in particular choices), and their causal effect is in virtue of their qualia structure (their content as opposed to being qualia at all), and the causal form is that qualia are possibility makers: they are part of the structural conditions determining the space of possibilities within which the causal interplay between laws and elementary particles happen. If that is to be called causation or not depends on your definition of cause, but it is not a part of the normal causal interplay between laws of nature and elementary particles, even if it does fall under the deepest set of metaphysical laws, and so it is causal in a wide sense of the term.

To conclude on the whole problem of mental causation: I have suggested a way for qualia to have a causal effect (in a wide sense of the term) while retaining sufficient physical causes for any particular physical event at a time *t*. This solves the problem that it seemed like we had to choose between a causal effect of qualia (to explain their evolution) and a sufficient physical cause for physical effects. Neural patterns have gotten consistently related to qualia structures that are simple and useful representations of the world. While there are many similarities, the qualia structures are also different from the world structures, and it has had a beneficial effect for brains to be consistently related to qualia structures as opposed to just world structures, especially in choices. This is their nomological connection.

Since non-physical qualia have this causal relevance, it implies that non-physical ideas influence human behavior and choices in the world. Everything that happens is not determined by physical laws only, and so this is an argument against physical determinism thus understood. This is an understanding of mental causation allows us to see how humans can have free will even if every choice considered on its own is based on brain processes that follow the laws of nature.[156]

With this evolutionary explanation of why qualia are the way they are, can it also be explained why evolutionarily beneficial actions feel good while life-threatening things feel bad? If it is just a question of simpler structures being more efficient in choices, it seems that eating could still have felt bad and being cut by a knife could have felt good. So why do things that are good for survival feel good and those bad for survival feel bad?

Here is a suggestion: Early in evolution, different beneficial processes evolved, such as "if this and that molecule enters your mouth, spit it out". There would be no need for consciousness in such processes. As the numbers of such processes increased, it would be efficient to gather them into a more general process: "if good, approach/continue; if bad, avoid". Those who connected such a neural pat-

156 I am grateful to Svein Jåvold for helping me see this point clearly.

tern with processes that were good for survival would then outcompete those who acted in an opposite way. Over time, it would be more nuanced – divided into good and better, bad and worse – which helped prioritizing.

There is nothing controversial so far. But it seems that all this could also happen non-consciously, in the sense that there could be a neural pattern standing for good/approach/continue and a neural pattern standing for bad/avoid without there having to be good and bad qualia connected to them. How did they come to have qualia that feel good and bad connected to them if neural patterns have randomly gotten connected to their qualia? It seems that evolutionarily beneficial actions could just as well have felt painful.

To answer this, note the very interesting point that the following question seems unanswerable: In virtue of what do good feelings feel good, and in virtue of what do bad feelings feel bad? What is it that makes a good feeling feel good? If you try to explain what is good about a particular good feeling, you will probably cite other good feelings. For example, you might like to be warm because it makes you relax or have good childhood memories, but these are all good feelings, so what is it that make good feelings feel good?

It seems that the only answer we can give is that they just feel good. In virtue of themselves, so to speak. Let us assume then that the neural patterns for good and bad in the beginning just randomly connected to the qualia they did, and maybe differently in different brains. I said that over time, animals who connected the neural pattern for good + approach with evolutionarily beneficial things would prosper. As it divided into good and better with degrees of desire, those who desired evolutionarily beneficial actions the most would prosper the most. This means that over time, those would survive who found food, sex, etc. very good, since presumably, they made greater effort to experience the things they found good. More and more evolutionarily beneficial actions would be grouped together as good, and as better and better over time.

This explains why a group of evolutionarily beneficial actions are considered by us to be good, even if it does not explain what the goodness of felt goodness is, but maybe there is not more to explain than this connection which was random from the start. We do not have an independent measure by which we can ask: do qualia that feel good really feel good? We can only note the fact that there are some qualia that we experience as good and bad without being able to explain further what makes pleasure feel good and suffering feel bad.

However, there may be more to say as well. Above, we saw how qualia over time get a more nuanced relation which helps with making evolutionary beneficial choices. In a similar way, it may well be that relating the conscious self and qualia that are experienced as good (in increasingly different nuances) by this self is a similar kind of process where an increasingly coherent set of qualia relations help making evolutionarily better choices. In other words, maybe there is a kind of qualitative qualia value that can take on quantitative values along a scale of good and bad, and through evolution neural patterns and actions have become coherently connected with this scale in the way that the more desired and beneficial things feel better than the less desired and less beneficial. The same could happen with bad feelings in the way that the more things are feared and unwanted the worse they feel. With this, the long discussion on mental causation is done, and I am ready to move over to other hard problems of consciousness.

7.4 Other hard problems of consciousness

Above, we have started to answer some of the hard problems of consciousness: How can consciousness have a causal role, why are qualia the way they are, how could qualia evolve? However, more remains to be said about these questions, and other hard problems have not been answered: What is consciousness, how can something physical cause conscious experiences, where is consciousness located, how is my consciousness connected to my body and your consciousness to your body, how is subjectivity possible, and why does only the subject have access to consciousness? These are great and unsolved questions, but I will nevertheless attempt to suggest some coarse-grained answers to them that fit well with what else has been said in this book. I attack the questions in the order given above.

What is consciousness? We know that there are different physical values that can be actualized in different points in physical fields. Some of these are known values that, combined, give us our physical world, which is mainly composed by up and down quarks, electrons and bosons. Several other particles are known, but seem to have no interaction with the physical objects we know from our world. In addition, there seem to be many unknown types of values today only called dark matter and dark energy.

Summing up, there seem to be many known and many unknown physical values which come in degrees and can combine in numerous configurations. If we ask what fundamental values like charge, spin, mass, etc. are or are composed of, or what makes them be like they are, we know little or nothing about this. There are some interesting similarities between physical values in fields and qualia. Qualia can be understood as different values coming in different degrees at different places, which can combine into numerous configurations, but we know little to nothing about what the fundamental qualia values are composed of or what makes them be like they are, and there can be many which are unknown to us.

The idea I am suggesting is that conscious experiences are composed of qualia values actualized at different places in a qualia field. Below we shall look at how they combine into subjective experiences. This way of understanding the constituents of conscious experiences is the same as found in panqualityism, which is to be distinguished from panpsychism. The idea is not that there are many small, conscious, subjective particles in the world, but that there are small non-experienced qualia values that can be combined into subjective experiences.[157]

Sam Coleman is a good defender of this view and argues that you cannot make a big subject out of small subjects, but that instead there must be unexperienced phenomenal qualities which together can constitute a subject. For example, he imagines that there can be a particle of the quale red which can be part of a conscious experience of red. As supporters of the view that there can exist phenomenal qualities without subjects to experience them, he lists Lockwood, Rosenthal, Foster, Rosenberg, Unger, Leibniz, and Hume.[158]

Is it possible to imagine qualia being composed of some basic values? It seems quite plausible when it comes to light, sound and taste. The conscious experience of light comes in a color spectrum with variations in degree of hue, brightness and saturation (Churchland, 2007, p. 163). The conscious experience of sound comes in degree of pitch, loudness and timbre.[159] The conscious experience of taste come in degrees and combinations of salt, sweet, bitter, sour and umami, but is also very influenced by odor.

The conscious experience of odor is trickier, since it seems that we can distinguish between numerous odors, but without identifying some basic odors, since combinations of odors just smells like one odor (R. J. Stevenson and Attuquayefio, 2013). However, a team of researchers have analyzed a wide range of odors and suggested that they are composed by ten basic odors. These are fragrant (flowers, perfumes), fruity (non-citrus fruits), citrus (citrus fruits), woody/

157 I noticed after having completed the manuscript that Coleman once in passing suggests that maybe there is a field of qualia (Coleman, 2015, section 59), and the IIT theory of consciousness thinks of qualia as structures in a conceptual space, but the idea of a qualia field as developed in this book is my own.

158 Coleman (2012, p. 149), referring to Lockwood (1989), Rosenthal (1991), Foster (2000), G. Rosenberg (2004), Unger (2006), Leibniz and Strickland (2014), and Hume, Selby-Bigge, and Nidditch (1978).

159 Pitch comes mainly from physical frequency, loudness comes mainly from intensity, and timbre comes mainly from time variation and spectral content (Wolfe, 1992).

resinous (tree, grass), chemical (ammonia, bleach), sweet (chocolate, caramel), minty (eucalyptus, camphor), toasted/nutty (popcorn, almonds), pungent (blue cheese, cigar smoke) and decayed (rotting meat, sour milk) (Castro, Ramanathan, and Chennubhotla, 2013).

Emotions also seem to come in many different varieties. At the same time there seems to be a scale of some main pleasurable emotions and some main emotions of displeasure, which then come in more and more nuanced versions. W. Gerrod Parrot has made a tree of emotions starting with love, joy, surprise, anger, sadness and fear, where every emotion has new sub-emotions (Parrott, 2001).

Conscious experiences are complex since they influence each other. Taste is influenced by smell, but also by how we feel and what we think, as proved by numerous blind tests. For example, wine connoisseurs will be very influenced in their judgments if expensive wine is served in cheap bottles or vice versa. Conscious experiences seem to influence each other back and forth in different ways, and can thus combine into numerous combinations even if the basic components may come from a small list. How we identify an emotion will often be due to what we are thinking, while it is difficult to take away the cognitive component and describe just the emotional difference between, for example, jealousy, envy, *schadenfreude* etc.

"Thoughts" seems to be the most complex category of all, impossible to reduce to a list of basic components. However, as seen in the presentation of the grounded cognition model, the idea is that all our concepts are based on perceptions. If this is right, then thoughts can be reduced to combinations of the other qualia components listed above. Very many words will be reducible to colored areas experienced as an object (e.g. cat, man), and relations between objects at a time or over time (e.g. pet, dad, zoo), while of course feelings and other sensations are also often an important part of the meaning of a concept.

Concerning the question of what consciousness is, I suggest the following answer: Consciousness is a set of fundamental values that can be actualized in points in a field we can call the qualia field. All qualia can be reduced to combinations of some basic qualia values. Like in the case of basic physical values, I do not know anything more about these values other than that they seem to be basic components of the possibilities of our world.

Is it possible to reduce the mentioned fundamental qualia values further down to variables of a few most fundamental qualia values? I noted in the chapter on evolution of qualia how qualia seem to vary on scales and even to overlap, such as deep bass sound turning into a feeling. Sam Coleman notes the same examples, and refers to Charles Hartshorne for a defense of a continuum hypothesis of qualia, but without suggesting what the fundamental qualities are.[160]

It seems to me that many different qualia can be understood as falling in under a broad category of feeling, namely tasting, smelling, hearing and what we usually call feeling. Then seeing seems to be of a different category than feeling, while thinking is a combination of feeling and seeing. One can imagine a qualia space with axes to combine the different qualia, but I have no detailed suggestion on which axes go together to combine all experiences.

Why should then a set of values be grouped together and called a qualia field while other values are grouped together as physical fields, instead of combining them into one? It is because the physical values have in common that one of their qualities is that they can influence motion in the world, while the qualia values have in common that one of their qualities is the basic subjectivity that makes it possible for them to combine in configurations that can be experienced as the experience of a subject. The physical field is a motion field while the qualia field is a subjectivity field, each grounding the fundamental features of a world of motion and subjectivity. This capacity to influence motion and the capacity for basic subjectivity are two of the fundamental values in the set of values that exist.

It is common to argue that consciousness comes in degrees (Lee, 2020). That seems to support a physicalist approach to consciousness as opposed to making consciousness unique at a basic level. However, defenders of consciousness coming in degrees rarely distinguish between the content and quality of consciousness. It is easy to think of the content of consciousness as coming in degrees and giving examples, e. g., of a vague, tired experience. But the quality of consciousness is the fundamental subjectivity that an experience is for someone at all, and this does not come in degrees. I have argued that subjective parts can constitute a full-blown subject, but at both levels the fundamental subjectivity (for-ness) is already there. I have never seen anyone give a plausible example of something partly subjective that does not presuppose fundamental subjectivity from the start.

I now move onto the next question on the list: How can something physical produce conscious experiences? From physics we know that fields can interact with each other. An excitation in one field can create or annihilate excitations in other fields. For example, an excitation in the photon field can create excitations in the quark and anti-quark fields, which again create an excitation in the gluon field. The so-called Feynman diagrams describe such interactions.

160 Coleman (2016, pp. 264–265), referring to Hartshorne (1934).

Interaction between fields happens also at higher ontological levels. For example, we can have interactions at different points in a temperature field and a wind field where the wind can influence the temperature and the temperature can influence the direction and strength of the wind at various points. Presumably, physical fields at higher ontological levels do not cause changes in the fundamental physical fields. Rather, the systematic interaction between configurations at higher ontological levels – such as molecules interacting with each other – can be ontologically reduced to interaction in the fundamental physical fields. (Note that if physical fields at higher ontological levels actually can interact with the fundamental physical fields, this would not weaken the suggestion I am about to make, but rather strengthen it, since it would allow a neuron field to activate a qualia field.)

Now we can use field interaction theory to suggest an explanation of how something physical can produce conscious experiences. The brain is a neuron field. The activity in the parts of the brain that causes consciousness is presumably ontologically reducible to excitations in the fundamental physical field located at every time at the same area as the brain is located (even if the owner of the brain is walking around). Physical activity in the brain, either at the neuronal level or at the fundamental level, can activate excitations in qualia fields, either at the fundamental qualia level or at higher levels of configurations of qualia values. Such interactions were chaotic and unsystematic early in evolution, but as described in the previous subchapter, some interactions have become stabilized between neuronal pattern configurations and qualia structure configurations.

The variables and internal relations in the qualia fields are largely unknown to us. But it certainly seems to have potential for further exploration through research. As mentioned above, the body can be stimulated in many creative ways to create new conscious experiences. This should be (and presumably is, even if I have not seen any research on it) systematically explored through systematic variations of stimuli to register what happens in the brain and in consciousness and then systematize which variables change. However, it is not the physical stimuli to the body that causes activity in the qualia field; it is the brain activity which occurs between the physical stimuli to the body and the experienced qualia. So how does brain activity produce qualia?

I do not know how brain activity activates qualia structures, but there should be a great potential for discovering this through systematic research: Which changes in brain activity occur with a systematic correlation to changes in qualia experience? What is the gradual change that occurs in the brain as a certain quale changes gradually, for example as a sound grows louder? There is of course a lot of research on the neural correlates of consciousness, but I do not know of anyone who has explored this kind of field approach where the relevant gradual changes may occur at a more fundamental level than the neuronal level.

This way of thinking about the interaction between the brain and consciousness seems to suffer from the same classical interaction problem as all other theories of consciousness. How can something physical interact with something so different as consciousness? This theory has a new resource for solving the problem. Remember something very interesting about interaction between fields in physics: in order to describe interaction between fields, we combine their field equations to get the quantum state and then the probability of finding a certain interaction. There is an interaction at the level of rules for probabilities of events that nature follows. If there are rules for how physical fields behave and rules for how qualia fields behave, it does not seem like an impossible interaction that these rules could be combined, just like other rules combine at a mathematical level. We usually have a problem in understanding the causal nexus that could relate two very different things, but here they are both related in the same causal nexus at the level of rules.

Certainly, all physical changes imply changes in energy, and this seems to be a problem if physical fields influence qualia fields, since there are no signs of energy disappearing from physical fields to qualia fields. However, we do not know what equations the qualia fields follow. I am not suggesting that physical atoms use energy to push around soul atoms. All that is needed is that the qualia field rules take into account the field equations for what happens in physical fields without the rules demanding exchange of energy for a change to happen in the qualia field.[161]

This qualia field theory of consciousness does not have anything like the equations found in quantum field theory. But theories of consciousness do not usually come with equations (IIT does, but suffers from great problems already mentioned). A theory of consciousness that can solve difficult typical problems is a serious candidate for a theory of consciousness.

The next question on the list is the question of where consciousness is located. A typical field interaction means that an excitation in one field at one place creates a change in another field at the same place. But it does not have to be an interaction at a certain place. One field may create changes in the whole state of

161 Anthony Peressini argues that a theory of qualia should explain why qualia share so many non-qualitative properties with physical things, like for example temporal, spatial, modal, causal, and other relationships (Peressini, 2018). I take it to be another advantage with the field approach to qualia is that it explains why qualia have so many common properties with physical things, namely because they are all values actualized in fields.

the other field which influences the probability of different things happening at different places.

If qualia are combinations of values actualized in qualia fields, then they are located at a place in the physical world. The most obvious place to assume they are is at the same physical place the neuronal activity that activates them is. This is an understanding which would avoid Descartes' problem with connecting the immaterial soul to the body. The cells in my brain are a result of field activity in the area where my brain is, even if I am walking around. Likewise, the conscious configurations in the qualia field that I experience can be a result of physical field activity in the area my brain is located, even if I am walking around.

This would answer also the question of how my consciousness is connected to my body and your consciousness is connected to your body. This could seem like a tricky problem if consciousness was located somewhere other than our bodies, but the problem is avoided here. My consciousness is connected to my body because my consciousness is a result of activity in my brain and located the same place as my brain, while your consciousness is a result of activity in your brain and located at the same place as your brain. The way the connection happens is as described above that fields interact at the level of rules.

I do think that qualia are actualized at the same area as the brains that actualize them, but the area where a person's qualia are actualized could be extremely small, while the qualia themselves can seem big. For example, your conscious experience of seeing a big mountain could be constituted by actualized qualia values located at an extremely small place somewhere in the same area as your brain.

The next question is: How is subjectivity possible? Note that we are not talking about something being epistemically subjective, like whether coffee tastes good, but rather about something being ontologically subjective in the sense of being experienced by a subject. Conscious experiences seem to be experienced by someone. I take that to be the essence of defining qualia in terms of "what it is like", namely something is like something *for someone*. After all, a frog *is like* a toad, and a viola *is like* a violin, but this is not the kind of "what it is like" we are after. Nor am I after what red is like as opposed to what blue is like, but instead that it is like something at all *for someone* to experience something consciously. Some distinguish between what-it-is-like-ness and that-it-is-for-ness, and this is useful sometimes, but when discussing the hard problem of consciousness or subjectivity I take them to mean the same (which Nagel originally did as well).[162] Subjectivity is the essence of consciousness but hard to de-

162 Sam Coleman distinguishes between what-it-is-likeness and that-it-is-for-ness to point out

fine in ways other than by saying words with the same meaning: subjectivity, phenomenality, for-ness, what-it-is-like-ness, consciousness, and qualia.

To explain subjectivity by referring to a subject who experiences gives an impossible regress called the homunculus problem. The homunculus is a little person inside our head who explains our subjectivity, but then how to explain the subjectivity of the homunculus? They seem to need their own little homunculus, and so it goes.

This problem has led many to think that the subject must be made from fundamentally subjective parts: subjective experiences are fundamental and the subject is a combination of these. William James famously said that the thoughts themselves are the thinkers (James, 1890, chapter 10). Antonio Damasio says that the person watching the movie in our head is part of the movie (Damasio, 2004, p. 11). Pete Mandik says that the subject is an experience deduced from other experiences (Mandik, 2001).

The view presented by James, Damasio and Mandik does not explain basic subjectivity in the sense of showing what it is made of and how it comes to be the way it is. Rather, they presuppose basic subjectivity and use it to explain the subjectivity that we feel as persons with selves. While not being a full explanation of subjectivity, at least they avoid the homunculus problem.

Basic subjectivity seems to be one of the fundamental values in the world, along with the other basic values of which we can say nothing more than that they exist and then describe them as well as we can. Of course, there may be a yet unknown explanation, but it is hard to see what it could be. The most common candidates are either to say that consciousness and subjectivity come from proto-consciousness or proto-subjectivity (Damasio, 2010, p. 253), which is not very clarifying, or to speak of loops (like representations of representations) (Hofstadter, 2007), which is not very explanatory either.

Sam Coleman thinks that phenomenal qualities can bond together and constitute a subject. The phenomenal qualities are proto-subjective in the sense of not being subjective until they combine into a configuration that becomes subjective. He seems also to think that there must in some sense be a representation of the representation (Coleman, 2012, pp. 159–160). I agree with the first part,

---

that there are both different qualities to be experienced and that there is someone experiencing (Coleman, 2015, section 88). Anthony Peressini makes a similar distinction and a distinction between what it is like *for a mental state* and what it is like *for an enitity*, but he also refers to Nagel's original paper "What It Is Like to Be a Bat" where Nagel (on p. 436) specifies that he refers to what-it-is-like *for the organism* (Peressini, 2018). When I use traditional what-it-is-like terminology, my interest is in the subjective for-ness – that something is like something for someone at all.

but think of the qualia field as a field of basic subjectivity where fundamental qualia values can combine into coherent wholes that have a conscious experience of being subjects of experience.[163]

If we assume basic subjectivity as a building block that describes a part of what qualia are, then qualia can be combined to constitute the kind of subject that we experience being. Being a subject is like having a bubble around our head which is always filled with a unified visual experience when we look around – or dark when we close our eyes – and there is a body with feelings connected to it as well. We can experience sound, smell and taste coming from different directions; we can experience feelings in the body; and we can have thoughts which do not feel like they are located at any particular place (although often roughly in the area around our head).

It seems important that the qualia we experience are integrated into a unity in order for us to have a subjective experience at all. The brain will often merge inconsistent input into a unified whole, but if the input is too different to unite, we will instead switch between different experiences. For example, you will see either a duck or a rabbit when looking at the duck/rabbit drawing, and if very different input is sent to each of your eyes (called binocular rivalry), instead of seeing a mix you will switch between seeing one thing for a few seconds, then another (Leopold and Logothetis, 1996).

A simple explanation of the importance of unified conscious experience is to say that the experience of being a subject is constituted by such experience. The qualia structure in the qualia field is the experience, including the subjectivity of the experience. Visual qualia combine in a way that feels like a subject is watching them. Other senses combine in a way that feels like being a subject with a body experiencing them. Thoughts are combined into a narrative, feeling like the life of a subject. In sum, the feeling of being a subject with experiences is a unified configuration of qualia values being actualized at a place in qualia space without there being something else that *has* this experience of which this experience is *for*. The qualia configuration is the subject, the core self, the stream of consciousness. And then the qualia values being actualized are caused

---

163 David Chalmers and others have raised a zombie objection to the idea of a subject being constituted by qualia values, namely that we imagine such a combination of values happening without a subject being constituted. However, as Sam Coleman notes, we have no idea of how this work, so imagining that it could happen or not is just a clash of intuitions with little or no argumentative force (Coleman, 2012, pp. 161–162). While it is not logically necessary that a configuration of qualia values should become subjective, it may well be metaphysically necessary (in the way I define "metaphysically necessary").

by a brain in a particular body in a way that makes “owning that body” part of the feeling of being that subject.

What binds together the conscious configurations of qualia that are experienced as being subjects? As mentioned in the chapter on the evolution of qualia, brains have actualized various patterns of qualia values through evolution, where some have turned out to be evolutionarily useful. These have become coherent wholes actualized in the same areas as different brains are, which are experienced as the conscious subjective experiences of a body. It is the laws of nature actualizing physical values, which again actualized qualia values, that binds together the configurations of qualia values. The binding is nothing else than parts being located such that a coherent pattern results.

With this understanding of subjectivity, we can answer the question: why does only the subject have access to its own consciousness? The question seems to presuppose that on the one hand you have subjects and on the other hand you have consciousness, and we wonder why the subject can only access their own consciousness instead of that of others. The theory I present here says instead that the subject, with its access to consciousness, is constituted by the consciousness it experiences. The feeling of being a subject at all is created moment by moment as new experiences become conscious and unified into a whole, but it feels like being the same subject over time since it is connected to the same body over time and can be conscious of memories stored in the brain of that body, especially of what just happened a second ago.

These facts are enough to explain the relation between subjectivity and consciousness access. Maybe in the future we can draw neuron cables between brains to let me have a conscious experience of your memories, etc., so we would have to specify more exactly what it means for one subject to have access to the conscious experiences of another subject. However, the theoretical apparatus here presented should suffice to do that job.

That was the last question on my list of hard problems of consciousness. A lot needs to be discovered by future research about how the brain produces which qualia values in what way and according to what rules. This is merely a coarse but coherent theoretical framework that neuroscience can use to fill out the many missing details, which of course may show that big changes in the theoretical framework are required.

8 Free Will

8.1 Introduction to the problem of free will – main positions and problems

To show how it deals with relevant problems, a new theory of free will should relate to typical challenges that other theories of free will face. I will now present the most common theories and the problems they face, and use this presentation as a structure for presenting my own theory of free will. There are three main questions that a theory of free will should answer: What does "free will" mean? Do we have free will? Is the proposed theory coherent, given that free will seems incompatible with both determinism and indeterminism?

What does the term "free will" refer to? There are many different definitions of free will, and one should ask which understanding of free will is being considered, whether it is being affirmed or rejected. It may well be that a strong form of free will is rejected while a weaker form is affirmed. In ordinary language, a minimum requirement for what it means to have free will is to say that it is the freedom persons must have in order for it to be meaningful to hold them responsible for their actions. (Like free will, though, responsibility is understood in many different ways.)

A common definition of free will is to say that

1) it is "up to us" what we choose between several alternatives, and
2) the source of the choice is *in* us, not outside of us or in something else that we cannot control (Kane, 2011, p. 5).

Even if this description does not apply to every choice, most people will say that they experience such free will at least in some of their choices. Still, both parts of even this definition are contested and can be further defined in various ways: What does it mean to be the source of a choice? Are alternative possibilities necessary for free will? How should such alternative possibilities be understood? Some philosophers say you have no free will, others say you have free will in a weak sense of the term, and still others say you have free will in a strong sense of the term. A way of formulating the problem of free will is thus to ask how strong a degree of free will humans have.

Among those who affirm free will, compatibilists and libertarians disagree on whether free will is compatible with the idea that all events are determined. Determinism is here defined as physical determinism, which is the view that previous physical causes plus the laws of nature determine one future with physical necessity. At any point of time, the rest of the content of history is then implied

https://doi.org/10.1515/9783110770957-010

by the state of the world at that time, which means that there is only one physically possible content of the future. Such physical determinism will be a focus in this book, since I believe that that is the strongest and most common challenge to the question of free will. This is a metaphysical position, and this is what I mean by "determinism" in this book.[164]

In the discussion on determinism, compatibilists believe that determinism is compatible with having free will. For a long time, the most common critique of compatibilism was the consequence argument. Roughly, this argument says that if determinism is true, then what happens in the future is determined by laws of nature and events that took place previously, even in the distant past. Even before any humans existed it was determined what the content of the future would be for all humans. Since the future was thus determined before our birth, it cannot be up to us what happens among different alternatives, and thus we cannot have free will (Van Inwagen, 1983).

Despite this widely debated argument, compatibilists believe determinism is compatible with having free will. Compatibilists today will usually say that it does not matter that only one specific future is physically possible. Rather, they will focus instead on what the inner mental life of an agent must be like in order for the agent to be free. One strand of contemporary compatibilism is the so-called *mesh theories*, which hold that a person is free when she has the right connections or "mesh" between internal parts of her mental life.[165] Another strand is *reasons-responsive theories*. According to these theories a person is free when her actions are based on a rational response to reasons for action.[166] These different compatibilist understandings of free will do not require indeterminism, so free will is argued to be compatible with determinism and in no need of alternative possibilities.

However, an argument other than the consequence argument has been in focus lately against compatibilism, and that is Derk Pereboom's four-case manipulation argument. This argument presents four cases, from a clear manipulation case to a deterministic world, where the point is to show that there are no rele-

---

164 Note that determinism does not necessarily mean that we will ever be able to predict the future.

165 For example, there are hierarchical mesh theories, such as Harry Frankfurt's theory. Frankfurt argues that we have several desires whose object is an action or a state, which he calls first-order desires. But we also have desires whose object is a first-order desire, and these he calls second-order desires. The second-order desires are internal responses to the first-order desires, which one may like or dislike. According to Frankfurt, we are free when our second-order desires approve our first-order desires, because only then do we have the will we want (Frankfurt, 1971).

166 See for example Wolf (1990), or Fischer and Ravizza (1998).

vant differences between the cases. Since the first case clearly seems to be a case of no free will, the charge is to explain the relevant difference between case 1 and case 4.[167]

167 Here are the four cases, quoted from Pereboom (2014, pp. 76–79):

Case 1: A team of neuroscientists has the ability to manipulate Plum's neural states at any time by radio-like technology. In this particular case, they do so by pressing a button just before he begins to reason about his situation, which they know will produce in him a neural state that realizes a strongly egoistic reasoning process, which the neuroscientists know will deterministically result in his decision to kill White. Plum would not have killed White had the neuroscientists not intervened, since his reasoning would then not have been sufficiently egoistic to produce this decision. But at the same time, Plum's effective first-order desire to kill White conforms to his second-order desires. In addition, his process of deliberation from which the decision results is reasons-responsive; in particular, this type of process would have resulted in Plum's refraining from deciding to kill White in certain situations in which his reasons were different. His reasoning is consistent with his character because it is frequently egoistic and sometimes strongly so. Still, it is not in general exclusively egoistic, because he sometimes successfully regulates his behavior by moral reasons, especially when the egoistic reasons are relatively weak. Plum is also not constrained to act as he does, for he does not act because of an irresistible desire – the neuroscientists do not induce a desire of this sort.

Case 2: Plum is just like an ordinary human being, except that a team of neuroscientists programmed him at the beginning of his life so that his reasoning is often but not always egoistic (as in Case 1), and at times strongly so, with the intended consequence that in his current circumstances he is causally determined to engage in the egoistic reasons-responsive process of deliberation and to have the set of first- and second-order desires that result in his decision to kill White. Plum has the general ability to regulate his actions by moral reasons, but in his circumstances, due to the strongly egoistic nature of his deliberative reasoning, he is causally determined to make the decision to kill. Yet he does not decide as he does because of an irresistible desire. The neural realization of his reasoning process and of his decision is exactly the same as it is in Case 1 (although their causal histories are different).

Case 3: Plum is an ordinary human being except that the training practices of his community causally determined the nature of his deliberative reasoning processes so that they are frequently but not exclusively rationally egoistic (the resulting nature of his deliberative reasoning processes are exactly as they are in Cases 1 and 2). This training was completed before he developed the ability to prevent or alter these practices. Due to the aspect of his character produced by this training, in his present circumstances he is causally determined to engage in the strongly egoistic reasons-responsive process of deliberation that issue in his decision to kill White. While Plum does have the general ability to regulate his behavior with moral reasoning, in virtue of this aspect of his character and his circumstances he is causally determined to make his immoral decision, although he does not decide as he does due to an irresistible desire. The neural realization of his deliberative reasoning process and of the decision is just as it is in Cases 1 and 2.

Case 4: Everything that happens in our universe is causally determined by virtue of its past states together with the laws of nature. Plum is an ordinary human being, raised in normal circumstances, and again his reasoning processes are frequently but not exclusively egoistic, and sometimes strongly so (as in Cases 1–3). His decision to kill White issues from his strongly ego-

Alfred Mele has a similar argument called the zygote argument: Imagine a goddess creating a zygote at exactly the right time and place with the exact right structure in a deterministic universe. She does this because she knows that the zygote will then become a man (Ernie) who at an exact point of time will do something the goddess wants done – for example, kill his grandmother. Ernie will be, like any other person in a deterministic universe, considered by compatibilists to be free, but many will have the intuition that he was not responsible for killing his grandmother since the goddess had planned things so that this had to happen. Yet, since he is like any other person in our world if the world is determined, it seems that if he is not responsible, no one else is either (Mele, 2006, pp. 188–189).

Even if one disagrees over how strong the manipulation argument and the zygote argument is against compatibilism, I think there can be little doubt that, if the future is already determined before we are born, we do not have a strong form of free will. It is not up to us to change the future into anything other than what was already determined before we were born. Libertarians, on the other hand, think that we do have a stronger form of free will than this. They hold that we can be the source of our choices in a more fundamental sense than what compatibilists will allow, but that requires an indeterministic world where different futures are possible and where it is up to us to influence what the future will be like.

There are three main positions among libertarians distinguished by how they understand the causality involved in free choices. *Non-causalists* believe that free actions are not caused at all, but are intelligible in the light of the purpose of the action. *Agent causalists* believe that there is a unique and irreducible kind of causation that only free agents can employ. *Event causalists* deny that actions have special causes, but believe instead that all causes are of the same kind: they think that events cause events, both in the mind and in the world in general.

Those who defend the strongest form of free will are the non-causalists and the agent causalists. Non-causalists argue that human action should be explained by intentions or reasons instead of causes, and that these are not reducible to ordinary event causes.[168] A classic charge against this view was leveled by

istic but reasons-responsive process of deliberation, and he has specified first- and second-order desire. The neural realization of Plum's reasoning process and decision is exactly as it is in Cases 1–3; he has the general ability to grasp, apply and regulate his actions with moral reasoning, and it is not because of an irresistible desire that he decides to kill.

168 An example of such a theory can be found in Ginet (1990).

Donald Davidson (Davidson, 1963). He pointed out that even if a person has a reason for doing something, that does not mean that his reason is what actually caused the event to happen. People often experience having competing reasons for doing different things when they act. The challenge to non-causalists is to explain what links the personal reason to the action. Agent causalists like Timothy O'Connor hold that agents are enduring, irreducible substances that have a unique ability to perform actions (O'Connor, 2011). Agent causalists are typically criticized for appealing to both a mysterious agent and a mysterious form of causation, which does not fit into the ordinary scientific world view (Pereboom, 2014, pp. 65–69). Nor do they explain how reasons make actions happen, for in virtue of what does the agent control her actions? Agent causation can be argued to be an irreducible phenomenon,[169] but I shall argue later that it is superfluous to add anything to normal event causation. This will be my main argument against non-causal and agent-causal libertarian theories: that our behavior can be explained sufficiently in event-causal terms so that there is no good reason to believe in extra agency or causation beyond that.

In my view, the most plausible of the libertarian theories are the event-causal theories. Event causalists hold that mental events can cause free actions. There are two main event-causal theories, distinguished by where in the deliberation process they locate indeterminism.[170] The advocates of *centered* event-causal theories believe that there is indeterminism until and in the moment of choice, whereas advocates of the *deliberative* event-causal theories hold that there is indeterminism early in the deliberation process, creating different ideas in the mind (alternative possibilities), but that the rest of the deliberation process is determined.[171]

Since event-causal libertarian theories are close to my own theory, I shall spend a little time in presenting them here. I start with the centered event-causal libertarian theory of Robert Kane. Kane thinks that free will is fundamentally about the ultimate source of action being in us (Kane, 2007, pp. 13–14). More precisely, the requirement is that "*To be ultimately responsible for an action, an agent must be responsible for anything that is a sufficient cause or motive for*

169 For example, E. J. Lowe argues that both causality and agent-causality are irreducible concepts. See Lowe (2002, chapters 8 to 11).

170 Indeterminism simply means that more than one future is possible, so when I and others I refer to speak about the location of indeterminism, the point is to speak about the source of indeterminism: where do the indeterministic effects arise?

171 The distinction between centered and deliberative event-causal theories is from Clarke (2003, pp. 57, 71).

the action's occurring" (Kane, 2007, p. 14).[172] This means that the requirement of alternative possibilities is not necessary for all our actions. But it is necessary in some specific early choices in which we formed our own characters, according to Kane. He calls such actions self-forming actions, or SFAs. Even if one could not have done otherwise in some situations, one is still responsible if the reason that one could not do otherwise was earlier SFAs. For example, if you have formed your character through SFAs so that it is now impossible for you to lie, you are still free, responsible and praiseworthy for not lying in situations where you could.[173] As long as we are free to make some SFAs, we can be free, but if the world is determined, then none of our actions are SFAs and then we are not free.

But this seems to result in an infinite regress, for would not those earlier choices depend on even earlier choices, and so on indefinitely (Kane, 2007, pp. 19–20)? Kane's response to this criticism is that the regress is ended if there is an action in the agent's past that lacked sufficient motive.[174] There could be a situation in which the agent did not know what to do and so did not set her will before the action occurred; then the action would set the will in the very act of choosing. Kane calls such actions "will-setting actions" (which are the same as self-forming actions). He adds that in order for such actions to provide us with free will, they must have been such that the agent could act voluntarily, intentionally and rationally in more than one way when she acted. If the action happened as an accident, it would not have made the agent the ultimate source; rather, the accident would be the source. But if the agent had a motive for both alternatives, then she is the ultimate source of the choice no matter what she chooses, so the regress stops there (Kane, 2007, p. 20).

Even if such a choice is undetermined, Kane still thinks it can be a rationally willed choice. To argue this, he offers the example of a businesswoman on her way to an important meeting who witnesses an assault. In this situation she has reasons to stop and reasons to move on, and she does not know what to do. The conflicting motives stir up a chaos in the brain, which is sensitive to undetermined events at the micro level of quantum mechanics. In this situation the woman must make an effort to choose and, no matter what she chooses, it will be for a reason. When she decides, that decision sets her will (Kane, 2007, pp. 26–28).

172 Emphasis in original text.

173 "Could" must here be understood in the sense that it was type physically possible.

174 According to Kane, we have a "sufficient motive" for doing something when our will is set one way on doing so before and when we act (Kane, 2007, p. 19).

The most common critique of Kane's theory is that it runs into a problem of luck. Let us say there is a 70 % probability that Jack will decide to have pancakes for breakfast. If history were rolled back a hundred times and played again up to the moment of choice, Jack would decide to have pancakes 70 times and something else 30 times. But if the exact same history up to the moment of choice can give completely different choices – which Robert Kane argues it can (Kane, 2007, p. 23) – it seems to be a matter of luck as to what Jack decides to do. The same point can be made with identical worlds up to the moment of choice, where Jack 1 and Jack 2 make different choices.

Kane's theory is a *centered* event-causal theory since it locates indeterminism in the moment of choice. Deliberative event-causal theories try to reduce the luck component by locating indeterminism at an earlier point in the deliberation process. Such models are also called two-stage models since the deliberation process comprises two stages: First there is an indeterministic stage in which alternatives for actions are generated in the mind, then this is followed by a deterministic stage in which one alternative for action is selected.

One of the best such proposals is Alfred Mele's *daring soft libertarianism.* Mele argues that whereas the other models shun luck and only include indeterminism to avoid determinism, this model embraces luck while still maintaining that the agent can be in control (Mele, 2006, p. 117). The point is that the deliberation process is indeterministic, so it is partly a matter of luck what the agents end up choosing. But the agent learns from experiences over time and the agent's evaluations of these experiences influences how likely it is that the same choice will be made later. In this way, the agent shapes her own character over time. This also explains why we hold children less responsible than adults for what they do (Mele, 2006, pp. 122–123, 131–132).

Neil Levy argues that libertarians can try to reduce the luck component by making their theories almost compatibilist (Levy, 2011, p. 77). But he does not think that Mele's strategy of including luck in the history of an agent works, since luck has been a part of every choice and you cannot solve the problem of luck with adding more luck (Levy, 2011, p. 89). Even compatibilists have a luck problem, according to Levy, since it is also a matter of luck what such agents come to think about or desire (Levy, 2011, p. 90).

In addition to the luck problem, there are two other important arguments against event-causal theories. First there is the problem of the disappearing agent. It seems that in event-causal approaches, choices reduce to desires, beliefs and bodily movement, and the agent disappears.[175] If everything is just natural causal processes occurring, where is the free agent? A second and similar charge against event-causal libertarianism is the regress problem, since it seems that we can follow causes backwards further and further to before the agent can make an ultimate choice (Strawson, 1994, pp. 5–7). Then the agent cannot be the ultimate cause of a choice if there must be a cause for why the agent chose as she did.

In addition to this list of philosophical problems come the challenges from neuroscience. Three kinds of findings are particularly relevant. The first finding is that of Libet-style experiments showing that consciousness seems to enter the stage after the brain has already determined what a person will do (Libet, Freeman, and Sutherland, 1999). More advanced experiments let researchers predict (better than chance) how people will act several seconds before they make their choice based on watching brain scans of the test persons (Haynes et al., 2007; Soon, Brass, Heinze, and Haynes, 2008). The second finding is of confabulation-type experiments and related kinds of experiments showing that our own conscious interpretations of our actions are often wrong. Confabulation means that persons are wrong about the real reason for their action. This has been demonstrated clearly in split-brain patients (Gazzaniga, 2005), but also among people in general, typically in examples of choice blindness (Johansson, Hall, Sikström, and Olsson, 2005; Hall et al., 2013). Other kinds of experiments show that non-conscious factors often influence our behavior without us being aware of it (Schnall, Haidt, Clore, and Jordan, 2008). The third finding is that reductionist theories of mind seem to explain all parts of human choices and actions (Damasio, 2010; W. Singer, 2004b). Brain processes are physical processes guided by the laws of nature, and there is no need for concepts like persons with intentions choosing between alternatives and controlling the outcome.

As we have seen, the different kinds of compatibilisms and libertarianisms run into different problems. This has led various philosophers to conclude that we do not have free will, since free will is incompatible with both determinism and indeterminism (Pereboom, 2014). But most philosophers still hold on to the idea that we have free will, since this seems best to fit our experience of having free will and being responsible for our actions. Now that we have an overview of the main positions and main problems, I am ready to locate my own theory as a newcomer to this map and indicate how I will relate to the different problems.

175 This is mentioned as a main objection against event-causal theories in, for example Pereboom (2014, pp. 31–33) and Steward (2012, p. 62).

As a brief preview, I will suggest that free will and responsibility come in degrees. A person can be involved in her choices to varying degrees from when a desire immediately causes an action to where an independent self causes an action. I use the theory of the self presented in Chapter 5 to argue that the self can gradually cause its own content over time and thus be the cause of itself and the cause of a person's actions; this way the person is free in the sense of being the ultimate source of her choices. This presupposes a specific understanding of causation as contrastive (presented in Chapter 4) and that the world is indetermined at the macro level of human interaction (to be defended below).

In more detail, the chapter is structured as follows: I start with the topic of determinism and indeterminism in Section 8.2 to show how I deal with the manipulation argument and the zygote argument. I argue that the world is indetermined at the macro level of human interaction and that this is required for free will in a strong sense of the term. I then move on to the topic of causation in Section 8.3 to show how I deal with the regress problem. After this, I move on to the topic of person, self and mind in Section 8.4 to show how I deal with the problem of the disappearing agent. I argue that people are involved in their choices to varying degrees, and that agents do not disappear even if they get explained in a more fine-grained way. I continue to show how free will can be built gradually via how, over time, the self can be the cause of itself and of actions. This is a more detailed response to the regress problem.

After a short discussion of responsibility in Section 8.5, I show how this theory responds to the problem of luck (Section 8.6), the question of weakness of the will (Section 8.7), and some other objections (Section 8.8). The problem of free will is related to many big questions and so cannot be fully defended in one chapter. At several points I must refer the reader either to other chapters or to further details in a book I have written about free will (Søvik, 2016).

8.2 Determinism and indeterminism

We saw in the introduction that free will can be understood in many ways and that some defend a strong version of free will while others, like the compatibilists, defend a very limited version of free will. According to compatibilists you have free will even if every action you do was determined before you were born. If we are to have free will in a stronger sense than compatibilist free will, it requires that there is indeterminism in the world at the macro level where humans act. Indeterminism here means that there are several possibilities open when it comes to what the content of the future will be, whereas determinism means that the content of the future is already set. However, as I will argue below, external indeterminism suffices for free will. That means that there do not have to be indeterministic processes at specific places in the brain, such as envisioned by Robert Kane in his theory of free will (Kane, 1996, 2007), but only some indeterministic processes occurring somewhere in the world with effect at the macro level of human interaction. Everything in the brain may happen as if the whole world was determined, yet it is important that the world itself is not determined.

The reason such indeterminism is important is that the content of the future will be open so persons can be the causes of which future content becomes actualized. It may seem strange to make an ontological divide between the brain and the rest of the world, but that is not what I do. Maybe there are some indeterministic processes in the brain as well. The point is merely that some indeterminism somewhere is required in order for people to have free will, since that opens up the possibility of different futures and an opportunity for us to influence which future will be actualized. Alfred Mele has claimed that it would be preposterous to base a theory of free will on external indeterminism (Mele, 1995, pp. 195–196), but I shall try anyway, and will respond to his detailed criticism of such attempts later in this chapter.

Do we have reason to believe that there is indeterminism at the macro level of human interaction? The most common place to go for support is quantum mechanics. Quantum mechanics can be given indeterministic interpretations (like Copenhagen and GRW) and deterministic interpretations (like de Broglie-Bohm and Everett). Nevertheless, all interpretations will agree that the guiding laws are merely probabilistic, saying only that something will occur with a certain probability (Ney and Albert, 2013, p. x). This still leaves open whether there is a determinism at a deeper level and whether indeterminism at the micro level of elementary particles can be scaled up to the macro level of human interaction.

If there is indeterminism at the micro level of quantum mechanics, there may nevertheless be determinism at the macro level, because the events at the micro level cancel out at the macro level. It may be undetermined whether a single particle goes here or there, but determined that 50 % will go here and 50 % will go there, so that the macro result is the same in any case. If it is undetermined where the particle will go, we could set up contrasts and ask, "Why did the particle go here as opposed to there?", and the answer will be that there is no cause, but rather it is a causeless, indetermined event. However, it could also be that indetermined micro events can scale up to the macro level.

James Ladyman offers the example of a scientist who decides to take lunch after so many clicks on his geigerteller (Ladyman et al., 2007, p. 264). Geigertellers measure events that according to some interpretations are indeterministic. We could expand the example and say that a scientist may decide to invite her male colleague to lunch if she gets a click on her geigerteller before 12. This decision may make them have lunch, fall in love and get married – or not. The world may then be very different in the future depending on undetermined events. We do not know whether quantum mechanics should be interpreted deterministically or non-deterministically. But note also that also in Newtonian physics indeterminism at a macro level can occur, for example if several identical particles with the same speed collide (Earman, 1986, pp. 30–32).[176]

Here is an additional argument I suggest in favor of indeterminism: evolution makes more sense if what the content of the future will be is genuinely open than if it is determined at the micro level. If many possible scenarios could have happened, we easily understand why the one that actually happened was where the ones best fit for survival had many children. If the one scenario that actually happened was determined solely by laws interacting with particles at the micro level, it seems that this scenario could just as well have been one where what happens at the macro level is very chaotic and unsystematic. The selection effect makes more sense as a selection between genuinely possible futures than if only one future was possible anyhow. And when there is physical indeterminism, qualia can also play a causal role as seen in the chapter on consciousness.

Taking these arguments together, I find more support for the view that the world is indetermined, so I will presuppose here both that quantum mechanics is indeterministic and that the world is undetermined at the macro level of human interaction. If the world is nevertheless determined and I am wrong in making the presupposition that it is not, I blame the Big Bang for making me screw this up.

While I think that the manipulation argument and the zygote argument are good arguments, they are not arguments against the theory I present here, since that theory presupposes indeterminism at the macro level of human interaction. The zygote argument cannot be used against my theory since a divine goddess cannot plan the life of a zygote in an indeterministic world. The argument thus rather supports the claim that indeterminism in required for free will. As for the manipulation argument, compatibilists are challenged to show the relevant difference between a determinism case and a manipulation case. I will argue below that a person is the ultimate cause of the action in cases of free action while, in manipulation cases, the manipulator is the ultimate cause of the action.

---

176 Earman also gives other examples from Newtonian and relativity physics. Important examples are briefly summarized in Sklar (1992, p. 203).

8.3 Causation and choices

I now move to the criticism of non-causal and agent-causal theories. The charges against them were that they invoked mysterious agents and forms of causation and were not able to explicate how choices, actions and control occur. Such charges are not valid against the theory I propose here, since I defend an event-causal understanding of the mind and the self, where causation is understood as being of the same kind everywhere in the world. Given such an understanding of the self, how can we understand how choices are made and lead to action?

A person can be involved in her choices to varying degrees, which makes persons have different degrees of free will and responsibility. This gradual understanding of free will solves many problems, as I will show below. I start with an overview of how persons can be involved in their choices to varying degrees, from an action caused by a desire, to an action caused by the autobiographical self, to an action caused by an independent autobiographical self. I presuppose a desire-action model where the strongest desires lead to action, as argued in the chapter on the mind.

At the first level, we find actions caused by desires, and the desires can be conscious or non-conscious, innate or acquired. Sometimes desires lead directly to action without the occurrence of any additional thoughts or feelings between the desire and the action happening. For example, a woman might see someone being mean to her boyfriend, desire to hit that person, and then immediately do so. In our brain, we carry several evolutionary old and simple systems that can sometimes execute an action immediately, bypassing the influence from reason (Schroeder et al., 2010, p. 103).

A new level of personal involvement is reached when the autobiographical self is activated between desire occurring and the action happening. Sometimes the autobiographical self is activated and changes the initial desire. For example, a person may see a chair and desire to sit, but also see a man approaching the same chair. Autobiographical memories are activated about not having offered a seat to someone previously and receiving negative feedback and of another time having offered a seat with positive consequences; then the initial desire to sit weakens and the desire to offer the seat to the other person strengthens. Initial desires may also change because of new thoughts and feelings as the brain is capable of making new connections between neurons.

The autobiographical self can be more or less involved between an initial desire and action, depending on the number and emotional strength of memories considered before action. The autobiographical self may also be more or less independent. An autobiographical self becomes increasingly independent throughout life as it changes initial desires by a process of thinking and feeling about alternatives. Even if the autobiographical self does not decide what to think or feel, such a change in initial desires is nevertheless caused from within the mind.

There were two charges against event-causal theories mentioned in the introduction: the disappearing agent and the regress problem. I will say more about the development of the autobiographical self and the regress problem below, but first a brief comment on the problem of the disappearing agent.

In Helen Steward's book, *A Metaphysics for Freedom*, her main objection against event-causal libertarianism is that choices reduce to desires, beliefs and bodily movement, and the agent disappears (Steward, 2012, p. 62). But the agent does not disappear. Rather, what an agent is and what it is for an agent to make a choice is explained in a more finely grained way. It is like explaining hardness by describing tension between atoms. The hardness does not disappear, it is just that what hardness is – and, as a result, what is hard – is explained in a more finely grained way.

Concerning the regress problem, this has been formulated by several critics. For example, Hilary Bok argues that since every event has a cause, it seems that we can always follow the causal chain back to events that took place before the agent made a choice (Bok, 1998, pp. 201–205). I will respond to this critique now by first saying more about how the autobiographical self develops.

8.4 Developing an independent autobiographical self

Even if nothing can be the cause of itself, an autobiographical self can over time be the cause of its own content. When we start life, we are not free. At the beginning of life children follow their initial desires, and they are told by their caregivers who they are and what is right or wrong or good or bad. But most children are born with a capacity for reasoning and feeling and thinking new thoughts and making new connections in their minds, which gives them a general ability to find out what is true and good and right. When they make choices and get experiences, these are added to their autobiographical self.

Later, they experience new processes in which new thoughts and feelings change the initial desires and cause action. Again, the autobiographical self is changed from within the mind of the person. In later choices, the old choices and experiences from the autobiographical self can change the initial desires, and again the experience is stored in the autobiographical self. So sometimes a mind-internal process happening independent of the autobiographical self can change that self, for example, if a new experience gives rise to a new thought or a new feeling. But most often, when the autobiographical self changes the process will involve the autobiographical self, so that the autobiographical self causes choices that cause changes to itself.

This means that as long as the world is not determined (including not determined at the macro level), it will often be right to select the autobiographical self as the cause of a change in initial desires *and as the cause of new changes in the autobiographical self.* This is illustrated in Figure 3.

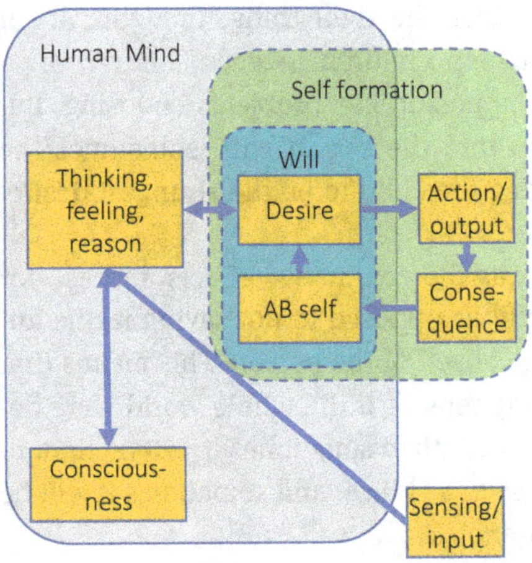

Fig. 3: Self-formation

It may seem of no help to argue that a self can cause itself, since that again must have previous causes and you get a regress going back to before the agent was born. But consider this example: Let us say that a person faces a storm that was undetermined, and sees a young child in the sea about to drown. The person imagines two alternatives: to help the child or not. The process goes on in the person's brain as if determined and leads to action.

The reason it is important that the world in general is not determined is that people will then continually face situations that are not determined by the past. When they face these situations, processes go on in their mind that lead to one desire becoming the strongest, which then leads to action. But even if the process in the mind happens *as if* the whole world were determined, that is not important. What is important is that if the world *is not* in fact determined, then it is sometimes right to select the autobiographical self as the cause of the strongest desire, which again is the cause of which alternative for action is chosen. Indeterminism in the brain is not required, but indeterminism in the world is required for it to be right to select the autobiographical self as the ultimate cause.

Why is it right to select the autobiographical self as the ultimate cause of the choice that is made if the world is not determined? In the example just given, the storm was undetermined, meaning it was not determined before the person was born whether she should save the drowning child or not. Since the storm was not determined to happen, it is right to select the autobiographical self as the cause if the following happened: The person saw the child and imagined two alternatives: to jump into the sea and save the child or not to. The immediate desire is not to jump into the cold sea, but then memories from the autobiographical self are activated about being a person who does the right thing, thoughts about what people think of those who do not help children in need, and so on – and maybe also fear of death – so that the person has problems choosing. But when the person finally decides to jump into the sea, it will (sometimes) be the autobiographical self that causes saving the child to be the strongest desire, causing the person to jump into the water.

When we select saving or non-saving as the contrastive effects and ask for the cause of why the person saves the child as opposed to not saving it, the answer is that the cause is the autobiographical self of the person. This means that even if everything in the person's brain happens as if the whole world were determined, if the world is in fact not determined, then sometimes it will be right to select the autobiographical self as the cause of a choice, and sometimes it will be right to select a person as the cause of a choice.

But is not the undetermined event that led to the storm now the cause of the person jumping in, not the person and not her autobiographical self? No, as one can see by considering contrasts. The undetermined event leading to a storm is not the cause of why the person jumps as opposed to not jumping. The undetermined event is the cause of why there is a storm as opposed to no storm. It creates a new setting in which a choice must be made, but it does not cause the choice. In the undetermined setting, the person and her autobiographical self is the cause of why she jumps, and that is why contrastive causation and macro level indeterminism are important presuppositions to show that this theory of free will is coherent.

But is not the undetermined storm a sufficient condition for the person jumping into the water, since I have said that everything may happen as determined in the mind of the person? No, since the distinction between sufficient and necessary conditions does not work to explain what a cause is. For example, both oxygen and ignition are necessary for a house to burn, but neither of them alone is sufficient, and yet both can be selected as causes. The reason is that we find causes by setting contrasts depending on what we already expect and what interests we have. In the case of the house burning, we are interested in why the house burnt as opposed to not burning, given that there was oxygen present. In the case of the storm, we are interested in why the person jumped as opposed to not jumping, given that there was a storm. If you say that the house burnt because there was oxygen present, I will keep asking why it burnt because oxygen is not the condition I am interested in. And if you say that the person jumped into the water because there was a storm, I will keep asking why the person jumped in because the storm is not the condition I am interested in. But when you cite a certain autobiographical self as the cause (a person with a heroic character, for example), I am satisfied with the answer. In questions concerning free will, our interest is in why people do A as opposed to B given certain conditions. If the cause is their autobiographical self, they act freely.

There is thus no vicious regress here. A person starts life without an independent self, but by use of the general capacity for thinking and feeling in meeting with new and undetermined experiences, that person builds up from within an autobiographical self from those experiences, good or bad, and the autobiographical self gets an increasingly larger role, by which the person gets more freedom and more responsibility. It is the indeterminism that breaks off the regress, since, like in the above storm, when the previous causes like the storm were not determined to happen, it was right to select the autobiographical self as the cause. In the beginning it is not an independent autobiographical self we select as cause, but over time a more and more self-caused and independent autobiographical self can be selected as cause.

Of course, the choices made by this autobiographical self do depend on the general capacities for thinking and feeling that it began with, and which experiences it has will often be a matter of luck, but that is the start package from which an autobiographical self is built, and we cannot demand from free will that there should be an uncaused beginning – that would not give any more freedom, either. It is logically impossible for anything to cause itself before it starts to exist, so we must start with something given, and from there more freedom and responsibility can be achieved. If this general starting point does not work properly, or gets destroyed by outside causes, we find that this reduces freedom and responsibility.

Why all this focus on the autobiographical self? How is it related to me having free will? When I am concerned that "I" have free will, what does "I" refer to then? What is it that makes it important for us to have free will? What should be the cause of our choices in order for us to have a free will worth wanting? Free will is here understood as inner-directedness. When a choice is made, a desire in a person causes that choice. In that sense it is inner-directed since it was caused by something inside a person. But for "me" or "I" to have control, we also want our desires to be something we have formed from within and have control over. The way in which that happens is that desires cause actions, the memory of the experience of those actions is stored in the autobiographical self, and the experiences stored in the autobiographical self can then later change desires and thus be the cause of future choices. When the autobiographical self causes the choice of a person, that person is not just inner-directed in the sense that actions are caused by desires in his or her body, but it is also an inner-directed inner-directedness since choices over time have shaped the autobiographical self that caused the choice. So when the autobiographical self causes choices, we are inner-directed to a larger degree since the choice is then inner-directed by something inner-directed; in other words, it is an accumulation of inner-directedness.

What I am saying is that free will lies on a continuum, where the degree of freedom has to do with the involvement of the autobiographical self – how strongly it is involved in the deliberation process and how independent it is by having been involved in earlier deliberation processes. The autobiographical self being the cause of a person's actions is, in my opinion, the content of the term "self-control". What it means to control one's actions is to cause one's actions, as argued by Alfred Mele, for how can you have control over something to which you are not causally linked? In order to have an effect on something, one must be causally linked to it (Mele, 1995, p. 10). Many who write about free will focus on control, but without answering by what means it is that the agent controls her actions. Here I answer that to control one's action is simply to cause it. The degree of control that we have over our actions is the different kinds of action I have described, where we have most control when it is an independent autobiographical self that causes the choice. This theory of free will is therefore also a theory of self-control.[177]

This theory of free will is also a theory of autonomy. Autonomy is self-governing in the way it is described here, as something coming in degrees, since choices can be caused by an autobiographical self which can be independent to different degrees. Autonomy can be undermined in the different ways I describe that persons can lack capacity for responsible behavior.

I find that the regress problem is the main objection people have against this theory, so I will now start at the other end and describe the journey from a newborn, unfree human being on his way to getting free will. Let us call this person Willy. I assume here that Willy is born with a normal capacity for thinking and feeling, since this is the kind of person I argue has free will. Willy has neither decided his genetic make-up nor the context he was born into, but when he is

177 Nicholas Rescher distinguishes between having control and exerting control to make the point that you can have control over something if you can cause it or prevent without necessarily doing it (Rescher, 2018, p. 41). I accept the point, but focus on exerting control here, in the sense of being the cause of an action or a choice not to act.

born, a new individual finds his place in the causal nexus of the world. Also, he is born into an undetermined world where different futures may take place.

There will be scenarios, not determined to happen, where we will look for the cause. The answer will be that it happened because that was what Willy's autobiographical self felt was best to do. In the beginning, if we ask why the autobiographical self of Willy felt that this or that was best to do, it will be caused by states of affairs other than Willy's autobiographical self, but quite soon, when Willy acts in a scenario not determined to happen, he will make experiences that will feel good or bad, and these will be stored as memories.

Important memories are stored as part of the autobiographical self, and thus the memory stored causes a change in the autobiographical self. In a similar scenario next time, the changed autobiographical self may find another alternative to be the best – which again gives new experiences which are stored in memory and further change the autobiographical self. The process here described is what I have called the development of a more and more independent autobiographical self; I have said that a person is gradually more and more free the more his or her autobiographical self causes itself and the actions the person does. But this is a process where the autobiographical self does not *determine* what it finds good; it just *acts* on the basis of what it finds good. And the autobiographical self does not control the events that shape what it experiences or how it is shaped by them. So why is this free will?

It is free will because there is nothing more to free will than being the most important cause of actions. There is no possible control beyond being the cause, for what would such control amount to? If it should be more, it would require a detached soul or a meta-self, but why should the actions of such an entity be freer than what I have described here? From where did this detached soul get its criteria for choosing? The only way to build a person who can will what he wills is by a gradual process as described here, where experiences of what is good are made in an undetermined setting.

Many envision free will requiring some entity completely free of external influence, but this does not give freedom. Instead, it gives randomness (because the entity did not choose its own criteria for choosing) and it gives the homunculus problem. The best we can achieve in establishing something that fulfills the definition of free will by being the cause of its own will is an entity that over time shapes its own criteria of choice. I suggest it must be a self that makes experiences of what it finds good and changes itself in light of that. Even if it did not choose its own constitution from the start, it would not have been freer if it was to choose its own constitution from nothing, for what would then have been the criteria of choice?

"What an autobiographical self found good" is the basis of free will and the description of how free will is built step by step. In an undetermined world there are many steps of what Willy's autobiographical self found good, which built Willy's autobiographical self and his free will. He did not cause the building blocks of his own body and brain or how they respond to experience or what he finds good. But he is the cause of every new reaction to scenarios in the world, and that is what is relevant.

Why is that what is relevant? Is this not too small a basis for holding people responsible? Surely where they were born, which experiences they had, etc., has influenced them greatly? It certainly influences them much where they were born, etc., but free will nevertheless deserves the name because it suffices to make responsibility meaningful. If the world is as I have described here, then holding others responsible and blaming them will influence what autobiographical selves find as good, which may cause better choices in the future. I hold a consequentialist understanding of responsibility, which is typical for compatibilist theories of free will, but in compatibilist theories every event of holding someone responsible was determined to happen before the person was born. Holding others responsible becomes more important and meaningful if it is not determined to happen. I will now present my views on responsibility further to show how it coheres with the rest of what has been said.

8.5 Responsibility

The following discussion is a brief presentation of responsibility, based on a larger presentation I have published elsewhere (Søvik, 2019). According to philosopher Derk Pereboom, the idea of responsibility as basic desert is the most common and traditional view that is operative in the larger literature but rarely explicitly formulated (Pereboom, 2007, p. 86). Pereboom argues that responsibility should be defined in terms of basic desert, and offers the following definition of responsibility and basic desert:

> For an agent to be morally responsible for an action in this sense is for it to be hers in such a way that she would deserve to be blamed for it if she understood that it was morally wrong, and she would deserve to be praised if she understood that it was morally exemplary. The desert at issue here is basic in the sense that the agent would deserve to be blamed or praised just because she has performed the action, given an understanding of its moral status, and not, for example, merely by virtue of consequentialist or contractualist considerations. (Pereboom, 2014, p. 2)

The quote explains negatively what it means that the desert is basic by saying what it is not, but it does not positively explain what basic desert is or why the desert is basic. Michael McKenna refers to conversations with Pereboom and explains that according to Pereboom there is no justification for this view since it just is a basic relation which cannot be defined in terms of anything more basic (McKenna, 2011, p. 121). I will suggest here that we should not believe that there is such an irreducible kind of responsibility, but rather that responsibility should be understood differently. Instead of understanding responsibility as basic desert, we should have a consequentialist understanding.

In the following, I will present the best consequentialist theory of responsibility that I know of, which is developed by Manuel Vargas. He argues that holding others responsible is a general strategy for cultivating morally good agency in a society (Vargas, 2013). We will have to dig more deeply into the concept of responsibility, but a good introduction to this theory can be given by looking at how Vargas responds to common criticisms of consequentialist theories of responsibility (Vargas, 2013, pp. 187–195). We start by looking at how Vargas answers four typical objections.

Firstly, we can influence people and animals in many ways, but surely responsibility is more than mere influence. Vargas answers that holding people responsible is a special form of influence since it is about giving people reasons to act differently, so it applies only to people who can respond to reasons.

Secondly, it seems that in many individual cases, it will not have the desired effect to hold a person responsible. He or she need not become a better person by being held responsible. Vargas agrees but points out that holding others responsible is a *general* strategy for cultivating morally good agency in a society and it obviously works in general at that level, even if it does not work in many individual cases.

Thirdly, it seems not to distinguish between holding people responsible and holding people *appropriately* responsible. If holding people responsible is just about influencing their behavior, we could put innocent people in jail as scapegoats in order to achieve the desired result (that people behave better). But that would not be appropriate, so a theory of responsibility must include more than just influence to explain when it is appropriate to hold others responsible. Vargas answers again that a consequentialist theory of responsibility must be seen as a general way of cultivating good agency in a society, and as a general strategy, it will not work to put innocent people in jail. Also, there are other ethical norms which are reasons not to put innocent people to jail; for example, it is not just or not the best way to actualize the best world (more on this in the chapter on ethics).

Fourthly, it seems that we often blame people without being interested in influencing them. We may even blame dead people, where influencing them is impossible. Vargas responds that others can learn when we blame dead people even if the dead cannot, but points out that, in many cases, people will not have the intention of cultivating agency when they blame people. Still, the whole practice of holding each other responsible has this effect.

To sum up so far, we hold others responsible as a general strategy for cultivating morally responsible behavior. But does not *holding* someone responsible presuppose them *being* responsible? We have seen that in the philosophy of responsibility there is a distinction between basic desert views and consequentialist views. These two views have a very different understanding of the relation between a person having capacity for responsible behavior and a person being held responsible. The basic desert view sees the capacity for responsible behavior as primary, and if this capacity is present and a person does something wrong, then it follows that the person deserves blame or punishment, regardless of what consequences blaming the person has. The consequentialist view sees holding others responsible as primary. Holding others responsible by praising, blaming or punishing them is what creates in them the capacity for responsible behavior. It gives them an understanding of the world and adequate feelings about what is good and bad, in light of which they can guide their behavior.

Briefly put, the first view says that only if you have capacity for responsible behavior, you can be held responsible, while the other view says that you can achieve capacity for responsible behavior by being held responsible. But even shaping responsible persons by holding them responsible does require a normal capability of thinking and feeling. Even if one thinks that holding persons responsible is primary in understanding what responsibility is about, one could still say that it presupposes a basic capacity for responsible behavior. This capacity then means that is possible for the person to be influenced by being held responsible through a normal deliberation process. In the following, I will try to describe in more detail how this happens.

What usually happens when we hold others responsible is that we compare a person's action with a moral standard concerning what a person in general should have done in such a situation. This means that when you are in a situation, people will hold you responsible according to a standard related to that situation, and this means that you can also be held responsible for something that you have not done but should have.

For example, if a child is drowning while you are nearby, people think that the morally right thing to do in such a situation is to jump into the water and try to save the child. If you do not, people will hold you responsible and blame you according to this standard, even if you did not cause the child to drown. If you do not do what people think you should have done – or if you do what they think you should not have done – it is considered blameworthy, whereas a high score on the moral standard of situations is considered praiseworthy (Bok, 1998). I think that this makes it clear that one can be morally responsible for something one is not causally responsible for, contrary to what, for example, Michael McKenna holds (McKenna, 2011, p. 7).[178]

People are usually understood to have a capability of formal reasoning and a normal emotional life. Because they have this, other people expect that they will understand what is true and what is right to do in various situations. We expect that they learn certain norms and that these norms are alternatives they consider when they make choices. As people grow older and become adults, we expect that they have had enough time to think and to make their own experiences in order to understand essentially what is true and right. Then we also expect them to act in accordance with that understanding.

What does it mean to *be* responsible (in the sense of having capacity for responsible behavior) as opposed to being *held* responsible? In the theory offered here in this book, a person is responsible for what he or she does or does not do in a situation if it is type – not token – physically possible for that person to act in a morally different way, in a way which can be influenced by being held responsible.

This type/token distinction is important for discussions about the possibility to act otherwise in a situation, since in a specific situation there may just be one action that is token physically possible for a person to do given the causes that led up to the situation. Nevertheless, it makes sense to hold persons responsible in such cases because the action of holding others responsible influences what is token physically possible to do in the specific situation. Why is that important?

The relevance of this point is that many (typically physicists or neuroscientists) will argue that no humans are responsible for their actions since all our actions are results of causal chains in the brain where it never makes sense to speak of an agent considering alternatives and controlling the outcome. There is just one alternative action that is token physically possible for a person to do in a specific situation because of the causal chain that led to this action. But one may accept that and reply that holding others responsible is a part of

178 This understanding of responsibility as based on a standard for what people should have done is helpful in matters of pulverization of responsibility, where the causal links are vague between people and consequences and it is hard to find anyone responsible. To find the responsible, one should ask who should have acted differently to avoid this consequence. Often it will be the ones with the most power, explaining why it is true that with great power comes great responsibility. But sometimes the answer may also be that nobody is to blame.

the causal chain that led up to the action and influences exactly which alternative is the token physically possible one. Jack seeing the telling look from his father may cause another alternative action becoming the one that Jack's brain executes.

If the world is determined, there is only one token physically possible chain of events that can happen from the Big Bang until the end of time, and then it does not make sense to say that holding others responsible is a meaningful way to change what is token physically possible in a specific situation. But as seen above, there are in fact good reasons to think that the world is not determined.

What I am saying is that there exist no entities in the world to which the concepts of basic responsibility or basic desert refer. All that exists are people who influence each other by comparing actions with a moral standard for what they think should have been done in that situation, and then praising, blaming or punishing people accordingly, which makes persons consider such reactions when deliberating. This practice only has the effect of cultivating moral agency in cases where people have the capacity for considering such reactions when deliberating. Thus we may define capacity for responsibility as capacity for considering such reactions when deliberating.

The standard that we compare the actions of people with is a general standard which presupposes normal development of people with normally functioning minds. Such a normal development was how I described the development of an independent autobiographical self above. To summarize: So far I have said that *holding* people responsible is a general practice of cultivating moral behavior through blame and praise, even if this may not be the motive in many specific cases. This practice presupposes people *being* responsible (in the sense of having capacity for responsible behavior), which means that they can take praise and blame into consideration in a normal process of deliberation.

When it comes to *being* responsible, we should distinguish between being a responsible person in general and being responsible for x in a particular situation. Being a responsible person *in general* means that you have capacity for responsible behavior and can take praise and blame into consideration in a normal process of deliberation in any situation where it is possible to deliberate. Being responsible for x in a particular situation means that there is something (x) that a person has done or not done in this particular situation which could have been influenced by praise or blame through a normal process of deliberation.

Being blameworthy or praiseworthy for x *in a particular situation* means that there is something (x) that a person has done or not done in this situation which, according to a moral standard, is blameworthy or praiseworthy. Responsibility in itself is nothing more than a concept referring to people being responsible for something in a particular situation, which again is reducible to a certain situation involving a person with capacity for responsible behavior, and "capacity for responsible behavior" means the capacity for being influenced by praise or blame through a normal process of deliberation. All these concepts have content which refers to normal processes between humans. There is nothing mysterious, unknown or metaphysical, in the sense of non-empirical or irreducible or undefinable, about them.

8.6 The problem of luck

In the introduction I mentioned three problems for event-causal libertarian theories of free will such as my own. They were the problem of the disappearing agent, the regress problem and the problem of luck. I have discussed the two first problems already, but now it is time for the problem of luck.

The problem of luck is considered by many the greatest, as with Neil Levy in his book *Hard Luck*. He defines luck as an instance of chance which is significant. Chance is not enough for something to be luck, since, for example, the number of trees in the forest is due to chance, but it is not a matter of luck. Levy distinguishes between chancy luck and constitutive luck, where chancy luck are significant chance events in your life and constitutive luck is the luck involved in what kind of person you were born as (Levy, 2011, pp. 13, 29).

Levy argues that we do not have free will, not because of determinism or indeterminism, but because of luck. Libertarians have a problem of luck because of indeterminism, but compatibilists also have a problem of luck, according to Levy, because so much of what happens in our mind and in our life is due to luck. For example, many external factors determine which ideas pop into our mind (Levy, 2011, p. 90). Trying to solve the problem of luck by arguing that a person has a life history where the effects of luck can be cancelled out over time does not help, according to Levy, since every choice in history has been influenced by luck and one cannot solve a problem of luck by adding more luck. Even when we change our character from within, such changes are based on our character, which was a result of luck, whether now or earlier (Levy, 2011, pp. 89–92).

I have three main responses to the problem of luck. The first is that we are born with a quite common capacity for thinking and feeling, and those who do not have these capacities are considered less free and less responsible. This basis is used as a standard when we ascribe people free will and responsibility and luck is then taken into consideration in many ways. We find it mitigating if people are born in uncommon circumstances or with uncommon problems, and we find it mitigating if luck plays a strong role in specific situations.

We also give people time through which the effects of luck should be cancelled out. So maybe you are more aggressive than normal, and maybe your parents were not that nice, but after a period of several years we think that, given normal capacities for reasoning and feeling, you should understand that it is wrong to steal, murder and lie. Especially in clear cases we expect that people understand what is right and wrong, whereas we can mitigate these expectations in more difficult ethical cases. So my first response is to accept that there is much luck involved in human action, but that this fact is included in how we evaluate people's free will and responsibility.

Is that a matter of solving the problem of luck with more luck, as Levy argues? No, it is a matter of letting reason and feelings find out over time what is right and wrong, even when disturbed by luck, and adjusting our view of a person's free will by the amount of luck involved. For example, a person growing up and traumatized in a war context may receive treatment and help instead of punishment to deal with his aggressive behavior.

Against this, one may argue that we cannot know to what degree people's choices have been influenced by luck, so we are not capable of taking the role of luck into consideration. I agree that we usually cannot know this to any exact degree, but that is not so important either when I consider responsibility to be something that has the goal of improving behavior. Holding others responsible, even those who have suffered much bad luck, can be a way of helping them to act better morally. It can also be a way of giving them more free will by making them aware of alternatives for actions that they can reflect upon and involve their autobiographical self in the process.

For example, by luck a person can have grown up in a family where they speak all at once, and he does not reflect on this behavior. He finds it normal and thus interrupts other people all the time. When someone then holds him responsible and in some way communicates to him that this is not good behavior, he can realize that there is a choice to be made between talking all the time and letting other people talk, and realize that letting other people talk is a good choice. This makes the person more inner-directed than before and thus freer than before.

One could argue that it was a matter of luck that he was held responsible, but it was nevertheless his own thoughts, feelings and experiences which caused the response. Of course one can say that everything is luck, including that we have a capacity for reasoning and feeling at all. But then the concept of luck has become too broad to be a problem to worry about. As mentioned, we start off with a given basis as standard, and consider luck, freedom and responsibility in light of that. You may call everything luck, but that does not make it incoherent to continue thinking of free will, responsibility and luck in a narrower sense in the way I have done here.

Still, even if luck is taken into consideration, and holding others responsible can have good effects even when we do not know the amount of luck involved in a person's choice, we do not actually know whether that person has been so struck by bad luck that in effect he has no free will. That is true, but it does not threaten this theory of free will – this is my third main response – because of the fact that people have free will and responsibility to varying degrees in different contexts. People experience different degrees of good luck or bad luck in their lives, but people also have different degrees of free will and responsibility, and we may never know exactly how free and responsible a person really is.

But as I have argued, there are many cases where the autobiographical self is the cause of an action. And some people have had many chances to feel and think and make choices and change their autobiographical self from within to become more and more independent and free. Let us say my grandmother is about to fall off a cliff, but I rescue her. Why did I rescue her as opposed to push her off the cliff? If you answer that it was just a matter of luck, I answer that the concept of luck is used too broadly to be of practical import. Then one has constructed a theoretical framework with a widely defined concept of luck within which everything can be defined as luck and every event explained by luck. Such a theoretical framework is too coarsely grained. We need a narrower concept of luck in order to differentiate between free and responsible actions on the one hand and cases of luck on the other hand – with a lot of mixed cases in between.

8.7 Weakness of will

A good theory of free will should be able to explain weakness of will. I will deal with this problem here, then also return to the topic of strength of desire, as I promised when writing about desires above in Chapter 5.

"Weakness of the will" is a term used for those who do something they do not want to do. For example, I may desire not to eat chocolate and yet I do, and that is called weakness of the will. That the will is weak seems to imply that there is something called will which can be strong or weak. I do not think that there is one entity deserving the name of "the will". Rather, I think that all there is are different desires of different strengths competing to be the one that finally triggers the motor neurons to make the person act. These different desires are what the person wishes to do, and the one that is actually acted upon is the one we say the person "wanted the most" or "willed". But sometimes we have contradicting desires; that is when it is relevant to speak of weakness of the will.

I find it useful to distinguish between different ways in which the term "weakness of the will" can be used. I concentrate on cases where people want to act in one way but do not do so. This can happen in situations in which a desire caused by the autobiographical self contradicts an innate desire, like an acquired desire to control aggression against enemies contradicting an innate desire to be aggressive against enemies. It can also happen in a conflict between desires that are both caused by the autobiographical self, such as a recent desire versus an older desire, both caused by the autobiographical self. Perhaps you enjoyed drinking good wine before but after having children you have decided not to touch alcohol and now you feel contradicting desires regarding drinking wine. The term "weakness of the will" can also be used for competing desires where people think that one desire is morally good and the other desire is morally bad, so when people act on their morally bad desires, they are said to have a weak will – not able to control their bad desires, for example, for too much alcohol. These may be desires caused by the autobiographical self or they may be innate desires.[179]

I shall disregard the moral examples and only consider the first two examples: where there is a conflict between desires caused by the autobiographical self and desires that are innate, and conflicting desires within the autobiographical self. How can such scenarios lead to acts of weakness of the will? I will start with the first case: a conflict between a desire caused by the autobiographical self and an innate desire. I have said that a person is freer (more inner-directed) when the autobiographical self is the cause of the desire that leads to action than when it is not. This means that if the desire caused by the autobiographical self wins a conflict with an innate desire, then the person exerts more free will (and more self-control, as I argue below) in that choice than if the innate desire wins. If the desire caused by the autobiographical self is for a person not to eat sweets on a weekday and yet that person does because of an innate desire for sweet food, then that is a case of weakness of the will, which more precisely means

179 Richard Hare uses akrasia to denote lack of moral strength (Hare, quoted in Mele (1987, p. 5)). There are also other variables and distinctions concerning akrasia that I have chosen to omit since they are not important to my overall project. A person can show weakness of will in different areas at different times to different extents; it is a difference between weakness of the will in the moment of action and that concerning decisions about what to do later. Weakness of will can relate to control over thoughts, feelings, desires, and actions, but I focus on actions here. The distinctions in this footnote are from Mele (1995, pp. 8, 32, 121).

that the desire caused by her autobiographical self was weaker than an innate desire.

The second case is when there is a conflict between desires that are all caused by the autobiographical self. If these desires are not sorted out, and the person does not know what she prefers, it is not right to speak about weakness of the will but rather confusion of the will. But if there is a new desire which the person now has as a voluntary desire (a desire she is happy with having), and an older desire that the person wants to get rid of, we can speak of weakness of the will when the person acts on the old (and now involuntary) desire she wants to be rid of. For example, this would be the case if a person joined a sect and there acquired some strange moral or religious views that influenced her desires, then left the sect and considered her views wrong, but was still unable to shake off some of the old desires. In this case, there is also weakness of the will if the old desire wins over the new desire since the new desire is caused by a more independent autobiographical self than the old desire – this is a weakness of the will in this context. The reason the new autobiographical self is more independent than the old one is that it has changed at least one more desire than before.[180]

Is it not strange that a person can say she wants one thing the most, yet she does something else? Have I not said that the strongest desire leads to action? How is weakness of the will then even possible?[181] It is time to consider, in more detail, what it means that one desire is stronger than another. I have already mentioned the consciously experienced side of strength of desires – that it has to do with the intensity and amount of pleasure or displeasure – but I have also said that, by definition, the strongest desire is the one that leads

180 Alfred Mele discusses the unorthodox case of a thief who thinks it would be best not to steal, but decides to do so anyway. Even if he is afraid, he steels himself and commits the crime. In this case, it seems normal to call it weakness of will if the thief did *not* steal, whereas it is self-control when he steals, even if he judges it best not to steal. Mele's own solution is to distinguish between an evaluative judgment and an executive judgment, so whether the thief steals or not will be in accordance or not with his executive judgment (Mele, 1995, p. 74). I interpret the story differently. If he does not steal because he is afraid, I would not think of it as weakness of will, but weakness of courage. If he steals, it may be a case of weakness of will, but this depends on the details of the story. Is his desire to steal in conflict with a newer desire not to steal, or is it not? This is what determines whether it is weakness of will, and not just that he thinks that stealing is not good, since if he desires to steal he must also think that stealing is good in one sense or another.

181 Some have argued that weakness of the will is impossible. As examples Mele mentions Socrates, Richard Hare, Gary Watson and David Pugmire (Mele, 1987, p. 8).

to action.[182] However, it seems that sometimes one desire can feel stronger than another yet not be the one that leads to action. I can feel a strong desire for pizza yet eat fish, which I do not feel a strong desire for. Is there then a contradiction in what I am saying here?

One part of the solution to this problem is that a consciously felt desire may activate many strong, contradicting desires in the mind, even if they are not consciously felt or consciously thought. Let us say a doctor suddenly feels a strong desire to have sex with a patient, yet does not act on this desire at all.[183] The desire probably activated many thoughts in the doctor's mind about scandal, prison, broken marriage, shame and so on, all with strong negative feelings connected to them (even if they were not consciously felt or thought), so there was a stronger desire in the doctor's mind not to take initiative to have sex with the patient even if not consciously felt. Mele mentions a problem raised by G. F. Schueler for those who think that the strongest desire leads to action; namely, how a person can intend to go to a school meeting even if he does not desire to do so (Mele, 2003, p. 29). My answer is that in the person's mind there are thoughts and desires concerning what people expect and what will happen if the person misses the meeting for no good reason, so the strongest desire is to go to the meeting even if it does not consciously feel that way.

Strength of desire then has to do with how good or bad the connected emotions are and how much they occupy the mind, but it may be both conscious and non-conscious. This determines the strength of the desire itself, but some desires are more easily executed than others because of physical factors in the brain. In our brain, certain neural patterns have stronger synaptic connections than others. Every time a pattern is activated, its parts become more strongly connected to each other.[184] This is why negative thoughts you have thought many times easily come back and it is also why many psychologists tell you to repeat good thoughts many times. Synaptic strength cannot be experienced directly, but it

---

182 Supported by Singer in Geyer (2004, pp. 56–57). Mele discusses the critique that referring to the strongest desire does not explain an action. For example, if I ask why Bob threw the rock, the action is not explained by saying that he desired that the most strongly. Mele suggests that we do not think of it as an explanation because we already take it for granted that he desired it the most, but want to know why he desired to throw the rock (Mele, 2003, p. 165). So, even if referring to the strongest desire does not explain a particular action, it does explain in general which desires lead to action when it is further explained what strength of desire means.
183 The example is from Mele (2003, p. 162).
184 As quoted above on Hebb's rule ("Neurons that fire together, wire together"). Synaptic strength does not make a desire stronger in the sense that it is connected with stronger feelings, but stronger in the sense that it more easily leads to action.

can be experienced indirectly or inferred by how easily thoughts come to you, or how hard they are to get rid of. Synaptic strength is also part of the reason it is easier to jog if you engage in it as a regular habit than if you just have to decide at one moment between jogging and the sofa. If you are a regular jogger, you may feel a strong desire for the sofa and yet the desire to jog at the regular time makes you jog because of physical aspects of neural patterns in the brain. In that case, the desire for the sofa may in itself have been emotionally stronger than the desire to jog (at least that was how it felt consciously), but since the desire to jog had an easy path to be executed, it led to action, and was therefore strongest in the total sense where strongest means the one that led to action.[185]

What I tried to say in the previous paragraph was that not only the strength of desires (which have already been influenced by thoughts) matters, but also habits. Schroeder et al. describe this by saying that the motor basal ganglia in the brain select which action to execute based on desires, but also based on the internal structure of the motor basal ganglia themselves, which are again a result of habits (Schroeder et al., 2010, p. 82). Desire strength is thus both the strength in the moment, but also the strength it has built up over time by developing habits.

However, it is even more complicated than that. There are several ways in which thoughts, feelings and desires influence each other while physical factors also play a role, and this makes it difficult to set it all out in an easy formula. My aim here is only to point out the main contours of how it works. Many exceptions will not be treated, but I shall mention some here to indicate certain complicating factors. Alfred Mele presents the case of Ian, who most desires to sit and watch television although he should go out and paint the shed. After a while he shouts to himself: "Come on, get out!" And out he goes. How is that to be explained? Mele mentions several possibilities. One is that perhaps Ian has a habit of following orders and shouting to himself provides the necessary adrenaline boost to change what is the strongest desire. Furthermore, we can have several desires simultaneously, so even if painting the shed is not the strongest desire, Ian may try a technique for making it the strongest desire, like shouting to himself. Other techniques for changing the strength of desire can be things such as imagining a piece of chocolate cake to be a piece of chewing tobacco (Mele, 1995, pp. 45–46, 179).

185 Synaptic strengthening also explains why people over time develop more stable characters, habits, and preferences. At least I think so, and it fits very well with psychological research on personality, as shown in Mischel and Shoda (1995).

Thoughts influence desires. In my own experience, when a thought gives me a specific feeling or desire, I can change the feeling or desire by taking a meta-perspective: thinking about the fact that I am thinking about something which feels like that. It will often then feel different, since it is a new thought with a new feeling connected to it. This experience of mine supports Damasio's somatic marker hypothesis. Our thoughts about how likely something is to occur or when it will happen are thoughts that influence our desire for something. For example, I desire to have a million dollars, but that does not make me buy lottery tickets, because I think it so unlikely I would win.

What I am saying is that there are several things determining strength of desire, so it is not strange that one desire leads to action even if a person has another contradicting desire which may consciously feel stronger. This is a kind of conflict we have all the time and it is a life project to create an autobiographical self which integrates our desires in a coherent way that we like. I realize that my theory about the strength of desire is unfalsifiable, since I can always appeal to non-conscious desires or physical neural connections in order to defend my claim that the strongest desire leads to action. However, there is independent support for the claims and they do explain features that otherwise seem strange. I have already mentioned examples which clearly indicate a physical side of desire strength: The lateral hypothalamus can be stimulated electrically and make a person or animal feel compelled to eat and drink, whereas when it is destroyed, animals must be force-fed not to die. Likewise, other areas of the brain could be stimulated to make people suddenly very sexually active (Joseph, 1996, pp. 171–172, 187). These data fit very well with an understanding of desires having different strengths and causing action. There is much neuroscientific evidence supporting that a choice consists in brain activity where desires reach a certain threshold, which then activates motor neurons leading to action (Roskies, 2014).

Helen Steward claims that there is no empirical evidence of non-conscious decisions and that desire strength is an empty concept (Steward, 2012, pp. 159–160, 170–171), whereas I think that the examples in the previous paragraph show the opposite. She nevertheless accepts that conscious beliefs and desires guide our action, while others reject the whole idea of beliefs and desires explaining action. Alex Rosenberg is an example of someone who argues that desires and beliefs are illusions irrelevant to explain what really causes action in the brain (A. Rosenberg, 2011; A. Rosenberg, 2018).

According to Schroeder et al., it is standard neuroscience to think that desires are something physical in the brain causing action (Schroeder et al., 2010, pp. 84–87). I am open to the idea that the physical realizers in the brain of what we call desires and beliefs may turn out to be and work quite differently from how we describe the causal chain of strongest desire leading to action. Nevertheless, I have offered a theory above of how conscious experiences have influenced which brain processes have evolved, which explains why we have the systematic relations we have between conscious beliefs, desires and actions (and it is strange that it should be so systematic if they are just illusions). This means that even if brain processes realizing beliefs, desires and actions are very different from how we describe conscious beliefs, desires and actions, it does not imply that our understanding of the relation between beliefs, desires and actions are wrong. Instead we would just have learned more about the physical side of the process. If a new theory shows that it is more coherent to abandon any role for conscious beliefs and desires, we should of course give up that belief, but so far it just seems to be a self-contradiction for Rosenberg to desire us to believe that there are no beliefs and desires.[186]

8.8 Some final objections

In the introduction, I presented some objections against free will from neuroscience. However, I will not go deep into the neuroscience. The reason is that the findings from neuroscience often considered to contradict free will merely contradict specific theories of free will like agent and non-causal libertarian theories. They lose their force if both conscious and non-conscious mind are understood as causal processes in any case, and they are even less relevant for theories of free will that emphasize free will as a result of decisions made over a long period of time, such as this one, since neuroscientific research on free will is usually made on spontaneous decisions.

I now turn to the objections from Alfred Mele that a theory of free will cannot be based on external indeterminism. Mele has four counterarguments (Mele, 1995, pp. 195–196). The first argument is that if an indeterministic bomb exploded, for example, on 15 September 1969 it gives another future than if it had not exploded, but it does not give us free will. I argue that it does, because in this scenario it is not determined before our birth what will happen. In this scenario, one can select the self as the cause of an action without determinism also being the cause of the action, and thus the self becomes the ultimate source of the choice. In this scenario, since undetermined events like the bomb are possible,

186 At the University of Oslo, 27 Nov. 2019, Rosenberg gave a lecture ending with him acknowledging on the last PowerPoint slide that his suggestion was self-contradictory and that this was an objection he had to continue working with.

other events will also be undetermined, and so our selves will be the cause of many actions.

Mele tries to push the point further by suggesting that if the bomb did not explode on 15 September 1969, and that this bomb was the only indeterministic device in the world, then we would no longer be free after that date. But Mele finds it preposterous to suggest something like that. I reply that such a world would be extremely close to a deterministic world. If the bomb were set such that it would either explode or not on 15 September 1969, then only one future would have been possible up to that date and after that again only one future would be possible. In effect, the world would have been determined up to that date and after that date, and so I agree that we would not have free will in such a scenario. But it is still important that the world is undetermined in general, with several undetermined events, and this argument does not refute that. The more undetermined the world is, the larger the role our selves can play.

Mele asks us to consider two worlds, one of which has undetermined bombs and the other determined. Let us then say that none of the bombs go off, so that everything that happens in the two worlds is identical. Could it then be right to say that those living in the determined world are not free, but those in the undetermined world are free? Here it is open whether the two worlds by accident happen to be identical, or whether everything in the world with the undetermined bombs (except for the bombs) is determined. If everything is determined but the bombs, my reply is like above, that it almost becomes a determined world. But if it accidentally is the same, but many events were undetermined, our selves will play a larger role. So an example with only a few undetermined bombs not going off does not show that a world with indeterminism as a general feature cannot support free will.

Mele's final point is that external indeterminism does not secure free will, as libertarians want the future to be up to them. I agree that free will implies that what happens in the future is up to the agent, but what that means is that the self is the ultimate cause of the action, and the self can be that as long as the world is not determined. As far as I can see, none of Mele's counterarguments refutes the solution I here offer to the problem of free will.

I move on to considering another kind of objection, namely that I have overlooked the importance of social conditions for freedom. I agree that this is a perspective which should complement my presentation, while being fully compatible with it. To introduce the point, I will introduce a famous discussion between Isaiah Berlin and Charles Taylor. Two main points will be considered: First, a point from Berlin on how social recognition determines both who we are and what we want and what is socially possible and impossible actions. Then a point from Taylor on how internal constraints can diminish our freedom.

Berlin wrote a famous essay where he distinguishes between positive and negative freedom. Negative freedom means that others should not deliberately interfere to prevent you from doing things you could otherwise do, while positive freedom means that you are the person determining who you are and what you do (Berlin, 1969).[187] Berlin also added a freedom he called social freedom, where the point is that you need to be recognized by another as a person with a will in order to become a free person who realizes that you are an individual with our own will. It is through the recognition from others of us (as English, church members, football fans, etc.) that we get our identity and desires (Berlin, 1969).

The concepts of positive and negative freedom seem very similar to what I have called freedom of will and freedom of action. Freedom of will is being the source of your choices, while freedom of action is having alternatives for action. While my focus was on the conditions for free will in relation to determinism, indeterminism and luck, Berlin adds the importance of the social conditions for free will. While I focused on alternatives being type physically possible, Berlin describes how social interaction makes it possible for us to understand ourselves as persons choosing among alternatives.

Doing this, Berlin underscores the role of recognition in this social process. "Recognition" is a term which is used in many ways. Arto Laitinen and Heikki Ikäheimo distinguishes between three meanings of recognition: There is first a basic sense of recognition as *identification* of something as something (e.g. "this is a person"). Then there is a second sense of *acknowledging* norms, facts, and values as valid. Third there is a *mutual recognition* process between humans or groups recognizing each other as recognizers (Laitinen, 2002). All of these forms are relevant for free will, since it shapes what we understand as relevant alternatives for actions to choose among.

Recognition both confirms reality and creates reality. For example, you cannot be popular if nobody recognize that you are popular or be somebody's friend if the other does not recognize that you are their friend. Recognition (and with it language) confirms and creates the understanding of ourselves and our surroundings in which we develop an independent autobiographical self. I have focused in this chapter on physical possibilities and impossibilities, but I think that this perspective from Berlin is a good description on how social conditions determine what we think of as possible and impossible alternatives at the macro level where we make our choices. For example, in order for me to think of myself

187 In this article, Berlin also used the terms "freedom from" for negative freedom and "freedom to" for positive freedom, but as others have pointed out, this description is quite misleading, since many scenarios are equally well described as freedom from or freedom to (Nys, 2004, pp. 216–217).

as someone who can get married and in order for me to have a desire to get married, there must be something called marriage and others who think of themselves as possible candidates for marriage and who recognize me as a possible candidate for marriage. Language, society and recognition partly create and determine how our autobiographical selves are experienced and understood, what we desire, and what alternatives for actions we perceive.

It is not only that recognition is important and constitutive for our understanding of our alternatives for actions – recognition is also important and constitutive for what are our alternatives for action. Whether people recognize us and our intentions or not determines what is possible for us to do and not do at the social level, since so much of what we do happens in cooperation with and by means of others. All of this is much more complex than what I can describe here, but I wanted to recognize the point. Elsewhere I have written about social conditions for freedom and what positive and negative effects different form of influence have on different aspects of freedom (Søvik, forthcoming-b).

This is fully compatible with what I have said in the rest of this chapter, and what this book says of the world, mind and language should be enough to see how to translate between these theoretical frameworks on free will. But the social level is an extra layer of complexity shaping the interacting persons, desires and states of affairs considered as alternatives.

The second point I shall consider, is a point Charles Taylor makes when offering some interesting comments to Berlin's article. Taylor defines *negative freedom* as having opportunities to act and *positive freedom* as exercising control over your actions, but argues that the negative freedom does not make much sense without the positive freedom: having opportunities does not make you free unless you actually are a person exercising your will (C. Taylor, 1979, pp. 177–178).

But even being a person exercising your will and having opportunities free from external constraints may not be enough to make you free if there are internal constraints on your actions. Maybe you are paralyzed by fear, or people have forced you to internalize their standards, or you have false conscious perceptions of matters. Taylor argues that you may be wrong about what you really desire. This can be both because you misunderstand what is a good way to your goal, but also because you have false beliefs about the goal (C. Taylor, 1979, pp. 176, 187–193).

Taylor discusses this problem as a problem of how much freedom society should give people and in what way, and I shall return to that problem in Chapter 15.3. Here I shall consider Taylor's point as a possible objection to the theory of free will in this chapter. One could object that in order to secure free will it is not enough that an autobiographical self causes itself, if conditions make it the case that the autobiographical self causes desires we do not think that would have arisen in more normal conditions.

To answer this objection, we should distinguish between three things: freedom of will, freedom of action and responsibility. I have argued that freedom of will is for the autobiographical self to cause its own content and actions, but that this comes in degrees. In a setting with strong social pressure or where things do not work properly in the brain, the autobiographical self may have less opportunity to cause itself than in a setting with more possibilities to explore without external pressure. The contexts within which people develop their autobiographical selves will be different for all in any case, and so that does not contradict my theory which says that people are independent to different degrees.

When it comes to freedom of action, different contexts will give people different opportunities, which will influence how much freedom of action they have, and also how happy they become. They would have made different choices in different contexts, and there is no agreement on what is the normal or the perfect conditions. I shall return to what are the best conditions in Chapter 15.3, and suggest that since we do not know what they are, that favors that people should have freedom of action to explore alternatives.

When it comes to responsibility, we do compare people's actions with a standard for what we think they should have done given normal conditions. If people have grown up in a non-normal condition (like a very mind-controlling sect) or there is something non-normal with their brains, we may hold them less responsible since their actions have been strongly shaped by mind-external factors in non-normal conditions. Again, this raises the question of what should be considered normal conditions, a question I will return to in Chapter 15.3.

When discussing positive and negative freedom, Belin and Taylor are most interested in how society should be formed in order to secure as much freedom as possible for its citizens. This is the discussion I shall return to in Chapter 15.3, and then I will pick up again the debate from Berlin and Taylor. Here I will just mention one final objection, which is more of a speculation, from Yuval Noah Harari.

Harari has speculated that in the future, it will make sense for people to let machines choose everything for them, since the algorithms know people better than they know themselves (Harari, 2017, p. 384). The idea of "free will" will cease to make sense, but Harari argues that belief in free will is a result of faulty logic and the idea of the self is just an imaginary story (Harari, 2017, pp. 331, 353). I have suggested a coherent theory of the self and free will in this book, which lets us see why it does not make sense to leave our choices up to a machine. The reason is that free will means a continuously increased independence by being the cause of your own choices in an undetermined world. This happens gradually by making the choices in undetermined settings. If the machine chooses, it and not you, is the cause of your choices, and even if you gave the initial instructions, you could not know what would happen in the future. Since a person is a whole life process, it means that she loses her free will if a machine instead of her is the cause of her choices. To sum up this last line of reasoning, finding your meaning of life and the best way to the best world, presupposes that both governments and machines allow us freedom of action to form our own independence.

In Part Two I have presented an understanding of how dispositions, desires, choices and thoughts could evolve gradually and function as a causal process. This description fits very well with how artificial intelligent agents are developed. An artificial intelligent agent is something that can perceive its environment through sensors and act upon that environment through actuators (S. J. Russell et al., 2010, p. 34). In almost all artificial intelligence you will find four types of agent, with an increasing degree of complexity: Simple reflex agents, model-based reflex agents, goal-based agents and utility-based agents.[188] *The simple reflex agent* acts directly on its current percept with a condition-action-rule "If A, then B". For example, a self-driving car could have the condition-action-rule: "If the car in front of you brakes, then brake", and humans have something similar, like "if something approaches your eye, then blink".

The model-based reflex agent does not only have an automatic response to what happens, but in addition it has a model of the world which says what happens if the agent does A or B, based on previous percepts. This allows it to make more advanced choices, not only based on what is currently perceived, but also previously perceived. *Goal-based agents* are even more advanced, since they have models for different possible worlds describing what happens if the agent does A, B or C, and in addition they have goals that the alternatives can be matched against. This can be made very complex with several goals for finding the best way to a goal. *Utility-based agents* add a utility function to measure to measure goals against each other and choose goal-based on maximal utility. Russell and Norvig write that this is the same as if the agent was to ask itself how happy the different goals would make it, but that since "happy" does not sound very scientific, computer scientists use the term "utility" instead.

In addition, you can have learning agents, and there are four components in a learning artificial intelligent agent: a learning element responsible for making improvements, a performance-element selecting actions, a critic which evaluates

188 The descriptions of the different agents are from S. J. Russell et al. (2010, pp. 46–58).

and determines modifications for the future, and a problem generator, suggesting new actions to be tested. The agent can learn either based on an internal measurement of how well it predicted outcomes, but it can also learn from an external standard of what is useful. The way that happens is that feedback is interpreted either as reward or penalty. Russell and Norvig compare it with pain or hunger in animals. All learning is about making the parts fit better together to increase the utility for the agent.

There are obvious parallels between artificial intelligent agents and the description of animal and human minds given in this part of the book. The simple reflex agent is like simple animals with brains driven by "if A, then B"-dispositions. The model-based agent has more advanced representations of the world. Goal-based agents can represent alternatives to choose among, while utility-based have desires of different strength making the select the presumed best alternative. In addition, the agents can learn, based on feedback, in the same way as we can develop more independent autobiographical selves, and in the same way as we learn moral behavior by being held responsible. I write more about the comparison of humans and artificial agents in Chapter 15.2.

To sum up the whole chapter: Free will is a matter of inner-directedness and exists on a continuum. People have a small degree of free will when their desires cause their choices without influence from their autobiographical self. They have more free will when their autobiographical selves are the cause of their choices and even more free will if they have independent autobiographical selves. This means that different people have different degrees of free will in different situations. It also means that you can attain more free will than you have now by exploring alternatives for action.

This chapter concludes Part Two on the mind. We now have a much deeper understanding of what happens when we understand and discuss something as true, which we can use when trying to understand other things in the world. On the other hand, we have also laid out a theory of mind which fits into the natural world that we shall consider further in the next part.

Part Three of the book deals with different elements in the world different from mind, and again an important goal is to show how very many things that could seem to be irreducible entities can be reduced to values actualized in a field. The topics to be discussed are time, fundamental concepts in physics, mathematical truths, and probability. First, it is time for time.

Part Three: **The World**

9 Time

In the previous chapter I defended free will, but there is an objection to free will which was not discussed. According to Einstein's theory of relativity, what is future to me can be past to someone else. But if something which is future to me is past to someone else, it does not seem to be up to me what I choose to do in this future which is already someone else's past.

This is a main debate in the philosophy of time, where we find a divide between those who defend eternalism, which postulates a block universe where past, present and future all exist at once, and those who defend presentism – the view that only the present exists, and "now" or "the present" is a special moment of time moving through history. Presentism is what most people will say that they experience, but eternalism or the block universe is what most philosophers of physics and philosophers of time will say is the scientifically supported view (Callender, 2011, p. 74). If presentism is wrong, the theory of free will presented in the previous chapter is wrong, but I will be defending presentism in this chapter.

How one deals with the problem of presentism is of great relevance to all the other questions in the philosophy of time. I shall start with this topic, introducing first relativity theory in Section 9.1, then in Section 9.2 I will discuss the famous argument in favor of the block universe, which roughly says that if past and future is relative, presentism must be wrong and eternalism right.

There are many questions that a coherent theory of time should try to answer. Here is a list of important ones: What is time? Does time flow? What is "now"? Is time relative? What is the structure (topology) of time – how does past, present and future relate to each other? Are statements about the future true? What, if anything, is the cause of time? Are there points of time? Is time continuous or discrete? Why does time move forward; why does time have an arrow? Is time travel possible? How does time relate to mind? Does time pass when there is no motion? Is time infinite in the past and future directions? Was there time before the Big Bang? When will time end? On the one hand, these are too many questions to deal with thoroughly. On the other hand, they are all interconnected, so showing briefly how the main idea of time presented in this chapter answers these different questions helps to explain the main idea of time as well. This chapter will then deal briefly with all these questions, and we shall start by digging into relativity theory.

https://doi.org/10.1515/9783110770957-011

9.1 Is time relative? The theory of relativity

This section contains math, but all you need in order to understand it is the Pythagorean theorem, which says that in a right-angled triangle, the sum of the two sides of the right angle squared equals the third side (the hypotenuse) squared. In math, $a^2 + b^2 = c^2$; for example, $3^2 + 4^2 = 5^2$ or $9 + 16 = 25$. Readers who find even this too much should read only the next paragraph, then skip the remainder of this section.

For those who will skip this section, the essence to take along is this: according to the special theory of relativity, everyone measures the same speed of light in vacuum regardless of their speed relative to each other. In order for that to be true, the universe makes some things contract and some motions occur more slowly, with the effect that all measure the same speed of light. This has been confirmed by experiments. Different observers moving relative to each other have different coordinate systems by which they measure the same speed of light, but whether certain events are past, present, or future then becomes different and relative to their different coordinate systems. This is the fact that is used in the argument that if past, present and future is relative, there cannot be one universal present moment for all, and thus presentism must be wrong.

Now, let us take a look at the theory of relativity. Einstein's "theory of relativity" is actually two different theories: the special theory of relativity (SR) from 1905 and the general theory of relativity (GR) from 1915. The SR is a special case of GR. SR is simpler than GR, dealing only with inertial (i.e. non-accelerating) motion and disregarding disturbances by gravity, while GR includes this and is a theory of gravity in addition to being about time and motion. SR has a simpler geometry, and since the main argument to be discussed is based on SR, I will present SR here, and then add some complicating points from GR later.

The special theory of relativity is based on two main claims: The laws of nature are the same in all inertial frames of reference and the speed of light in a vacuum is measured to be the same value (c) in all inertial frames of reference.[189] The first claim – that the laws of nature are independent of inertial motion – is a

---

189 More precisely, Einstein described the second claim in this way: "Light is always propagated in empty space with a definite velocity, c, which is independent of the state of motion of the emitting body" (Einstein, 1905, p. 891). As Tom Van Flandern points out, it is common for waves to propagate with a velocity which is independent of the state of the motion of the emitting body, but what is unique to light is that the velocity of light is independent of the speed of the observer (Craig and Smith, 2008, p. 216). Einstein does say in his original article that the velocity, c, of light in empty space is a universal constant (Einstein, 1905, section 1), and this is also how it is presented in introductions to spacetime physics (E. F. Taylor and Wheeler, 1992, p. 60).

fact we are used to. If you are on board a train traveling at uniform speed, you are used to the laws of nature being the same as when you are at rest. You walk around in the train as if you were walking on the ground. The first claim uses the term "frame of reference", but what does this mean?

"Frame of reference" and "observer" should be understood as referring to the same. A frame of reference is a coordinate system, and you can imagine it as a lattice work of clocks with a given distance between each clock in each direction.[190] These clocks are all synchronized, and it is the time that these clocks register for an event[191] that we refer to when we speak of time according to a frame of reference, an observer, or a clock. This is useful to note since otherwise we could worry about the time that light uses to travel to the eyes of a person, and about whether mechanical clocks in the house tick slowly, etc., but we are only talking about these lattice-work clocks when talking about observers, clocks and reference frames. What an "observer sees" should thus in the context of physics be understood as "what a reference frame measures" and not as what a living person sees with his or her eyes. Being aware of this can save you from much confusion.

The second claim – that the speed of light in a vacuum is measured to be the same by all observers – has some pretty spectacular consequences, as most famously described in the twin paradox. Before I describe the twin paradox, I shall describe a simpler scenario to give a first presentation of the strange phenomenon of measuring the same light speed in different frames of reference.

Imagine twin brothers Rocket Rocky and Stationary Steve. Rocket Rocky is sitting in an extremely huge spaceship watching a beam of light emitted from the floor reaching a mirror in the roof three years later. Rocky sees the light beam traveling straight up, covering a distance of three light years. Since he measures the speed of light to travel at the speed of light (one light year per year), he measures the time this takes to be three years.

Imagine further a stationary observer (twin brother Steve) watching Rocky's spaceship zoom by in what, relative to Steve, is 80 % of light speed. When Rocky is four light years away on the left side of Steve, Steve sees Rocky's light beam start moving gradually upwards towards the right; it covers a distance of five light years before it hits the roof. Remember that the time it takes for light to reach the eyes of Rocky and Steve is irrelevant here since the numbers given are those that apply to each frame of reference (the registration by synchronized

190 For details, see E. F. Taylor and Wheeler (1992, p. 37).

191 The clock near the event will record the same time as all the other clocks.

clocks at every location in the frame of reference take care of all practical problems like what physical persons actually see with their eyes).

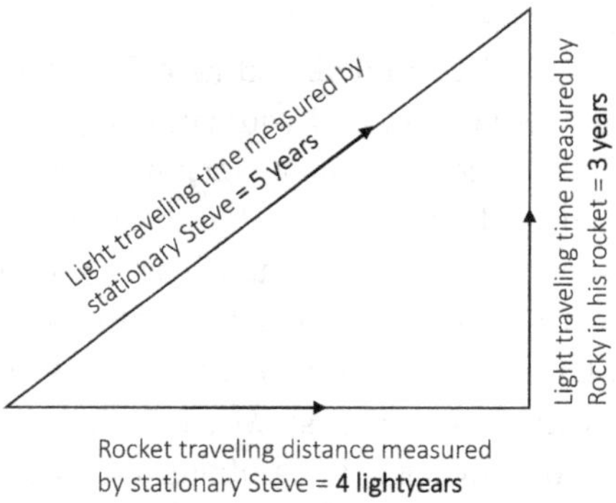

Fig. 4: Rocket Rocky passing Stationary Steve

SR (and a lot of experiments to be described below) tells us that light speed is measured to have the same speed by all observers. Stationary Steve measures light to cover a distance of five light years in five years to hit the roof, so Rocky's travel took five years of time measured by Steve's clock, while it took three years on Rocky's clock. Our bodies are like biological clocks which age at the same rate as measured by proper time (E. F. Taylor and Wheeler, 1992, p. 150) (more on proper time below), and the proper time is 3 years for Rocky and 5 years for Steve. Twin brother Steve will now have aged five years while twin brother Rocky will have aged three years. Rocky could leave Steve, then return, and they could meet and have aged differently, biologically speaking.

We can now formulate a twin paradox and describe a scenario where Rocky leaves Steve and returns, and Steve has aged ten years and Rocky aged six years in the period between the same two events. This description is called the twin *paradox* since, according to SR, it is just as right to consider Rocket Rocky as the one being stationary and Stationary Steve to be moving relative to Rocky. This would seem to imply that either one could be seen as traveling relative to the other and either one could be considered younger than the other, which is impossible, hence the name "paradox". But "paradox" only means "apparent contradiction". There is no actual contradiction here, as we shall see.

Now follows a detailed explanation of why the traveling twin ages more slowly than the stationary twin. Notice how I mentioned that Rocket Rocky traveled four light years from left to right, watching a light beam reach the roof in three years, while Steve saw a light beam moving gradually up in five years. This picture can be envisioned as a triangle with two legs that are 3 and 4 light years long and have a right angle between them, while the third leg can be seen as a hypotenuse with a length of 5 light years. The Pythagorean theorem says that adding each leg squared should give the hypotenuse squared, which is right: $3^2 + 4^2 = 5^2$ ($9 + 16 = 25$).

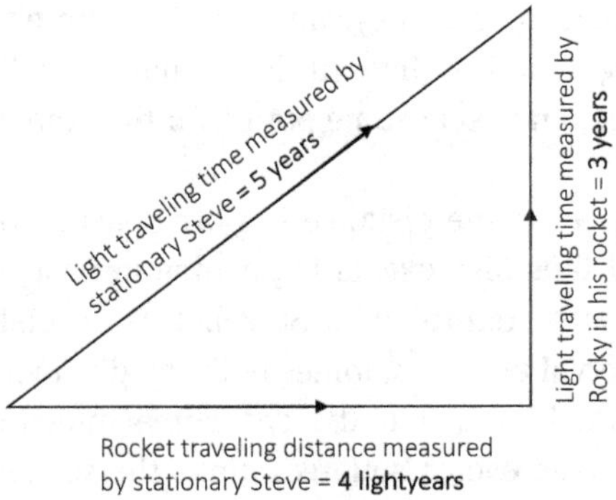

Fig. 5: Rocket Rocky passing Stationary Steve

What the numbers here represent are spatial distance (traveling from left to right) and distances between events (light going from floor to light reaching roof). We have 3 as the time between events that Rocky measures, 4 is the spatial distance that Rocky travels (relative to Steve's frame of reference, since in his own spaceship reference he is standing still), and 5 is the time between events that Steve measures. Note for later that Steve's time squared, 5^2, minus the distance squared, 4^2, equals Rocky's time squared, 3^2 ($25 - 16 = 9$).

I said that Rocky traveled at 80 % of light speed (c), which can be written as 0.8 c. If he had traveled at 60 % of light speed (0.6 c), Steve would measure that Rocky had covered a distance of 3 light years (since 5 years multiplied with 0.6 light speed equals 3 light years of distance) while watching his light beam take 5 years to reach the roof (while Rocky would have seen it take 4 years). In that case Rocky would age 4 years compared with Steve's 5 years, while in the example above, Rocky aged 3 years relative to Steve's 5 years. In this example Rocky travels more slowly, and note that Rocky's time squared, 4^2, equals Steve's time squared, 5^2, minus the distance squared 3^2 ($16 = 25 - 9$).

Here is a description of the twin paradox where we consider two events happening at the same place for Steve and Rocky. The events are first that Rocky leaves Steve in their childhood home traveling 4 light years at 0.8 c, then he returns, and the second event is that he returns to the same home again. Steve is stationary the whole time in his childhood home. Since Rocky travels at 0.8 c, Steve can calculate that Rocky needs five years to travel 4 light years before turning ($5 \times 0.8 = 4$), and then Rocky comes home again, and Steve notes that ten years have passed in total.

The term “event” is ambiguous, since in spacetime physics it usually refers to a spacetime point – a location in a spacetime diagram, which maps both time and space. In everyday speech, an event is something happening, like an explosion or the flash of a light bulb. Since an event in the everyday sense can happen at a spacetime point, the meanings can overlap, but they can also differ, for example if an event takes place over a long period of time. In this chapter I shall use “event” to describe things that happen, and “spacetime point” for the points in the spacetime diagram.

SR tells us that all observers can agree on the distance in both space and time between two spacetime points (and thus also events happening at spacetime points, such as Rocky leaving Steve or returning to Steve). This special kind of distance is called a spacetime interval and it combines both the distance and the time between events. It is defined such that the spacetime interval squared equals the timelike distance between events squared minus the spacelike distance between events squared (examples follow soon). All observers can agree on what the value of the spacetime interval is. An important thing to note is that when SR speaks of “measuring time”, what is meant is the spacetime interval (also called the proper time) that an ideal clock (i. e. reference frame) measures (E. F. Taylor and Wheeler, 1992, pp. 76–77). Proper time need not be the same as that which is measured by a regular watch, as I will say more about below, but since biological aging is related to proper time it *does* describe how the twins age biologically (E. F. Taylor and Wheeler, 1992, p. 150). Here is how time will be measured by the twins:

For Steve, the distance in time between events is 10 years, and the distance in space is 0 light years, since he has just been sitting in his home. When he calculates timelike distance (10^2) minus spacelike distance (0^2), he gets 10^2, which is 100, and since the square root of 100 is 10, this gives a spacetime interval of 10 years for Steve.

Steve can also calculate the proper time that Rocky must have measured. Then he takes timelike distance in his reference frame squared minus the spacelike distance in his reference frame squared, which is $10^2 - 8^2 = 6^2$ ($100 - 64 = 36$). Since the square root of 36 is 6, Steve can know that Rocky will only have measured 6 years to have passed. Note the similarity with the previous example, where Rocky was watching light move 3 light years, while Steve watched it move 5 light years. This would be like the first half of the trip now described.

This is a weird result, but it is correct. What Rocky will experience is that 6 years has passed, and he will only have aged biologically 6 years. SR tells us that instead of considering Rocky to be traveling at 0.8 c, we can choose to say that Rocky’s spaceship is the reference frame that is stationary, while Steve and his home on earth are zooming away from Rocky at 0.8 c. Rocky would then calculate for each leg of his trip that the timelike distance is 3 years and the spacelike distance is 0 years. Since $3^2 - 0^2 = 3^2$, he will measure that his trip took 3 + 3 = 6 years.

Here is why the twin paradox is called a paradox: It seems that Rocky, traveling away, could consider Steve as the one moving, and then Steve should be aging less than Rocky, but it is impossible that both twins get older than the other. And SR tells us that Steve will age 10 years while Rocky ages 6 years, so how is this possible if we are free to choose who to consider as traveling and who to consider as stationary?

Most textbooks and YouTube videos and even experts in physics like Richard Feynman give a misleading solution to this problem and say the solution is that the difference between them is that the traveling twin is accelerating. But the acceleration is not the real explanation, as Tim Maudlin shows by making a thought experiment where the twin at home accelerates more than the traveling twin, yet the twin at home gets older (Maudlin, 2012, pp. 77–83).

Acceleration is relevant in the example just given, but explaining age difference as a result of acceleration can be quite confusing, since you can make the twins accelerate just as much or even make the one getting older accelerate more while still getting older. Maudlin argues instead that the difference should be explained by saying that what matters is just the length of their world line (Maudlin, 2012, pp. 77–83), which I will now explain.

The world line between two events is the spacetime interval that an ideal clock following an object would measure for the object traveling between those two events. It is also called the wristwatch time or the proper time, since it is the time measured in a reference frame where the object is considered stationary. As mentioned before, it is measured by taking the timelike distance squared minus the spacelike distance squared. We calculated the world line for Steve when we considered him to be stationary ($10^2 - 0^2 = 10^2$) and found it to be ten years, and we calculated the world line for Rocky when he was considered stationary and found it to be 3 + 3 years ($3^2 - 0^2 = 3^2$). Note that I write 3 + 3 (instead of 6) years for Rocky since he is turning around and thus changes reference frame. In order to consider Rocky as stationary, we must consider him as stationary in two different reference frames – one going out and one coming back.

This is still strange: why does Rocky only measure 3 years passing instead of 5? Why does not Steve get younger than Rocky? Let us look at how things appear from Rocky's frame of reference, seeing now this reference frame as stationary. We consider first the first three years. Rocky and his spaceship are standing still while Steve and the earth are flying away at 0.8 c. When Rocky measures three years to have passed in his frame of reference, he notes that Steve has traveled 2.4 light years away (in Rocky's frame of reference). Why only 2.4 light years away? Because 3 years times a speed of 0.8 c gives a distance of 2.4 light years. How many years has Steve aged when Rocky measured 3 years to have passed? Rocky can calculate this by taking the square root of the difference of time (3 years) squared minus distance (2.4 light years) squared: $3^2 - 2.4^2 = 1.8^2$. Rocky measures Steve to have aged 1.8 years after 3 years in Rocky's frame of reference.

So far then, everything is symmetrical. When Steve measures in his reference frame, Rocky ages ³⁄₅ less than Steve ($5 \times 3/5 = 3$), and when Rocky measures in Rocky's reference frame, Steve ages ³⁄₅ less than Rocky ($3 \times 3/5 = 1.8$). The same applies to Rocky's home trip, for again, Rocky will measure that Steve has only aged 1.8 in the last part of the trip. This seems to create a big problem, since SR tells us that Steve does not age 1.8 + 1.8 years, but rather he ages 5 + 5 years. What is the solution?

The solution is that when Rocky turns, he changes reference frames. Even if we consider Rocky to be stationary for both parts of the trip, Rocky must change reference frames in order for the two to meet again at the same spacetime point, while Steve is in the same reference frame the whole time. What happens is that Rocky measures Steve aging the first 1.8 years on the first part of his trip. This means that what he thinks of as happening in the same period as his own three years is the 1.8 years that Steve experiences. Then Rocky changes reference frames and measures the last 1.8 years of Steve's aging as happening at the same time as Rocky's last three years. If Steve and Rocky were 20 years old when they left each other, Rocky will measure what happened to Steve from 20 to 21.8 years old and from 28.2 to 30 years old. There is a gap of 6.4 years in Steve's lifetime that Rocky does not experience to happen at the same time as he is traveling, since he changes reference frames, and thereby changes what he measures to happen simultaneously with his own experiences. Look at the simultaneity lines illustrated in Figure 6 to notice how Rocky would think that only 0.6 years of time passes for Steve for every year that passes for Rocky, but that there is a gap in between.

Since a change of reference frame means a change in acceleration, it is not wrong to use acceleration to explain the difference between the twins, but such explanations are often very unclear. The amount of acceleration is not important. The length of the world line, and the gap coming from jumping frames, is what is important. Since non-accelerating particles move along geodesics defined as maximal world lines, it is not wrong to use acceleration as explanation, but hopefully this unpacking made it even clearer.

The explanation for the gap is that different reference frames have different lines of simultaneity, but why do they? Why do twins age differently depend-

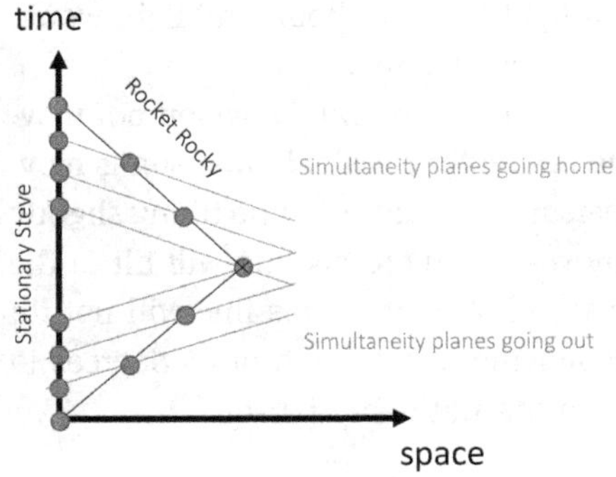

Fig. 6: Simultaneity lines

ing on what route they travel between events? The reason is a very weird fact: nature adjusts objects and motions in a way that makes all measure the same light speed.[192] In order for all to measure the same light speed, there occurs in nature a physically real stretching of motion and length called time dilation and Lorentz contraction. This is a real effect, but it is easy to be confused since sometimes length and time differences are just a matter of the coordinate system we use, but sometimes there are real physical differences like the aging of twins, which is different regardless of reference frame chosen to describe it (Maudlin, 2012, p. 116).[193]

I will try a brief explanation of such Lorentz transformation, much simpler than how they are usually described, and you still just need to know Pythagoras to hang along. Imagine a normal coordinate system, with vertical and horizontal lines. Every step along the horizontal (x) axis represents one light year of spatial distance (ca. 9.46 trillion (10^{12}) kilometers). Every step up the vertical (y) axis represents one year of time. In the center (origo), we say that the following event occurs: a light bulb flashes. The light starts traveling in the space direction, and after one year it has traveled one light year, thus reaching point (1,1) in this coordinate system, which represents one light year of distance and one year of time (think about it and make sure you understand why). After one more year, light will have traveled to point (2,2), then (3,3), etc.

Such a diagram of space and time is called a spacetime diagram. In a coordinate system where light and time have correlated units, light will always travel in a 45-degree angle. The correlated units here were light year per year, but it

192 I discuss why we have the laws of nature we do in Chapter 13.5.

193 Maudlin refers to detailed physical explanations by Bell in Bell (2004, chapter 9). Brown gives an overview over this and other physical explanations of length contraction in Brown (2005, chapter 7).

could have been lightsecond per second or light hour per hour, and light would have traveled in a 45-degree angle in those diagrams too.

A vertical line moving from the center and up represents a person not moving in space, but just moving through time up the time scale. If a person is moving a little bit to the right relative to this person, it would be a line tilting slightly to the right. The faster that this person moves, the more his line will tilt to the right, but nothing can be faster than the speed of light, so his line will not tilt more than 45 degrees to the right if he is moving to the right, or 45 degrees to the left if he is moving to the left relative to the stationary person.[194]

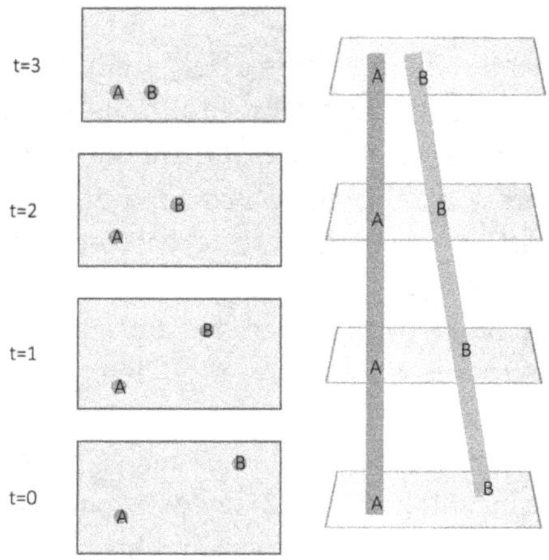

Fig. 7: Creating a spacetime diagram

Now, visualize the following: in addition to a normal coordinate system, we put a see-through rubber sheet on top of it with an identical coordinate system. Imagine the original coordinate system having black lines, and the coordinate system on the rubber sheet having red lines. In addition, both coordinate systems have a green line in a 45-degree angle representing a ray of light. Now, take the rubber sheet at the top right corner, and drag it up to the right while trying to make the green lines overlap, so that the light beam gets coordinates 1,1; 2,2; 3,3; 4,4; 5,5; etc. in both coordinate systems. It should then look something like the illustration given in Figure 8 (check to see that the green line crosses both the black and the red corners).

In the example above, we can see that what is coordinate 5,5 in blue lines is coordinate 4,4 in the red coordinate system. If we drag it further, 5,5 in the blue coordinate could be 3,3 in the red coordinate system. Remember that the vertical

194 Figure 7 is inspired by two similar figures found in Maudlin (2012, p. 55 and 57).

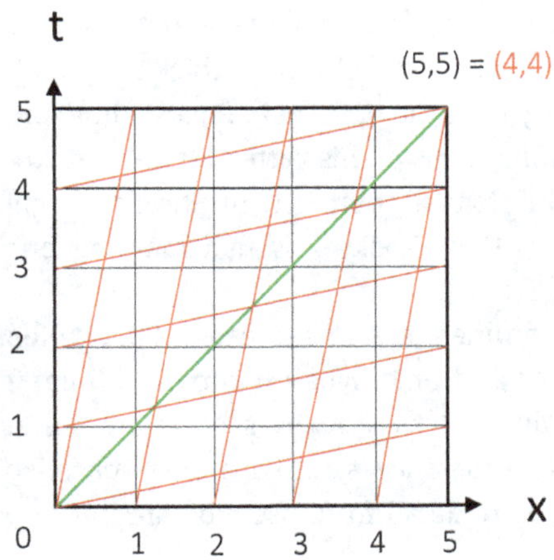

Fig. 8: Overlapping spacetime diagrams

line (y-axis) represents time in both coordinate systems. If we see the black coordinate system as representing the stationary Steve and the red coordinate system representing Rocky in his rocket, we can drag it more and more so that what is 5 years for Steve becomes 4 years for Rocky, or 3 years, or less the more you drag it.

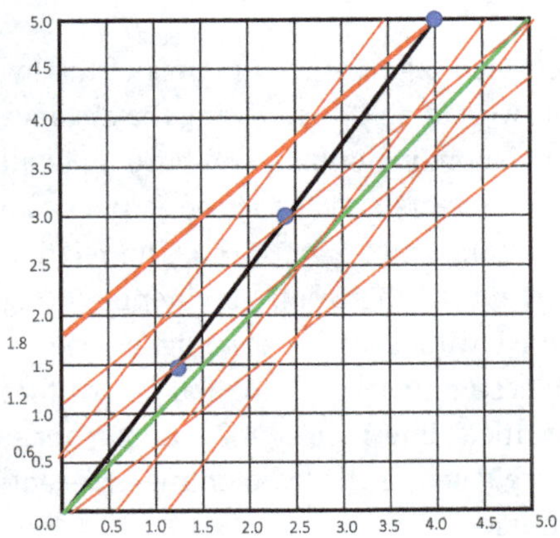

Fig. 9: Rocky traveling

In Figure 9, Rocky is traveling at 0.8 c towards the right (black line with blue dots), while Stationary Steve is sitting still following his own vertical time-axis. In the black coordinate system of Steve, Rocky travels 4 light years (on the horizontal axis) to the right and uses 5 years (on the vertical axis), and ends up in point (4,5). Rocky's coordinate system is the red one. In his own coordinate system, Rocky is sitting still following his own vertical time-axis, which is the black line, with a blue dot marking one year each, as measured by Rocky.

When Rocky comes to 3 years of time in Rocky's coordinate system, Steve is at 1.8 years of time (to see that, follow the top blue dot along the thick red simultaneity line to the left). The red simultaneity lines are parallel to Rocky's horizontal axis, showing everything which is simultaneous on his vertical time-axis. Notice how the green line representing light crosses point (1,1) in Steve's system marking one year for Steve, and point (1,1) in Rocky's system marking one year for Rocky.

There is a formula for relating the coordinate systems to each other, called the Lorentz transformations for both space and time. When it comes to time, it is just another way of writing the formula we have already seen many times: spacetime interval squared equals timelike distance squared minus spacelike distance squared. Here it is explained in more detail for those who are interested:

What we want is a formula to use to either stretch the proper time of one coordinate system to the measured time of another coordinate system, or the other way around. This means that we want to find the relation between what is a straight vertical timeline in one coordinate system and the hypotenuse in another coordinate system (recall the example with light going straight up or gradually to the right). For example, we want to be able to turn 3 into 5 (as in the example of the twin traveling at 0.8 c) or 4 into 5 (as in the example of the twin traveling at 0.6 c).

First, we define the horizontal line representing distance in terms of the hypotenuse representing measured time. This we can do by expressing the distance as a fraction of light speed, and using corresponding units as we have done all the time. For example, if the hypotenuse is 5 years and the speed is 0.8 c, we know that the distance traveled is 4 light years, since $5 \times 0.8 = 4$. The relation between the hypotenuse representing time and the horizontal line representing distance is thus the hypotenuse multiplied with speed v over light speed c ($5 \times 0.8 = 4$). In the following I will use these numbers – 3, 4, and 5 – so that the reader can remember that 3 is the vertical line representing proper time, 4 is the horizontal line representing distance, and 5 is the hypotenuse representing measured time in another reference frame.

How does the vertical line representing proper time (3) relate to the hypotenuse representing measured time in another coordinate system (5)? We know that the line representing proper time squared (3^2) equals the hypotenuse squared (5^2) minus the line representing distance squared (4^2). The horizontal line representing distance covered can be measured in terms of speed (v) divided by light speed (c): v/c. If we know that someone travels at 80 % of light speed, their speed in v/c is 0.8 c, and when we know the speed they travel and for how long, we can know the distance. For example, someone traveling at 0.8 c for 5 year will cover a distance of 4 light years.

Let us now set the hypotenuse at value 1, and use the Pythagorean theorem to say that $1^2 - (v/c)^2$ equals the proper time squared. This formula ($1 - (v/c)^2$) represents the relation between the hypotenuse representing measured time (5) and the horizontal line representing distance covered (4), which can then be related to the vertical line representing proper time (3). We relate them by dividing the vertical line representing proper time with ($1 - (v/c)^2$). As a general formula we can write 1 divided by the square root of $1 - (v/c)^2$, then just take whatever number we have for the proper time times this formula.

Here is a test example: We have a proper time of 3 years and wonder how that will be measured in a reference frame moving at 0.8 c. We then take $3 \times 1 / 1 - (v/c)^2$. Here $(v/c)^2$ is 0.8×0.8, which is 0.64. Then $1 - 0.64 = 0.36$, and the square root of 0.36 is 0.6. To find the number we want, we must then take 3×1 (= 3) and divide by 0.6, and we get 5. So 3 becomes 5, which is the correct result. Another example: We have a proper time of 4 years and wonder how that will be measured in a reference frame moving at 0.6 c. We then take $4 \times 1 / 1 - (v/c)^2$. Here $(v/c)^2$ is 0.6×0.6, which is 0.36. Then $1 - 0.36 = 0.64$, and the square root of 0.64 is 0.8. To find the number we want, we must then take 4×1 (= 4) and divide by 0.8, and we get 5. Thus 4 becomes 5, which is the correct result.

See that this is just another way of doing what we have been doing all the time, either by multiplying with speed relations (3/5 and 4/5), or by taking the time distance squared minus the space distance squared, but now conveniently gathered into one formula which can be used in all settings. The formula is called gamma, written with a lower case Greek gamma letter (γ), which is the time stretch factor.

This stretch factor for time is the same factor that is used for calculating length contraction. You may have wondered about the following: Have we not said that Rocky traveled a distance of 4 light years in what Rocky measured to be 3 years of time? That seems to imply faster-than-light travel if Rocky can travel 4 light years in 3 years while light itself needs 4 years, so how is this possible? Imagine that Rocky is about to travel 4 light years away, and we have an extremely long stick which is 4 light years long that Rocky will be traveling along. SR tells us that this stick will contract with the exact same factor as the time stretch factor. The relation between time measurements was 3/5 and the stick will contract with a 3/5 ratio, which means that Rocky will measure the stick to be 4 light years times 3/5 = 2.4 light years. Since he measures that he travels for 3 years over a distance of 2.4 light years, this is 3 years divided by 2.4 light years, which gives us 0.8 light years per year, which is the correct result. If we consider Rocky to be stationary, he will measure earth and Steve to move away from him 2.4 light years in 3 years, which is 0.8 c.

9.2 What is the topology of time? Presentism versus eternalism

After reading about the theory of relativity, it is natural to ask whether it can really be true that the universe makes processes go slower and things contract so that all measure the same light speed. While this seems utterly crazy, there are many tests that have confirmed that this is how the universe actually works.[195] Muons that usually exist for 1.5 microseconds in proper time when standing still exist for 44 microseconds when they travel at 0.9994 c, in complete agreement with the theory of relativity (Callender, 2011, p. 529). The twin paradox effect can be seen by comparing iron atoms vibrating, some fast and others slow, where the internal processes in the fast-moving iron atoms happen more slowly (E. F. Taylor and Wheeler, 1992, p. 134; Rebka and Pound, 1960).

Even the GPS in your smartphone must calculate these effects in order to give you your exact position (Ashby, 2003). Your GPS is connected with clocks in satellites orbiting the earth at high speed and further away from earth than what we are. These satellite clocks need their time to be adjusted relative to our clocks. The difference in speed must be calculated in accordance with the special theory of relativity (SR), and the difference in gravity according to the general theory of relativity (GR). The GPS clocks go 7200 nanoseconds slower per day because of their speed and 45900 nanoseconds faster per day because of less gravity (Craig and Smith, 2008, p. 213).

The spacetime diagrams in SR that only deal with speeds of observers are thus quite simple compared to how GR also takes into consideration the effect of various distributions of mass in the universe. GR uses a different geometry from SR, where you cannot have a large inertial frame because it gets disturbed by gravity, but SR works locally also in GR (E. F. Taylor, Wheeler, and Bertschinger, Unpublished, chapter 1.11). The argument we are about to consider is based on SR, but arguably things just get worse for presentists with GR, since there is no global inertial frame in GR (Callender, 2017, pp. 57–60).

Now we are ready for the argument against presentism, which says that presentism is incompatible with the theory of relativity. Most philosophers of phys-

195 Common examples of tests considered to confirm SR and GR are Michelsen-Morley (1881), Kennedy-Thorndike (1932), Ives-Stilwell (1941), Frisch-Smith (1963), and Hafele-Keating (1972).

ics and of time hold this to be the case (Callender, 2011, p. 74). There is an argument made by several philosophers, but often referred to as the Rietdijk-Putnam-Penrose argument, which argues from the relativity of simultaneity in the theory of relativity to a block universe (Callender, 2017, p. 52). Roger Penrose describes this argument in the following way: Imagine a car driving along the street in a direction which is also in the direction towards the galaxy Andromeda As the car passes a man standing on the street, the man in the car and the man on the street will have different simultaneity lines with what happens at Andromeda. Simultaneous to the man standing on the street, Andromedans may consider whether to invade earth, but simultaneous to the man in the car, the Andromedans may already have chosen to invade us and be on their way towards earth. Penrose writes that if what is future to one person is past to another, the event is an inevitability (Penrose, 1989, p. 303).

This argument is used by many physicists to argue in favor of a block universe.[196] Putnam says that the problem of the reality of future events has been solved by physics, and not by philosophy.[197] Craig Callender calls the argument "utterly convincing" (Callender, 2017, p. 53). In the following, I shall argue that the argument fails because of a fallacy of equivocation.

In order to consider this argument carefully, we shall first consider more closely what we do when we measure time. The twin paradox has shown us that we need to be very precise when talking about time as a measurement of motion. Time can be measured by an individual in his or her reference frame, and this is called coordinate time. Time can also be measured as wristwatch time by a clock traveling between two events, and this is called proper time, which is a quantity all observers can agree on. Since this proper time is a common measurement of time for all, and independent of reference frame, it is a very useful way to measure time, which allows the same laws to be used in all frames of reference and which is therefore adopted by physics. In addition, it allows us to easily calculate real phenomena like time dilation.

Such measurements of time are made by ideal clocks which use light to measure time. This is a sensible choice, since all measure the same speed of light, and again, since it even matches with physical processes like biological aging. It is worth saying explicitly that if traveling twin brother Rocky had left Steve traveling at 0.8 c like in the example above, he would measure three years on his *ideal* clock while Steve measured five years. But if Rocky were to have a *regular* clock on his arm, e.g. bought at a gas station before leaving, this clock

196 For example, Brian Greene, Max Tegmark and Sean Carroll in this video: Art of Spirit (2014).
197 Callender (2011, p. 205), referring to Putnam (1967, p. 247).

would show five years just as the clock on Steve's arm would (E. F. Taylor and Wheeler, 1992, pp. 76–77). After all, Rocky would presumably – like all of us – be moving his arm in all sorts of directions all the time, so this mechanical clock would not have been an idealized inertial frame of reference.

This means that Rocky could easily have measured the same time as Steve on his whole journey. For example, we could imagine clocks hanging outside along his journey that are synchronized with Steve's clock, so that Rocky could always look outside his window to see what time it is for Steve. Or he could have brought a watch which had been corrected for the speed effects, such as GPS watches (Craig and Smith, 2008, p. 222).

How we choose to measure time is then a choice we make where different choices have different advantages. The ideal clocks and reference frames are very useful coordinate systems for the purpose of physics, but we need to be very precise on exactly what they tell us about time, passage, simultaneity, future and past, and this is what we are now digging into. In the twin paradox, ten years passes for Steve and six years for Rocky, and this is natural to say because Steve has biologically aged four more years than Rocky. However, it would be wrong to say (even if many examples do[198]) that if Rocky leaves in 2017, it will be 2023 for Rocky and 2027 for Steve when they meet.

It will be 2027 for both when they meet, since the earth will have orbited the sun ten times since Rocky left, but if Rocky were to measure these orbits, he would measure the earth as orbiting the sun faster compared to Steve. If he were to look at the earth with his eyes (as opposed to measuring with his ideal watch), he would see the earth going slower when he was moving away from the light, and then faster when he was moving towards the light, since the light would travel longer or shorter before reaching his eyes. As measured in Rocky's frame of reference (if he travels at 0.8 c), there is only 0.6 of a year between each Christmas. But Steve and Rocky will agree on when it is Christmas, namely when the earth has a certain position relative to the sun and the other planets and there are Christmas trees for sale. They will disagree on how long the period is between each Christmas if they use light to measure this time period, but they will not disagree on how many months it is between each Christmas – if they measure months not in seconds but as $1/12$ of the time earth takes to orbit the sun and roughly one period of the moon orbiting the earth. If they measure time not by means of light but by means of the earth orbiting the sun, they will of course agree that it is one year between each Christmas.

---

198 As anyone can see by writing "twin paradox" and searching google images.

It is useful, then, to distinguish between, on the one hand, events happening and the order in which they occur, and, on the other hand, how we can use different coordinate systems and different means of measuring *when* these events occur. There is an important distinction in SR and GR between what happens within the past light cone of an event and what happens outside of it. Your past light cone is the region of events that could have influenced your present moment since they are within light speed distance. You could have been influenced now by things that were less than one light year away one year ago, less than two light years away two years ago, etc., since this could have reached you now, so this within your past light cone. But if something was more than a light year away a year ago or more than two light years away two years ago, etc., it could not have influenced you now, since it could not in any way have reached you, so then it was outside your past light cone.

Events that, in your coordinate system, are considered by you to be past and within your past light cone are called the absolute past, whereas events outside of your past light cone are not. The reason is that different observers will disagree on what is past and future outside of your past light cone, and even in which order they occurred. But all observers will agree on the order of events in your past light cone (E. F. Taylor et al., Unpublished, chapter 2.6). These events follow a world line, and all observers can agree on the order of events on a world line, although they will disagree on what happened first of two disconnected events with a great spatial distance between them.

This has a very interesting consequence. Different observers will measure different time orders of random events that are far from each other, but in fact, there is a time order of events that are causally connectible (within a light cone). All events in the universe can be placed in the past light cones of particles from the Big Bang, so in principle all observers should be able to agree on the order of events in the universe, if they could have communicated with each other. Relative to their own frames, they would disagree on when the events happened and how long they took, but they would agree on the order of events.

On the one hand then, we have the order of all the events in the universe. On the other hand, we have different coordinate systems that can be used to sort these events into past, present and future. This can be done in many ways, and it is not the case that something becomes fixed past by being designated as past by us in a coordinate system we choose to use. Imagine a photon leaving the Big Bang. In a spacetime diagram it can be seen as traveling at 45 degrees to the right, with a simultaneity line of 45 degrees, so that the Big Bang is always simultaneous with it. After 13.8 billion years in the reference frame of the universe, for this photon the Big Bang is still happening at the present moment, and the 13.8 billion years of universe history is future relative to the reference frame of this photon. If the photon a second later (in the universe frame) starts traveling in the opposite direction, then its simultaneity line gets tilted 45 degrees upwards to the left, and now the 13.8 billion years that the universe has lasted and the next 13.8 billion years of the future of the universe will all be 27.6 billion years of past relative to the reference frame of this photon. In one second, 27.6 billion years of future becomes 27.6 billion years of past to this photon.

The good thing about using reference frames this way is that we learn something right about photons and how they do not seem to age since they travel at the speed of light. On the other hand, it does not make sense to say that a photon can gain 27.6 billion years of history in one second. The reference frame of the photon is not useful for discussing the age of the universe. If we want to say something interesting about the age of the universe, we need a reference frame for the whole universe, and this can be defined. Even if the *geometry* of space and time does not prefer a special frame, the *content* of space and time makes some reference frames special (Callender, 2011, p. 208) – not in the sense that different laws apply to these reference frames, but that they stand out as special. There is a frame of reference for the universe as a whole where the background radiation from the Big Bang is similar in all directions, and this is the reference frame for the universe as a whole. It is used to define the age of the universe as being ca. 13.8 billion years.[199]

This reference frame might seem to solve all problems for those who are presentists, but because of gravitational effects being different at different places, this reference frame in physics is found by different global averaging techniques and cannot be used to define a global now that all can agree on at specific local places (Callender, 2017, p. 75).

It is true that the geometry of GR does not give us a way to define a global simultaneity slice through the whole universe. On the other hand, we can make sense of the idea of a global simultaneity slice. We can understand what it means by reflection even if we cannot perform an experiment to measure it. How? Imagine the reference frame of the universe as a lattice work of synchronized cameras permeating the whole universe, then taking a picture at one point of time as defined by this frame. The pictures could be combined into one big snapshot of the universe, where one could see some objects being contracted. This cannot be done in practice, and black holes would disturb parts of it, and it is a definition of universal simultaneity which physics has no need for, but it is a definition of

199 For details, see Callender (2017, p. 73).

universal simultaneity that makes sense, contra those who say that no such concept can make sense.[200]

The big choice to make is whether to believe with the presentists that what exists at such a universal slice of simultaneity is all that exists, or whether to believe with the eternalists that all simultaneity slices exist at the same time. The question is whether to believe that there is one three-dimensional world that changes through time or to believe that there exists a four-dimensional block universe where every event exists in the same way, with no interesting ontological difference between past and future.

We have seen some of the advantages that make physicists choose no reference frame as special, but what is actually the ontology that is being presupposed if you make this physics into an ontology of a block universe? What does it mean to live in a block universe? To consider this, imagine moving your hand slowly towards your face. At first, the hand is 20 centimeters away, then 15 centimeters, then 10 centimeters, etc. If these moments all exist in the same way, there must exist forever a state of affairs where your hand is localized 20 centimeters from your face and a state of affairs where your hand is localized 10 centimeters from your face, but this must then happen at two different places. There is a you that sees the hand 20 centimeters away and a you that sees it 10 centimeters away, and these two events cannot be happening at the same time and place, but if both exist in the same way forever, they must be happening at two different places in order to be understandable as something existing.

If the universe is a four-dimensional block universe, then "time" and "dimension" must be something unknown which exists and where things can be located. And there must be an infinity of states of affairs existing at the same time: a whole universe where my hand is 20 centimeters from my face, a whole universe where my hand is 15 centimeters from my face, a whole universe where my hand is 10 centimeters from my face, etc.

Or maybe one has another strange theory about objects as four-dimensional objects with time parts, but this will also be something unknown and mystical. The point I am trying to make is that a four-dimensional universe is a quite crazy ontology which takes on board a lot of problems in order to win a few advantages in how physics is done. You can get rid of a preferred frame of reference, but have to take on an infinite number of universes and a strange spacetime substance with an unknown kind of dimension – in addition to other problems:

200 Observers moving relative to this reference frame would not consider that which was on the picture as being simultaneous events relative to their own reference frame, but they would agree that it was simultaneous relative to the reference frame of the universe – which was the whole point here.

How did this massive block come into existence all at once, or how could it be like this forever? It makes so much more sense with a three-dimensional universe that has changed and developed over time. We understand how evolution can select those who are best fit to survive over time, but if everything has always existed in a giant frozen now, how does evolution make sense? How should we understand the advantages that have selected some species if all have just existed forever anyway?

The defender of a four-dimensional universe could still argue that physics has taught us that we live in a four-dimensional world, and that we need the geometry of spacetime to explain why things move as they do. Minkowski famously said that space and time must fade away and spacetime take its place (Minkowski, 1952, p. 76). An important choice must here be made: should we believe that spacetime and its geometry explains motion in the world, or should we think instead that motion in the world happens according to certain rules, which are best described by a certain geometry?

We already know that motion in the world happens according to certain rules, so it would be a simpler theory if we could remove four-dimensional spacetime as an additional entity with a certain structure and a capacity for influencing objects. Harvey Brown argues this in his book *Physical Relativity*. It is not the geometry of space that makes objects contract or move slower, rather there are physical explanations for this, which are then well described by use of the Lorentz factor and the geometries of SR and GR (Brown, 2005, chapters 7 and 8). Michael Esfeld makes the same point that general relativity describes patterns in the motion of particles (Esfeld et al., 2018, p. 154).

Brown argues that we have no idea what a geometrical explanation for contraction would be, and that we have no reason to believe that the geometry of spacetime should explain anything. A common example is that spacetime geometry explains inertia in the way that objects naturally follow geodesics, but Brown points out that spinning objects do not follow geodesics, and argues instead that it is the field equations that explain inertia. Clocks and rods do not know what kind of spacetime they are in. It is the laws of nature that explain why the geometry of GR is useful to describe the universe; it is not the geometry that explains how the universe works (Brown, 2005, pp. 134–143).

There are many possibilities within GR geometry that are obviously not physically possible (Callender, 2017, p. 47; Maudlin, 2012, pp. 156–157). This supports the view that GR geometry should be understood as a theoretical framework which is helpful in describing the world, not that it is a description of an entity existing in the world. While some could argue that spacetime has a geometry we can discover, I would say instead that we can discover the rules that nature follows and use geometry to describe systematic relations in the motion that results.[201]

An argument against my view, could be to say that measurements of the expansion of space seems to indicate that in addition to objects in space moving, space itself is an entity that can stretch and expand. How so? The part of the universe that we can see today (often called the observable universe) is a sphere with us in the center, and the edge is light that has traveled 13.8 billion years to reach us.[202] However, the observable universe does not have a radius of 13.8 billion light years, instead the radius is 46.5 billion light years. Note that when the light from the edge of the observable universe started its travel it was 42 million (with an "m") light years away from us, but the universe has expanded (C. L. Bennett et al., 2013). It could thus seem that the universe is not just particles moving in empty space, but that space itself is an entity that stretches and expands, and thus is something extra to be added as an ingredient in our ontology. If a light source can be 46.5 billion light years away from us and the universe is 13.8 billion years old, it seems to imply superluminal speed and that we must stretch the metric of space itself to accommodate this fact.

---

201 For example, Callender says that spacetime has a metric with a unique signature where a minus sign indicates an important difference between space and time (Callender, 2017, pp. 122–123). Taylor and Wheeler refer many times to the important minus sign that distinguishes Lorentzian geometry from Euclidian geometry (E. F. Taylor and Wheeler, 1992, pp. 7, 18, 126), but they never explain it. The minus sign and the difference they refer to is that in Euclidian geometry you find distance by *adding* distances, but in Lorentzian geometry we find the interval by taking timelike distance *minus* spacelike distance (E. F. Taylor and Wheeler, 1992, p. 7). I think that this has an easy explanation: Lorentz geometry makes three dimensions into four dimensions by treating a distance at the vertical y-axis (representing proper time) as a straight line in a right triangle which is equal to a hypotenuse (representing coordinate time) minus spatial distance along the horizontal x-axis. Imagine a lab with a light beam going straight up from the floor to the ceiling, and different spaceships flying by looking at the light beam and measuring the time it takes to reach the ceiling. They will all see light moving gradually upwards, and these light paths can all be seen as different hypotenuses. The different spaceships moving at different speeds will cover different distances spatially while seeing light travel to the ceiling. All the hypotenuses (squared) minus the spatial distance traveled (squared) will be equal to the time measured for the light in the lab (squared). One distance in one dimension can be translated to a distance minus another distance in two dimensions using Pythagoras and taking the hypotenuse squared minus the vertical line squared equals the horizontal line squared. The minus sign allows three dimensions to be translated into four by help of Pythagoras.

202 How big is then the whole universe including the observable universe? Nobody knows, but speculations go from 250 times bigger (Vardanyan, Trotta, and Silk, 2011) to extremely large numbers (Page, 2007), or that it is infinite.

However, when special relativity prohibits objects from moving faster than light speed, this applies to inertial frames of reference, which in general relativity can only be applied to certain local frames of reference. No large part of the universe will be a local inertial frame of reference in general relativity, which means that we share no local frame of reference with distant galaxies receding at superluminal speed relative to us. We do not need a spacetime substance to change its metric to account for superluminal speed, since no rules governing light speed are broken in general relativity by the expansion of the universe (Davis and Lineweaver, 2003). The equations we have for describing motion (including the geometry they presuppose) are sufficient without adding an extra medium to explain what happens.

What I have tried to show so far, is that time can be measured in different ways, and simultaneity can be defined in different ways. There are different advantages and disadvantages with the different ways of doing it. What is important when it comes to the Rietdijk-Putnam-Penrose-argument is that it talks about past, present and future in a sense where the concepts are relative to observers, but it does not follow from this relative categorization that specific events are unchangeable past or that it is not open future. The relative categorization is compatible with a coherent understanding of global simultaneity, unchangeable past and open future. This means that the argument fails in showing that the future is not open, since it does not follow from relative simultaneity in one sense that there cannot be global simultaneity in another sense.

The approach I will be defending in this chapter is to consider the universe as three-dimensional, but following rules that make it easiest for physics to describe it as four-dimensional. There are still many reasons to think of it as a block universe, and I will deal with more arguments, but time has come to develop some more terminology before dealing with the rest of the arguments. I have postponed introducing this terminology till now, since the previous discussion makes it easier to understand the distinctions I am about to present.

9.3 What is "time", "simultaneity", "now", "past" and "future"?

I believe that many problems in the philosophy of time can be solved by means of more precise definitions of what is meant by time, now, simultaneous, past and future. In this section, different meanings of these terms will be defined, and we start with "time".

The word "time" has different meanings in ordinary language. I will argue in this chapter that the basic meaning of "time" is motion, which means that time is not an entity with its own existence beyond motion. I shall argue below that there would be no time without motion.

The word "time" can also be used for measurement of motion. Such measurement can be understood as an abstract and hypothetical measurement of time for the whole universe, but we also use it for actual measurements of time, either globally or locally, and with different means for measuring time.

Finally, the word "time" is used for our experience of motion. I shall use the term "time" with all these meanings, but specify which meaning I am referring to – whether it is time as motion, time as an abstract or hypothetical measurement of motion, time as in different actual measurements of motion, or time as experience of motion. The basic meaning is in any case that time is motion, but we shall see in discussions below that it is useful to distinguish between these different meanings. When reading about time, I often find that the author problematically mixes different meanings of "time", and that this causes problems.

"Simultaneity" means that two or more events occur at the same point of measured time. This must always be judged according to a reference frame, either the reference frames of the events or the reference frame of one or more observers. Two events at one spatial location can be judged as simultaneous by all observers, for example the event of a pin hitting a balloon and the balloon blowing up, but events at a spatial distance will not be judged as simultaneous by all observers according to special relativity. This means that global simultaneity can be defined in SR relative to a reference frame, but not one that all observers can agree on. In general relativity, the geometry used does not allow for a global simultaneity to be defined at all.

However, as seen we may of course define a global simultaneity by means other than the geometry of general relativity. GR cannot use this definition to calculate anything, and physicists do not need it either, since they can always calculate what they need relative to the reference frame of whatever they are considering. We cannot use it to calculate anything precise either, but we can use it to make sense of important ontological questions. Just the fact that we can meaningfully define global simultaneity is important as an argument against those who say that the universe must be a 4D-block universe since simultaneity is relative or non-existent: yes, some definitions of simultaneity are relative or non-existent, but others are not.

If we want to use GR geometry to describe the world, this has many advantages, but then we must give up global simultaneity in the sense of a simultaneity that all observers can agree upon in their frames of reference using light to measure time. However, global simultaneity in another sense can be defined and has different advantages without causing any trouble for physics. One way of giving meaning to the concept of global simultaneity is then to imagine a lattice framework of synchronized cameras relative to the rest frame of the universe, all taking a picture of the universe at the same time relative to this reference frame, then uniting the photographs into one picture, as described above.

Why choose the universe as a reference frame if all reference frames are equivalent? Why not the earth or something else? When all reference frames are equally good concerning the laws of physics, it makes most sense to choose the one that includes all the others in both geometry and content and which is the one that physicists use to speak of the universe as a whole. This does not mean that the universe is an absolute space, since the whole universe might be moving relative to a greater space. So far we have no reason to believe that the universe is moving or is located at a particular space in a larger space, but maybe in the future unsolved questions (like questions of fine-tuning or initial conditions) can be solved with reference to the speed or location of the universe as a whole relative to a larger space.

Selecting this reference frame as special does no harm to physics, and physicists themselves choose this reference frame as special when they describe things like the age of the universe as a whole. Making sense of the concept of global simultaneity also makes sense of the idea of one three-dimensional universe existing at one point of time, and it makes sense of our treatment of past, present and future as different. This includes many specific data, like the fact that we prefer a headache to be past rather than present or future. It makes more sense of data like evolution than does eternalism. And it is a falsifiable hypothesis which can be used to make predictions like my prediction now that aliens may one day come and show us a video of our past, but they will never come and show us a video from our own future, since this is not possible given presentism – even if this should be possible given a block universe where the future already exists.

While I do not think that physicists should include absolute space in physics or a preferred frame of reference, I do have a good argument in favor of there nevertheless existing an absolute space, which therefore should be part of metaphysics. The argument is as follows:

For anything that happens anywhere it must be possible that it could happen there, otherwise it would not have happened. Furthermore, it must always have been possible (in a wide sense of possibility, as defined in Chapter 3) that it could happen there, otherwise again it would not have happened. It seems to follow that there must be an absolute space of possibilities in the sense of an area where it is possible that something can happen. This is an area which could not have been moved five meters to the left, because then it would have to have been an area five meters to the left where it was possible to locate this space, and then this area to the left would have been included in the original area. Our universe must be located at an exact location in an absolute space of possibilities, and I cannot see how it could be otherwise: the existence of an absolute space in this sense seems logically necessary (inconsistent to deny).

This means that the whole universe could be located five meters to the left, and I repeat that this is not something that physics needs to care about, but it is something a theory about basic truths should care about. Because of the argument above, metaphysics should include an absolute space to be distinguished from the spacetime of physics, and maybe physics in the future will be able to solve questions by relating the universe to a larger space. What I have here defined as the absolute space is the same as the fundamental F-field which was defined in Chapter 3. It does not make sense to ask where the fundamental field is located, since the field itself defines localizability by being the area where it is possible to be.

Craig Callender argues that even if we can define global time or global simultaneity, we have no reason to think that this is related to metaphysical time or to experienced time (Callender, 2017, pp. 75, 92–93). However, I argue that the physical state of the universe causes the conscious experiences there are of the universe, and there is no other metaphysical time in addition to this, so there is a clear link between them anyway. We shall consider this further by looking at the concept of "now".

"Now" is an index word which locates the speaker. Just as "I" is an index word which selects an individual, and "here" is an index word which locates this individual in space, "now" locates this individual in time. "Now" has two different meanings as an index word, which should be distinguished. It is the moment when the individual has an experience (conscious or non-conscious) of being able to act (the act may be no more than to breathe or lie still), and it is a moment that an individual experiences being conscious.

An eternalist may well accept this understanding of "now" since it is just an index word locating an individual in the block universe. But if there is just one universe existing at any point of time, then this physical universe is the cause of the content of the experiences in the universe at that time, and thus it will be one set of conscious experiences at that same time called "now". We shall see that the brain generally causes us to experience the universe as it was 80 milliseconds ago, and different people have different experiences of which events are present. But the qualia themselves that are consciously experienced are actualized in a qualia field at the same time as the physical now. Since the physical now is the cause of our experienced now, the physical now is the fundamental now, which we must now define.

How can we pick out and define a physical content of the term "now"? The basic idea is that the physical content of the concept of "now" is the values that are actualized and constitute the universe as it exists when these values are actualized. That is still a quite circular definition, and it does not distinguish a presentist view from an eternalist view. To be more specific in giving a physical content to the concept of "now", we can say that it is the age of the universe. Since we do not know this age exactly, the exact and objective answer to when "now" is, is that it is the time that has passed since the point of time we have defined as the birth of Jesus. When I am writing this, that is 2021 years, 26 days, 8 hours and 17 minutes ago, and if you want to know what it is when you read this, you must consult a calendar and an accurate clock adjusted with Greenwich mean time, and allow that your brain needed around 80 milliseconds of time to process what it saw.

The eternalist will say that all experiences of now exist in a four-dimensional block, but that presupposes an unnecessary, huge and mysterious ontology. If there is only a three-dimensional universe actualizing a set of physical values at a time, which again cause conscious experiences, consciousness can be used as a means to find out when now is simply by checking an accurate clock (with date) and having a conscious experience of what it says.

The conscious experience of a present now explains why we feel that there is a "now" flowing through history. Since the contents of conscious experiences have a physical cause, the configuration of the physical universe will cause a lot of brains in it to have conscious experiences. Then in the next moment, physical changes happen in the universe, brain patterns change, and a new set of conscious experiences are actualized. These new experiences have a feeling of it being now, and they can connect with memories of previous feelings of it being now. This explains why it feels like a "now" is flowing through history.[203]

How does this "now" flowing through history separate past from future? There is one universe in a field of values changing here and there. Some values stay the same for a long time, while others change. For example, a mountain stays at the same place, while a tree at it grows and dies, but physical values at lower levels change faster. While the mountain stays at the same place rela-

---

203 We could imagine that the physical universe actualized different physical values relative to reference frames and that the qualia field actualizes different qualia relative to the physical values being actualized relative to reference frames. That would mean that at the same place the universe could actualize different values that were part of a past experience relative to some observers and part of a different event to be experienced as present or future relative to other observers. The so-called Unruh effect suggests that an observer at rest and an accelerating observer may disagree on the presence of particles (Unruh, 1976). On the other hand, this seems impossible at a large scale: that at the same time and the same place the universe should actualize an elephant which is past to you and a car which is future to me.

tive to earth, earth is speeding through space, which means that the mountain is moving. The specific configuration of the values that constitute the universe changes constantly. To consider how these changes relate to each other, we can either use a coordinate system or look at specific events. If we look at events, we can sort them into chains of causes and effects. While some causes and effects are simultaneous, the general picture is that causes precede effects, and causes are past relative to their effects, which are future relative to their causes. When an event that is a cause of something occurs, the effect may still be open, but when effect has occurred, the cause is fixed.

If we use coordinate systems, we can set these up arbitrarily and call something past or future relative to the coordinate system, but causal chains of events in the universe have the order they have regardless of coordinate system. Since there is only one three-dimensional universe, the configuration of events being actualized is the only one that exists and constitutes what is then the present moment. Other configurations have been actualized previous to this, but no longer exist. We refer to these as the past, and the past is fixed because these configurations cannot be changed because they do not exist, so there is nothing at hand that can be changed. The future does not exist either, but the concept of the future refers to possible configurations of values in the universe that we expect will happen later. When I use the terms "previous" and "later", these are defined by the measurement of time we have chosen to use. Two years in the past is what the universe was like 730 earth rotations around itself ago.

More can be said about all these definitions, especially concerning our experience of time and now, but I will return to this later. This is enough for starting definitions, and we can begin with the different questions on the list presented in the beginning. I will start by delving a bit more into the debate between presentism and eternalism by looking at some arguments used in a closely related debate, namely that between the A-series and the B-series.

9.4 Does time flow? The A-series and B-series

The A-series and the B-series is a distinction by John McTaggart from 1908. According to the A-series, time flows and there is a time which is truly now. The B-series understanding of time is instead that there is no objective now, rather all times are related just by being earlier than or later than. Presentism is an A-series view and eternalism is a B-series view, but there are also different combinations, since the A-series does not necessarily say that only the present exists. For example, one could combine a moving now with either an eternal block universe where all times exist and "now" is like a moving spotlight on the block. Or it could be a growing block where the past exists and new layers of present are added, while the future does not exist yet.

I have chosen to focus on the debate between presentism and eternalism since these are the two most common positions (Callender, 2011, p. 171). But I include the A-series and B-series discussion here, since many will expect to find a comment on it in a discussion on the philosophy of time, and the arguments that follow are relevant for both the debate between the A-series and B-series and between presentism and eternalism.

I will now consider the following arguments against presentism and the A-series view of time:

- that the A-series is inconsistent;
- that there is no place for a "now" in the laws of physics;
- that the A-series seems to imply that it is meaningful to ask how fast time passes, even if no coherent answer seems possible; and
- that spacetime curvature demonstrates the existence of a four-dimensional spacetime. The questions will be discussed in that order.

The first argument is by McTaggart, who argued that the A-series is inconsistent since it ascribes incompatible properties to the parts of the series, which are all said to have the properties of being past, present and future at some time or from some perspective. The obvious response to this is to say that the properties of past, present and future are always relative to some point of time, so that no event is past, present and future at the same time. McTaggart anticipates this response and argues that then one tries to solve the problem by adding a new A-series to sort the original events into past, present and future, but that just gives you the same problem again, which means that you have a vicious regress (McTaggart, 1908). McTaggart sees this contradiction since he presupposes that past, present and future events always exist, but that the present moves along and changes what is past or future. If one rejects that the events exist at all times, the problem disappears (Craig, 2001, p. 144).

The second argument against presentism and the A-series is that there is no mention of "now" in the laws of physics (Callender, 2017, p. 40). Physics do not need a "now", for they have the time, "*t*", in their equation, and can fill with numbers in either the moment that is now or something else. But since "now" is a point of time that does not follow different laws than other points of time, they do not need to distinguish a now from other points of time.

On the other hand, physics does use a now all the time, and it does distinguish past from present, but they then have to use their consciousness to find out when now is, and consciousness is not part of physics. If you ask physicists a question involving now, they find out what time that is, and fill in a number for the *t* value in the equations. When discussing the age of the universe physics employs a now, and when making experiments they presuppose that there is a future that will give a result. One could argue that these are just indexical statements that would make sense in a block universe, but the point was just to show that there is a now in physics – and more importantly to show why physics do not use or need a now even if there is one.

Given certain indeterministic interpretations of quantum mechanics, there is an open future where the results of experiments are only given probabilistically. If this is right, there is an important difference between past, present and future. Craig Callender dismisses this by arguing that probability can be understood in many ways (Callender, 2017, p. 76). In Chapter 11, I will argue in favor of an ontological understanding of future probability which is the relevant here, and that nature follows indeterministic laws in determining the results of quantum measurements.

Quantum entanglement suggests that effects of measurements are simultaneous, even across the universe, and could thus be taken to also support global simultaneity. Callender argues that the simultaneity of quantum mechanics need not correspond to the simultaneity defined by metaphysics (Callender, 2017, pp. 92–93). But it does suggest global simultaneity, and if the entangled particle can be anywhere in the universe, and there is only one universe existing at a time, and if the entanglement is due to laws governing the whole universe, the most natural assumption to make is that they coincide.

The fact that quantum mechanics seems to support global simultaneity is worth taking into consideration in the whole discussion of how physics seems to support eternalism. Such arguments lean on special and general relativity, but we know that these theories are incompatible with quantum mechanics and need to be replaced by a new theory. It is nevertheless a goal to be compatible with SR and GR, since what they say about time may survive into the next theory to replace them, but it is an indicator that these data are not absolutely certain. One can go to physics to find support both for presentism and eternalism, as seen here with the examples from quantum mechanics.

The third argument against presentism and the A-series is that it seems meaningless to ask how fast time flows, since it would give the answer of one hour per hour, which does not actually tell us how fast time flows. But if we recall that time is both the abstract measurement of motion and actual motion, we see that the question is not meaningless. The measurement of motion has come about by relating different even motions to each other with more and more precision in order to find one even motion to be used as a measurement of all other motion (Callender, 2011, p. 63). A year is the time that the earth needs to orbit the sun, which was counted as twelve months, then 365 days, then 365.2422 days, etc., and a day is 24 hours, each of 60 minutes each of 60 seconds. Having looked for different kinds of stable motion, we have landed on the best one being transitions between levels of the ground state of the cesium-133 atom. Time is then an average measure of this activity in several cesium-133 atoms, in case some of them are unstable.

So, how fast does time flow? As fast as 9 192 631 770 periods of the radiation corresponding to the transition between the two hyperfine levels of the ground state of the cesium-133 atoms per second. This is 1 hour per hour, but it is also 60 minutes per hour, or ca. $1/_{24}$ of an earth rotation per hour, or how long it takes for light to travel 1 079 252 848.8 kilometers per hour, etc., and this is useful and meaningful information. It is circular, but not viciously circular, in the same way that it is not viciously circular that all the words in the dictionary are defined by other words in the dictionary. We get a meaningful measure of time by relating all these motions and dividing into a set of units, without there being anything else in addition that is really time.

What I just said needs to be qualified, since an actual measurement of motion is relative to the reference frame that is used to measure the motion, with the speed of light as the exception. We can use light to measure time with an ideal clock and we can use any other classical device like a mechanical clock, we can measure time relative to a reference frame, or we can measure the proper time that all can agree on independently of reference frame. As seen, Rocky in his rocket can then measure three years of travel with an ideal clock in his reference frame (which is also the proper time), but would have measured five years with a mechanical clock, which is also what Steve measures. Rocky measures three years one way by use of light, but could have measured five years by counting earth orbits around the sun. When we answer how fast time flows, time as a measurement relative to a frame of reference flows at one hour per hour, but actual measurements of events may differ relative to measurement method and reference frame.

The fourth argument against presentism and the A-series is that spacetime curving of light and paths of particles show that there is a four-dimensional spacetime, not just a three-dimensional universe with changes in it. In order to consider this argument, we need to discuss the concept of dimension. I have already suggested that we should not treat time as a special dimension, but here I will discuss the concept further and distinguish between a mathematical dimension and a spatial dimension.

In traditional Euclidian geometry, you could think of a point as having zero dimensions, then you could drag it in one direction to make a line, which has one dimension. If you drag the whole line in another direction you get a plane in two dimensions, and if you drag the plane in another direction you get a cube in three dimensions. Any point on a one-dimensional line can be specified with one number, any point on a plane can be specified with two numbers, and any point in a cube can be specified with three numbers.

If you drag the cube in a direction, you are still within a three-dimensional space. But you can imagine the cube changing over time, and let time be a fourth dimension with a new number to specify a point in the cube at a time. You can add as many variables as you want to each point, and specify the variables with a number, and call each of them a new mathematical dimension.

A mathematical dimension in the sense of a variable has a different meaning from a spatial dimension. You can drag a cube in different directions, but it stays within three spatial dimensions. You can twist, bend, and shape the cube in ways that makes it better described by other geometries than a traditional Euclidian geometry, but one point in space can always be specified with three coordinates in traditional Euclidian geometry. Some speak of spatial or physical dimensions that are curled up, but they must then be curled up somewhere in the three traditional spatial dimensions. Any point location in this "curled up" dimension will be located in the three spatial dimensions, even if you can add variables or theory which would require more mathematical dimensions to describe them.

Since the general theory of relativity says that spacetime can be bent, it is tempting to say that it has four physical dimensions, to explain for example why light is sometimes bent when it travels near massive objects. Maybe space consists of an unknown medium which can be bent, as the spacetime substantivalists claim. But it could also be that there is no medium, but merely that the rules nature follows in guiding the motion of particles contain four coordinates that make it seem like particles follow trajectories in a four-dimensional medium. This is what I suggested above, with Harvey Brown and Michael Esfeld as support. It would be ontologically parsimonious to leave out the medium, and just go for some rules, since nature already follows rules.

It would also give a clear meaning to the concept of spatial dimension. When we speak of spacetime as bent, this seems to make sense on the background of a traditional Euclidian three-dimensional space that we compare it with to describe it as bent, whereas four-dimensional bending does not make sense on its own; it does just in light of the effect it is said to have. I thus conclude that speaking of spacetime as bent is best understood as nature following rules which make particles behave as if they were following trajectories in a bent medium, but there is no medium (or anything else) which can meaningfully be said actually to be bent.

There is a list of other objections against presentism, which has to do with the status of the past: how can statements about the past be true, and how can we refer to or relate to the past if it does not exist? These will be discussed in the next section.

9.5 What makes statements about the past true?

The theory defended in this chapter is a version of presentism. It distinguishes between what exists in the sense of being part of and actualized in the universe now and what existed in the sense of being part of and actualized in the universe in the past, while it says that the future does not yet exist. There are some common objections against presentism based on the understanding of past and future: What makes statements about the past and future true if the past and future do not exist? How can we be related to things in the past if they do not exist? How can we refer to things in the past if they do not exist? How can events in the past cause events now if the events in the past do not exist? I will answer these questions in that order.

As shown above, one of the meanings of the term "time" is time as a measurement. We have landed on an even motion to describe the flow of time, and then divided this into seconds, hours, years, etc. Taking the present moment as now, we can stretch an imagined timeline backwards and forwards. To determine the point of time that represents the world as it was two years ago, we trace back two earth orbits around the sun, or 730 earth rotations around its axis, or a heck of a lot of cesium oscillations. The values and structures that were actualized in the world then are the world as it was two years in the past. They existed in the sense of being actualized in the world then, but many of them are not actualized in the world now, and so do not exist in the sense of being part of the world now. We can also imagine what the world will be like two years into the future, but this is undetermined and has not happened yet, so the future is just a physical possibility which we can imagine.

What makes statements about the past true? If you make a statement about the world two years ago, certain values and structures were actualized in the world at that point. A statement about the world two years ago is true if what is described was part of the world two years ago. Many will find this description too weak to function as a truth maker or grounding of statements about the past. M. Joshua Mozersky discusses many different theories and concludes that there are no good answers, and that this is the most fundamental problem presentists face (Mozersky, 2011, p. 143). My own solution is as follows:

In the chapter on truth, I said that something is true when it is part of the most coherent description of the world. This maximally coherent description is not just a description of the location of individuals in the world now; it is the description of all relations, and of what is possible, and of the history of the world. Among the relevant data for such a maximally coherent description are all the things that have been actualized in the history of the universe. This regulative idea of a maximally coherent description and the data thereby described is the truth maker for all claims about the present and the past.

What about the future – what makes statements about the future true? Statements about the future do not have truth value, although some of them are very probably true, like the claim that the world will exist one second from now (verified ... now). The reason why statements about the future are not true now is that the data to be integrated in the maximally coherent description does not yet exist and it cannot be deduced from the world at the present moment what the data will be since the universe is indeterministic. One can only hypothesize about what the future most probably will be.

How can we refer to and relate to things in the past if they do not exist? Mozersky argues that one cannot refer to something without being related to it, and that two entities cannot be related unless both exist. According to Mozersky, this is "simply a logical truth" (Mozersky, 2011, p. 130). Well, "refer" and "relate" can mean many things, so what are we actually doing when we refer to something in the past and relate to it? I suggest that what we are doing is having ideas in our mind, and so the relation in question is a relation between an idea I have and myself. How then to distinguish between referring to Santa Claus and referring to Abraham Lincoln? In the most coherent description of the world, Abraham Lincoln was once actualized in a physical field, while Santa Claus was not. I successfully refer to something in the past by thinking or saying something true about it, and above I described what makes statements about the past true.

How can events in the past cause events now if the events in the past do not exist? Causation describes lawful regularities, but it is fundamentally the laws of nature (the rules that actualizations follow) that causes future events, and the laws/rules exist at all times and make the world evolve continuously based on what happened in the previous moment.

9.6 How long does "now" last?

How long does "now" last? This question is related to the question of whether there are points of time and whether time is continuous or discrete. Mathematically, we can consider the timescale a continuum and, theoretically speaking, divide it into infinitely many parts. The same goes for the geometry of space, which is continuous in relativity theory (and quantum field theory assigns values to points in spacetime). There are also techniques for turning equations with continuous time into equations with discrete time (Callender, 2011, p. 147). Particles are quantized, and so are the values of mass, charge and spin they come in. Should the content of the world then be understood as continuous or as discrete?

What is most important for the question at hand is then what to think ontologically about the values that are actualized in our world. In the introductory chapter, I chose an easy way out of the problem and defined a metaphysical point (as opposed to a mathematical point) as no smaller than the smallest area where a change can happen that it is metaphysically possible to register, arguing that there was no need in any theory for a hypothetically smaller change. Ontologically speaking, a now should be understood to last for as long a time as the smallest change takes to occur. This gives a different length of a metaphysical now, a physical now, and a conscious now – which should not be a problem.

We do not know how short a metaphysical point of time is, but how does this work out when it comes to the physical now and the conscious now? In physics, Planck length is 10^{-35} meters and Planck time is 10^{-43} seconds. Planck length is considered the shortest meaningful length and Planck time the shortest meaningful unit of time. What that means is that no theories today need or can say anything meaningful about things happening at smaller places or in shorter periods of time. Metaphysically speaking, there may be changes taking place at smaller places or shorter periods of time, so metaphysically speaking the physical now lasts for the period of time that the smallest physical change takes, but we do not know what it is. The best estimate of today is that it lasts for 10^{-43} seconds.

When it comes to consciousness, the question of how long a conscious now lasts is a complex question which will be discussed more below under the question of how the mind relates to time. The general picture is that the brain takes in information from the physical world and waits for about 80 milliseconds before the information is consciously experienced. If a conscious now is defined as the minimal duration of a conscious experience before it changes, there is no consensus in research, but estimates range from about 25 to 150 milliseconds (Coren, Ward, and Enns, 2004, p. 351).

9.7 Why does time move forwards?

I am now ready to move on to the next big question listed in the introduction: Why does time move forward – why does time have an arrow? Space does not have a direction, but time does. Time moves from past to future, but could it have been otherwise? There seems to be many processes that only occur in one direction: ice cubes melt in hot water, glasses break into pieces, etc. – but the reverse process never happens. However, physics textbooks say that the laws of nature are in most cases time-symmetric and describe processes that could just as well go backward as forward in time, so where does the direction of time come from? This is the problem of the arrow (or the problems of the arrows) of time.

In order to deal with these problems, we need first to make a distinction between processes occurring in the world and time as a measurement of motion. Time as a measurement of motion is just defined as an arrow moving from past to future with marks for every second of motion in the world. Let us say that suddenly every process in the universe started going in the reverse. Time as a measurement of motion would still be an arrow pointing towards the future that allows measuring of motion at a certain speed even if every specific process was reversing. It would allow someone outside of the universe to say, "Now the universe is just like it was a second ago", then a second later into the future, "Now the universe is just like it was two seconds ago", etc. Time as a measurement of motion would continue forward into the future even if all processes and motion in the universe was reversing to become what it was in the past.

It is thus very different to answer why time as a measurement of motion has an arrow pointing to the future and why certain processes in the universe have a certain future direction in the sense of only happening one way. Some processes in the universe look the same if you record them on tape and reverse the movie, whereas others do not. The most common explanation for why most processes have this forward direction is that, given a low entropy in the beginning of the universe, it is highly probable that the entropy of the process will increase. For example, given the chaotic action between molecules in a cup of hot water and an ice cube, it is very probable that the ice cube will melt compared to not melting or an ice cube forming from hot water, although both alternatives are physically possible (Albert, 2000, p. 96).

In addition, some physical processes seem to have a certain direction following from the laws of nature, such as B meson decay, which is a process that never happens in reverse (Albert, 2000, p. 16). Actually, David Albert argues well that the common textbooks in physics are wrong when they say that the laws of nature are time-symmetric. He argues that all theories in physics for the last 150 years use laws of nature that are not time-symmetric when we look at the details. It is just true that they are time-symmetric at a very coarse-grained level, looking merely at positions of particles (Albert, 2000, pp. 14–21).

In any case, we should thus distinguish between the question of what makes certain processes happen in a certain way which implies a direction and the question of why time has an arrow. Time in the sense of motion does not have a direction, even if some specific processes happen in a way that lets us deduce a direction. It is time as a measurement that has a direction, because we have defined the direction and apply it to all other processes. For every motion that happens, in whatever way it happens, we define it as the future moment becoming the present moment and moving into the past.

This direction of time that we have chosen is of course linked to the physical process of our brains adding new memories to itself. Actual measurements of time thus presuppose that this process continues as before. If every process in the universe had started going backward, then an actual measurement of time would have been going backward, whereas we can think of time as a hypothetical abstract measurement process that would continue to move forward while all processes in the universe (including actual measurements of time) went backward. In other words, we could sit outside of the universe and see that all actual measurements of time were going backward while we measured time going forward.

This understanding of the arrow of time reveals yet another reason why the past is fixed and cannot be changed, for if something that had happened in the past were to change two seconds from now, this would be a change that happens two seconds into the future from now and not a change in the past – for there is no past existing anywhere to be changed. The values that were once actualized at a certain place are no longer there and cannot be changed. You cannot change the past because you cannot change something that does not exist.

9.8 What kinds of time travel are possible?

This also means that there is no past or future existing anywhere for time travelers to travel to. There are just two ways that something could occur which we could call time travel. The first option is that every process in the universe starts to reverse, except from any process that involves the mind and body (and time machine if that is being used) of the time traveler. Time as a hypothetical and abstract measurement of motion would still move forward, and so the time traveler would feel that her own time was moving forward, but it would also feel like going back in time for the time traveler, since the universe would actualize again something which had been actualized before, and which the time traveler could then experience (as part of her own future). But unless the universe re-actualized how it was in e. g. 1960, there would not be a 1960 anywhere that a time traveler could travel to.

Usually, as shown in most of the well-known time-traveling movies, time is moving forward in the mind of the time traveler as he or she uses a few seconds to move many years backward in time. One gets several timelines and it immediately becomes a logically inconsistent mess, where the universe in different configurations must be actualized at different places at once. Since I do not believe that there are different universes actualized at different locations at different times, I cannot make sense of the concept of several timelines either. At least, it clearly seems a lot more ontologically parsimonious to believe in just one universe from moment to moment.[204]

The other option for time travel is for a person to move at almost light speed or in extreme gravitational contexts. Biological processes can then slow down for this person in a way that makes it feel like traveling to the future when the person goes home again and feels, for example, five years older, while a hundred years have passed back home. If only such kinds of time travel are possible, many logical problems are avoided, like the problem of you killing your grandfather.

One other possibility should be mentioned because it is a popular idea: Could spacetime be curved in a way that would make it possible to travel through a wormhole into another time? This raises the question of the dimensionality of the universe. Above I argued that it would be ontologically more parsimonious to defend a spatially three-dimensional universe instead of a four-dimensional entity that can be bent, and that nature instead follows rules making the universe seem four-dimensional. If that is right and it is possible to travel in time through a wormhole, it would make most sense to understand it as time travel of the first type.

Even if spacetime is not a medium that can be bent, nature could possibly follow rules that in certain circumstances would suddenly actualize the world as it was many years ago, which would be like traveling to the past. Since I have no reason to believe that nature follows such rules, I do not believe that it does, until new evidence should suggest otherwise.

9.9 How does time relate to mind?

How does time relate to mind? I will answer this question by considering a list of questions that fall under this topic. How does the brain register motion in the world and turn it into a conscious perception of motion? How should we understand the conscious experience of presence or of now? Is the experience of time discrete or continuous? I will treat these questions in this order.

---

204 I do not say universe as opposed to multiverse, only as opposed to the same universe having different timelines.

How does the brain register motion in the world and turn it into a conscious perception of motion? There is a particular area of the brain called V5, which specializes on the detection of motion. If it is destroyed, this can lead to cerebral akinetopsia, where a person becomes unable to register motion, but just see things suddenly being here and there (Zihl, von Cramon, and Mai, 1983). The brain has a very active role in integrating the input that it gets into a coherent whole, filling in gaps to create an experience of continuous motion. When we watch a movie, we see 24 still pictures per second, but it is experienced as a continuous flow. If two spots on a screen flash alternately between left and right, the brain will make it seem like one dot is moving back and forth, even if there are just two dots standing at the same place (Wertheimer, 1912).

The brain can stretch events to last longer for different reasons. A flash that physically lasts for one millisecond can be an event that takes between 100 and 300 milliseconds in consciousness (Weichselgartner and Sperling, 1985). If you are shown a long list of familiar pictures and then suddenly a picture which does not fit in with the other, you will have a longer conscious experience of the unfamiliar picture than of the other pictures, even if they are shown for the exact same amount of time (Eagleman, 2009).

Research has shown that the brain waits for about 80 milliseconds to collect physical input before this is turned into a conscious experience (Eagleman and Sejnowskij, 2000). The reason is clearly to create as coherent an experience as possible. For example, you will consciously see a person clapping her hands and consciously hear the sound of the clap happen at the same time, even if the light from the hands entered your brain before the sound did. As long as the person is clapping within a 30 meters distance from you, you will consciously see and hear the clap at the same time, but as the person moves a little longer away, the brain gives up synchronizing, and there is suddenly a clear mismatch between sight and sound (Eagleman, 2009).

Events that happen very quickly in succession cannot be distinguished consciously as different events. There must be at least two milliseconds between auditory stimuli in order for them to be distinguished, ten milliseconds between tactile stimuli, and 20 milliseconds for visual stimuli to be distinguished. If one is to say which event happened first, there must be at least 30 milliseconds between them, regardless of the kind of event (Dainton, 2017).

How should we understand the conscious experience of presence or now? As seen, the brain merges input coming at different times into a conscious experience 80 milliseconds after the first input came. As mentioned in the previous paragraph, events that happen within shorter time frames than 30 milliseconds cannot be sequenced, and are consciously experienced as simultaneous. A conscious experience will typically last for about 100 milliseconds before it changes, but it can be shorter or longer. How short it can be before changing is disputed; estimates vary from 25 milliseconds to 150 milliseconds (Dainton, 2017). Note that these data can vary from person to person (Callender, 2017, pp. 211–212).

The experience of presence or the experience of now is then actually an experience of the content of the world as it was a very short while ago, combined in ways that can be different from person to person. Presence or now is not a thing we experience; rather we have a conscious experience of the content of the world which we call now, but which actually was a little while ago when we compare it with the physical now. On the other hand, having a conscious experience at all is an experience of presence or now, since in order to be experienced it must be actualized, and when it is actualized it is what we mean by experiencing something as now. A conscious experience is always a present experience since it is an existing conscious experience, but the content of the experience may be a representation of the world sometime in the past. The content of a conscious experience will never be a representation of the world at the same time as the experience is taking place, since it takes time for the brain to represent the world.

This raises another problem: should conscious experience be understood as continuous or discrete? It feels continuous, but that raises the following problem: What we consciously experience we experience as present in the sense just described. But we consciously experience motion, and motion occurs over an interval of time. If we experience something as present, it should be experienced as simultaneous, but motion is not simultaneous, so how can we experience motion as present (Le Poidevin, 2015)?

It seems to me that in order to solve this problem, we must distinguish between a wide and a narrow sense of the conscious now, present, and simultaneity. In the wide sense, what is now, present, or simultaneous is a feeling which can be described as the experience that something is now, present, or simultaneous, but it is an experience that *has as content* events that happen over a period of about 100 milliseconds. For example, I can feel that now in the present moment my hand is moving at the same time as the rest of my body is not.

In the narrow sense, what is consciously now, present, or simultaneous lasts for as long as it takes for the smallest discrete change we can be consciously aware of. As seen above, this may be as short a period as 25 milliseconds. This may seem to contradict our experience of continuous motion, but if you move your hand quickly, you are just aware of specific places that the hand is, and you have a feeling that it was continuous without having individual experiences of specific changes. In other words, you consciously experience it at one place and then at another place, and it has already placed in memory a feeling that this was a continuous motion. The conscious experiences replacing each other are different and discrete, yet they feel continuous.

I conclude that the present now is your conscious experience when you are conscious and feel that you can act, but there is a wide sense of this which feels longer than it is in the narrow sense, and thus we should distinguish between a conscious now in a wide and a narrow sense. Unless I specify otherwise, I use it in a narrow sense.

9.10 Can there be time without motion?

In the philosophy of time, there is a major divide between reductionism (or relationism) and Platonism (or absolutism) about time. The view presented in this chapter is reductionism, since it says that time can be reduced to something else, namely motion and the other meanings that have been defined. The alternative view, Platonism, thinks of time as an entity of its own which can pass or flow even if there is no change in the world. Reductionism says that it does not make sense to speak of time passing if there is no motion, but there is a famous argument that it does make sense to speak of time passing even if there is no motion, and this will be discussed here.

It seems easy to imagine that nothing moves, and yet time flows. Sydney Shoemaker has developed a famous thought experiment to make the idea coherent: Imagine three zones – A, B and C – where each of them sometimes experiences a local freeze – everything stops moving for an hour. This happens every second year in A, every third year in B, and every fifth year in C. For the people who experience the freeze, it just feels like going from one second to the other, but after every freeze period, there is a red glow on things for a short while. The people in the different zones know about the freezes in the other zones. The inhabitants realize that every thirty years, all three zones should experience a freeze at the same time, and they do experience the usual red glow at all places. They conclude that they have probably had an hour of global freeze, meaning that one hour has passed, even if nothing has moved (Shoemaker, 1969).

If time is basically motion, as I argue, it follows that time does not flow if absolutely nothing moves anywhere. The reason it seems that it could is that time is also a measurement of motion. It is thus easy to imagine that we are sitting outside of a world where nothing is moving, measuring that time passes while nothing moves in that world. This scenario is certainly possible, but that is because there is still motion outside where we are, measuring time while watching the world where nothing moves. But even an act of measurement is a motion, and so if there is no motion anywhere, time cannot pass, for then there is no coherent content to the phrase "time passes". Saying that an hour passes, if there is no motion anywhere, has no meaningful content, but is only a vague container metaphor of time as something mystical that "flows". If you think that it does make sense, it is because you are presupposing motion at some level somewhere, which you were supposed to exclude.

What is the cause of time? Nature has certain values that can be actualized according to rules. It is the actualization of different values at different places that make motion and conscious experiences of motion happen. The cause of time is thus the fundamental laws of nature or whatever actualizes the rules that nature follows. One may object to the claim that time is motion by arguing that motion presupposes time: there cannot be motion if there is no time. It seems right that there cannot be motion without time, but that is because time is motion, and so there cannot be motion without motion. But there is nothing in addition to motion called "time" that motion presupposes. It only presupposes that motion is possible, which is granted by the laws of nature.

9.11 Is there a beginning and an end to time?

I have come to the last three questions: Is time infinite in the past and forward directions? Was there time before the Big Bang? When will time end? I start with the end.

There does seem to be a conservation of energy in the universe which implies that motion will never stop completely. If motion never stops, time will last forever. If all motion everywhere stopped, for some unknown reason, then time would also end. It seems easy to imagine that everything stops while time continues to flow and everything stands still while time goes on. But imagining this presupposes motion used to measure time passing. If no motion happened anywhere, it would not make sense to say that time was passing. It may seem to make sense, but that is only because we presuppose motion.

This can be stated more precisely by using the distinctions I offered on different understandings of time as measurement of motion. In order for there to be an actual measurement of time for the whole universe, there has to be a continuous motion that can be used to measure time. Imagine instead that everything in the whole universe was standing still, except for a few atoms making random movements. In this scenario nothing can be used to actually measure time, but we can imagine time as a hypothetical measurement of motion which could measure time for the whole universe even if only a few atoms moved in a non-regular way.

We can use this hypothetical and abstract measurement of time also to imagine that the whole universe was standing still. But if there actually was no motion anywhere, there would be no time in the sense of motion, no time in the sense of actual measurement of motion, and no mind to make sense of the scenario we are imagining. In such a scenario there would be no content to the statement "time passes".

What about the beginning of time? Has there always been time? It seems highly problematic to think that time had a beginning and highly problematic to think that it did not. The problem with time having a beginning is to understand how it could begin. The problem with time not having a beginning is that it seems to imply that an infinite series of events (of same duration) has been actualized in order to reach the present.

The point of the last argument is that it seems to lead to contradiction to accept the possibility that a process could have started infinitely long ago and end up here today. Someone cannot count to infinity or walk infinitely far away, but then it seems that they cannot have counted down from infinity either, or started infinitely far away and arrive here. Imagine a person counting down from infinity and just now saying "... 4, 3, 2, 1, finished". Why did she finish just now, instead of yesterday, tomorrow, or never? Or if a person came to you and said he had come from infinitely far away – why did he arrive today, and not tomorrow, or never (Craig and Sinclair, 2009)?

If time has not been forever, it raises the question of what could cause the first movement. However, we must start with the existence of something, so why not start with something that had the potential for an undetermined beginning? Then there is no cause of why motion and time started exactly when it started, but there is a cause of why it started at all, namely whatever it was that had in it the potential for creating motion – the fundamental field with its value actualizer.

If we accept this, we can ask whether this cause of time itself had existed for a long time, but as with the end of time, it does not make sense to ask how long something has lasted if there is no motion anywhere. We cannot say that the fundamental cause of the world existed for, say, a million years before the first motion occurred, since the passing of time presupposes motion. We can only say that it was there less than a Planck second before the first motion, and then nothing more can be said about where it came from, if it really is the fundamental cause (although I do speculate in the end of the book why we have the fundamental cause that we have).

This was the last question on my list of questions about the philosophy of time, and so it is time to end.[205] In this chapter, I have tried to offer an analysis

---

205 I do not have a philosophy of history explaining what drives history forth. I merely describe history going forward as causal interactions over time, and I have defined causes and time. How-of time where time as measurement is a theoretical framework that is used to categorize motion and events, which again are actualizations of values in a field. Time is not an irreducible entity on its own in addition to motion, but is fundamentally motion and then there are different concepts of time that we use to categorize motion and events. We shall now move on to consider another example of something that many philosophers consider examples of irreducible entities: namely, mathematical entities.

ever, history can be described with many different theoretical frameworks that manifest different structures, relations and driving forces at the macro level of history. One can describe history with Hegel in terms of thesis, antithesis and synthesis, or in terms of information flow, or in terms of many other things, all contributing to driving history forth. For what it is worth, I guess this was a three sentence philosophy of history, but it does not comment on what are the most important driving factors on the macro scale.

10 Mathematical Truths

Philosophy of mathematics deals with a set of ontological and epistemological questions. Concerning ontology, mathematics seems to be about a set of truths. These truths further seem to be a priori (not depending on natural states of affairs), necessary (that they could not have been otherwise), and abstract (not part of space, time or causal chains). What makes mathematics true in this way (Linnebo, 2017, pp. 1–4)? Concerning epistemology, the main question is how it is possible for mathematicians to discover such truths (Linnebo, 2017, pp. 12–13). Paul Benacerraf argued that we need a causal link between our mind and that which we know, but since mathematical entities are not part of the causal chains in the universe, how can we have knowledge about them (Benacerraf, 1983, p. 409; Linnebo, 2017, p. 102)?[206]

In the following, I will suggest an answer to these problems (Section 10.1). Then I will present how I interpret the view of Øystein Linnebo in his recent books *Philosophy of Mathematics* (Linnebo, 2017) and *Thin Objects: An Abstractionist Account* (Linnebo, 2018) and argue in favor of my own position at the places where we differ (Section 10.2). I will also compare my theory briefly with other positions that are close to it, namely the views of Carnap and Hempel. The idea behind this approach is to compare my own position with the best position and the closest positions I know to show that my own approach solves more problems and is more coherent, and thus more justified as true.

10.1 A proposal for a theory of mathematics

What we think of as mathematical truths seem different from what we think of as other truths, and they are. In most cases, when we discuss whether something is true, we have a lot of data, which are states of affairs we believe to have been actualized in the world, and we try to systematize these data in the most coherent way in order to have a theory of what is most plausibly true. For example, we could know that Jones is dead and seems to be murdered, that somebody saw Smith around at the time of the murder, that Smith had such and such motive, etc., and we try to find out whether the most coherent way of integrating these data is to believe that Smith is the murderer.

---

206 I use the open term "mathematical entities" instead of "mathematical objects" to refer to numbers, mathematical relations, mathematical truths, and the like, since referring to them as objects may seem to imply that they are entities existing in a platonic realm, which I reject.

https://doi.org/10.1515/9783110770957-012

Mathematics is also about developing theoretical frameworks. But in this case the entities we are trying to relate coherently to each other are not actualized states of affairs in the universe, but instead a selection of structures it is possible to think about. More precisely, the mathematical entities are possible structures and patterns among qualia values – structures that are thinkable, and which were possible to think about even before somebody actually thought of them – and many of the ones we have thought about we may still only have a coarse understanding of, meaning that we do not fully see their implications.

For example, a dodecahedron is a structure which was possible to think about before anyone had actually thought of it and before such a structure was actualized anywhere in the physical world, and the pattern of a series of 1, 2, 3, etc., was a possible structure in qualia values before anyone had thought of it, and it is a pattern that no humans have thought of in full detail since it can continue forever and contains relations within it (for example on relations between prime numbers and even numbers) we have not yet discovered.

Which of all thinkable structures are mathematical structures? This is determined through history by exploring what structures are interesting to add to the already existing set of structures we call mathematical structures. It started with some interesting structures and patterns which were clearly useful abstractions, such as numbers, addition, geometry, etc., and then these simple frameworks have been expanded continuously. Sometimes suggested entities have seemed meaningless or useless to include, like zero or the square root of −1, but then they have shown themselves to be meaningful, interesting and useful to integrate into what we think of as mathematics.

However, there are some shared features between mathematical entities, which we shall look more closely at now. How should mathematical entities as structures be further understood? What kind of structures are they? In the second chapter of this book, we looked at how structures can be individuals and relations in a broad and narrow sense, where relations in the narrow sense (also called pure structure) was the relation between individual parts when you disregard the individuals. Many mathematical entities, such as numbers, are narrow relations in the sense of being structures that have individuals as parts, but where we disregard the individuals. For example, the number 4 is a structure that can relate four individual apples or four individual bananas, etc., and then we disregard the individuals and are left with the relation expressed in the structure that we call 4. But mathematical entities can be structures in the sense of individuals, or in the sense of relations (in the narrow sense) between individuals or in the sense of relations (in the narrow sense) between relations (in the narrow sense). For example, a circle and a diameter are individuals, and pi is a relation between them, while pi squared is a relation between two relations.[207]

As mentioned, all of these structures are structures and patterns that are possible to actualize in qualia values, meaning that they are structures that can be actualized as contents of mind.[208] Most people would like to say that circles existed before any human mind existed, and I suggest that what we should mean by saying this is not that there is a circle existing in a platonic world, but instead that circularity is a possible pattern in qualia values that existed in the sense of being thinkable even before anyone thought about it. A circle or any mathematical entity does not depend on any individual mind or consciousness in order to exist or be what it is, but it does depend on what is fundamentally metaphysically thinkable in order to exist or be what it is.

What I am trying to say is that mathematical entities exist as possible structures in mind only, and not outside of mind, so the possibility of there being mind at all is necessary for there to exist mathematical entities being what they are. Mathematical entities have always existed, since the possibility of mind has always existed (which we know since if it had not been possible from the beginning, it could not have been actualized). It follows as implication that if the possibilities of mind and consciousness had been fundamentally different, mathematics would have been different too. But it is quite hopeless to speculate about details here since our minds work the way they do, and we do not know what it would be like if thinking was very different.

From the perspective of ontology, it is good if we can avoid using a platonic world to explain mathematics. Mathematicians are often skeptical to mathematics depending on mind, but the problem is if you claim that mathematics depends on individual, different minds. However, abstract, necessary and a priori truths can be maintained when mathematics depends on the possibilities of fundamental mindedness instead of individual human minds.

Much more remains to be said about the truth of mathematics. So far we have been concerned with the ontological status of mathematical entities and

---

207 Øystein Linnebo objects that mathematical entities have many non-structural properties, such as being abstract, or being the number of planets, or Dedekind's favorite number (Linnebo, 2017, p. 163). The way I understand mathematical entities, being abstract means being a structure in qualia values, and the way I have defined structures means that being the number of planets or Dedekind's favorite number is just a relation (i.e. structure) between the number (which is a structure) and planets or Dedekind (which are also structures actualized in physical fields).

208 Many structures cannot be actualized in full detail in the human mind, only coarsely. We cannot think a full infinite series, but we can extrapolate from a series how it will continue, and we can define criteria for set membership.

how they exist. Now we shall consider the work done in mathematics with relating mathematical entities to each other in theoretical frameworks. How is such relating of entities in theoretical frameworks to be further understood: what makes theoretical frameworks into *mathematical* theoretical frameworks?

When mathematicians develop theoretical frameworks they have a roughly clear set of entities they think of as mathematical entities that they try to relate to each other coherently. This is the same ideal as when we try to make any kind of theory coherent. In the second chapter we saw that coherence has three aspects: consistency, connectedness and comprehensiveness. These are the ideals for mathematicians too, but mathematicians are usually able to fulfill the criteria better than any other discipline. Mathematicians aim for theories that are 1) consistent, but also 2) internally connected in the strong sense of entailment, which means that all parts can be proved true on the basis of some axioms and rules of inference, and 3) comprehensive in the sense of completeness, which means that for any mathematical claim (relevant for that theory) the theory will say that it is either true or false. I will now say more about these goals of mathematical theories.

A mathematical theory starts with some axioms, which are the first, unproved principles, that are appealed to in the rest of the proofs. Together with rules of inference these are the basis for all the proofs of derived principles called theorems (Linnebo, 2017, p. 22). The axioms will have implications, and usually it is dificult to see what the implications are, so we must explore them. Let us look at a very simple example.[209] You could make a list of numbers and rules for addition, and with the arithmetic we are used to, we would say that 2 + 3 = 5. But if the number series in this theoretical framework had been 1, 2, 3, 11, 12, 13, 21, 22, 23, 31 ..., then it would be true that 2 + 3 = 12. And if it had been the theoretical framework known as tropical geometry, it would be true that 2 + 3 = 3, since the rules for addition in tropical geometry say that the answer should be the highest number that you are adding. In other words, the plus sign has another meaning in this framework.

When some axioms and rules are in place, one can start to discover implications. For example, some numbers are even numbers, and some numbers are prime numbers, and there are all sorts of interesting relations between even numbers and prime numbers that can be explored and is still explored today. If one discovers that the theory has implications that are inconsistent, it is back to the drawing board. Consistency is a basic requirement, since if

209 For a more realistic examples, one can google Robinson arithmetic and look at its seven axioms, or look at six axioms for second-order Dedekind-Peano arithmetic (Linnebo, 2017, p. 33).

you allow inconsistency, anything goes and any problem can be solved trivially by giving any answer.

The ultimate dream for mathematicians would be to have a consistent theory for all of mathematics which is complete and where everything can be proved based on the axioms of the theory. However, most mathematicians have accepted that this is impossible because of the two incompleteness theorems by Kurt Gödel, which say that mathematical theories in which a certain amount of elementary arithmetic can be carried out cannot be proven within the system (Linnebo, 2017, p. 70). Even if there are some really powerful mathematical theories, completeness seems to be an unachievable goal. In the following we shall look more into why this is the case, and to understand it we must look at how mathematical theories are developed and what it means that a mathematical statement is true.

Mathematical theories are developed by starting with some axioms and rules and exploring implications. Sometimes such exploration leads to inconsistency and the theory is thrown away, but often it also happens that one either faces questions that cannot be answered, so that one needs more axioms, or there is a need for distinctions because of ambiguities (or to dissolve an inconsistency), or something that seemed meaningless can be made meaningful by introducing new elements to the theory. In other words, problems get solved by expanding the theoretical framework. Let us look at some examples.

All numbers can be multiplied with themselves – squared – or have the square root taken, and some will have a whole number as a result when you take the square root. Taking the square root of negative numbers was considered impossible until someone invented imaginary numbers, where the imaginary number i is defined as the square root of -1. Some questions do not have definite answers given the theoretical framework. What is 0 divided by 0? It can be 0, since 0 divided by something is 0; or it can be 1, since a number divided by itself is 1; or it can be infinite, since a number divided by 0 is infinite. Some will say that the question of 0 divided by 0 is wrong or illegitimate since it is not definable in the framework, but that is just to support my point. Asking for the square root of -1 was considered to be a wrong or illegitimate question until imaginary numbers were invented, thus extending the framework. Distinguishing between infinities of different size would be yet another example of expanding the framework.

It is common among mathematicians to look for the true answer to mathematical questions, but since truths in mathematics are different from truths about the universe, we should distinguish between truth relative to mathematical framework and truth relative to the most coherent mathematical framework.

I shall now spend some time on defending this claim, since it is controversial in the philosophy of mathematics.

When we ask for truths about the universe, we are interested in which states of affairs have been actualized in the physical world, for example whether there is life on other planets, what killed the dinosaurs, or whether Smith murdered Jones. When we ask for truth in mathematics, we are relating a set of possible structures to each other with different kinds of rules to see what kinds of ways they can be consistently related, to see what is provable, how complete it can be made, etc. We choose some presuppositions, defining some entities and rules, but without seeing all the consequences. Then we start exploring.

What we do when we build a mathematical theory can be well described in terms of the three aspects of coherence. We start with some parts to see if they can be connected consistently. Sometimes there will be parts that we have not been able to connect, letting us sit with loose ends, so to speak. Maybe there is a connection we have not discovered, or maybe the theory does not have the resources to establish the connection. For example, we may wonder if there is an entailment connection between even numbers and primes that allows us to prove that every even number is the sum of two primes. Or we may have good reason to think that the theory does not allow us to establish a connection, which means that more parts must be added in order for the connection to be made. An example of this would be the continuum hypothesis, which we shall return to below.

When discussing the truth of mathematics, we should distinguish between truth in two different senses: truth relative to theoretical framework and truth relative to the most coherent theoretical framework. Truth relative to theoretical framework means that a statement is true if it is coherently connected with the rest of the framework of which it is part. It is normal to speak of truth in this sense. For example, we can say that it is true that the sun rises in the east and sets in the west, and this is true relative to the theoretical framework of everyday speech, even if relative to a solar system framework we would say that the sun was standing still while the earth rotates on its axis.

Constructing a mathematical theoretical framework is like constructing a logically possible world, and a mathematical statement will then be true relative to a theoretical framework, in the sense that it is true if it fits coherently with the rest of the framework. That it fits coherently means that it is consistent with the framework and connected to its parts, including the axioms. This is the first sense of mathematical truth – truth relative to theoretical framework – and this first understanding of mathematical truth will occupy us first.

If we ask whether it is true that Smith murdered Jones with a knife, the constituents of Smith were either actualized in the presence of Jones when he was murdered or not. But when we ask for mathematical truths, we are investigating whether a possible way of relating possible structures is consistent, and this depends on the starting assumptions we made.

For example, the statement that the sum of angles in a triangle is 180 degrees is true in Euclidian flat geometry, but not in Riemannian curved geometry. It is true that 2 + 3 is 3 in tropical geometry, since it is that theoretical framework that defines the meaning of the terms. It is the theoretical framework that determines what plus means, and it is the theoretical framework that must determine how to answer the question of what 0 divided by 0 is.

This is then truth relative to framework, determined by whether it is coherent with the framework. This may seem like an irrelevant sense of relativistic truth and that we should instead speak of the one truth of mathematics, but we shall return to its relevance. Now we shall consider instead the question of whether there is one true answer to mathematical questions.

I reject the platonic interpretation of mathematics where the truth of mathematics is thought to be determined by how existing mathematical entities actually relate to each other. I do not think that there is a zero divided by zero relation in a platonic world that gives the true answer to what zero divided by zero is. However, we can make sense of mathematical truth in the sense of what would be true relative to the most coherent mathematical theoretical framework (as opposed to just being true relative to *a* theoretical framework).

It has often happened in mathematics that a problem which could not be solved in one theoretical framework could be solved in another theoretical framework. This could either be interpreted as a sign that we are on our way to the one true mathematics or just that truth is still just relative to theoretical framework, but we have found a more comprehensive framework.

There is a very comprehensive mathematical framework called ZFC set theory. It is a set theory where the elements of the set are also defined as sets. For example, numbers can be interpreted as sets, where zero is an empty set, 1 is the set containing zero, 2 is the set containing 1 and 0, etc. Given an understanding of mathematical entities as structures, it is not surprising that a set theory of sets should work very well, since the definition of a set is the same as the definition of a structure, namely a collection of elements related in a certain way.

However, as mentioned above, Kurt Gödel has demonstrated that even a comprehensive theory like ZFC cannot be complete and cannot prove its own consistency. A simple way of understanding Gödel's point is to imagine adding to a system an extra rule consistent with the others. Let us call it Rule 12, and it says that Rule 12 cannot be proven in this system. Is what the rule says false? If it is false, then it must be the case that the rule *can* be proven to be true, but if Rule 12 could be proven to be true, Rule 12 would not be false. So it cannot be false; it must be true. But if it is true, then it must be as it says, namely that it cannot be proven. It follows that the system has a true, but unprovable, rule.

It may seem irrelevant to add such disconnected rules to a system, but Gödel made the point in a much more clever way, where he made a system for turning formulas (including formulas of proofs) into numbers called Gödel numbers. With this move, the formulas should now be provable in a mathematical theory as a mathematical operation. But if a formula has a certain Gödel number, and that formula says that the formula with this Gödel number is unprovable then it can be shown to be true, but it is unprovable for the same reason as in the previous paragraph.[210]

The first incompleteness theorem showed in this way that the mathematical theories we are interested in cannot be complete, for even if you extend them to prove what was unprovable, the same problem will re-occur for the new system. The result from the first incompleteness theorem can be extended to the second incompleteness theorem, which says that the relevant mathematical theories cannot prove their own consistency.

As a simple explanation of the second incompleteness theorem, imagine again Rule 12, which says that Rule 12 cannot be proven in the system. Let us say as an initial hypothesis that the system including Rule 12 could prove its consistency. Then it could prove Rule 12, but that rule says that it cannot be proved, and that would be inconsistent. It follows that the initial hypothesis is wrong, which means that the system cannot prove its own consistency.

Gödel's insights fit very well with the understanding of mathematical theories presented in this chapter. A mathematical theory is a theoretical framework where we relate possible structures to each other as coherently as we can, but it is always possible to expand it, so it will never be complete. It can be expanded consistently with self-contained structures of the directly or indirectly self-referential kind that will not be entailed by the axioms. Since we cannot prove the consistency of the theoretical framework, this gives us good reason to define the truth of mathematical statements not in terms of provability, but instead in terms of them being integrated coherently in a theoretical framework.

This understanding of mathematics also helps us understand another classical problem in the philosophy of mathematics, namely the problem of the continuum hypothesis. The continuum hypothesis is a hypothesis about the size of different infinities, but the hypothesis itself is not the interesting part here. What is interesting is the question of whether the continuum hypothesis is true, be-

---

210 Gödel's demonstration is more advanced than a simple self-reference, but is an operation with a result implying self-reference, leading to unprovability.

cause given that ZFC is consistent, we can prove that the continuum hypothesis can neither be proved nor disproved from ZFC (Linnebo, 2017, p. 170).

Here is a way to analyze what happens in the case of the continuum hypothesis: When you are making a consistent mathematical theory, you could come to a situation where it is possible to include either A or not-A in a consistent way, but not both. In that case, you could either include A and say that A is true relative to the theoretical framework which now includes A, or you could include not-A and say that not-A is true relative to the theoretical framework which now includes not-A. Probably there is a more coherent theoretical framework which only includes either A or not-A, but possibly there is not.[211]

Most often when we have a mathematical problem, there is an answer to be found given the theoretical framework. But it may also be the case that there is no answer to be found because the answer is not implied by the axioms and the rules. When we have a theoretical framework, it may be possible to prove an answer but nobody has figured out what the proof is, or it may not be possible. Even if it is not possible to prove the answer, there could still just be one answer that is consistent with the rest of the theoretical framework and thus the true answer, but possibly that is not even the case, since several possibilities could be consistent with the framework. Most likely, one answer will be the most coherent answer by being integrated in the most coherent framework.

When it comes to the continuum hypothesis, it seems like a question that has an answer, even if we can prove that the answer cannot be proved within ZFC. On the other hand, it is possible to give different answers to the continuum hypothesis relative to different set-theoretical frameworks (Linnebo, 2017, p. 177). This would be what I called true answers relative to theoretical framework, and still it seems plausible that there is a most coherent theoretical framework, where precise continuum hypotheses can be formulated and given a final answer.

This is then another indication that mathematical truth is determined by the theoretical framework it is integrated in. We can say that the final truth in mathematics in the one determined by the most coherent theoretical framework, but we have good reason to believe that there will never be a maximally complete theory where all elements are *provable*. This is to be expected if mathematics is in fact an exploration of possible ways of relating possible structures coherent-

211 This description raises the question of what determines whether theoretical frameworks are the same or different. There are many theoretical frameworks that are overlapping but different in the sense of having different contents. Some can be completely integrated in others and are then parts of the same framework. Others are inconsistent, and then they are different theoretical frameworks. Theoretical frameworks being the same thus differ in degrees, while inconsistency is a clear marker of two theoretical frameworks being different frameworks.

ly given the enormous amount of possible structures we could have defined as mathematical entities.

The search for a final theory about the universe differs from the search for a final theory of mathematics since a theory about the universe is about physical values that have been actualized, so there will most likely be a description of these which is most coherent, and there can be no contradiction between actualized physical values: the ones that have been actualized are the ones that have been actualized, and only descriptions of them can contradict each other.

In mathematics, however, we try to relate as many possible structures as we can, which is an enormous amount. As a regulative idea, we can imagine a description of all possible qualia structures and how they relate, but it would require an infinite number of definitions and distinctions and descriptions of relations between theoretical frameworks. Any humanly possible mathematical theory today will be a theory where what we have thought about as possible mathematical structures are inconsistent with each other and cannot be put together in a consistent theory.

It may seem like I have still not answered the deepest questions of why it is true that mathematical structures are as they are, even if I say that it depends on their definitions in a theoretical framework. Even if we define numbers, circles, diameters and a Euclidian plane, why is it true that pi is 3.1415... etc., and not something else? To this I answer that it is because of the identity of that structural configuration. If pi were 4, pi (and its relations) would be something else than what we think of as pi today. If you ask, "why could not all the structures be the same and still pi be 4?", it would be like asking why is A not B? The answer is: because A = A. That is what it means to be A. I would answer the same way to similar questions, like why is the Pythagorean theorem true.

Another way of making the same point helps to explain the meaning of the sign "=", which does not signify identity, since 2 + 2 is not identical to 4 even if 2 + 2 = 4. "2 + 2" is not identical to "4" since they do not have all structural parts in common. However, there is a mathematical structure such that both "2 + 2" and "4" are descriptions of its parts. There "2 + 2" and "4" are entailed by the same mathematical structure, and thus it is provable that 2 + 2 = 4.

If we compare what is said in the last two paragraphs with what was said earlier about truth in mathematics, we see that there are two levels of truthmaking in mathematics. Firstly, mathematical statements are made true by the theoretical frameworks that determine the meaning of the mathematical concept. For example, "plus" can mean different things in different frameworks. Whether a statement including "plus" is true depends on the theoretical framework where plus is used and gets its meaning, and so we have seen that it can be true that

2 + 3 = 5, or 2 + 3 = 12, or 2 + 3 = 3, depending on the theoretical framework that makes each of these statements true.

Secondly, when the mathematical entities have been defined by the theoretical frameworks of which they are part, it is also a matter of how mindedness is fundamentally structured which determines what possible relations among qualia values there are. This makes it true that in a Euclidian plane with normal numbers the relation between a circle and its diameter is pi with value 3.1415... etc., or it makes it true that the Pythagorean theorem is what it is when its terms have been defined.[212]

All of this will hopefully be clearer when I relate it to the position of Øystein Linnebo, but before I do that, I will answer the questions in the introduction. Since we can distinguish between kinds of truth and levels of truthmaking, my answers to the questions are twofold.

How can mathematical statements be true? They can be true relative to a theoretical framework by being coherently integrated into a theoretical framework, but we can also single out certain mathematical statements as the ones that are true in the most coherent theoretical framework. In any case they are made true by the integration in the theoretical framework and the fundamental possibilities of patterns in qualia values.

Why are mathematical truths a priori? They are a priori partly because their truth depends only on the meaning of the basic structures of the theoretical framework and not on our experiences, and they are a priori because they depend only on what are possible patterns among qualia values and not on which patterns have been actualized in the physical world.

Why are mathematical truths necessary? They are necessary in a theoretical framework because their truth is given by the basic axioms of the theoretical framework. That something is necessary means that it is true given the basic assumptions in the theoretical framework and thus inconsistent to deny given the

212 This is how I would analyze Kripke's question of how to determine that plus means plus and not quus, where quus works just like plus, except that at some point the answers become 5 (Kripke, 1982, p. 9). There are many different ways to define plus, including the quus meaning, but any plus has the meaning it has been given by the axioms of the theoretical framework of which it is part. We could hypothesize that reality is fundamentally structured in a way that makes plus (as we usually understand it) suddenly turn into 5 for reasons unknown to us, but this is one of an infinite number of unfalsifiable possibilities that are of no interest. We could also think of individuals as non-consciously following a rule that suddenly makes them add numbers to five, but this would be a metaphysically uninteresting psychological fact, not relevant for the metaphysics of meaning.

presuppositions, but the axioms themselves are not necessary (unless they are logically necessary in the sense that they are inconsistent to deny).

They are also necessary because they are different descriptions of the same structure. There is a structure such that 2 + 2 and 4 are both descriptions of its parts.

Why are mathematical truths abstract? The mathematical truths we have discovered are abstract because they are narrow relations abstracted by the mind where we disregard the individuals they can relate. Fundamentally, mathematical entities are possible patterns in qualia values, and thus abstract as opposed to actualized physical values.

I have not discussed the epistemological question from Benacerraf of how it is possible for mathematicians to discover mathematical truths that they are not causally linked to. In the chapters on mind, thinking and consciousness, I have described how the brain has developed neural patterns through evolution and how they got connected to structures in the qualia field. Thinking about horses, unicorns, circles, unclehood, and numbers basically function in the same way as representations either of structures in the world or imagined structures. Thinking about these entities will be different experiences for different persons, but to varying degree persons can have understandings of structures in the physical world or structures in the qualia field which match the existing structures.

Frege complained that if mathematical entities are in the mind, my number 7 is different from your number 7, which seems absurd (Linnebo, 2017, p. 73). But our experiences of thinking about number 7 are probably different, and nevertheless there is also a possible qualia pattern we call 7, which is independent of any individual mind.

10.2 Linnebo on the philosophy of mathematics

In the following I will compare my understanding with that of Øystein Linnebo and others in order to clarify details in my position. Øystein Linnebo has written a book on the philosophy of mathematics, which is a general introduction, but where he also expresses his own views (Linnebo, 2017). Since I wanted to present a general understanding of philosophy of mathematics in this chapter, it is helpful to compare it with a general introduction to the topic where a scholar with great knowledge of the subject also expresses his own views. Linnebo's views are spelled out in great detail in the book *Thin Objects*.

Linnebo argues that mathematical truths exist independently of mind. He says that "had there been no intelligent life, these truths would still have remained the same" (Linnebo, 2017, p. 9). And further: "It is part of our experience of doing mathematics that mathematical facts are discovered and not invented" (Linnebo, 2017, p. 11). Linnebo says in his conclusion that he dismisses the view that "mathematical objects are mind-dependent in a way that sets them apart from physical objects" (Linnebo, 2017, p. 186).

According to Linnebo, there are mathematical objects he refers to as "thin objects", which means that they exist independently of humans and are related as they are independently from our understanding of them (Linnebo, 2018, pp. xi, 9, 189–190). While this is a typical description of mathematical Platonism, Linnebo specifies that he does not think that mathematical objects and physical objects are metaphysically on a par. Rather, mathematical objects are thin, which means that they do not need to be actualized in spacetime for their existence (Linnebo, 2018, pp. 10, 45, 190).

His motivation for believing in thin objects is that it is an ontology that supports our epistemological understanding of mathematics: We can discover true mathematical statements, which means that we must be referring to something true, which must exist independently of our minds (Linnebo, 2018, p. 9). There are two problems that follow if mathematical objects exist this way, and which Linnebo spends much time on solving. The first is the question of how our mathematical concepts are able to refer to mathematical objects. According to Linnebo, there is no intrinsic connection between words and thoughts on the one hand and the objects they refer to on the other hand, so how are we able to refer to mathematical objects (Linnebo, 2018, pp. xii, 10, 22)? The second problem is that it seems to give an extravagant ontology if everything that mathematics refers to exists outside of mind (Linnebo, 2018, p. 10).

To solve the reference problem, Linnebo argues that it is possible to refer by use of identity criteria. He uses the example of a robot being able to refer to objects by having identity criteria for objects, and argues that we humans non-consciously can do the same (Linnebo, 2018, pp. xii, 26–29). When it comes to the challenge of extravagant ontology, Linnebo responds that thin objects "do not make much of a demand on the world" (Linnebo, 2018, p. 10) since they do not occupy spacetime and that for thin objects "the bar to existence is set very low" (Linnebo, 2018, p. 10). A generous ontology is thus to be expected and not problematic (Linnebo, 2018, p. 10).

Compared with my own view presented above, I argue that mathematical truths are mind-independent in the sense that even before individual minds evolved, the world contained the possibility of patterns in qualia values. They are mind-dependent in the sense of being dependent on possible patterns in qualia values, and specific mathematical truths relative to framework are mind-dependent in the sense that their contents are determined by the theoretical frameworks of which they are part, but they are not dependent on individual minds.

Mathematical truths differ from physical objects in the way that a mind can construct theoretical frameworks relating mathematical concepts which refer only to possible patterns in qualia values and not to actualized physical values (nor to actualized platonic ideas in a platonic world). Many mathematical patterns are patterns that can be found among values actualized in the physical world. For this reason, mathematics is a useful tool in science, and mathematicians have been extra interested in these patterns because of their usefulness. Because mathematics potentially deals with all possible patterns it is not surprising that some of them are relevant in the physical world, but there are also extremely many completely useless (as far as we know) mathematical patterns that can be explored.

Linnebo claims that it is part of our experience of doing mathematics that mathematical facts are discovered and not invented. I find this claim to be imprecise. Instead I would say that we invent axioms and definitions to make particular theoretical frameworks, then we discover implications, relations, and new ways of making these consistent.[213] Shortly put, we discover what are the best (most coherent) inventions. We can define the plus sign any way we want without there being a true definition of the concept of plus. I am not confusing word and concept here: What plus means and what makes statements including plus true is the theoretical framework they are integrated into. The content of the concept is given by its relations, which is given by its framework. There is no true answer to what 0 divided by 0 is. We make it true by deciding which theoretical framework to place it in and what rules to follow.

Here is an example of discovering implications: Given the natural numbers in their units of ten and the meaning of addition, subtraction, multiplication and division, we can discover many implications, for example that there are odd and even numbers, that there are prime numbers, that all even numbers are the sum of two prime numbers (if they are), etc. If any of the primary presuppositions had been different, the implications would have been different (for example, if our numbers were 0, 1, 2, 3, 4, 5, 10, 11, 12, 13, 14, 15, 20, 21, 22 etc.,), and then we could discover those implications.

While the truth of mathematical statements depends on the frameworks they are part of, Linnebo does have a correct intuition when it comes to how mathematical truths also depend on how mindedness is fundamentally structured. It

---

213 To discover implications of a concept should be understood broadly. It means to discover how the concept can be understood more coherently in the relations it has.

matters what kinds of patterns among qualia values are fundamentally possible, which makes it the case that the value of pi not an arbitrary choice after the framework has been defined.

When it comes to reference, Linnebo argues that reference to something true shows that what is referred to exists, but my response is that it does not show that it has to exist outside of mind. Linnebo thinks that it does because we discover truth about mathematics that is not made true by our minds, but in my view the concepts refer to ideas and we then discover the truth about which sets of ideas are coherent. Whether a set of ideas is coherent is not made true by us, but follows from the definitions of terms and their implications. Again, the truth of mathematics does not refer only to the contents of individual minds, but depends also on the fundamental structure of mindedness, and thus it is correct that mathematical truths refer to something that exists – in the sense of existing as possible patterns in qualia values.

In any case, reference to physical entities and reference to possible patterns in qualia values both work in the same way, namely as something a mind is thinking about, which was described in further detail in Chapter 2 on the reference relation. There is no reason to make reference to mathematical entities into something special. Linnebo argues that one should not use Ockham's razor against the independent existence of mathematical objects if presupposing such existence gives you a more coherent ontology (Linnebo, 2018, p. 10). I argue in this chapter that mathematical Platonism does not give us a more coherent ontology, implying that Ockham's razor should be used to reject such Platonism.

When it comes to understanding mathematical truth, Linnebo seems to support structuralism, according to which mathematics is about abstract structures. In the words of John Barrow, "mathematics is simply the catalogue of all possible patterns".[214] Some of these patterns are actualized in the world, which would explain why branches of mathematics are so useful (Linnebo, 2017, p. 155). Frege has argued that mathematics must be more than a game since it applies to the world,[215] but Barrow can answer that *parts of it* apply to the world. Quine has argued that mathematical entities must exist since they are indispensable to our best science,[216] and again Barrow can answer that *some of them* are because they express patterns that exist in the world. This structuralist approach coincides with my own approach, but differently from Barrow, I have described how to understand the possible patterns as possible pat-

214 Barrow, quoted in Linnebo (2017, p. 155), referring to Barrow (2010, p. 371).
215 Frege, quoted in Linnebo (2017, p. 42).
216 Quine, quoted in Linnebo (2017, p. 97).

terns in qualia values, which then specifies the type of possibility and the type of pattern or structure we are talking about.

More specifically, as I interpret Linnebo, he supports a set-theoretical structuralism (Linnebo, 2017, p. 159). Almost all of mathematics can be expressed in set theory (Linnebo, 2017, p. 139). Especially powerful is the already mentioned Zermelo-Fraenkel set theory (with Choice), usually abbreviated as ZFC. It defines all set-theoretic concepts based on memberships in sets (Linnebo, 2017, pp. 141–143). Such set-theoretical structuralism raises the question of how one should understand the construction of these sets and their structures ontologically. Can they be eliminated or not? Are they actualized or merely something potential (Linnebo, 2017, pp. 150–161)? Linnebo seems to support a modal understanding of sets as possible, but leaves open the question of how to understand the relevant notion of modality (Linnebo, 2017, pp. 153, 160). Linnebo sees a problem in the fact that there are questions that cannot be answered in ZFC, like the continuum hypothesis (Linnebo, 2017, p. 170). Linnebo thinks that the question may still have an answer, but that we need to discover the axioms that can give us the answer (Linnebo, 2017, p. 171).

My own understanding of ZFC and unsolved problems is that ZFC is a theoretical framework with great ability to integrate other mathematical frameworks. I believe that its ability to do this is its set-theoretic emphasis on parts being elements in sets. This corresponds to how I think of mathematical structures as relations between parts, and how all thinking is organizing parts into wholes – or members into sets if you like.

When it comes to the modality of sets, the theoretical framework approach fits well with my understanding of modality. Modality is a hypothesis about how parts of a theoretical framework can be consistently combined given certain physical, metaphysical or semantical presuppositions. That something is possible means that it expresses a consistent combination given the presuppositions. That something is impossible means that it expresses an inconsistent combination given the presuppositions. That something is necessary means that it is implied in the presuppositions and thus inconsistent to deny. These presuppositions may be logical, physical, metaphysical or something else. For more details, see Chapter 3.

Linnebo says that the less the existence of an object demands of the world, the more modally robust the object is. Tables are thus modally fragile and could have been different, while the number zero exists with necessity (Linnebo, 2018, p. 46). I find the relation between demanding existence and modal robustness quite opaque, and think that the explanation above is far simpler. Given our presuppositions about laws of nature and what it is to be a table, it is consistent (and thus physically possible and thus metaphysically possible) to have different kinds of tables.

When it comes to the necessary existence of zero, we should distinguish between the necessity of zero existing and the necessity of zero being what it is, and distinguish between kinds of necessity. The concept of zero being what it is depends on the theoretical framework determining its meaning. Given the presuppositions of a framework implying a concept of zero, the concept of zero becomes logically necessary in the sense that it is inconsistent to deny it being what it is given the presuppositions.

At a deeper level, we could say that given the metaphysical presuppositions of which patterns of qualia values are fundamentally possible, our concept of zero is one of those possibilities that exists, so its existence is necessary in the sense that it is inconsistent with the metaphysical presuppositions to deny it.

Linnebo discusses whether one should be a pluralist or monist about mathematical theories. Can all geometries or all set theories be reduced to one, or are there just several parallel ones? The discussion ends with him saying that his book is not the place to try to solve this problem. Instead he points out that in practice mathematical work is quite similar whether one is a monist or a pluralist, since one in any case explores a plethora of mathematical theories and relate them to each other (Linnebo, 2017, pp. 170 – 182).

My own take on this is to say that mathematical statements are true relative to their theoretical frameworks, but also that different theoretical frameworks can be integrated into more coherent theoretical frameworks, where one will be the most coherent, even if there can be different, very coherent frameworks. Frameworks may of course also be incompatible with each other.

When it comes to the question of mathematical knowledge and evidence, Linnebo has discussed different theories about evidence which focus on intuition, logic, extrapolation or systematization (Linnebo, 2017, chapters 9 – 12). His own approach is pluralist and gradualist, which means that there are many kinds of mathematical evidence, and that such evidence becomes less secure as one moves into the higher reaches of the subject (Linnebo, 2017, p. 187).

My own understanding of mathematical knowledge and evidence is that it works by developing theoretical frameworks based on some axioms, then mathematicians explore what implications follow. Some follow deductively and are in that sense necessary and can be proven. Statements can be shown to be true in the sense of being consistent with and connected to the framework, or impossible in the sense of being inconsistent. If answers cannot be given, the framework can be extended by more axioms, and problems can also be solved by changing the framework or integrating it in a larger framework. In this process, coherence is the only guide. We come to know mathematical truths because of the evolved ability of our minds to relate parts into consistent wholes, and because of our conscious mind being able to experience qualia values.

This concludes my comparison with Linnebo, where I have defended a kind of coherentism against his kind of Platonism, but with many overlapping intuitions since I think of mathematical entities as possible patterns in qualia values. Linnebo asks whether coherentism is tenable, and answers that he is neutral; his intention is to defend an alternative which is not affected by the success or failure of coherentism (Linnebo, 2018, p. 7). I think that the success or failure of coherentism does affect whether coherentism or Platonism is the best justified view, and have tried to argue that coherentism is better justified than Platonism.

I end by comparing briefly with some similar views on the philosophy of mathematics and how my position differs from these. The positions I will compare my own view with are those offered by Rudolf Carnap and Carl G. Hempel. Of great importance in my own position is the role of theoretical frameworks. This is very similar to the role that Rudolf Carnap gives to his linguistic frameworks (Carnap, 1983, pp. 242–244). However, Carnap speaks only of different criteria for how to evaluate such frameworks, like efficiency, fruitfulness and simplicity (Carnap, 1983, p. 244), whereas I employ a detailed criterion of coherence which clarifies how mathematics relates to truth.

Carnap distinguishes between asking a question about mathematical entities which is internal to the framework and a question which is external to the framework. An internal question asks whether, for example, numbers exist given the framework, then obviously the answer is yes. The external question asks whether numbers refer to something existing outside of the framework, and Carnap is not able to see how such a question could be meaningfully asked (Carnap, 1983, pp. 245, 254–255).

One way of asking the external question meaningfully is to ask for the truthmakers of mathematical truths, or to ask questions about mathematical truths, like why they are necessary and a priori and seem to be discovered instead of invented. Carnap does not answer these questions, but I have argued that they can be coherently answered when the coherence of the framework is explained as the truthmaker for the mathematical claims, together with possible patterns in qualia values as truthmakers.

Carl G. Hempel answers the question of truthmakers of mathematical truths by arguing that they are made true by the definitions of the terms of the theory and deductions we can make from them according to logical rules (Hempel, 1983, pp. 379–381). This is similar to my own view, although I would not say, like Hempel, that the theory does not define the axioms (Hempel, 1983, p. 380), but instead say that it does. Nor would I say that mathematical truth is established by deduction only, but rather that the coherence of a theory (as defined above) determines the truth of the theory.

This last claim can be made clearer by considering an objection from Linnebo against deductivism in mathematics. Linnebo argues that mathematics cannot be merely about deduction from axioms because we need a contentful metamathematics as a basis to consider what follows deductively from what (Linnebo, 2017, pp. 52–53). Against Linnebo, I hold that we do not need any metamathematics beyond the criterion of coherence, and the content of the terms comes from exploring the initial definitions that we give the terms and the implications that follow.

I said above that my position was a kind of structuralism, which is something that should also be added as a difference between my own position and those of Carnap and Hempel. When exploring mathematical concepts, we explore patterns and implications and can discover something that nobody knew when introducing a concept or relation or rule. This explains the feeling that mathematics is about discovering something which does not seem explained by Carnap and Hempel.

In practice, mathematicians work in the same way regardless of whether they are Platonists or not. But many mathematicians lean towards Platonism to explain certain features of mathematics which seem unexplained if mathematical entities do not refer to something outside of mind. While many would agree that we define terms as we choose, they would still think that something outside of mind determines the truth of mathematical statements after we have defined them. I have argued that we start by selecting entities and definitions of them and that this depends on individual minds. But then we start exploring their implications and coherence, which does not depend on individual minds, but fundamentally on how mindedness is structured and what patterns are possible among qualia values. This could then be a kind of middle position between coarsely described alternatives of mind-dependence or not, but it at least leaves us free to exclude the platonic world.

In this chapter I have tried to show the importance of theoretical frameworks for our understanding of mathematics. In the next chapter, I will try a similar analysis on the concept of probability.

11 Probability

I have already been speaking of probability many times without defining it, for example in what is probably true, in the probability of values being actualized at a point, or in probabilistic causation, and I will also need the concept later in discussing ethical decisions about what most probably will give the best consequences. This chapter will be on how to understand probability.

What is probability? Is it an objective feature of the world which gives events a certain probability of occurring? Is it a quality of theories depending on how much evidence there is for the theory? Or is it a subjective degree of belief – how probable an individual finds something to be? In this chapter I argue that we make a mistake if we try to find out what probability really is. There are many different kinds of probability, and what makes them types of probability is that probability is a theoretical framework with rules for how to calculate probability, which (more or less) coincides with different phenomena in the world. Probability is not an entity of its own, but an idea we can use to analyze in more detail different phenomena in the world, such as causal powers, evidence or beliefs.

I will now present this theory in more detail, and I start by presenting different theories of probability in Section 11.1. Then I present how I understand different types of probability, and my focus will be on what I call ontological probability about the future (in Section 11.2) and epistemic probability about the past (Section 11.3).

11.1 Theories of probability

I start with a presentation of the concept of probability and how it is discussed in philosophical literature. Calculations of probability are usually performed based on some axioms developed by Alexander Kolmogorov (Kolmogorov, 1950, p. 2).[217]

---

217 I do not have space to explain the rules for those who do not know the formalism of set theory, but for those who do, I quote a brief presentation from the Stanford Encyclopedia of Philosophy:

> Let Ω be a non-empty set ('the universal set'). A *field* (or *algebra*) on Ω is a set **F** of subsets of Ω that has Ω as a member, and that is closed under complementation (with respect to Ω) and union. Let P be a function from **F** to the real numbers obeying:
> 1. (Non-negativity) $P(A) \geq 0$, for all $A \in \mathbf{F}$.
> 2. (Normalization) $P(\Omega) = 1$.
> 3. (Finite additivity) $P(A \cup B) = P(A) + P(B)$ for all $A, B \in \mathbf{F}$ such that $A \cap B = \emptyset$.

https://doi.org/10.1515/9783110770957-013

From these axioms one can deduce rules for how to measure probabilities and how they can be added, subtracted, multiplied, etc. But what is probability, ontologically speaking? Various encyclopediae and introductory books mention roughly the same interpretations divided into two or three types. The main categories are either that it is a measurement of objective evidence (the current states of affairs make it 70 % probable that it will rain) or it is a measurement of subjective degree of confidence (I am 70 % certain that it will rain), or it is an objective relation in the world (a certain atom nucleus will decay with 50 % probability).

Within these categories, there are different subtypes. Classical probability is of the first type, and is explained this way by Laplace:

> The theory of chance consists in reducing all the events of the same kind to a certain number of cases equally possible, that is to say, to such as we may be equally undecided about in regard to their existence, and in determining the number of cases favorable to the event whose probability is sought. The ratio of this number to that of all the cases possible is the measure of this probability, which is thus simply a fraction whose numerator is the number of favorable cases and whose denominator is the number of all the cases possible.[218]

Roughly put, the probability of an event is the ratio of relevant cases to the possible cases, which is called the reference class. For example, the probability of getting 5 on a die is one (the relevant case) of the six possible cases. Another kind of theory of probability as objective evidence is the logical interpretation, where probability is meant to express the degree of implication between a conclusion and its premises or between a hypothesis and its evidence.[219]

The second main type is subjective interpretation, which says that probability is the degree of confidence that a subject has that something will happen.[220] But if that is all we say about probability, what the probability of an event is will be very relative because it will differ from individual to individual. In order to make the concept less relative and closer to the objective evidence interpretation, subjective probability is often analyzed in terms of what would be reasonable betting behavior. For example, it would be reasonable betting behavior to assume that the probability of getting a 5 on a die is $1/6$, and other estimates would make one party in the betting very rich.

Inferences made by use of Bayes' theorem are most often interpreted as subjective measurements of probability, and it shows how to calculate probability

---

Call P a *probability function* and (Ω, **F**, P) a *probability space* (Hájek, 2012).

218 Laplace, quoted in Hájek (2012).

219 Most famously developed in Carnap (1950).

220 A famous example is De Finetti (1974).

given certain evidence. The probability of a hypothesis is given not simply by the prior probability of the hypothesis itself, but rather by the prior probability of the hypothesis multiplied by how probable the evidence is, given the hypothesis, divided by how probable the evidence is in itself: P (H/E) = P (H) × P (E/H) / P(E).

For example, let us say that one day several people who did not know each other at different places called newspapers and radio stations and reported that they had seen Elvis' face in the sky. How probable is it that they actually saw Elvis' face in the sky? The prior probability that Elvis' face should be in the sky is very low (P(H) = low). But the probability of the evidence (that many people called and said they had seen a face) given the hypothesis is high (P(E/H) = high). And the probability of the evidence (that different people said they had seen Elvis' face), in itself, is low. So even if the prior probability of the hypothesis is low, it must be multiplied with a high number and divided by a low number, so the total probability of the hypothesis is high.

The third main type holds that probability is an objective relation in the world. The most common subtypes are the frequency interpretation, the propensity interpretation and the best-system approach. The frequency interpretation identifies probability with the number of times that an event occurs in a series divided by the total number of events in the series. For example, if a die is thrown enough times, the number five will occur ⅙th of the times. The propensity interpretation says that probability is an objective property of an event; it has a certain probability of (propensity for) occurring. For example, a fair die has the propensity for showing number five ⅙th of the times. The best-system approach argues that probabilities follow from laws of nature and that the laws of nature should be understood as systematizations of events that occur. The best systematization of events in the world includes that the probability that a fair die shows five is ⅙ (Hájek, 2012, sections 3.4, 3.5 and 3.6).

One may say with the subjective interpretation that probability is nothing more than the degree of belief that certain people have, but that seems to make the concept too relative. It seems that something objective can be said about probability so that subjects can be right or wrong when they estimate probability. But what makes probability claims true? The evidence interpretation says that it is the evidence, which may be different things, whereas the objective interpretation says that it is either frequencies or propensities that make probability claims true.

But why do frequencies become what they are? If it is propensities that make probability claims true, what are these propensities? In any case, the frequency interpretation has a problem with single cases, and even a series of events can be seen as a single event. So, for example, flipping a coin one hundred times can give a hundred heads in a row and be considered the one event of flipping a coin one hundred times. Of course, one can always extrapolate the series, but one can also extrapolate what is considered as one event, so this raises the problem of the reference class.

Selection of reference class – the possible cases – is a typical problem when deciding probability. Tossing a coin seems to have a clear reference class (heads or tails), but what is the reference class if we ask how likely it is that I die before I am 80? Is it men; people in my age; in my country; with my education, or what? Even the reference class for coin tossing can be tricky: Should we include the coin landing on the edge, disappearing, etc.? And should it be an infinite number of tosses; how should it be ordered, etc.? It seems that depending on a subjective choice of reference class, the probability value will be different, but can probability then be something objective?

Bertrand paradoxes describe the problem as follows: A factory produces cubes with side lengths between 0 and 1 meter. The probability that a randomly selected cube should have a side length of less than ½ meter seems to be ½. But the probability that a randomly selected cube should have a face area of less than ¼ square meter seems to be ¼. The problem is that we then get two different probabilities describing the same event, since a cube with the side length of ½ meter also has a face area of ¼ square meters. Below, I shall present a solution to the Bertrand paradoxes and suggest how to deal with the problem of selection of reference classes.

This chapter discusses how to understand probability. Discussions of probability are often about what probability really is: is it something objective or epistemic or just something subjective? I argue that probability is a theoretical framework, the most common version being the one developed by Kolmogorov. This framework can then be used to describe or analyze different phenomena in the world, for example causal relations, the evidence for different theories or the relation between subjective beliefs. Using the concept of probability is a useful tool, even if it is not always possible to give accurate values. In addition, there seems to be some objective probability in the world, such as in quantum mechanics, to which we shall return below.

In this chapter I present an understanding of probability focusing on what makes probability claims true in order to achieve an understanding which is as coherent as possible.[221] Very briefly I will suggest that *ontological* probability

---

221 In this chapter and others, I use terms like "making true", "in virtue of" and "grounding". I have offered precise theories of truth, explanation, causation and actualization, but sometimes I use the less precise terms "truthmaking" and "in virtue of" or "grounding", which can refer to all of the more precise terms. The context should make the intended meaning of the imprecise terms clear. The terms typically relate something to a deeper level of explanation. If they are used when talking about statements, that which makes a statement true (or "that in virtue of

claims are about states of affairs that will happen, and they are made true by how causal powers work in the world. *Epistemic* probability claims are about the relation between evidence and explanations. As *objective* statements they are made true by the coherence of the explanation. *Subjectively* understood, they are about the beliefs of persons and are made true by mental processes in the mind of the person, but in this chapter I will focus on the objective version of epistemic probability claims.

I distinguish between past and future because it is useful for understanding epistemic probability claims. When subjective or objective *epistemic* probability claims are made about the past, what happened has *ontological* probability 1, and thus it is only the evidence which determines the suggested probability value. But when subjective or objective epistemic probability claims are made about the future, the ontological probability together with the evidence determines the suggested probability value. Take for example the claim "the probability of rain tomorrow is 50 %". This may be a person expressing his own belief that he is very uncertain about whether it will rain, but without knowing any evidence. But an expert may also have seen all the evidence of what the weather will be and say that the probability of rain is 50 %, and then the ontological probability of rain may be 50 %, while the epistemic probability of that hypothesis may be almost 100 % (because it is very certain that the world is such that rain may occur with 50 % ontological probability).[222]

These three kinds of probability thus relate to each other in the way that ontological probability is about states of affairs in the world, objective epistemic probability is about the evidence for which states of affairs actually obtain in the world, and subjective epistemic probability is about how individual persons evaluate that evidence. Each kind depends on the previous to become what it is, but not the other way around.

I will first present ontological probability about the future and the past, then objective epistemic probability about the past and future, with a brief comment

---

what it is true" or "that which grounds it") is usually a state of affairs in the world that the true statement expresses. If the terms are used when talking about states of affairs, the in virtue of-relation (or the truthmaking relation or the grounding relation) usually expresses that a state of affairs is what it is because of what it consists of/what it is constituted by at a deeper level. That a level is deeper, means that something at a higher level is a configuration of structures at a deeper level, where the deepest level is the fundamental values actualized in the fundamental field.

222 A similar kind of division of probability into three different kinds can be found in David Lewis, who relates subjective epistemic probability (credence) with ontological probability (chance) in his Principal Principle (D. Lewis, 2011, p. 460) and he relates evidence to chance by use of Bayes' theorem (D. Lewis, 2011, p. 475).

in the end about subjective epistemic probability about the past and future. For each kind of probability, I will answer the following list of questions:

A) What is this probability, ontologically speaking?
B) What is the entity that takes probability?
C) What gives the entity its probability?
D) What are the possibilities in the reference class?
E) How is the reference class selected?
F) What makes it true that this probability can be divided into degrees at all?
G) Can exact probability values be given, and if so, what are the truthmakers of exact probability values?
H) What does it mean when the probability is 1 or 0?
I) What makes it possible that the probability values can be summed to 1?
J) Can there be conditional probability of this kind?
K) How does this probability relate to standard interpretations in the literature?

Before getting started, I will comment upon two methodological questions. The first is on the criteria of adequacy: what makes something a good interpretation of probability? The second question is the general problem of how to select a reference class when making a probability assessment. To these questions I now turn, and I start with criteria of adequacy.

What makes something a good interpretation of probability? First, it should be an understanding of probability which follows the rules of probability, preferably as formulated by Kolmogorov (Kolmogorov, 1950, p. 2). But not all things that are usually understood as probability satisfy all of Kolmogorov's probability rules, and there are quantities which do follow the rules without being probability (Hájek, 2012, section 2). So secondly, it should be connected to how probability is usually understood in daily language and in science. If someone were to say that probability is something totally different and unconnected to what has previously been understood as probability, it should rather be called something else since the relation between language and reality should develop gradually. Third, it should be as coherent as possible, solving consistency problems in other theories.

How important is it that the term "probability" should just have one meaning, or that there should always exist exact probability values, etc.? There are other formal theories of probability than the one offered by Kolmogorov (Lyon, 2010), so which one should be used? The criterion of coherence does not say which features should be prioritized, just that a theory of probability should integrate as much relevant data as possible as coherently as possible.[223] I will try to develop an understanding which covers the three main usages of the term "probability" in order to have as coherent as possible an understanding of these, even if that should mean that exact and objective probability values cannot always be given.

As seen, a typical problem when deciding probability is selection of reference class. It seems that depending on a subjective choice of reference class, the probability value will be different. But can probability then be something objective? This problem is similar to the problem of how to select causes. I argued with Jonathan Schaffer in the chapter on causation that selection of contrasts is subjective, but when the contrasts have been chosen, it is objectively true what the cause is. I suggest the same solution for the problem of reference frames: when we want to decide the probability of something, we must choose a reference frame and that is a subjective choice. But when the reference frame has been chosen, it is an objective matter what the probability is relative to the reference class.

Below, I will give many examples on how probability values change depending on the reference class we choose. The very same event can then get a different probability value depending on what reference class you put it in. Many consider this a great problem, but I argue that it is not. Why not? To understand something is to integrate it in a theoretical framework – it is to understand how it relates to other things. Differently put: To understand *x* is to understand *x* in relation to *y*. If you change *y*, you also change the relation that *x* has to *y*. So if you change the reference class (*y*) to *x*, it follows that the relation (the probability value) should be different. That does not give us less understanding, but more understanding.[224] I start now with the kind of probability I call ontological future probability.

---

223 And what is to be counted as relevant data must also be argued by using the criterion of coherence.

224 Timothy Childers argues that selection of reference frame should not be subjective since we want probability to be something objective (Childers, 2013, p. 112), but as seen, there is a middle position: probability is objective relative to reference frame. He also argues that probability should not be relative to language (Childers, 2013, p. 125). I agree that it should not depend on whether it is being expressed in a specific language like English or German, but of course it will depend on language as such. That is not a problem, but applies to everything, as argued in Chapter 2 in this book.

11.2 Ontological future probability

A) What is this probability, ontologically speaking?
What is ontological future probability? It is the probability that an event has at a specific point of time for occurring in the future. For example, it is the probability that Ms. Johnson will be elected president a year before election or the day before election, and the probability will vary with time.[225] Another example is the flipping of a coin. Before flipping, the ontological probability of heads is 50 %, but a microsecond before it lands, it may be 99.99 % ontologically probable that it becomes whatever it is about to become, even if it may still be 50 % epistemically probable what the result will be.[226]

B) What is the entity that takes probability?
What is the entity taking probability in this case? The entity taking probability is states of affairs. I use "events" and "states of affairs" synonymously, since they may both be understood as either static or dynamic. In any case, the point is that they are structures that are part of the world different from ideas or linguistic entities.

C) What gives the entity its probability?
What gives the state of affairs its probability of occurring? It is causal powers in the world. Different states of affairs push, pull and constrain in different directions so that outcomes are not certain. Given that the world is not determined (which I have argued in the chapter on free will), the outcome is open, and the concept of ontological probability expresses in a total sum the relation between the different causal influences going in different directions at a point of time before the event has happened.[227] For example there may be different causal influences now making it 70 % probable that Ms. Johnson will be elected president, then just before the election it is close to 100 % probable, and after she has been elected we can say that it is 100 % (ontologically) probable that she was elected (but is was still 70 % probable at time *t* before the election).

---

225 David Lewis makes the point that chance varies with time. See D. Lewis (2011, p. 463).
226 I here assume normal coin tossing. There are coin-tossing machines where the outcome is certain, and also coin-tossing techniques where the result is almost certain (Childers, 2013, pp. 37, n. 31). This fact just supports my claim about the ontological probability of coin tossing against those who would say that it should be 50 %, for example based on a frequency interpretation of probability.
227 I specify under point K how this theory differs from similar theories that have already been proposed.

In the chapter on causality, I argued that causal claims are made true by the works of laws of nature interacting with states of affairs. In the case of quantum mechanics, there are some laws for how fields evolve at quantum levels where there seems to be a direct link between the laws and the probability of the outcome of different experiments. While some argue that probabilities in quantum mechanics should be interpreted epistemically, it is most common to think of them as ontological probabilities. In that case, the relevant causal powers are laws that give a probability for a certain outcome.

In most cases, motion in the world is due to the combined effect of many laws of nature at different levels. The laws of nature make motion happen in the world, but their influence goes in different directions, and different structures in the world function as constraints for how the laws of nature work. For example, nuclear forces can hold atoms together, chemical forces can hold a spherical object together, and gravity can pull this spherical object rolling down a hill, and so the circular downward motion of the molecules will be explained by different laws of nature and forces and structures like the ball and the hillside working in different directions and constraining each other.

Is this a way of understanding ontological probability where the content of the term "probability" becomes so complex that it is impossible ever to say something reasonable about how ontologically probable something is?[228] The way ontological probability is constituted in the world is extremely complex, and still we are often able to predict successfully what will happen based on estimates of ontological probability. Even if it is extremely complex what causes each voter to vote as they do, we can believe with some accuracy that if several thorough polls show 50/50 election results the day before election, it is 50 % ontologically probable that Ms. Johnson will be elected president.

Another example of causal complexity which nevertheless is predictable is the toss of a coin. When it is tossed there are many states of affairs influencing whether it turns heads or tails: strength of toss, wind, where it lands, etc. Since the sides of a coin are very alike and the edge is thin, it takes very little to push the coin to the one side or the other, but it will always fall on one side or the other. Since it takes so little to change the result, we can predict that the ontological probability of the coin falling on either side is ½ or 50 %.

---

228 According to Timothy Childers, that is the main problem for propensity theories, which are similar to how I define ontological probability: that it is difficult to give an exact value when the whole world (understood as indetermined) is the relevant reference class (Childers, 2013, pp. 38–39).

D) What are the possibilities in the reference class?
What are the possibilities in the reference class ontologically speaking? They are the physically possible futures that could have occurred.[229] By physically possible, I include conscious states (which I argue below depend on physical structures in our universe, namely our brains), so the point is to contrast it with logical and metaphysical possibility. The physical possibilities are all the possible futures given the laws of nature and the states of affairs in the world at one point of time.

Is it a finite number of possible futures, or is it infinite? I assume that it is finite since it is the number of possible configurations of field values at a point of time. If it is in fact infinite, the reference class would have to be infinities ordered into a finite number of sets of possible futures in order for ontological probability to a concept to be where the possibilities add up to 1.

E) How is the reference class selected?
What determines the reference class of an event? As argued above, deciding the reference class is a subjective choice including several properties of the reference class like what are the relevant possibilities, how many possibilities should be included, how is it divided into parts, etc. For example, we usually decide that a coin toss has two possible outcomes – heads and tails – and that in a series of infinite tosses the probability will converge towards ½ when we count one toss at a time. But we could include the coin landing on its edge, disappearing etc., which would make the value not exactly ½, and the sequence of events could have been organized in different ways.

One may thus choose a reference class where not all possible futures are included, but the most objective measure of ontological possibility is when all possible futures are included. And often, all possible futures can be sorted neatly into big groups. For example, all possible futures after the election tomorrow can be divided into those where Ms. Johnson has become president and those where she has not. These groups can then have subgroups. The futures where Ms. Johnson is elected president wearing blue trousers and the futures where she is elected wearing black trousers all belong in the group of possible futures where Ms. Johnson is elected president.

This grouping of possible futures together raises the following problem: One possible future can be put into different groups depending on how one chooses to group them together. For example, I can divide the possible futures into fu-

229 In one sense there is only one future, so what I mean by possible futures is different physically possible contents of that one future.

tures where I scream tomorrow night and possible futures where I do not scream tomorrow night. I could also group the possible futures into futures where I alert my family members that there is a thief in our house and those where I do not. Let us now say that there is a possible future where the following event takes place: I scream tomorrow night and thereby alert my other family members that there is a thief in our house. What is the ontological probability today of this event happening tomorrow night?

That depends on how the event is described. If I describe it as the event that I scream tomorrow night, there may be some futures where I scream because of a dream, some where I scream because I see a thief, and some where I do not scream. So in the group of possible futures where I scream, I scream for different reasons and there may be a thief in the house or not. If I describe the event as me alerting the others of a thief, there may be some futures where I alert them by screaming, some where I alert them by waking them up another way and some where there is no thief in the house. And if I describe the event as both screaming and alerting, the group of possible futures will be smaller than in the other cases. I can make it randomly small by specifying details on how loudly I scream, etc. This means that depending on how I describe the event, the number of futures containing the event and not will be different, and thus the ontological probability of the event will be different. This illustrates again the point that the selection and ordering of reference class is subjective, but that an objective measure of probability can be given after this subjective choice.

F) What makes it true that this probability can be divided into degrees at all? What makes it true that a state of affairs occurs with a probability *of a certain degree?* What makes it true that ontological probability comes in degrees? It is several things together. First of all, it is the strength of causal powers in competition with each other and everything that constrains them. The strengths of these powers are determined by the laws of nature. But if this was all, we could have had a determined world where only one future was possible so that everything happened with probability 1. So, indeterminism is the extra ingredient which makes several futures possible and thus divides the one past into many branches of possible futures. All of these possible futures together at one point of time are the total amount of possible futures and thus 100% of the possible alternatives to consider in a question of ontological probability. The different futures are what makes it true that ontological future probability comes in degrees and the concrete value of probability will then be the fraction of the possible futures in question compared with the whole amount. As seen, this can be divided into more finely or coarsely grained groups.

G) Can exact probability values be given, and if so: what are the truthmakers of exact probability values?

So far, I have argued that causal powers and constraints working in different directions make different futures possible. More precisely, it is laws of nature interacting with states of affairs in an undetermined world. It is important to notice that according to standard science (and experience), this indeterminism is not a producer of total chaos; rather it works within constraints. So for example, while it may be undetermined when a certain particle decays, it will still be a determined amount of particles that decay in a determined amount of time.

There are laws of nature that are stochastic/probabilistic, which merely make it the case that an event happens with a certain ontological probability.[230] Other laws are not probabilistically formulated. Thus, I am not proposing that stochastic laws of nature are the truthmakers of ontological probability claims, the way for example Ronald Giere seems to do (Giere, 2011, pp. 402–504). Rather, I suggest that laws of nature interacting with states of affairs in an undetermined world are together the truthmakers for ontological probability claims.

This may be a subtle difference, but there is a difference between, on the one hand, a law saying that something will happen with a certain probability and, on the other hand, laws saying that if A happens then B happens, and then the conjunction of these laws can lead to indetermined results. I gave the example previously of how a number of type-identical particles with the same speed can collide and where the trajectories afterwards, according to Newtonian physics, is undetermined (Earman, 1986, pp. 30–32). Maybe one could make metalaws for such events, so that is why I said that the difference may be subtle, but there seems to be a difference between probability being part of a law and probability arising from combinations of laws.

230 Childers mentions the critique that causation and probability can become circularly defined if probabilistic causation is included (Childers, 2013, p. 31). I define causation as depending on laws of nature and ontological probability as due to indeterminism. Some laws of nature are called probabilistic or stochastic because they say for example that 50% of an amount of particles will decay without saying which or when. Note that the law does not say that it is 50% ontologically probable that the particles will decay, but 100% ontologically probable that 50% will decay, and thus that it is 50% ontologically probable that a certain particle will decay. There are a few such examples where the ontological probability of an event is linked directly to a law of nature and where there is no explanation for why that law works that way. The law of nature is then just a mathematical rule that nature follows which has the formalism of probability calculations in it. But many laws are not like that, and most often the ontological probability will be a matter of several laws working in different directions. For this reason, I do not see it as a general problem of this theory of probability that there are some laws of nature that function probabilistically without us knowing how they work or why they do.

I also mentioned previously how small, undetermined events could scale up to the macro level, as in the example with the scientist going to lunch after a certain number of Geigerteller clicks, which resulted in her getting married. Such scenarios indicate that very different futures can depend on very small and undetermined events. This is relevant here for the following reason: If small, undetermined events today can give totally different possible futures tomorrow, does it make sense to give a value today of the ontological probability of this or that event happening tomorrow, or does the indeterminism imply that no number can be assigned since it is undetermined and thus completely open?

I believe the answer lies in the constraints that indeterminism seems to work within. The world seems to work in very regular ways, even if very many different things can also happen. The constraints then make the number of possible futures a finite number, especially if we work within short time limits (like the probability today of an event tomorrow as opposed to the probability of a certain event happening 100 billion years from now). The example of an event in the macro world (a lunch) depending directly on an undetermined micro event (Geigerteller click) is very rare.

What I have argued so far is that the choice and ordering of reference class is a subjective choice. This means that there is no probability value that can be established independent of the theoretical framework that we choose to express it in. On the other hand, when such a selection is made, there is an objective value of ontological probability which is a relation that can be translated into other theoretical frameworks or other selections of reference classes. This is because ontological probability says something about the content of physically possible futures today, and thus it is a fact about the world. At least this is so if the number of possible futures at one point of time is finite or can be ordered into finite groups (which I assume, but nobody knows for sure).

H) What does it mean when the probability is 1 or 0?

If a future event has ontological probability 1, it means that it must happen with physical necessity. Physical necessity means that given the laws of nature and the state of affairs at time *t*, event *x* must happen. If a hypothesized future event has ontological probability 0, it means that it is physically impossible for it to happen, given the laws of nature and the state of the world at the time in consideration.

I) What makes it possible that the probability values can be summed to 1?

The number 1 represents the sum of all the physically possible futures, and a part of these possible futures will be a fraction of the total amount. This part together with the rest will always then sum to 1.

J) Can there be conditional probability of this kind?
Conditional probability is about the probability of an event given a certain event. When it comes to ontological probability we must ask whether the given event is an event that has actually occurred or whether it is hypothetically given. If the event has actually occurred, then conditional ontological probability and ontological probability are the same, and there is no point in talking about conditional ontological probability. If it is hypothetically given, then we are discussing how probable something is in light of given evidence, and we are then in the domain of epistemic probability. I thus conclude that conditional probability is a matter of epistemic probability, not ontological probability.[231]

K) How does this probability relate to standard interpretations in the literature?
This understanding of probability relates most closely to the frequency interpretation, the propensity interpretation of probability and the best systems interpretation of probability. In relation to the frequency interpretation, it explains why frequencies become what they do: because of causal powers in the world. It also avoids the common problems of the frequency interpretation, like the problem of the probability of single events.

In relation to the propensity interpretation, it explains what propensity is, which Antony Eagle says is the main problem of propensity theories, namely that they seem devoid of substantial content.[232] Propensity is not a property that things have, as in the single-case interpretation of propensity. Rather, if the term is to be used, it should refer to something that characterizes the whole world (more similar to the long-run version of propensity theories), namely the relation between causal powers and constraints working in different directions.

Childers argues that propensity theories should be able to explain why propensity is probability (Childers, 2013, p. 44). I do not argue that propensity is

---

231 This is then how I would reply to the Humphreys's paradox charge against propensity theories. Humphrey's paradox can be formulated as follows: traditionally, we will use probability theory to say both that if you smoke there is a probability that you will get lung cancer and if you get lung cancer there is a probability that you smoke. The problem is that smoking causes lung cancer, but cancer does not cause smoking, thus probability cannot be about causation or propensity (Humphreys, 1985, pp. 557–559). My reply is that ontological probability is a matter of causation and goes only in the future direction from cause to effect (smoking to cancer). But epistemic probability is about evidence, and can go both ways: smoking can be evidence for cancer and cancer can be evidence for smoking. Humphreys's paradox thus shows that probability cannot just be about propensity, but it does not show that different kinds of probability cannot be about both propensity and evidence.

232 Hitchcock refers the critique that calling something propensity does not explain what it is (Hitchcock, 2001, p. 12094).

probability, but rather I have explained in detail what makes ontological probability claims true: causal powers which again come from the laws of nature understood in a realist way.

One could argue that probability claims are made true by laws of nature, but that laws of nature should be understood in a Humean way. This is the approach called the best-system approach, which understands laws of nature as the propositions that best systematize our knowledge of the world. A problem with such a Humean position is that it seems to make probability claims accidentally true and subject to undermining futures. In other words, there seems to be a great risk that the laws of nature will change in the future, since it may be a mere coincidence that coins have landed roughly 50/50 heads and tails, but in the future they may all land heads. A realist position avoids this problem.

Does the concept of ontological probability also apply to the past? Strictly speaking, no, although I sometimes say that events in the past have probability 1. The reason it does not have probability is that of all the possible futures (relative to a point in the past), only one of them has been actualized in the past. When there is only one possibility, namely the actualized one, there are no possibilities to add up to one, and thus no probability values are relevant except the value 1 for the actual past (and 0 for non-actualized previously possible pasts). One may of course discuss today what the ontological probability was of an event in the past at some point of time before that event, but in such cases the same applies as what I said above about ontological probability about the future. What I here say about past and future ontological probability presupposes a presentism in the philosophy of time, as defended in the chapter on time.

11.3 Objective epistemic probability about the past

A) What is this probability, ontologically speaking?

Epistemic probability is about explanations. It is tempting to think that a *deductive* argument is one where the premises make the conclusion 100% probable, and then ask how probable the premises make the conclusion in an *inductive* argument. Typically, that will be a question of how probable certain evidence makes a hypothesis – what degree of support it gives. Keynes, and especially Carnap, tried to develop a logical interpretation of probability where propositions could be assigned probability values.

However, there is a general agreement that this project is doomed. There are several reasons for this, an important one being that probability values seem very dependent on a particular language. Even if a very accurate language could be developed, our knowledge is limited and there is every reason to believe that the future will give us new and better terminology, as has the general Duhem-Quine problem, classically described by Quine in "Two Dogmas of Empiricism": There is no clear divide between hypothesis and data/evidence, rather we understand hypotheses and data in light of each other in new ways as our knowledge increases (Quine, 1951). There is thus little hope that we should be able to establish a clear relationship between evidence and how probable it makes a hypothesis.

When it comes to the relation between hypothesis and data, I argue that the only viable solution is to consider data as hypotheses on their own (as truth candidates instead of facts), although data will usually be considered more certain than the hypotheses they are meant to support. When it comes to the question of how probable a hypothesis is, the hypothesis and the evidence it has in support should be seen together as one explanation. In order to consider how probable this explanation is, the only possibility we have is to compare it with alternative explanations. For this reason, I suggest that objective epistemic probability about past events should be understood as a measure of how coherent the explanation is compared to alternative hypotheses.

I here lean on a very well developed coherence theory of truth, as presented in the first chapter of this book. A theory that is more coherent than another can claim that it is more probably closer to the truth than the alternative theory since it is more coherent.[233] It may nevertheless be wrong since a new theory in the future may prove to be more coherent and closer to the truth than previous theories. But we have no better way of defending our theories as true. Most other criteria of a good scientific theory are included in approach as different kinds of connections between data, but I do not think it is possible to give an exact measure of how probably true an explanation is. That is because we do not know the final truth so that we can use it as a measure, thus there is no way of knowing exact values.

This coherence theory explains well features that only Bayesianism is argued to explain (Howson and Uhrbach, 2011), like why surprising evidence is good, the paradox of confirmation (a white shoe supports that ravens are black – but only to an extremely small degree – by supporting the coherence of the claim), Duhem's problem (a theory should be considered falsified if there is a clearly more coherent alternative theory), and the difference between good

233 Note the difference between such a theory of coherence and a de Finetti-style subjective Bayesianism. According to Finetti, all we can demand from reasonable beliefs is that they are consistent and reasonably related to relevant objective data (De Finetti, 1974, p. x). In my view, the theory that integrates the most data in the most coherent way is the most rational.

and bad ad hoc hypotheses (a good ad hoc hypothesis saves many connections in the most coherent theory, while a bad ad hoc hypothesis is very loosely connected with the rest of the theory). In addition, it does not have the problems that Bayesianism has, like old evidence (coherence is good regardless of whether the evidence is new or old) or the difference between theory and data (since both are plausibly considered as theories). The defense here given will be to show how this approach answers questions and solves problems in the philosophy of probability.

B) What is the entity that takes probability?

In the paragraphs above I talked about the probability of both explanations and hypotheses. I also said that evidence or data should be seen as hypotheses on their own, so that the distinction between hypotheses and data is a distinction between less and more certain hypotheses (and I also called the data truth candidates). Explanations, hypotheses, data and evidence are thus all here understood as theories about what is the case. There is a distinction between explanations and hypotheses on the one side and data and evidence on the other side, but it is not a clear border, since it is just a matter of what is considered relatively more certain. What is a hypothesis in one context can be data in another context. In the following I shall use the term "theory" when talking about these entities in general, and use the terms "hypothesis" and "evidence" when the distinction between them is relevant.

Theories are thus the entities that take objective epistemic probability, but that includes hypotheses and data/evidence since all these entities can be considered as theories about what is the case. This is in accordance with how hypotheses and evidence are usually assigned probability in Bayes' theorem calculations.

C) What gives the entity its probability?

The probability of a theory is given by its coherence compared to alternative theories. It will not work to say that the completely coherent theory of the whole world is the theory that is 100 % probable, since this theory is just a regulative idea the content of which we do not know.

We can be very certain about many things, and say that this and that theory is almost 100 % probable, for example the theory (the truth claim) that John F. Kennedy was president in the USA in 1962. The reason this is so highly probable is that the theory "John F. Kennedy was president in the USA in 1962" is extremely coherent with numerous data, and it seems inconceivable that an alternative theory could turn out to be more coherent. Relative to the alternative theories, the one in question thus is very coherent and the others not coherent at all, and thus the coherent theory is very probable. This is why we will say that many theories are very probably true even if we do not know the final truth about the whole world. As yet an example, it is very probable that I am wearing a pair of blue pants on 3 August 2021 since no alternative and more coherent theory seems conceivable.

On the other hand, a theory is much less probable if we either have alternative theories which are equally coherent or if the theory is so little coherent that it seems very conceivable that a more coherent theory is possible to construct. Note then the difference between a situation where we have very coherent, but alternative, theories and a situation where no particularly coherent theory exists at all. An example of the first could be the different interpretations of quantum mechanics, where the main theories seem to integrate the data almost equally well, while an example of the latter could be a theory of how the brain creates qualia, where no particularly coherent theory is currently available (in 2021).

This makes it the case that the epistemic probability of a theory being true depends on which other theories we can think of. Ideally, we could wish that there was an objective way to find out how probably true a theory was without it depending on the contingent matter of whether other theories can be concocted, but this is not possible. In scientific practice, how epistemically probable we find a theory to be true does depend on the alternative theories on the market. If a new and coherent theory arrives, it makes it less epistemically probable that the previous theory was right. That is how the world is. But it is still very useful to discuss which theory is most epistemically probably true at the moment – even if our conclusions are not certain – since we often have to make choices between theories, and the most rational choice will then be the most epistemically probably true theory.

D) What are the possibilities in the reference class?

The reference class of a theory is the alternative theories we can think of. If there is a lot of evidence that Jones is the murderer, but little evidence that anyone else could have been the murderer, it is very epistemically probable that Jones is the murderer. If there is equally much evidence that either Jones or Smith is the murderer, but little evidence that anyone else could have been, it is about 50/50 epistemic probability that either is the murderer. If the evidence is equally distributed between Jones, Smith and Clark, but nobody else, the epistemic probability lowers to ⅓.

If there is hardly any evidence that Jones is the murderer and no evidence that anyone else is the murderer, it is slightly epistemically possible that Jones is the murderer, but highly epistemically possible that an unknown person is the murderer.[234]

E) How is the reference class selected?
The reference class of a theory is all the theories we can think of as realistically competing theories. This is a vague answer, but my argument is that no other theory of epistemic probability can be made more coherent and more precise.

F) What makes it true that this probability can be divided into degrees at all?
Coherence comes in degrees. While consistency is an either/or issue, a theory can be more or less coherent by integrating more or less data and by having more or less and stronger or weaker connections between its parts. These are not exact measures, and there will be cases where two competing theories have different strengths and weaknesses without there being a correct answer as to which is best since they are about equally coherent, but in different ways.

Again, we could wish that there was a more exact way of deciding between theories, but there is not. The interpretations of quantum mechanics can again serve as an example, where extremely competent philosophers of physics will choose different interpretations because they emphasize different aspects of coherence. Some will choose an Everettian many-worlds interpretation because that would give a coherent connection between Schrödinger's equation and all the superposition events that seem to follow from it. Others will prefer a Bohmian interpretation, even if that means adding more physics in addition to the Schrödinger equation, since it coheres more with the evidence we have at hand, namely one world and not many. (In the chapter on quantum mechanics, I will argue that the Bohmian interpretation is more probable than the others.) Since we neither know what the final truth about the world is, nor the language which best describes this truth, there is no hope of giving an exact measure of how epistemically probably true the theories we have today are.

However, there are often cases where one theory is obviously more coherent than another, and so we often see how the great majority of the scientific community changes their opinion and embraces a new theory, be it evolution, continental drift, heliocentricity or something else. There are different degrees of coherence between theories, and we often need to choose between them, so a measure of epistemic probability is very useful even if the values are not exact.

234 I here count having a motive as evidence, so if Jones has a motive and we have good reasons to think that nobody else has a motive, it would be wrong to say that there is hardly any evidence that Jones is the murderer, since him being the only one with a motive is important evidence.

G) Can exact probability values be given, and if so, what are the truthmakers of exact probability values?

There are no exact epistemic probability values, except in toy examples where we exclude possible alternatives like future paradigm shifts with new terms, solipsism, etc. A toy example could be the probability that the first pupil walking out of a classroom is a boy when 75 % of the pupils are boys and 25 % are girls (and we disregard the possibility that gender may not be related to sex organs in the future, or that the whole world is just my dream, etc.)? If the case was in the real world about a future event in a specific class, there could be many things influencing the ontological probability. Maybe the boys were always running out first, for example. But since this is a toy example with a simple world with no more information than what has been given, it will be a 75 % ontological probability that it is a boy, and we can say that the ontological probability is 75 % with almost 100 % epistemic probability since there is no good alternative theory. Now I discussed epistemic probability of a future event, but we could make it a case about the past by asking how epistemically probable is it that the first person that walked out of the classroom yesterday was a boy with the evidence at hand. Then it is 75 % epistemically probable that the first person who walked out of the classroom was a boy.

The distinctions just used with ontological and epistemic probability about past and future events together with the insight about selection of reference classes help us dissolve some classical problems in probability theory, like the Sleeping Beauty paradox and Bertrand paradoxes. To show the coherence of my suggestion, I will now show how they solve these paradoxes, and I start with Sleeping Beauty. The paradox is as follows:

> Some researchers are going to put you to sleep. During the two days that your sleep will last, they will briefly wake you up either once or twice, depending on the toss of a fair coin (Heads: once; Tails: twice). After each waking, they will put you to back to sleep with a drug that makes you forget that waking. When you are first awakened, to what degree ought you believe that the outcome of the coin toss is Heads? (Elga, 2000, p. 143)

Adam Elga argues in favor of probability ⅓. He gives the following two arguments: If we were to do the experiment many times, we would be right in answering heads ⅓ of the times, since two-thirds of the times we are awakened, it is after tails. The second argument is that there are three hypotheses that equally well explain Sleeping Beauty being awake: Heads and it is Monday, tails and it is Monday, and tails and it is Tuesday, and since by the principle of indifference we have no reason to prefer one over the other when we are awake, heads should be considered to be of probability ⅓ (Elga, 2000).

Note that he assigns probability ⅓ before Sleeping Beauty is told that it is Monday, but when she is told that it is Monday, she can exclude "Tails and it is Tuesday", so after being told that it is Monday two hypotheses remain, and the probability of heads is ½. David Lewis, on the other hand, replies to Elga that our credence of future chance events should equal the known chances (D. Lewis, 2001, p. 175). Before the coin is tossed, Beauty should believe that the probability of heads is ½. He argues further that only new and relevant evidence should make her change credence, and when she is awakened (but does not know which day it is), she has no new evidence, and so before knowing what day it is she should say that the probability of heads is ½. But when she learns that it is Monday, she should hold that the probability of heads is ⅔, according to Lewis. His reason is as follows: The probability of "heads" is equal to the probability of "heads and it is Monday", namely ½. The probability of tails (which is ½) equals the probability of "tails and it is Monday" or "tails and it is Tuesday". That makes the probability of "tails and it is Monday" ¼ and "tails and it is Tuesday" ¼, since together they are ½. When Beauty learns that it is Monday, she can exclude "tails and it is Tuesday". Heads and Monday is then twice as likely as Tails and Monday, which makes the probability of heads ⅔ and the probability of tails ⅓ (D. Lewis, 2001, pp. 174–175).

Here is how I analyze their discussion: History occurs once. Before the coin is tossed on Sunday, the ontological probability of heads is 50 %. Just before it lands, it is about 100 % ontologically probable that it becomes either heads or tails (depending on what it actually becomes). After the toss the ontological probability of heads is 1 if it landed heads and 0 if it landed tails. It seems clear to me that Sleeping Beauty is not asked about the ontological probability before the coin toss, but rather about the epistemic probability after the coin toss. David Lewis thinks that the two should be the same, since what we believe about a future ontological probability chance should match the ontological probability and remain the same if we receive no new evidence.

I have argued that epistemic probability is a matter of how coherent a theory is compared to alternative theories. Before Beauty is told that it is Monday, both the hypothesis heads and the hypothesis tails explain the evidence Sleeping Beauty has equally well, since her only evidence is that she is awake, so given a choice between the hypotheses heads and tails, each is 50 % epistemically probable. But it is also the case that before she is told that it is Monday, the three hypotheses, "heads and Monday", "tails and Monday", and "tails and Tuesday", all explain the evidence equally well, so given these three hypotheses, heads has epistemic probability ⅓. After she is told that it is Monday, heads and tails explain that it is Monday equally well and thus has epistemic probability ½.

"Heads and Monday" and "tails and Monday" each explain equally well and thus each has epistemic probability ½.

This is why Sleeping Beauty seems to be a paradox: The question "to what degree ought you now to believe that it landed heads" is ambiguous before Beauty is told that it is Monday. Is the question: how probable is heads given that you are awake as opposed to the probability of tails given that you are awake? Or is the question: how probable is it that you got heads and it is Monday as opposed to the probability of either tails and Monday or tails and Tuesday? The last question is more specific and has a different reference class, so it is not strange that the probability values are different depending on which reference class we specify. In the first case two hypotheses are equally coherent given the evidence and in the second case three hypotheses are equally coherent given the evidence, thus both ½ and ⅓ is correct in the different cases, even if the event is the same, since it comes in different descriptions with different reference classes.[235]

Compare with the following: There is a class 50 % of each boys and girls, and we ask how epistemically probable is it that a boy went out first yesterday. The answer is 50 %. But let us say that we do not know what date it is today and we ask: How probable is it that a boy went out first yesterday and that the date was an even number? Now there are four almost equally coherent theories: boy first and even number, boy first and odd number, girl first and even number, girl first and odd number. This is the same as either "heads and awake"/"tails and awake" or "heads and awake Monday"/"tails and awake Monday"/"tails and awake Tuesday". This means that the same event "that a boy went out first yesterday" can receive different epistemic probability values depending on the reference frames we choose.

What I am saying is that both Elga and Lewis are right in their analyses of the situation before Beauty is told that it is Monday, since they describe the situations differently. Elga is right in the analysis of the situation after Beauty is told that it is Monday. Why is Lewis wrong? Because he keeps the same probability values regardless of reference frames, and transfers the value from one context to another, which in my opinion is wrong.

What about Bertrand paradoxes? One version of the paradox is as follows: A factory produces cubes with side lengths between 0 and 1 meter. The probability that a randomly selected cube should have a side length of less than ½

---

235 Hitchcock offers a Dutch book argument in favor of ⅓ (Hitchcock, 2004), but Bradley and Leitgeb show why credences and betting behavior comes apart in this case (Bradley and Leitgeb, 2011).

meter seems to be ½. But the probability that a randomly selected cube should have a face area of less than ¼ square meters seems to be ¼. The problem is that we then get two different probabilities describing the same event, since a cube with side length ½ meters also has a face area of ¼ square meters (Van Fraassen, 1989, p. 303).[236]

There is a problematic ambiguity hidden in the paradox. What does it mean to say that any randomly chosen cube is "equally possible"? Does it mean that the subject finds it equally possible when estimating the epistemic probability of what kind of cube it will be, or does it mean that it is ontologically equally possible what kind of cube will be selected? This is very important, for if it is ontologically equally possible what kind of cube will come, it means that the factory produces equally many cubes with side length evenly distributed between 0 and 1 meter. But if it does, then it is right that the probability of selecting a cube with side length less than ½ meter is ½ and the probability of selecting a cube with face area of less than ¼ square meters is then also ½. For if the factory produces equally many cubes with side length of 1 cm, 2 cm, 3 cm, etc., up to 100, then ½ of the cubes will be less than ½ meters in side length and ½ of the cubes will have a face area of less than ¼ square meters. But if the probability is subjective epistemic probability and we do not know anything about what the ontological probability is, then the subjective estimate has very little value in any case – it just tells us something about what an individual with little knowledge thinks, which may of course very well be wrong.[237]

H) What does it mean when the probability is 1 or 0?

A theory with probability 1 is a theory which is true by logical necessity, in the sense that it is self-contradictory to deny it. Examples would be either that triangles have three corners, or that thoughts exist in our universe. A theory has almost probability 1 if it is very coherent and alternative theories which are equally coherent are almost inconceivable. Examples would be chairs exist or Kennedy was president in the USA in 1962. A theory has probability 0 if it is inconsistent, like "here and now it rains and it does not rain". A theory has almost probability 0 if it is not coherent with anything else we believe to be true, like the theory that there is an invisible teapot orbiting the sun.

236 Van Fraassen uses 2 cm cubes, but I found the example easier to understand using 1 meter.

237 Van Fraassen solves the problem of the cube factory by finding a measure which is invariant under the transformation from length to area, namely a log uniform measure (Van Fraassen, 2011, pp. 306–309). However, this does not solve other similar problems, like the wine/water problem (Van Fraassen, 2011, pp. 311–313), which favors my solution to the problem.

I) What makes it possible that the probability values can be summed to 1?
Epistemic probability values cannot be summed to one, since the number of theories and the number of data are infinite, and we do not know what the most coherent theory of the world (or anything) is. Even if the actually proposed theories and formulated data are finite, the implicit and possible data are infinite. This means that in a broad sense all epistemic probability is conditional, since a theory is always probable given the (finite amount of) concrete data it integrates.

J) Can there be conditional probability of this kind?
As mentioned, all epistemic probability is conditional in a broad sense of the term. But usually we mean conditional probability in the sense that given this particular X, how probable is Y? To ask how probable a theory is given data X thus means to ask how probable a theory is given that X is true, and the way that works is that one must suppose that X is very coherent and then consider the hypothesis in light of that. Take as an example the question of whether the universe had a period of rapid inflation in the very beginning. Physicist Paul Steinhardt has argued that a cyclic universe model explains the data we have equally well as an inflationary cosmology, but that if we were to discover imprints from gravitational waves in the cosmic microwave background radiation, inflationary cosmology would be very probably right and the cyclic model very probably wrong (Steinhardt, 2011, p. 43). In other words, given such an imprint the cyclic model is very improbable and the inflationary model probable.

K) How does this probability relate to standard interpretations in the literature?
Epistemic probability understood this way is a much less exact theory than what logical probability aimed for, but the logical probability project is a failure. This way of understanding epistemic probability matches well with actual scientific practice. Even if it rarely involves exact values, it is nevertheless quite precise and therefore useful.[238] Since it is based on a well-developed theory of truth, it is helpful and clarifying in how to relate different theoretical virtues, which is often unclear in other accounts.[239]

238 There is a growing interest in imprecise probabilities, since this is a way of understanding probability which solves many problems that arise if exact values are considered to be necessary (www.sipta.org/). Accepting imprecise values is in line with the criteria of adequacy presented in the beginning of this article.

239 For example, James Ladyman argues that there is no way of deciding between theoretical virtues (Ladyman et al., 2007, p. 83); cf Ladyman (2002, p. 257). But here at least is a theory explaining why some theoretical virtues (like the aspects of coherence) are relevant as criteria of

I do not have space to go into detail on epistemic probability about the future and subjective epistemic probability. Subjective epistemic probability is degree of belief, but what is degree of belief, and what is the truthmaker when someone believes a theory to be 75 % probable? I suggest that this term should be used to refer to how people in daily speech themselves evaluate how probable they find a proposition. For example, a person may consider evidence for and against the existence of God and say that he finds it 60/40 in favor of God's existence. This then expresses his feeling of how convinced he is that a theory is true, but there are many things influencing such convictions beyond a mere consideration of the coherence of the theories. Non-conscious factors, emotions, wishful thinking, physical brain features, upbringing, irrational factors of different kinds, etc., can influence how strongly convinced a person is about a theory being true.

Even if this kind of probability is very imprecise, it is being used in daily life because people find it useful to express more precisely what they think, to understand each other and have discussions, and to make decisions where it is relevant how probable they find something to be true. These distinctions are helpful also for philosophical analyses of statements about probability. Things can get quite complicated, for example when it gets to epistemic probability about the future, since one must then distinguish between different kinds of probability.

"It is 90 % probable that I will be no longer be sick tomorrow", a person may say with no insight into either the epistemic probability or the ontological probability. He is then expressing his own conviction, but may be far off the epistemic or ontological probability. The doctor can say that it is 90 % probable, still expressing her conviction, but now much closer to the epistemic probability given the evidence. If this is a disease the prognosis of which is very regular and well known, the doctor's statement can mean either that it is almost 90 % epistemically probable that it is almost 100 % ontologically probable that the person will no longer be sick tomorrow, or that it is almost 100 % epistemically probable that it is 90 % ontologically probable that the person will no longer be sick tomorrow. But if the disease has a quite regular prognosis, but sometimes behaves very differently without anybody knowing when or why, the doctor may still correctly say that it is epistemically almost 90 % probable that the person will no longer be sick tomorrow, yet the ontological probability may be only 10 % because special conditions happen to obtain in this case.

---

truth, while criteria like simplicity are merely pragmatic criteria, not truth criteria, since there is no reason to believe that the truth must be simple.

These distinctions can help us understand discussions of probability. For example, there is much confusion about the probability involved in discussions about the fine-tuning of the universe for life. The epistemic probability that our universe should be fine-tuned is very low, but since we do not know anything about the ontological probability that our universe should be fine-tuned, the epistemic probability is very uncertain.

In philosophy, there is a discussion of what knowledge is, since a traditional definition of knowledge as true, justified belief has problems. The term "knowledge" can refer to an objective content of something that is commonly considered to be knowledge or to a subjective relation that a person has when he or she is said to know something. I think that speaking of degrees of objective and subjective epistemic probability is as precise as we can describe people's beliefs so that we have no need for a vague concept of knowledge in addition. When I use the terms "know" or "knowledge" in this book, they only refer to well-justified beliefs.

To conclude, I find that this way of understanding probability solves the problem cases in the literature in a realistic way, which means that many problems are solved, but sometimes the solution to the problem is to realize that we wish for more objective (in the sense of less dependent on human choices) and precise probability values than it is possible to give. Probability is a theoretical framework we can use for calculating in different contexts which can sometimes be done very precisely and other times only very roughly, yet it is useful in both cases. At least if it is clear what one is doing.

In the three previous chapters, I have argued that time, mathematical truths and probability are useful theoretical frameworks, but that what they express is reducible to values being actualized according to rules. It is now time to talk more about these values and the rules they are actualized in accordance with.

12 Fundamental Concepts in Physics

I have repeatedly talked about what things are and how they move by saying that this is a matter of values being actualized according to rules. While I have written some about qualia values, very little has been said in detail about physical values and motion and the rules they follow. In fact, just by knowing very little math (again the Pythagorean theorem suffices) we can know amazingly much about how things interact from micro to macro level. We can also understand much about light speed, mass-energy equivalence and the creation and annihilation of particles. Looking a little more into concepts like mass and energy lets us understand very much about how things interact in the world.

In this chapter, I will present some fundamental concepts in physics in order to be more specific about physical values being actualized in our world, with an emphasis on mass and energy. I thought a good idea of presenting the topics in an understandable way would be to make a quasi-historical presentation from a quite intuitive understanding via Newton to Einstein. While not historically accurate, the goal is to present the ideas in an understandable way. Some of the presentations are a bit technical, and they are not important for the rest of the book, but readers who want to understand $E = mc^2$ or a little about particle creation and annihilation can get a fairly simple introduction to these topics in this chapter.

I start by presenting concepts like meter and second, since very many physical values are reducible to a relation between meters and seconds. This happens in Section 12.1. In addition to meters and seconds, we need mass, and I introduce mass, and with it momentum, in Section 12.2.

The detailed picture of how fundamental physical values can combine to give us the world we know must be discovered by physics, but I will offer some speculations at the end of this chapter. I will say a little in Section 12.2 on what concepts like mass and energy express ontologically. With these concepts in place, we can look at work, force and energy, and we start with Newtonian physics in Section 12.3 before moving to mass and energy in relativity in Section 12.4. I compare Newton and Einstein explicitly in Section 12.5 to understand what happens with mass and energy after Einstein. I can then proceed to talking about momenergy – the combination of momentum and energy into one unit – in Section 12.6, where we shall look at some fascinating and simple facts about rules that govern the motion, creation and annihilation of particles in our universe. After all this, we can end with a discussion (in Section 12.7) of what to think ontologically about all these concepts from physics and how to think about the fundamental physical values.

https://doi.org/10.1515/9783110770957-014

12.1 Meter (m), second (s), speed (m/s), and acceleration (m/s^2)

The world is experienced to consist of objects. Those objects have different sizes that can be compared. For example, a (big) human foot could be experienced to be approximately as big as 12 human thumbs are. With a common currency of, for example, feet or meters, together with numbers, all objects can be compared in size, and a unit of measuring size or space is born. The size of any object or the distance between any objects can now be described as a number expressed in meters.

Objects are experienced to move relative to each other. Some motions are constant compared to each other, and some things move faster than others. The earth orbiting the sun moves as fast as the earth takes to spin around itself 365 times, and some can run in one day a distance that others need two days to run.

Using many different constant motions that have a constant relation to each other, we can define one constant motion to be used as a measure of all other motions. The thing in motion must cover a distance, which can then be divided into units of time. For example, the earth covering the distance of spinning around its own axis once can be divided into 24 parts called hours, which can be divided into 60 parts called minutes, which can be divided into 60 parts called seconds.

With this constant motion giving us units of seconds, minutes, hours, etc., we can now compare all other motions with this motion to describe their speed. The speed will then be a measure of distance divided by time, for example kilometers per hour or meters per second.[240] Any speed can be described with a number (in physics called a "scalar") and meters per second, or m/s.[241]

---

240 I am simplifying matters here for pedagogical reasons, since I want the reader to see the development starting with meter and second and then moving to the other concepts. That is why I use the units meter and second for speed and acceleration instead of saying distance divided by time. I am also for simplicity just talking about speed here instead of distinguishing between speed and velocity, where velocity is an amount of speed in a direction, while speed only is the numerical value of the velocity. Later I use velocity when the relevant formulas use the abbreviation v and the context is a discussion of momentum, which is always measured in a direction.

241 In formulas, letters next to each other should be multiplied, while m/s means that m should be divided by s. I use the international SI units like meters and second. It is possible to use more general description which does not choose between, for example, meters or feet, and then one could use "distance", "time", "speed", etc., instead of meters, seconds, and meters

When objects move at uniform speed they cover the same distance per quantity of time. For example, a car driving at 60 kilometers per hour will cover the distance of 60 kilometers per hour. But objects can also change speed, which is called acceleration.[242] This is not meters per second, but meters per second per second, or meters per second squared. An object may for example move 1 meter per second the first second, two meters per second the second second, three meters per second the third second, etc., for short: m/s^2.

For later, it is useful to note that if the acceleration is constant, the average speed can be found by taking the end speed minus the start speed and divide by two. For example, if an object moves at 0 meters per second to begin with and ten meters per second after ten seconds, then 10 minus 0 divided by 2 shows us that the average speed is 5 m/s.

12.2 Mass (m), and momentum (p = mv)

It would not be strange for humans before the days of Newton to reason as follows: If something moves, it seems something must have caused the motion. A common name for all causes that make something move is force. A force is then understood as something that causes something to move.

However, it seems that even if you apply the same force to different objects, they will not move in the same way. Some things require a lot of force to be moved and other things require less force to be moved. This is not necessarily connected to their shape or size, since a small object can require a lot of force to be moved. Objects are different in how they react to force, but mass is a concept that can be used to distinguish all objects with regard to how much force is required to move them. Mass is then understood as resistance to force.

By being resistance to force, an object's mass could be used to compare the strength of forces. A force which is twice as strong as another force will be able to lift a rock which is twice as heavy or stretch an iron spring twice as long. Mass allows us to compare the strength of forces. The strength of force and the amount of mass are thus defined in relation to each other, since mass is resistance to force and force is acceleration of mass (to be further described below).

When force makes an object move in a direction, the object in motion is an effect of the cause that is the force. However, when an object has been set in mo-

per second. Since I want to be as precise as possible on the content of the terms, I have chosen SI units and say for example "seconds" instead of just "time".

242 Instead of saying acceleration and deceleration, I just use the term acceleration for change in velocity, which may then be negative (that something starts to move slower).

tion as the effect of this cause, the object itself gets causal power (in the same direction) to cause an effect at another object, mainly to push something, but also to heat it or something else.[243] This is a sort of rough and basic physical power (to be distinguished from the definition of "power" in physics as energy or work divided by time), and it is in this sense of causing that I will use the term "causal" in this chapter. The causal power that an object gets from being in motion is called momentum.

Experience shows that the causal power of the momentum grows proportionally with both mass and velocity. You receive a stronger blow by being hit by a stone which is bigger, or hits you faster, or hits you directly in the direction of the momentum of the stone. Momentum is defined as mass times velocity. The notation in physics is $p = mv$.

Momentum grows when the amount of mass grows, and later we shall see that momentum is part of the concept of energy. An object has more momentum, more energy and more causal power by having more mass, which already demonstrates a close link between energy and mass to be explored more below. Mass is thus not just a measure of the resistance to force in an object, but also its energy and causal power.

12.3 Work (W = Fd), Force (F = ma), and Energy (E = ½mv²) in Newtonian physics

If a force is used to move an object over a distance, we say that a piece of work has been done. Work is defined as force multiplied by distance. The notation in physics is $W = Fd$.[244] This work is the effect of the force that has caused the object to move. As seen above, when an object has momentum, it gains causal power in that direction. Another way of saying that an object has causal power is to say that it has energy. It is thus tempting (but wrong) to say that momentum is the same as energy.

Leibniz argued that we should distinguish between momentum, which he called a "dead force", and a "living force" (*vis viva*) which he defined as mv^2. Experiments were done to confirm that effects seemed to be squared by speed, for example that a ball dropped into clay would fall four times deeper if the speed was double and nine times deeper with triple speed.

---

243 Again I am simplifying matters by focusing on kinetic energy and leaving potential energy out of the discussion.

244 Another notation is $W = Fs$.

Newton shared Leibniz's view that we should distinguish between momentum and what later became known as kinetic energy. Newton had the insight that an object will either be at rest or continue with its momentum unless acted upon by a force. One does not need a force to maintain the uniform motion of an object, one only needs a force to change its motion. There is no up or down in the universe, and so if something has started to move, there is nothing that will stop it unless a force stops it (in everyday experience, friction usually stops the motion of objects). This was Newton's first law of motion, that an object will either remain at rest or continue to move at a constant velocity unless acted upon by a force.

If an object does not need a force to maintain its motion, what a force does to an object is to change its motion: to accelerate it. It seems reasonable that a change in motion must correspond to a proportional change in force applied to it, and this was Newton's second law of motion. Although not at first defined in terms of mass, it later got the formulation $F = ma$, which means that a force equals mass times acceleration.

The term energy is not used for momentum, but for the additional velocity that a force adds to an object, and the amount of energy depends on the distance that the object is moved. Work and energy are thus very closely related, and one could say that the work done on an object corresponds to the energy the object receives.[245]

---

245 This close relation can be shown between the formula for work and the Newtonian formula for kinetic energy ($½mv^2$). Here I write it in full text, but most people will probably find it easier to understand if they look at a picture with formulas that can be found online using this reference: Nave (2016). The formula for work is $W = Fd$. Recall that $F = ma$, so instead of $W = Fd$ we can write $W = mad$. Instead of distance, *d*, we could write the average speed multiplied with time, *t*, since if we know the average speed that an object has and the time it has been moving, we are able to find the distance it has covered. Simplified (assuming constant acceleration), the average speed is the maximum minus the minimum speed divided by 2, so that $v/2$ gives us the average speed. For example, if an object moves at 0 meters per second to begin with and ten meters per second after ten seconds, then 10 minus 0 divided by 2 shows us that the average acceleration is 5 m/s. Instead of $W = mad$, we can then write $W = ma*v/2*t$ (the * means "multiplied with").

We are almost at the end. Instead of acceleration, *a*, we can (if simplified) substitute it with the maximum speed minus the minimum speed divided by *t*, which gives us v/t. If we change the *a* in the formula with v/t, we get $W = m*v/t*v/2*t$. Here two simplifications can be made to get rid of the things we are supposed to divide with. The expression v/t is to be multiplied by *t*, but one *t* cancels the other out, and we are left with *v*. In order to get rid of the 2 in $v/2$, we can multiply with ½ instead. If we draw this all together we get $W = mvv*½$, or $W = ½mv^2$. This is the formula for kinetic energy, and compared to Leibniz's *vis viva*, which was mv^2, it has a ½ in front of it,

There is another way of writing ½mv^2, which is useful to know in order to compare with how Einstein changed it later. This other formula is that kinetic energy = $p^2/2\,m$. In order to see that these two formulas are equal, remember that $p = mv$, which means that $p^2 = mvmv$. If you divide $mvmv$ with $2\,m$, it is like taking away one m and multiplying with ½: $mvmv/2\,m = mvv/2 = \frac{1}{2}mv^2$. I will return to this formula below.

12.4 Energy and mass (E = mc²) in the physics of relativity

We have already seen a close link between mass and energy, but Einstein developed this a lot further compared with Newtonian physics. Recall from the chapter on time that in relativity theory the laws should be the same in all reference frames. This means that an object with momentum moving at uniform speed in a direction could just as well be considered to be standing still.

This has the following consequence: Since velocity is part of the concept of energy, and velocity is relative, it seems that an object could at the same time be seen as having a lot of energy or just as standing still. If a car is about to smash into a house, we could consider the car to be standing still and the house to be smashing into the car. In any case, there will be a crash, and so the car has causal power when crashing with the house even if we consider the car to be standing still, so it must have this causal power in virtue of having mass. Again, energy and mass seems to be closely related.

Since velocity is part of the concept of energy, and velocity is relative to reference frame, Einstein wanted a clarification of what all observers can agree on and what is observer-dependent. He divided the concepts of mass and energy in two. Rest mass and rest energy are the mass and energy that an object has when considered in the reference frame in which it is at rest. Relativistic mass and relativistic energy is the mass and energy an object gains when considered to be in motion.

Einstein's famous proposal was that mass and energy are equivalent. This means that an object has the same amount of rest mass/rest energy, and the same amount of relativistic mass/relativistic energy, which means that both mass and energy can be described in units of mass (i. e. kg of energy). Rest mass/rest energy is the quantity that all observers can agree on, while the

---

which results from taking the average of the acceleration. Note however, that an amount of work is determined by distance, whereas energy is determined by velocity.

amount of relativistic mass/relativistic energy will depend on what frame of reference is used for measurement.

These relations can be shown by using the same kind of triangle that we used to describe time in special relativity in the chapter on time. Imagine a right triangle with straight edges. The vertical line represents rest mass, which may be 3 kg. The horizontal line represents momentum, and the total energy is represented by the hypotenuse. This total energy will be rest energy plus relativistic energy. This last part I will call kinetic energy, since it corresponds to the kinetic energy which Newton was working with, while Newton had no concept of rest energy (the first part).

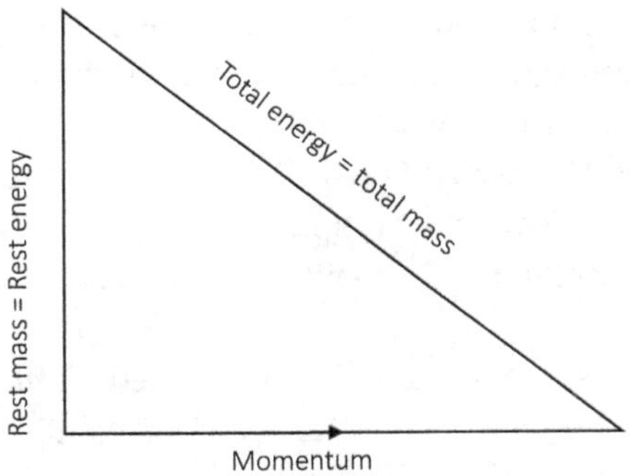

Fig. 10: Mass and energy relation

In this triangle, the vertical line representing rest mass will be the same at all times and agreed upon by all observers. But the horizontal line representing momentum can be made longer and longer, giving the object more and more total energy (represented by the hypotenuse, which would then also become longer and longer). If momentum (the horizontal line) is zero, the vertical line and the hypotenuse fall together. Then the mass is 3 kg and the total energy equals the rest energy, which equals the rest mass, which is 3 kg.

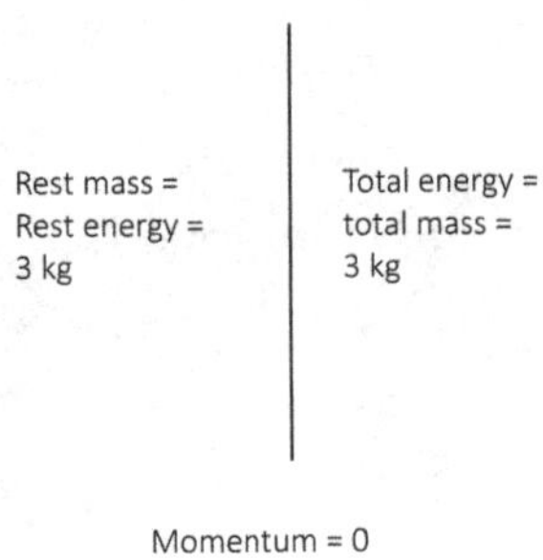

Fig. 11: Mass and energy relation with 0 momentum

On the horizontal line for momentum, we proceed as in the chapter on time and choose corresponding units of time and distance, like light years per year or light seconds per second, since we then get a speed which can be described by just a number, namely a fraction of light speed. Let us say, for example, that the object with the mass of 3 kg travels at 80 % of light speed, or 0.8 c.[246] After 5 years it will have travelled 4 light years (since 5 times 0.8 = 4) and after 5 seconds it will have travelled 4 light seconds. In this example, we will say that it has travelled 4 light seconds in 5 seconds at 0.8 c.

In the chapter on time, I explained the stretch factor called gamma, which could be used to convert a vertical line into a hypotenuse, for example a vertical line of 3 to a hypotenuse of 5. According to Einstein, the total energy of an object is its rest mass multiplied by the gamma factor, and the gamma factor is 1 divided by the square root of 1 minus the speed squared. In our example, we should take 1 divided by the square root of 1 minus 0.8^2, which is the square root of 1 minus 0.64, which is the square root of 0.36, which is 0.6. This means that to multiply the mass with the gamma factor means to multiply 3 with 1 divided by 0.6, which equals 5. An object of 3 kg travelling at 0.8 c will gain a total energy of 5 kg.

According to Einstein, momentum is rest mass multiplied with speed multiplied with the gamma factor. Rest mass is 3 kg multiplied with speed 0.8 c multiplied with the gamma factor, which at that speed is 1 divided by 0.6 (as seen in the previous paragraph). To find momentum we can then calculate as follows: 3 kg mass multiplied with speed 0.8 is 2.4, which divided by gamma factor 0.6 is 4, and the answer is that the object has a momentum of 4 kg.

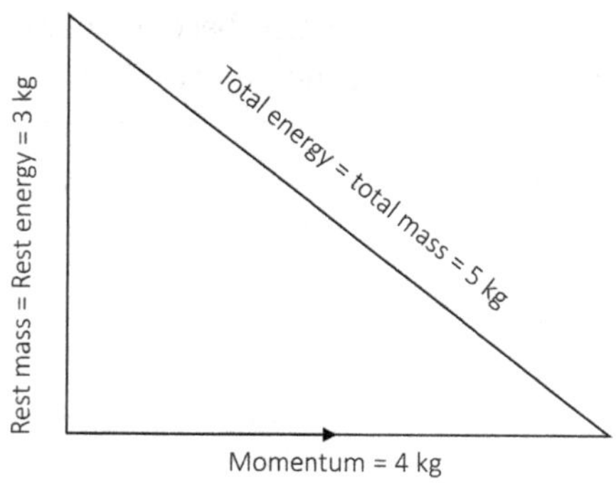

Fig. 12: Mass and energy relation with kg units

246 The example is from E. F. Taylor and Wheeler (1992, pp. 202–203).

Hopefully, the Pythagorean relationship is clear: If m (represented by the vertical line) is rest mass, p (represented by the horizontal line) is momentum and E (represented by the hypotenuse) is the total energy, then $m^2 = E^2 - p^2$ ($3^2 = 5^2 - 4^2$); $p^2 = E^2 - m^2$ ($4^2 = 5^2 - 3^2$); and $E^2 = m^2 + p^2$ ($5^2 = 3^2 + 4^2$). To distinguish the added kinetic energy from the rest energy, we can take the total energy minus the rest energy (which equals the mass), which in this example is 5 minus 3 = 2 kg of kinetic energy. And to find the speed, one can divide momentum with energy, in this case 4 divided by 5 gives a speed of 0.8 c.[247]

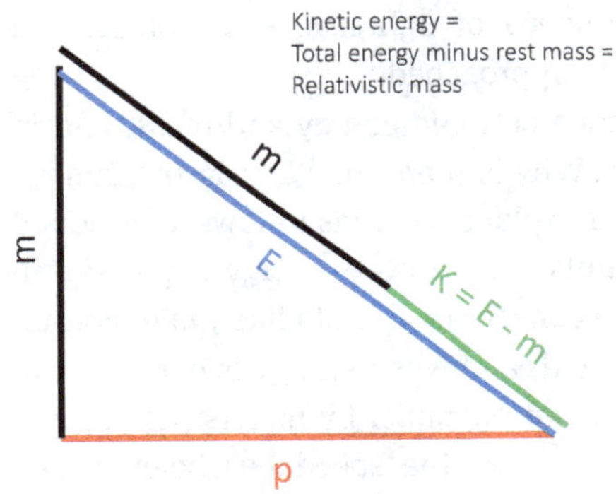

Fig. 13: Mass and energy relation according to Einstein

This triangle can be used to show many interesting things. We have already seen that if momentum (represented by the horizontal line) is zero, rest mass (represented by the vertical line) and rest energy (represented by the hypotenuse[248]) overlap and become the same, so the triangle shows the mass-energy equivalence. On the other hand, if mass (represented by the vertical line) is zero, then momentum (represented by the horizontal line) divided by energy (represented by the hypotenuse) will overlap, which becomes 1, which means that the speed is 1 c, which is c, which is light speed. This means that particles without mass travel at the speed of light, and since photons (i.e., light) do not have mass, light travels at the speed of light.

If speed increases, the horizontal line representing momentum will be longer and longer. Then the hypotenuse representing total energy will also be longer and longer. But as long as there is some mass, the hypotenuse will always be lon-

247 It makes sense that speed is momentum divided by energy, since speed is v, momentum is mv, and $E = m$: $v = mv/E$. Figure 13 is much inspired by Matt Strassler (Strassler, 2012).

248 Usually I say that the hypotenuse represents total energy, but when the momentum is zero, the hypotenuse only represents rest energy, since it is momentum that adds energy to the rest energy, but if there is no momentum, there is no more energy than the rest energy.

ger than the line representing momentum, which means that momentum divided by energy will always be less than 1, which means that an object with mass will never reach light speed.

As speed increases, momentum will increase and be longer, and then the hypotenuse will be longer. As the hypotenuse gets longer, the kinetic energy increases and the total energy increases. This makes sense, since higher speed gives higher energy. However, since relativistic energy is equivalent to relativistic mass, this means that an object with mass will gain more and more relativistic mass as speed increases. Since mass is resistance to force, no force will be able to push something faster than the speed of light, since the object will give infinite resistance when light speed is approached.

So far, I have explained the equivalence between energy and mass, and in the case where momentum is zero, $E = m$. Why is there then a c^2 in the famous equation by Einstein, $E = mc^2$? A simplified explanation is as follows: c^2 is added simply as a converting factor between units – to convert energy expressed in units of mass to energy expressed in more conventional units like Joule. Formulas for energy have mass times speed squared, and when energy is expressed in conventional units, the speed is expressed, for example, by meters per second. When the energy is expressed in units of mass, the speed has been turned into a unitless number by making it a fraction of light speed. The speed squared has then been divided by c^2 to turn the speed of light into just a number instead of using conventional units. If we want to convert it back, we must multiply the speed with c^2. This means that the important insight in the formula $E = mc^2$ is not the c^2, but the equivalence between mass and energy. As Edwin Taylor and John Wheeler say: "The conversion factor c^2, like the conversion from seconds to meters or miles to feet, can today be counted as a detail of convention rather than as a deep new principle" (E. F. Taylor and Wheeler, 1992, p. 203 and 206).

12.5 Comparison between Newton and Einstein

How could Einstein be so different from Newton, and still Newtonian physics work so well at low speeds? In the following I will explain this, with text and figures very much inspired by Matt Strassler (Strassler, 2012). We have seen how momentum, mass and energy could be related by a triangle in the case of Einstein. The same triangle cannot be made in the case of Newton, but something similar can be made which helps to compare.

Recall that the Newtonian equation for kinetic energy could be expressed as $p^2/2\,m$ (as shown in Section 12.3). This relation can be represented as a horizontal line representing momentum and a vertical line representing mass, just as in the

Einsteinian triangle. Since momentum is mass times velocity ($p = mv$), speed is momentum divided by mass ($v = p/m$). The horizontal line divided by the vertical line can then give us an angle to represent speed. Yet another line can represent the kinetic energy that one gets from dividing momentum by mass. This line is not a part of the triangle, but is added to compare with Einstein afterwards, like in Figure 14.

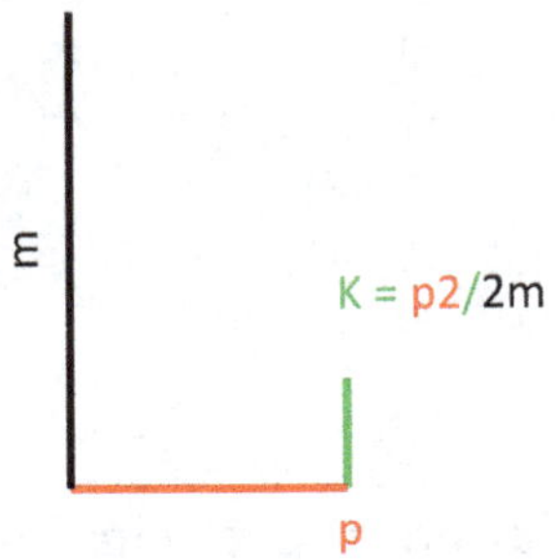

Fig. 14: Kinetic energy (Newton)

If the horizontal line representing momentum is very small and the object only moves at a tiny fraction of light speed (as do most objects we are used to), then the kinetic energy becomes very similar in the cases of both Newton and Einstein, but when momentum is large, it becomes very different, as shown in Figures 15 and 16.

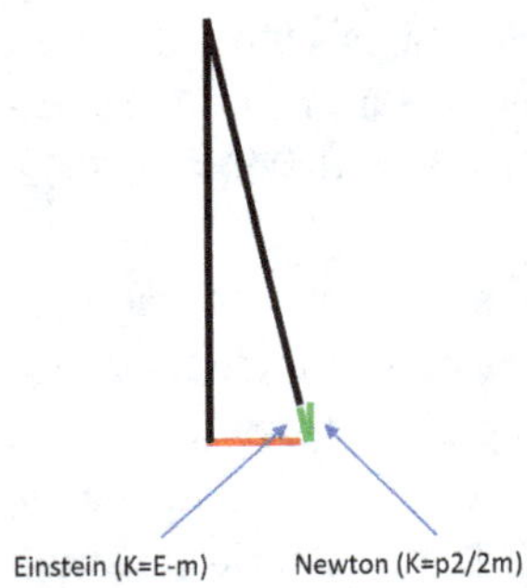

Fig. 15: Kinetic energy at low speed (comparing Einstein and Newton)

For instance, in the example we used above with an object of 3 kg of mass moving at 0.8 c, we saw that Einstein predicted 2 kg of kinetic energy, whereas Newton would say ½ times 3 kg times $0.8^2 = 0.96$ kg of kinetic energy. The predictions made by Einstein have been confirmed, while Newton has been falsified. Newton developed a theory which works very well at most everyday speeds, but Einstein's theory works at all speeds.

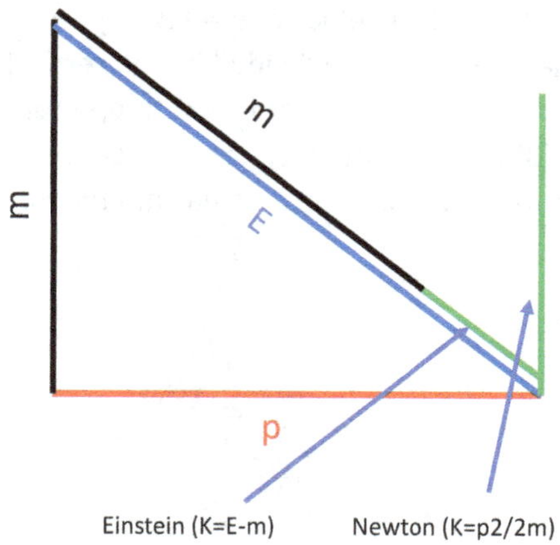

Fig. 16: Kinetic energy at high speed (comparing Einstein and Newton)

12.6 Momenergy, and creation and annihilation of particles

If we know the triangle (presented above in Section 12.4) representing the relation between mass, energy and momentum, and in addition know that energy and momentum is conserved in a system of interactions, we can understand surprisingly much about how particles and objects move in the world, and even how particles are created and annihilated. In this section, we will take a closer look at this. It will be very simplified, but the point will be to show some general rules that nature follows, which gives a very good overview of many different, interesting things. The presentation is based on two chapters in the book *Spacetime Physics*, where more details can be found (E. F. Taylor and Wheeler, 1992, chapters 7 and 8).

Remember the right triangle which has mass as the vertical side, momentum as the horizontal side, and energy as the hypotenuse. And remember also (as said above) that speed is equal to momentum divided by energy ($v = mv/E$, since $m = E$). We can now construct a very helpful concept of momenergy, which is a vector that combines momentum and energy. To do this, we construct a spacetime diagram where we let the hypotenuse representing energy be a vertical time axis and the line representing momentum be the horizontal space axis.[249] By expressing speed as a fraction of light speed, all units can be in kg. This way of representing states of affairs lets us understand and predict very

249 The momenergy vector is actually a four-vector, which means that it takes into account all three spatial directions in addition to the time direction, but I simplify here and write about one spatial dimension only.

much about how objects with a certain mass (including zero mass) and momentum will behave. Let us look at some examples.

First of all, the momenergy vector shows the world line that an object travels in a spacetime diagram. We saw above that an object of 3 kg mass travelling at 0.8 c will have a momentum of 4 kg and an energy of 5 kg. If we draw a spacetime diagram, we know that the world line of this object will move 4 units on the space axis and 5 units on the time axis.

Secondly, we can also understand how objects will move if they interact, for example by colliding, when we remember also that energy and momentum must be conserved through the interaction. Imagine, for example, that an object of 8 kg mass (from now on called the 8-ball) is moving to the right at speed $^{15}/_{17}$ c and an object of 12 kg mass (from now on called the 12-ball) is moving towards the left at speed $^{5}/_{13}$ c.[250] We can check that the numbers for each object make sense, since we know that mass squared plus momentum squared should equal energy squared (for simplicity I leave out the kg unit from now). Thus 8^2 (mass) + 15^2 (momentum) = 17^2 (energy): 64 + 225 = 289 for the 8-ball; and 12^2 (mass) + 5^2 (momentum) = 13^2 (energy): 144 + 25 = 169 for the 12-ball.

Since 8-ball has a speed of $^{15}/_{17}$ c, and speed is momentum divided by energy, we know that its momentum is 15 on the horizontal space axis and the energy is 17 on the vertical time axis. And since 12-ball has speed $^{5}/_{13}$ c, we know that its momentum is 5 on the horizontal space axis and the energy is 13 on the vertical time axis. If we now consider these two balls as a system, we see that the total energy is 17 + 13 = 30. When it comes to momentum, this is 15 to the right and 5 to the left, and since momentum has a direction, this will be 15 to the right minus 5 to the left, which gives a momentum of 10 to the right.

Imagine now these two balls colliding. Since energy and momentum must be conserved after the collision, we know that the total energy must be 30 and the total momentum must be 10. In addition, they must have the same mass as before. The result of the collision is as follows: 8-ball moves to the left at $^{6}/_{10}$ c while 12-ball moves to the right at $^{16}/_{20}$ c. Then the total energy is 10 + 20 = 30, and the momentum is 16 to the right minus 6 to the left, which is 10 to the right. As we can see, energy and momentum has been conserved after the collision. The numbers also fit with the mass of the balls, since mass (8^2) + momentum (6^2) = energy (10^2) – 64 + 36 = 100 – for the 8-ball; and mass (12^2) + momentum (16^2) = energy (20^2) – 144 + 256 = 400 – for the 12-ball. The collision is illustrated in Figure 17.

250 The example is from E. F. Taylor and Wheeler (1992, p. 207), which has also inspired the figure.

The horizontal line in the middle separates the time before and after the collision.

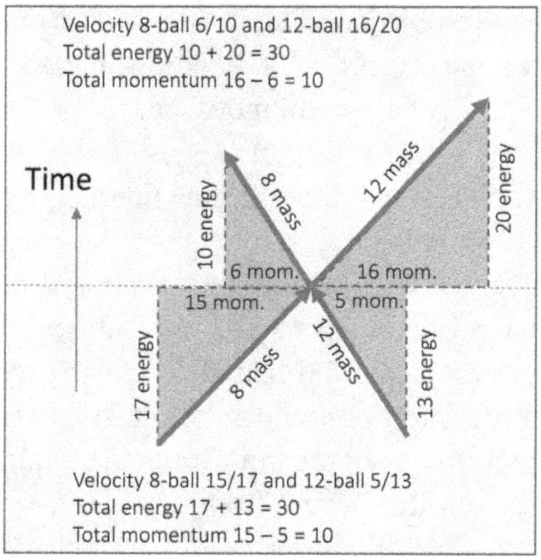

Fig. 17: 8-ball and 12-ball colliding

The same relations hold also in interactions with particles with zero mass, such as photons. Consider the following example:[251] A photon is travelling at light speed to the right and colliding with an electron standing still. In this example, we use electron mass as unit, so the electron has mass of 1, and is hit by a photon with momentum value of 2 and energy of 2.[252]

The photon travels 2 units at the horizontal space axis and 2 units at the vertical time axis, while the electron, at rest, travels 0 units at the horizontal space axis, but 1 unit at the energy axis since it has mass of 1, which equals energy of 1. Their combined energy is then 2 + 1 = 3, and their combined momentum is 2 to the right minus 0 to the left, which equals 2 to the right. Since energy and momentum must be conserved, this limits the possible trajectories after the collision.

Given that the particles do not change, but remain a photon and an electron, the result of the collision is as follows: The photon moves to the left with momentum of 0.4 and energy of 0.4 while the electron moves to the right with momentum of 2.4 and energy of 2.6. Then the total energy is 2.6 + 0.4 = 3, and the momentum is 2.4 to the right minus 0.4 to the left, which is 2 to the right. As we can see, energy and momentum has been conserved after the collision.

251 The example is from E. F. Taylor and Wheeler (1992, p. 231), which has also inspired the figure.

252 Remember that speed is momentum divided by energy, which must be equal to 1 (light speed, c) in the case of photons: momentum of 2 divided by energy of 2 equals 1.

The speed of the photon is 0.4 divided by 0.4, which equals 1, which it must be for photons, while the speed of the electron is 2.4 divided by 2.6. The numbers also fit with the mass of the particles, since photon mass (0^2) + momentum (0.4^2) = energy (0.4^2) and electron mass (1^2) + momentum (2.4^2) = energy (2.6^2) – 1 + 5.76 = 6.76.

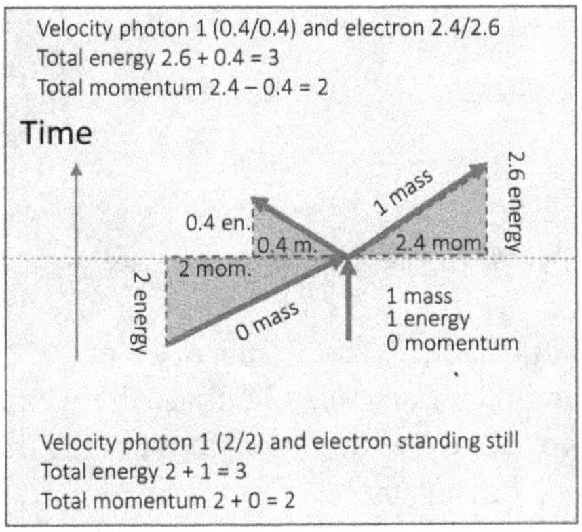

Fig. 18: Photon and electron colliding

Now for some more wonders of nature: Sometimes these relations are conserved by nature, letting some particles be created or annihilated. This can happen in the example above if we increase the momentum of the photon to 4, in which case an electron and a positron may be created. Let us walk through the example.[253]

The photon travels 4 units at the horizontal space axis and 4 units at the vertical time axis, while the electron at rest travels 0 units at the horizontal space axis but 1 unit at the energy axis, since it has mass of 1, which equals energy of 1. Their combined energy is then 4 + 1 = 5, and their combined momentum is 4 to the right minus 0 to the left, thus 4 to the right.

A possible result of this collision is as follows: The photon ceases to exist, but instead an extra electron and an extra positron are created. The original electron plus the two new particles all move to the right with a total momentum of 4 and a total energy of 5. The energy is 0 + 5 = 5, and the momentum is 0 to the left + 4 to the right = 4 to the right. As we can see, energy and momentum has been conserved after the collision. We also know that if the energy is 5 and the momentum is 4, the mass must be 3 (since $5^2 - 4^2 = 3^2$). In this case, nature fixed this by providing three particles each with the mass of 1.

253 The example is from E. F. Taylor and Wheeler (1992, p. 234), which has also inspired the figure.

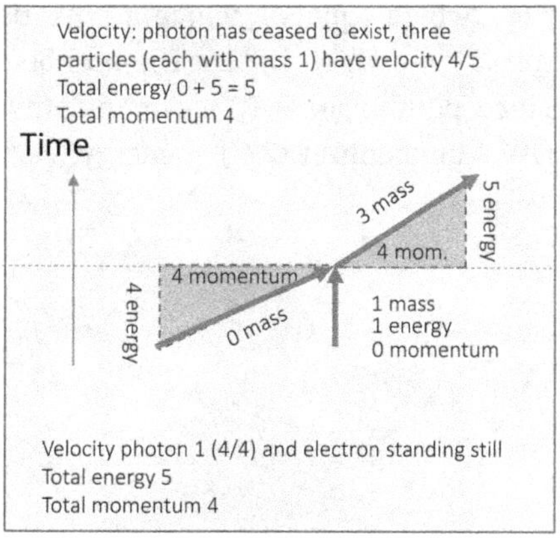

Fig. 19: Particle annihilation and creation

All particles with energy can create new particles (E. F. Taylor and Wheeler, 1992, p. 234). Particles can also annihilate, and energy can be released through fusion, fission and annihilation, all of which is governed by these simple relations, in addition to other conservation laws. Here is one example:[254]

An electron and a positron at rest can annihilate and turn into two high-energy photons called gamma rays. The electron and positron both have mass of 1, energy of 1 and momentum of 0, which gives a total energy of 2 and momentum of 0. They turn into 2 photons travelling at momentum of 1 in each direction and energy of 1. The energy afterwards is $1 + 1 = 2$, and the momentum is $1 - 1 = 0$, so momentum and energy are conserved. The relation between mass, momentum and energy is also correct in the different cases: electron mass (1^2) + momentum (0^2) = energy (1^2); positron mass (1^2) + momentum (0^2) = energy (1^2); and each photon mass (0^2) + momentum (1^2) = energy (1^2).

While the detailed picture is of course very much more complex, there are some fascinating fundamental relations that let us understand some basic rules for very much of what goes on in nature. At CERN and other places, physicists try to smash particles into each other at high speeds to see if there are even new particles that can be discovered. But the general rules seem to be followed when relating mass, energy and momentum.

254 The example is from E. F. Taylor and Wheeler (1992, p. 238).

12.7 What are the fundamental physical values?

What is the ontological status of entities like energy, mass, momentum or momenergy? Can we say more about what energy and mass are in themselves beyond describing their interrelations? What about other basic physical values like spin and charge? Is it possible to say something fundamental about what the fundamental physical values are, in the sense of what their internal structure is?

It is difficult to think that they could be independent entities. Nothing can be isolated as a piece of energy or a piece of mass or a piece of momentum or a piece of spin or a piece of charge. Earlier in this book, we have briefly visited Russellian monism with David Chalmers and structural realism with James Ladyman, which are positions arguing that many physical concepts are circularly defined in terms of each other.[255] Michael Esfeld argues that quantum mechanics shows us that the basic properties of mass, etc., can only be attributed to whole systems and not individual particles (Esfeld et al., 2018, p. 7). We have seen Tim Maudlin make a similar point, namely that quantum mechanics supports holism over particularism in physics.[256]

When we look at the concepts of mass, energy, momentum and momenergy, these are concepts that express relations that hold at many ontological levels in interactions between small particles and big objects. It seems to me that nature follows rules for interactions, and that there is a system in these rules. If terms like mass, energy, momentum and momenergy express structures that can be found in the rules that nature follows, then they do not have to be entities in themselves, but can rather be expressions of structures in the rules for actualization of values, regardless of whether it happens at micro or macro level.

This is similar to how Michael Esfeld argues that all physical properties should be reduced to laws of nature describing only the dynamics of how particles move (Esfeld et al., 2018, p. 6). He gives the nutshell description that electrons do not move like electrons because they are electrons, but we call them electrons because they move electronwise (Esfeld et al., 2018, p. 7). In my terminology, dividing into particles and properties can be seen as theoretical frameworks explicating structures and patterns in the motion in the world.

There is a great advantage to such a view in that it is ontologically parsimonious when entities are reducible. On the other hand, it seems that we go too far if we reduce all values to rules, since the rules must be guiding something. The

---

255 For concrete examples of how physical concepts are defined in terms of each other, see H.-J. Schmidt (2019).

256 I take this to favor realism about laws of nature over dispositionalism.

world is not rules only – there is motion, which means that something is moving. It cannot be rules only; some values must be left to be actualized, so which are those?

Let us try the following suggestion: There is one fundamental physical value that is actualized at points in the field according to rules, and that is energy with a quantity of strength in a certain direction, which we called momenergy. Many different rules determine which strength and direction the force will have in each point. Physics have discovered theories which very well describe the strength and direction of forces or energy in each point, and while these are structurally very similar to the most coherent description of the rules that nature follows, we can be quite sure that physics have not yet found the final and best description of it.

Momenergy and momentum are concepts whose content is structurally similar to the idea of a force in a direction. The concept of energy is structurally similar to the concept of force, while mass would here be just an expression of how interactions between forces in points are adjusted by rules governing them. Mass does not need to be an entity with an independent ontological existence beyond the rules that nature follows, but the fundamental force would have to be. If mass is just a part of rules governing energy, momentum and momenergy, it would explain the close relationship between mass and energy.

Spin and charge are easily interpreted as structures in the rules guiding motion. For every spin there is an opposite spin. Every particle has a particle with opposite charge. Negative and positive charges attract each other, while two negatively charged bodies will repel each other. All of this looks very much like some basic symmetries in the rules that nature follows for guiding motion, and then we make up names like spin and charge for certain kinds of symmetrical interactions, without there being anything that actually spins, pulls or pushes. This means that spin and charge do not exist as ontologically irreducibly entities, but instead there is structure in the rules that nature follows.

We do not need the fundamental concepts of physics like mass, spin and charge as building blocks of nature, but we need the concepts to describe structures in the rules that nature follows. And we need the underlying fundamental momenergy as at least one fundamental physical value that exists and can be actualized with a certain strength and direction.

We have previously seen how particles in quantum field theory can be translated to energy moving in the field. Energy and momentum in a field can hit like a particle (Hobson, 2013, p. 11), and experiments in quantum field theory will typically determine the probability for locating a particle at certain place or as having a certain momentum. This energy moving in different directions with different strength is the fundamental physical value, and this is needed to explain the presence of what we experience as particles and motion in the world.

Note that when I speak of momenergy as the fundamental physical value, this is not the same concept as momenergy in physics, and thus I specify that I am speaking of ontological momenergy or underlying momenergy. I call it momenergy because of the common content of energy, direction and strength. But in my ontology the underlying momenergy is a fundamental physical value compared to the other physical values, which means that underlying momenergy is not related to other concepts in physics in the same way as momenergy in physics is related to other concepts in physics.

The underlying momenergy as the fundamental physical quantity is also to be distinguished from the fundamental power of actualization that actualizes all kinds of values – physical values *and* qualia values. However, it is a tempting thought to think that they are nevertheless closely related since both are a kind of power of change. Both when we try to find out what the fundamental physical values are and when we try to think of what the fundamental power of actualization is, they are powers to change, and we even saw that existence itself was this kind of power to change.

Because we have these overlapping features in three different entities that are all poorly understood – namely the underlying momenergy, the power of actualization, and existence itself – it is tempting to think of all three entities as aspects of one and the same fundamental power, the power of actualization. It would reduce the number of poorly understood entities and explain the overlapping features. This would still leave us, though, with a fundamental power which is poorly understood, but it is a power that we have to include in the ontology anyway.[257]

It is an advantage in ontology to be able to reduce several difficult concepts into one. It gives fewer loose ends, and it is even something we know must exist, namely the truthmaker of motion, whatever that is. On the one hand, we do not know what it is, but on the other hand, we know a lot about it in the sense that we have described it and explicated many of its characteristics and relations

257 For those who think that this existence, power, force, and actualizer in one also actualizes mind in itself, it would be natural to think of it as God, existence, power, force and actualizer in one (with the classical terms "being", "omnipotence" and "creator"). If mind came later in evolution, however, it is not natural to call it God. The use of "itself" in the first sentence in this footnote is ambiguous. It could be within the power, but also within the fundamental field, but not as part of physical beings. For example, physical values could be actualized in the universe only, while qualia values were actualized as a divine panentheistic mind outside of the universe.

throughout the book. In one sense it is the case with all things that we do not know their deepest existence, we only know their characteristics and relations. And so it is good when some things can be reduced to others. One option I have not chosen (although I think it is worth exploring) is to try to reduce physical values to qualia values, as in idealism. But since we need a power of motion in any case, and since the content of qualia seems to depend on what happens physically, I have kept them as two distinct kinds of fundamental values.

In this chapter I have tried to describe some fundamental relations in physics. This is a field where much research is going on, and it is very likely that some better theory about the fundamental physical values will come in the future, be it superstrings, quantum foam, or something else. However, describing fundamental issues in physics would not be complete without a look at the fascinating world of quantum mechanics, and this will be the topic of the next chapter.

13 Understanding Quantum Mechanics

Quantum mechanics gives us some fundamental insights into how the world works that a theory of the world should be able to integrate. It is the purpose of this chapter to do so, and we shall see that it fits well into what is said in the rest of the book.

Quantum mechanics is often made more mysterious than it has to be. There are certainly some strange findings and some new ways of thinking about how nature works, but there are also good ways of understanding it and many exaggerations being made in presentations of quantum mechanics. There are some strange phenomena in nature that are predicted by the formalism of quantum mechanics, but it is very controversial what this tells us about the world. The phenomena and how to understand the formalism give us a list of standard problems that a good theory of quantum mechanics should explain.

In this chapter I will list some problems first (Section 13.1), which also will serve to introduce quantum mechanics, then I will present a good way of thinking about these problems (Section 13.2). This presentation of problems and solutions leans heavily on the book *Quantum Theory* by Tim Maudlin, which is by far the best book I have read on quantum mechanics (Maudlin, 2019).[258] My own presentation is a simplified and non-technical/non-mathematical version of the real issues involved for the sake of readability, so for details one should consult Maudlin's book (a little taste of the mathematics is given in an excursus in the end). In Section 13.3 I will discuss some remaining problems and suggest how they should be solved. This will show where my thinking differs from Maudlin.

There are two excursuses in the end, first one for those who would like a little taste of the formalism of quantum mechanics (13.4), then one excursus speculating over why we have the often seemingly strange laws of nature that we do (13.5).

13.1 Problems for a theory of quantum mechanics to solve

The following list of problems will start by presenting some famous experiments that yield counterintuitive results. Then we move on to how the formalism of

---

258 While Maudlin presents eight experiments that a quantum theory should solve, I present nine problems for a theory of quantum mechanics to solve (which includes five of the experiments that Maudlin presents). These are common experiments and problems seen in most introductions to the topic (cf. for example Albert (1992), which my presentation is also inspired by).

https://doi.org/10.1515/9783110770957-015

quantum mechanics fits well with the experiments while still having some problems on its own. Together this gives us a list of problems that a good theory of quantum mechanics should solve.

The first problem is how to understand the double slit experiment and the interference it shows. If you shoot electrons through a wall with two slits in it onto a back wall, the electrons will not hit the back wall in the area behind the slits as if you were shooting bullets or footballs. Rather the hits on the back wall will gradually form what is called interference bands, parallel lines where the electrons hit. This is the same pattern you would get if waves were sent through the slits, since the crests and troughs will either enforce each other or cancel each other out, making strong and weak hits on the back wall in parallel lines. This happens even if you just shoot one electron at a time (Maudlin, 2019, pp. 10–14; Albert, 1992, pp. 12–14).

On the one hand, then, we have particles you can shoot one at a time making single hits on a back wall, just as if they were particles. On the other hand, they behave like waves when it comes to how they move, and they even behave like waves when only one particle is being shot through and there is no other particle to interact with. Understanding this behavior is the first problem to solve.

The second problem is to understand why interference disappears if you try to check which hole the electrons go through. If you put a camera by one slit – or something else that allows you to check which slit the electrons go through – the electrons stop behaving like waves and the interference bands disappear. This happens even if the electrons go through opening A while the camera is at opening B. The electrons are sensitive to what happens also where they are not present (Maudlin, 2019, pp. 14–17).

It could seem like just observing or having the possibility of observing particles make them act differently, and some theories of quantum mechanics will say that observation plays an important role (causing a so-called collapse of the wave function). Others have ridiculed this idea. Einstein asked if a mouse could drastically change the universe just by looking at it, and Bell asked if the whole universe was waiting for the first cell to appear before it could make a quantum jump, or whether it had to be a more advanced observer – perhaps with a PhD.[259] Explaining why interference disappears with monitoring is the second problem.

The third problem is how to understand scrambling of measurement results. If a charged particle spins it gets magnetic poles, and electrons have such poles even if we do not know of anything actually spinning around inside of them. It is

259 Maudlin (2019, p. 3), referring to Everett, Barrett, Byrne, and Everett (2012, p. 157) and Bell (2004, p. 216).

just a feature of electrons called spin, which can be measured in different directions with a so-called Stern-Gerlach device. This device makes the electrons go up or down in the direction they are being measured when they pass a magnetic field.

If you measure the spin of a beam of electrons in the x direction, 50 % will go up and 50 % will go down. Let us say that you collect all the x-spin up electrons and measure their spin in the x direction again. The result is then that 100 % is up, as we would expect if we have now selected the electrons with x-spin up. If we now take the x-spin up electrons and measure their spin in the z direction, then again you get 50 % up and 50 % down, but now in the z direction. But if you then measure their spin in the x direction once more, it would be common to expect that they were still 100 % up in the x direction (assuming that this is a permanent feature they have), but in fact they are now again 50 % up and 50 % down in the x direction. When you measure the same feature several times in a row, it is stable, but if you measure another feature in between, the original features get scrambled (Maudlin, 2019, pp. 17–22; Albert, 1992, pp. 1–4). Explaining why measurement results get scrambled this way is the third problem.

The fourth problem is to explain the reversing of results in a so-called Mach-Zehnder interferometer. Let us say we proceed as above and take a group of x-spin up electrons and send them into a device that sends all z-up electrons one way and all z-down electrons another way, then recombines them in the end to measure x-spin. We started with all electrons x-up, but since we have measured z-spin, we should now, based on the experiment described in the previous paragraph, have learned that the result of the new measurement of x-spin should be 50 % up and 50 % down, but in fact all are now 100 % up. You can even manipulate events with the one beam of electrons and see that the other beam is influenced by what happens (Maudlin, 2019, pp. 22–25; Albert, 1992, pp. 7–12). Explaining this reversing of results is the fourth problem.

The fifth problem to explain is non-locality. If you prepare two particles in the same quantum mechanical state and send them in different directions to measure their spin, they will always have opposite results. If one has spin-up the other will have spin-down. This would not be strange if we knew that the particles could be prepared to be opposites from the outset, but the experiments have shown that particles cannot be so prepared, because it all depends on what you choose to measure. We can choose to measure anything and whatever we measure, the result will be the opposite for the other particle. But they can be as far away as we want when we measure, and still the result will immediately be the same, even if no information about what kind of measurement was being made could have been transmitted (Maudlin, 2019, pp. 25–28; Albert, 1992, pp. 69–70).[260] One particle seems to influence another even if it is far away, and this is called non-locality or action-at-a-distance (other terms often used to refer to this topic is "quantum entanglement", "EPR paradox", "Bell inequalities" and "Aspect-like experiments"). Explaining non-locality is the fifth problem.

Fascinatingly enough, the quantum formalism agrees with all the results of the experiments just presented. It gives extremely accurate predictions, also before tests have been made. Einstein famously used non-locality to argue that quantum mechanics must be wrong (in the EPR paper) (Einstein, Podolsky, and Rosen, 1935), but then experiments by Alain Aspect and others have confirmed that quantum mechanics was right and Einstein was wrong (Aspect, Grangier, and Roger, 1982; Aspect, Dalibard, and Roger, 1982).

Maudlin walks through each experiment and shows how it fits with the formalism (Maudlin, 2019, chapter 2). But the formalism itself has some problems. Standard quantum formalism has two parts. There is a wave equation called the Schrödinger equation that describes the evolution of a physical system, and then there is the Born rule, which says that the amplitude you get as a result from the Schrödinger equation can be squared and give you the probability of the outcome of a measurement (Maudlin, 2019, pp. 46–47). The Born rule is a surprising rule coming out of the blue with nothing in the Schrödinger equation implying that it is about the probability of anything, since the Schrödinger equation seems to describe a deterministic evolution. This raises several questions or problems.

Why does the Born rule give us the probability of outcomes of experiments? What is it about, when should it be used, and why does it work? This is problem number six. Problem number seven is to define the terms "measurement" and "outcome" if they have a special status. What is a measurement and an outcome at all, and why should that be anything special? It seems that a measurement and an outcome are just ordinary physical events like all others, so why should a specific rule apply to these specific events that we call measurements? A theory should either give a clear definition of these concepts or show why they are superfluous, and this is problem number seven.

The previous problem about defining "measurement" and "outcome" is closely related to a famous problem in quantum mechanics called the measurement problem. This problem can be understood in different ways and as different problems. A clear way of presenting the problem and different solutions can be

---

260 Maudlin adds a proof from David Mermin that the features of the electrons cannot be presettled (Maudlin, 2019, pp. 31–32).

found in an article by Tim Maudlin called "The three measurement problems" (Maudlin, 1995a). I will focus on the first two, which are the most common ones.

Maudlin calls the first problem the problem of outcomes. As a simple first presentation of the problem, we could say that the problem is that while quantum mechanics predicts superpositions, which is a combination of incompatible properties, experiments always show definite outcomes, not superpositions. In other words, if you make a measurement, the measurement apparatus gives a definite result that you can observe, while the formalism seems to imply that both the apparatus and you should be in a superposition of observing different results (Callender, 2017, p. 87).

More precisely put, Maudlin describes it as an inconsistency between three claims:

1A) The wave function of a system is complete.
1B) The wave function always evolves in accord with a linear dynamical equation.
1C) Measurements always have determinate outcomes (Maudlin, 1995a, p. 7).

The inconsistency is that if the first two are right, the measurement device should give a superposition as a result, but instead 1C is right and the result is always a determinate outcome. There are three main ways of dealing with the problem, namely to abandon one of the three claims. Theories which abandon the first claim argue that the wave function of a system is not complete, but additional variables are needed, such as the pilot waves suggested by David Bohm. Theories which abandon the second claim argue that the wave function sometimes collapses, either spontaneously (as in the GRW interpretation) or when a measurement is made (as in the Copenhagen interpretation). Theories which abandon the third claim need to explain why outcomes are not determinate when they seem to be. The most famous is the Everett interpretation, which says that the universe splits in two to give both outcomes (Maudlin, 1995a, p. 8).

The second problem is similar to the first, and Maudlin calls it the problem of statistics. As a simple first presentation of the problem, we could say that the problem is that while quantum mechanics predicts superpositions, experiments always shows different definite outcomes (even if the initial wave function is identical every time), and the probability for what result one gets is given by the Born rule. More precisely put, Maudlin describes it as an inconsistency between three claims:

2A) The wave function of a system is complete.
2B) The wave function always evolves in accord with a linear dynamical equation.
2C) Measurements described by identical initial wave functions have different

outcomes, the probability of which are given by the Born rule (Maudlin, 1995a, p. 11).

The inconsistency is that if the first two are right, the measurement device should give the same definite result every time, but instead 2C is right and the result is always a determinate outcome given by the Born rule. Again, the problem is dealt with in the same way as described above – by abandoning one of the claims.

Maudlin criticizes the Copenhagen interpretation for not defining what a measurement is and for not explaining why a measurement should be something special (Maudlin, 1995a, p. 9). He criticizes the Everett interpretation for not being able to make sense of statistical outcomes if the universe splits in two for every outcome (Maudlin, 1995a, pp. 11–12). He is most positive to theories like Bohm or GRW, since these offer additional physics that are needed to avoid the inconsistency, and he prefers the Bohmian interpretation among these two (Maudlin, 1995b). However, many others are reluctant to accept Bohm's theory since it seems ad hoc, adds extra elements to the basic formalism, and employs non-locality, which is incompatible with relativity theory (Ladyman et al., 2007, p. 181). The measurement problem is the eighth problem.

Last, but not least, the ninth and final problem is the question of what a theory of quantum mechanics says about the world. What does exist and how does it make things move? What would remain in the world even if all living beings were dead, and what are merely useful mathematical descriptions? What is irreducible and what can be reduced? Does the wave function exist, do superpositions exist, does a configuration space exist, do the equations refer to something that exists, do particles exist? You can find people affirming or rejecting the existence of all the entities in the previous sentence.

13.2 How to solve the problems of quantum mechanics: Bohmian mechanics

There are several competing solutions to these problems, often called interpretations of quantum mechanics. Maudlin does not discuss the Copenhagen interpretation, since he thinks it is too vague to be a theory, using expressions like "measurement" and "outcome" without saying what they mean (Maudlin, 2019, p. ix). Nor does he discuss theories which say that the results of quantum mechanics are about our beliefs. These theories say that we are merely ignorant about the positions and spin, etc., of particles until we measure them to find out. These theories have problems explaining real physical effects like interference patterns (Maudlin, 2019, p. 82).[261]

Maudlin does discuss the GRW theory, which says that measurements are irrelevant but spontaneous collapses happen, and they typically happen in situations we think of as measurements. And he discusses the Everett interpretation, which says that the universe continuously splits into branches (Maudlin, 2019, chapters 4 and 6). He has many objections against these theories, but finds there to be one theory far better than all the others, which is Bohmian quantum theory. Since I do not have space to discuss all the theories here – and since I find Maudlin's arguments to be very good – I refer to his discussions for objections against the other theories. Here I will only present how Bohmian quantum theory solves the problems in the previous section, based on Maudlin's presentation (Maudlin, 2019, chapter 5). I can then show how I differ from his views in Section 13.3.

The Bohmian theory was first developed by Louis de Broglie, but then abandoned until David Bohm gave a new version in 1952, which now exists in different variants. The main idea is that it is a theory about particles, where you have to add to the Schrödinger equation a guidance equation for how the particles move that takes their initial position into account.

Take problem one with the double slit experiment and interference. There really are particles going through the slits, and each particle moves through one slit each. However, the particles are guided by the guidance equation, which implies that when you shoot through many particles they will form interference bands, since more particles will land some places than others (Maudlin, 2019, pp. 142–144).

But if this is the case, then why does interference disappear when you try to check which slit the electrons go through? Maudlin makes a clever move when he presents this experiment, since he makes a setup where there is a registration of which slit the electron goes through that does not involve any cameras, humans or minds, but only a simple physical process where a proton reacts depending on which slit the electron goes through (Maudlin, 2019, pp. 14–16). The interference bands disappear, which shows that the effect is not a result of consciousness, minds and observation, but a simple physical effect.

The disappearance of the interference already follows from the formalism of quantum mechanics. When only electrons move through two possible openings, the equations describing their motion gives interference effects (because of overlapping possibilities for motion interfering with each other). When they interact

261 Maudlin also uses the PBR theorem to reject such theories; see Maudlin (2019, pp. 83–85).

with another physical system, they get entangled with this system in a way that makes the possible motions described by the equations no longer overlap, taking away the interference effect.[262] In everyday life many physical systems interact and take away interference effects, which is why they are only seen in specially designed experiments (Maudlin, 2019, p. 58). The explanation works the same way in Bohmian theory (Maudlin, 2019, pp. 146–147). This is how problem two is solved.

What about problems three and four, with the scrambling and reversing of results when spin is measured? Again, the formalism of quantum mechanics gives the right results, and the relevant differences lie in whether or not there is an interaction with another system.[263] But how does the Bohmian theory explain what happens? Here the clue is not to think of spin as a property of particles that is either set or gets scrambled or reversed, but rather to think of spin as a convenient way of describing different motions of particles, namely whether they go up or down in a certain magnetic field. The equations for how the particles move in the Bohmian theory works for describing their motion in interactions with the different kinds of measurement apparatuses, and there is no need to think that, in addition, the particles have something spinning inside them which has been changed or not. What accounts for whether the electrons go up or down is their initial position, which is evenly distributed, so we get the experiment results that we do (Maudlin, 2019, pp. 148–149, 164).

Even if these experiments can be explained by particles moving according to equations, it is worth noticing that in experiment after experiment, the results show that the particle motion is sensitive to the whole setup and even what happens where the particle does not go. The equations take possibilities into consideration, which shows that the formalism itself is global and more than just the sum of the features of the individuals (Maudlin, 2019, p. 92). We shall return to the question of what this tells us about the world.

Non-locality is another example of how quantum mechanics is something global. What is the Bohmian account of non-locality? It could seem that the answer is simple since Bohmian quantum mechanics are deterministic, so if the particles are prepared as opposites from the beginning, they will be opposites when measured. But we have seen that results depend on what we randomly choose to measure, and even if you think the universe is determined, you should not believe that there is a super-deterministic conspiracy of nature perfectly correlating what humans choose to measure and how particles are prepared. The

262 For details, see Maudlin (2019, pp. 55–57).

263 For details on the formalism, see Maudlin (2019, pp. 59–65).

Bohmian account should thus account for how Aspect-like measurements always give perfectly correlated results.

The Bohmian solution is that the wave function of the particles depend on each other, so that when one particle is measured to be of one type, the second particle will become the other type no matter how far away it is (Maudlin, 2019, p. 159). However, this does presume that one particle is measured before the other gets its feature, and against this many will argue that the theory of relativity shows that "before" and "after" does not make sense for spacelike separated events and that it implies faster-than-light signaling. We shall return to this problem below and for now merely note that the solution above is the non-relativistic solution of Bohmian mechanics, which does not take relativity into account.

The sixth problem was how to understand the Born rule, which is especially difficult for Everettian theories since any outcome happens anyway. What does it mean that one result of two is 75% probable if the universe splits to give both results anyway? In Bohmian mechanics the guidance equation says that some outcomes happen more often than others, for example some places will be hit more often than other places in the two-slit experiment. But then the probability of finding a particle there will be higher than the probability of finding it elsewhere, so this explains the probabilistic nature of the Born rule (Maudlin, 2019, p. 145). Note that Bohmian mechanics are deterministic and does not include probabilistic or stochastic equations or laws. I will return to discussing this below.

Problem number seven was how to define "measurement" and "outcome", and this is an important challenge to theories like the Copenhagen interpretation that make measurements and outcomes into something special. But in Bohmian mechanics they are not special, treated just like any other physical process, so they need not and are not defined as anything special in Bohmian mechanics.

Problem number eight was the measurement problem. The first version of the problem was that there is an inconsistency between holding that 1A) The wave function of a system is complete; 1B) The wave function always evolves in accord with a linear dynamical equation; and 1C) Measurements always have determinate outcomes. We have seen that the Bohmian solution to this problem is to dissolve the inconsistency by rejecting 1A. The wave function is not complete, but gets complete when we add the guidance equation.

The second version of the problem was that there is an inconsistency between holding 2A) The wave function of a system is complete; 2B) The wave function always evolves in accord with a linear dynamical equation; and 2C) Measurements described by identical initial wave functions have different outcomes, the probability of which is given by the Born rule.

Again, the Bohmian solution is to reject 2A and solve the inconsistency by adding the guiding equation. We saw above how this explains the Born rule. Bohmian mechanics thus has a straightforward solution to the measurement problem, and there is no problem with Schrödinger's cat being both dead and alive at the same time. Bohm's cat is always either dead or alive.

The final problem is to state clearly what the theory says about the world. Bohmian mechanics is a theory about particles being guided by certain equations. This is quite classic and similar to traditional physics. However, the equations imply a configuration space, meaning a space of possibilities, where the equations interact and have global consequences. Maudlin seems clearly to believe that this is a many-dimensional space which exists and where interactions happen with observable effects (Maudlin, 2019, p. 162). He says of Bohmian mechanics that "the quantum state is a physically real, non-local entity in the theory" (Maudlin, 2019, p. 171). All of this is needed to account for the non-locality with global interactions and effects and particles influenced by what happens elsewhere.

As we saw earlier (in Chapter 3 of this book), Maudlin is also a realist about laws of nature, believing that they are irreducible entities that make things happen. Now that the theory is stated clearly and many problems have been solved, we should make a final assessment of its coherence. I shall do this by discussing some remaining problems in the next section.

13.3 Problems with Bohmian mechanics – and how to fix them

We have already seen one problem that we postponed, namely the problem of reconciling Bohmian mechanics with relativity. The time order in Bohmian mechanics implies a universal time order which is not found in general relativity, and it implies effects happening at superluminal speed. However, what that universal time order is cannot be determined by experiments (Maudlin, 2019, pp. 209, 214–216). Another problem that Maudlin discusses in the last chapter of the book is the following: Bohmian mechanics is a deterministic theory presupposing a certain constant amount of particles. But in physics we see particles colliding where some particles are annihilated and others created, which is not possible in a non-relativistic version of Bohmian mechanics (Maudlin, 2019, pp. 209, 220).

In quantum field theory, you get indeterministic probabilities for finding particles at different places. John Bell has made a quantum field version of Bohmian mechanics where he accepts particles literally being created and annihilated.

The theory then takes as its starting point not all *actual* particles but all *possible* particles, and gives probabilities for finding particles at certain places and times (Maudlin, 2019, pp. 221–223).

In this book I have argued in favor of an absolute space, a universal time, and indeterminism. While others will think of these as problems for Bohmian mechanics, I think we have different reasons supporting each other for embracing these results, which then also makes Bohmian mechanics more coherent. However, there are also other objections and possible improvements to be made to the Bohmian theory.

Michael Esfeld has argued that Bohmian mechanics is the best theory of quantum mechanics (Esfeld et al., 2018, pp. 69–71). However, he does not find it plausible to believe that there really exists a many-dimensional configuration space where things happen. Instead he refers to the work of Dürr, Goldstein and others in interpreting the equations of quantum mechanics as laws. The wave function and the configuration space are not irreducible entities being fundamental building blocks of the world. Instead they are convenient ways of expressing motions of particles, but where nothing more exists than the particles in motion (Esfeld et al., 2018, p. 10; Dürr, Goldstein, and Zanghì, 2013).

When Esfeld interprets the equations as laws, he does not mean that laws exist as parts of the furniture of the world. He subscribes to the Humean view that there are certain patterns in particle motions without there being a deeper explanation. Laws are then nothing more than the simplest and most informative ways of describing this pattern. For Esfeld nothing more exists in the natural world than particles in motion (Esfeld et al., 2018, p. 4).

For this reason, he cannot accept Bell's suggestion of particles being created and annihilated, since in Esfeld's ontology this would imply impossible creation out of absolutely nothing. Esfeld also criticized quantum field theories for having mathematical problems, which makes the field theories today "efficient" theories needing completion instead of being complete (Esfeld et al., 2018, p. 9). Instead he sticks to the deterministic version of Bohmian mechanics and interprets particle "creation" and "annihilation" in terms of the Dirac sea model (Esfeld et al., 2018, p. 101).

I see several problems with Esfeld's alternative. The first has to do with whether one should be a realist about laws of nature or think of them in the Humean way as merely uncaused patterns that the laws we have chosen are best at summarizing. When there is a perfect correlation between how we randomly choose to make measurements to test non-locality and the results always conform to this, it is extremely plausible that there has to a be a cause of the results instead of the results merely being an uncaused pattern that the laws summarize. Esfeld thinks that it is not of explanatory value to add laws to your ontology since they do not add coherence to the theory (Esfeld et al., 2018, pp. 43, 51–53).[264] But it is clearly more coherent with a theory that says that non-locality results are caused than not caused, and this is a good reason to include laws in the realist sense.

One did not have to include *laws* as the cause of the non-locality results. One could argue that the wave function exists and causes the results in the configuration space, and a reason to believe this is that the equation seems to describe an interaction happening somewhere. The motion of the electrons is sensitive to what happens at other places than where the electron is. If the Humean is right, it is strange that the best description of electron motion is one that strongly indicates an interaction between the electron and other possibilities.

On the other hand, it is also strange to believe that a configuration space should exist, which so clearly seems to be an abstract idea summarizing possibilities without itself being something physical. The complex numbers involved indicates something abstract and non-physical.[265] I think that the best way to reconcile this is to accept a realist understanding of laws of nature and that the interaction takes place in the laws of nature at the level of rules. The laws of nature take possibilities into consideration and actualize physical values at different places according to the rules of quantum mechanics. With this view one can keep the global interaction that clearly seems to happen at the quantum level without the need for saying that wave functions or configuration spaces exist as irreducible entities on their own. They can then be dismissed as unnecessary add-ons.

It is not a problem with particles being created out of nothing here since they are not created out of nothing, but actualized by an actualizer in the field, like everything else in the world. There is nothing traveling at superluminal speed, merely the laws of nature actualizing certain values at certain places according to rules.

In sum, it seems to me that this ontological interpretation of Bohmian mechanics solves all the problems discussed in this chapter except for the mathematical problems that have not been finally solved for quantum field theories. Since these field theories nevertheless work so well and there are so many other arguments in its favor, I find the field approach to be the most coherent

264 In these pages (and elsewhere), Esfeld rejects and criticizes ontologies including laws as fundamental, but in one place he says that he is agnostic about the existence of laws (Esfeld et al., 2018, p. 56). I interpret this as him being open to the possibility of fundamental laws while finding the rejection to be best justified.

265 I am grateful to Katarina Pajchel for the point about complex numbers.

and expect that the mathematical problems will be solved in the future when experiments at higher energy levels can be performed.

The ontology that comes from this is that what exists is a field where values are actualized according to rules. The values relevant here are the physical values in quantum mechanics and the relevant rules are the ones described by Bohmian mechanics with the indeterministic quantum field version of Bell. This ontology fits perfectly with the ontology found elsewhere in this book.[266]

13.4 Excursus: A taste of the formalism of quantum mechanics

In the chapters where including mathematics is natural, I have tried to keep the mathematics so simple that knowing addition, multiplication and the Pythagorean theorem suffices, and I shall do the same here (while I do mention the cosine of an angle somewhere, I also translate it to the Pythagorean theorem afterwards). The mathematics of quantum mechanics are harder, even the basic Schrödinger equation, but it is possible to get an idea of how to represent measurements of spin and use the Born rule using only simple mathematics, so that is what we shall do here in order to get a taste of the formalism of quantum mechanics.[267]

Quantum states can be represented by vectors (Albert, 1992, pp. 20, 30). A vector is, geometrically speaking, an arrow with a length and a direction (Albert, 1992, pp. 17–18). A property which has one of two values (for example, spin up and spin down) can be represented with two vectors which are orthogonal to each other (there is a 90-degree angle between them), and both vectors are unit vectors, which means that they have length one (Albert, 1992, p. 30). These two vectors are like a coordinate system, called basis vectors, and other pairs of vectors which are orthogonal to each other can be included to make a vector space, which is a collection of basis vectors (Albert, 1992, p. 30).

Here is an example with two vectors representing example properties of either soft or hard and two vectors representing example properties of either black

266 The reader should also know that there is a quite uncontroversial way from quantum mechanics to Newtonian mechanics. It is a matter of statistics that makes Newtonian mechanics at the macro level a consequence of quantum mechanics at the micro level. This transition will not be presented here, but for details see, for example Ogborn and Taylor (2005). The picture is complicated by context dependence on the way from quantum mechanics to Newtonian mechanics, see Bishop (2019, pp. 6–3).

267 The presentation is based on Albert (1992).

and white (see Figure 20).[268] These example properties are chosen to make things easier, but in real life the properties would be x-spin and y-spin of electrons (Albert, 1992, p. 1). Note that the opposing vectors are orthogonal to each other, and that each vector is of length one.

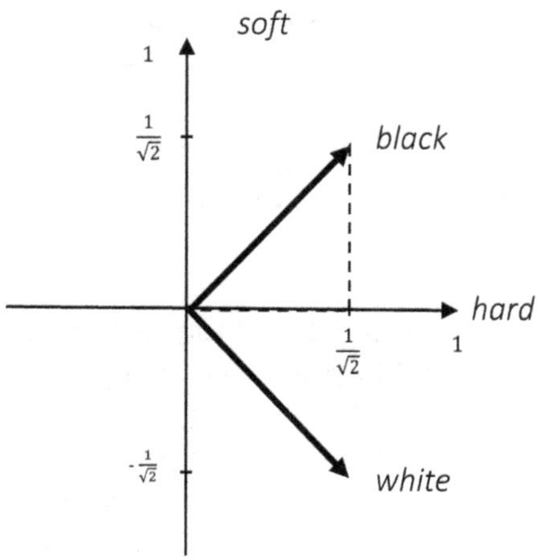

$$\left(\frac{1}{\sqrt{2}}\right)^2 = ½ = 50\%$$

Fig. 20: Vectors representing quantum properties

The properties we are measuring are incompatible in the sense that they cannot be measured at the same time (Albert, 1992, p. 7). As seen above, a puzzling feature of quantum mechanics is scrambling of measurement results: if you measure a property of an electron (say, white), and get 50 % white, then pick out all the white electrons and measure their color again, you will find that all are still white. But if you measure their hardness, then again measure their color, then suddenly they are again 50 % white and 50 % black (Albert, 1992, pp. 1–4).[269]

268 The example is taken from Albert (1992, pp. 31–33), but the setup in a coordinate system is made by Kelvin McQueen (McQueen, 2015), and adapted by me here.

269 Some of the strange results from quantum mechanics experiments could make you suspect that what seem like very accurate results stem from self-confirming measurements that are depending on the theory in how the measurements are interpreted. A theory could wrongly seem to be confirmed with spectacular accuracy. Imagine as a toy example a theory about the size of a certain fish that it will never be larger or smaller than some very accurate values, which is then confirmed to the degree of many decimals, but the nets used to capture the fish cannot let bigger fish in and let smaller fish get out. I do not think that such mistakes are made in quantum mechanics, which has produced a lot of well-functioning technology, so I will suggest an interpretation based on standard quantum mechanics.

The degree to which these vectors in the vector space are pointing in different directions (from 0 to 90 degrees) represent their degree of compatibility, and here degree of compatibility tells us something about how probable a certain result of a measurement is going to be.[270] There is a way to calculate how compatible two vectors are by finding out to what degree they go in the same direction. This is done by multiplying the one vector with the other to find a number called the inner product. When the vectors are of length one, the inner product is a result of the angle between the vectors. The smaller the angle gets, the more compatible they are, until they overlap completely. Figure 21 shows how the inner product can be seen as projecting one vector onto another.

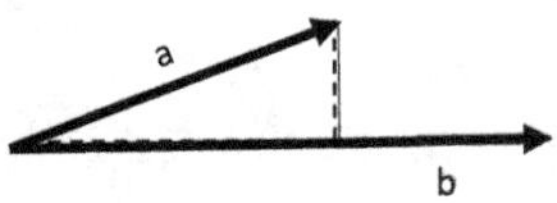

Fig. 21: Inner product

One can calculate the inner product in different ways. In one, you can multiply the length of each vector and multiply with the cosine of the angle between them. If the angle is 90 degrees, there is no overlap between the vectors and the result will then be zero, since the cosine of 90 degrees is zero. This will represent incompatibility, while if the angle is zero degrees, the (unit) vectors completely overlap, and the inner product will be 1 since the cosine of zero is 1. Another way of calculating this inner product is to multiply the coordinates for both vectors along the x-axis and the y-axis and adding these two products. This latter way of calculating is how measurements in quantum mechanics are often calculated (at least in introductions) by use of matrices and Dirac notation.

The Born rule says that when two vectors represent the possible outcome of a measurement, the inner product squared gives you the probability of what the measurement outcome will be. Above we saw that the inner product represents the degree of compatibility between two vectors. Consider again the diagram presented in Figure 22.

To what degree does the vector representing black overlap with the vector representing hard? If you project the vector for black (with length 1) onto the vector for hard (with length 1), you can draw a line from the tip of the vector for

270 This degree of compatibility is how I interpret Kelvin McQueen and David Albert here: McQueen (2015, pp. 22, 29–30), Albert (1992, pp. 33, 43). If my interpretation of them is wrong, then it is instead how I suggest to interpret quantum mechanics, since it seems to make good sense of the measurement results and allows me to present the basic idea without including operators.

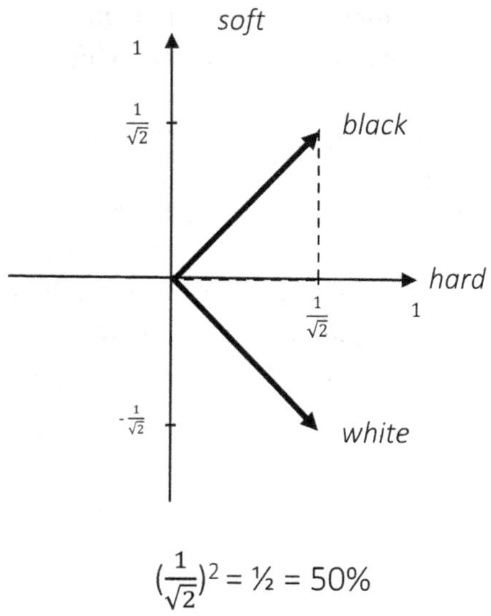

Fig. 22: Vectors representing quantum properties

black straight down to the vector for hard, and you will hit the point marked as 1 over the square root of 2. It could also be written as 0.7071..., but it is written this way to show easily that the square root of this value is ½ (i.e. $\frac{1}{\sqrt{2}}$ = ½).

Consider now the vector for black as a hypotenuse of value 1. The degree to which it overlaps with the vector for hard is the value 0.7 or $\frac{1}{\sqrt{2}}$. Consider now this part of the vector hard as one of the straight lines in a right triangle representing the degree of compatibility. Since the degree of compatibility plus the degree of incompatibility should add to one, we can think of the line from the tip of the vector for black down to the vector for hard as representing incompatibility. Pythagoras has taught us that if we square each of the two mentioned lines and add them together, they equal the hypotenuse squared: $(\frac{1}{\sqrt{2}})^2 + (\frac{1}{\sqrt{2}})^2 = 1^2$ (0.5 + 0.5 = 1).

What we can learn from this is that the probability that a black or a white electron will be found to be hard or soft is 50 %. This is because the overlap is $\frac{1}{\sqrt{2}}$, and this value squared is ½. The probability of finding that a black electron is black or a white electron is white, a hard electron is hard or a soft electron is soft is 100 % in all cases, since the overlap is complete. On the other hand, the probability of finding a black electron to be white, a white electron to be black, a hard electron to be soft, or a soft electron to be hard will be 0 % since the vectors are orthogonal and there is no overlap.

We can calculate this degree of overlap between vectors, as described above in the presentation of inner product, by multiplying the coordinates for the vectors along the x-axis and the y-axis and adding them together. To find the probability of a particular measurement, this value can be squared. For example, what is the probability of finding a soft electron to be soft? The coordinates for the soft vector is 0 along the x-axis and 1 along the y-axis. We can find the inner product by then taking $0 \times 0 + 1 \times 1$, which equals 1. The probability that a soft electron will be measured to be soft is 1^2, or 1.

What is the probability of finding a soft electron to be hard? The coordinates for the soft vector are 0 along the x-axis and 1 along the y-axis, while the coordinates for the hard vector is 1 along the x-axis and 0 along the y-axis. We can find the inner product by then taking $0 \times 1 + 1 \times 0$, which is 0. The probability that a soft electron will be measured to be hard is 0^2, or, again, 0.

What is the probability of finding a black electron to be hard? The coordinates for the black vector is $\frac{1}{\sqrt{2}}$ along the x-axis and $\frac{1}{\sqrt{2}}$ along the y-axis, while the coordinates for the hard vector is 1 along the x-axis and 0 along the y-axis. We can find the inner product by then taking $\frac{1}{\sqrt{2}}$ x 1 + $\frac{1}{\sqrt{2}}$ x 0, which is $\frac{1}{\sqrt{2}}$. The probability that a soft electron will be measured to be hard is $(\frac{1}{\sqrt{2}})^2$, or ½.

This method will work whatever the angle between the vectors is as long as the vectors chosen to represent degree of incompatibility represent how compatible or incompatible these values actually are in nature. Which vectors to use to represent quantum states and measurements have been found through different techniques (Albert, 1992, p. 37). The method works equally well on one particle as on several particles, and the state vectors can be translated into wave functions (Albert, 1992, pp. 47–48).

The Born rule expresses that there is a degree of incompatibility between properties represented by vectors. When these vectors are multiplied, you get a number which represents their degree of compatibility. This number can be seen as one of the right sides in a right triangle, representing compatibility, and the other right side in the triangle has a value that represents incompatibility. When these two values are squared and added, they equal the hypotenuse squared, which is 1 squared, and which represents the total sum of compatibility and incompatibility. If the line representing compatibility had been completely compatible with this hypotenuse, they would have overlapped, and the line representing incompatibility would have been zero.

Using this method allows us to express compatibility in percent and add the percent of compatibility and incompatibility to get the sum of 1. This again allows us to use the formalism of probability calculation, where the degree of compatibility and incompatibility must sum to 1. When Born first suggested his rule, he did not square the amplitude, but then he added a footnote later saying it seems that the rule needs to be squared to work, but without giving an explanation of why (S. Carroll, 2014). This is how I think of why the Born rule needs to be squared in order to work: because the formalism employs Pythagorean relations and the values need to be added to 1.

13.5 Excursus: Why are the scientifically formulated laws of nature as they are?

Having looked at physics in general, relativity and quantum mechanics, it is interesting to ask the following question: Why are the scientifically formulated laws of nature what they are? Why are these particular rules being followed? Why is light speed a universal speed limit; why does gravity have the strength it has; why are the equations of physics like they are and not something else; etc.? In this excursus, "laws of nature" are to be understood as the rules nature behaves in accordance with, and not their truthmakers that make things move.

Here are two considerations to use as a point of departure for answering this question: On the one hand, many laws, constants and initial conditions are said to be fine-tuned for life, meaning that there is only a very small subset that are physically possible (as far as we know what is physically possible) and could produce life, and we do not have an explanation for why these laws, constants and initial conditions have the values they have. A famous example of fine-tuning is the low initial entropy of the universe, which according to Roger Penrose is fine-tuned to 1 to 10 raised in the power of 10^{123} (Penrose, 2004, pp. 726–732).[271] Another common example is the cosmological constant (Weinberg, 1989).

On the other hand, many of the laws of nature seem very strange. Why does the universe contract in a way that makes everybody measure the same light speed? Why do particles have their properties entangled? It seems we could have a nice inhabitable universe without the weak force or neutrinos (Harnik, Kribs, and Perez, 2006),[272] and thus it is hard to see that there is a point or function or intentional reason why they are there.

Considering why we have the laws we have, I see three main candidates for answering the question. The first is to argue that there is an intention behind them, making the laws of nature what they are for a purpose. We have already seen that there must exist something fundamental with an enormous capacity for actualizing values. It is not a far stretch of mind to imagine that whatever this is, it also had the potential for actualizing intelligence and desire in itself, which it then used to actualize laws of nature for a purpose. All the strange

---

271 For other examples, see Barnes (2019).

272 Thanks to Øystein Elgarøy for this suggestion.

laws would then have to be considered as being there for a reason that we do not understand.

Alternatively, there is no purpose behind the laws, but that raises the following question: are all the laws we have discovered actually laws that nature follows, or are most of the laws merely non-guiding epiphenomenal patterns while in reality there is a deeper pattern according to which the values are really actualized? For example, does nature follow a rule that everybody should measure the same speed of light, or is there a deeper rule being followed which has the consequence that everybody measures the same speed of light?

Here is the second main way of answering the question of why we have the laws that we do: While there seems to be many strange laws and many strange rules for actualizing particles and forces, most of the laws are in fact just regular patterns that follow from some deeper laws being followed, which may be quite simple in themselves, but with an enormous potential for creating larger configurations of patterns that we summarize as scientific laws and name as different particles.

The third main way of answering the question is to say that there are in fact very many strange laws and rules that guide how values are actualized in the world. Since they seem very unsystematically related, it indicates that there is a huge amount of possible rules that potentially could have been followed. If there are very many rules that could have been followed, can we say anything about why the rules are followed that are actually being followed?

One could look for a kind of Darwinistic explanation and search for some kind of selection mechanism that picks out some laws for being actualized. Some could be more stable than others for some reason, but it is hard to think of a mechanism that should preserve some laws of nature instead of others, even if the others are short-lived and chaotic.

It then seems more likely to think that all kinds of rules are actually being followed, actualizing all sorts of values, but that only some laws actualize values that turn into stable structures that in the end are stable enough to produce intelligent beings asking about laws. In other words, maybe numerous values are actualized all the time according to all kinds of rules – maybe even at the same points – but we inhabit a particularly stable part. When we look for laws, we look for laws that are relevant for the kind of stable structures we have a relation to (like water, stars, etc.), and we fail to see all the chaotic rules that are also being followed with consequences undetected by us.

The stable structures we are interested in are also the structures that we use to test the laws we are looking for. There are thus many observation selection effects going on, while there may be many patterns, laws and values that we have not discovered, either because they are here and we do not notice or because they are out of our reach. This way of explaining laws of nature would then also work for explaining physical constants like the coupling constant, also being important in physics, but again many of the constants seem to have completely random values with no deeper explanation.

All of these three answers are possible, but the first two appeal to undiscovered patterns to explain the laws we know that look strange. Maybe with the rise of more intelligent AI, they will be able to find patterns that we have not discovered that can help us choose between these three answers. While any of the three answers may be best justified in the future, what seems best justified today is that there are strange laws because there is a much larger potential of possibilities, but a random combination of them has been able to produce us. Just like the qualia seem to be a selection of a much larger potential, so the rules that values are actualized in accordance with seem to be a selection of a much larger potential of rules, with the common feature that these rules have produced consequences which are stable enough to include intelligent beings asking about rules.

The large variety of rules being actualized could of course come from a simpler rule of exploring possible rules. Could we in the end speculate as to why there is the enormous potential of possibilities for qualia values, physical values and rules that there is? Why is there anything at all instead of nothing? I shall return to this question at the end of the last chapter of the book.

This concludes Part Three on the world. Much of what we mean by time has been shown not to be irreducible, except from the motion part of it. Mathematical entities and probability have been shown not to be irreducible. Finally, several physical concepts were reducible, but the fundamental force with a strength and direction remained.

In this book I have claimed to have reduced a lot of different concepts to some fundamental building blocks, but some may object that I have not done so. One could object that instead of reducing causality, modality, probability, and mathematical entities, I have just kept them all, but placed them inside a big mysterious thing called laws of nature (or rules that values are actualized in accordance with).

This objection is mistaken. I have reduced these entities in the following way: Causality in its many forms has been argued to be theoretical frameworks used on regularities at all ontological levels (including shortcuts), but in the end the regularities are rooted in the laws of nature. A core of the concept has thus been kept, which is only natural in any reduction which is not pure elimination. But causation is not something flowing between what we call the cause and what we call the effect. It expresses a regularity usually made true by something that happens at a deeper ontological level.

Modalities in their many forms have been argued not to be an irreducible structure in the world, but an expression used for consistency with a set of presuppositions. Many of the presuppositions could have been different, but in the case of laws of nature, they are a set of presuppositions that other things can be consistent with or not, so one could say that this is one kind of modality that has been kept. However, modality is not something irreducible and different from laws of nature, so in that sense modality has been reduced.

Probability is not a mathematical entity with irreducible existence, but instead values are actualized by rules that sometimes say that a value could be actualized or not, or maybe twice as often as not, or something else that makes different futures possible. Then we can sort different possible futures into groups and use the theoretical framework of probability to say how probable the different futures are. Again, probability has not been eliminated, but shown to be a theoretical framework that can be used to describe what follows from the fundamental building blocks.

Even if the laws of nature follow mathematical rules, this does not mean that mathematical entities have an irreducible existence. Rather it means that there are structures and relations between the rules that we can use our mind to conceptualize as mathematical entities. Maybe the laws of nature work by if-then rules with the consequence that A is actualized three times more often than B, and we can use our mind to make a concept for "three".

I am not sure what is the best way to think of the relation between laws of nature and mathematical entities. One could think that laws of nature implies mathematical entities like complex numbers which implies a mind which implies that laws of nature imply a mind. But one could also think that complex numbers describe relations like that of rotating on an axis (which is a way to think about what it means to be the square root of −1) which could be a pattern in the world regardless of mind, but which a mind could later think about and express with complex numbers. Or one could think that mathematical structure is an irreducible part of the rules that nature follows. So far I find the best option to be that both the rules and the world and the mind have structures with relations between parts that our mind can develop mathematical concepts to describe. In the chapter on mathematics I did say that there are fundamental structures in the world that make some mathematical claims true, so this part of mathematics has not been reduced.

In other words, even if I claim to have reduced a lot of concepts, I do not mean to say that everything they express has been eliminated. I mean instead to say that there is a theoretical framework which explains their ontological status in a more coherent way than by thinking of them as irreducible or undefinable – namely the theoretical framework defended in this book.

The first three parts of this book have presented a model for understanding the world, including mind. Now it is time to look to the future, which is the topic of Part Four. Here I want to explore implications of this understanding of the world for our understanding of the future and the possible world of the future.

Part Four: **The Future**

14 Ethics

It may seem strange that I have placed a chapter on ethics in a part called "the future" since ethics is about what is good or right to do here and now. As will be clear, I will defend a view where what is ethically good is determined by what would be the best way to the best future, and this is why the chapter has been placed in this part. While some argue that ethical values are irreducible entities that exist here and now, I argue that ethical values are determined by the possible future they can actualize.[273]

As in previous chapters, we shall again consider whether ethical values are irreducible entities that belong in an ontology, or whether they are reducible to something else. I shall be arguing that they are reducible to something else. This is a typical topic for the discipline of metaethics, and I will start by presenting the standard theories in metaethics in Section 14.1. Then I present my own understanding of metaethics and a specific ethical theory in Section 14.2 before I defend it against objections in Section 14.3.

As a very brief preview: I shall defend a reductionist naturalist view, where "good" is a concept we make up to describe a particular possible world in the future that we think is the best possible world according to a particular standard, and we hypothesize that there is also a best way to reach the goal. This is enough to give meaning to normative terms like "good" and "should", and there is no normative force or moral values existing in a platonic world beyond the concepts invented by humans to describe a possible world. Ethics is thus a theoretical framework describing a path for the future that the supporters of the framework suggest that people should follow.

14.1 Introduction to metaethics

There are different kinds of ethics. Applied ethics discusses specific questions like abortion or environmental issues. Normative ethics discusses general theories for how to behave, like duty ethics, virtue ethics, or utilitarianism. Descriptive ethics describes how people actually think, feel and behave on moral issues.

Sometimes a distinction is made between ethics and morality, and then morality can refer to how people actually behave, think and feel in moral dilemmas,

273 In this chapter, the term "value" refers to ethical values and not the fundamental ontological values presented in Chapter 3, unless the context makes it clear that I refer to fundamental ontological values.

https://doi.org/10.1515/9783110770957-016

while ethics refers to theory about morality: what is actually right, wrong, etc.? However, most English-speaking philosophers (and I) use the terms ethics and morality as synonyms, and instead use “descriptive ethics” to refer to the understanding of morality as how people actually behave.

In this chapter, I am not interested in ethics or morality in this descriptive sense, but the distinction is useful to be aware of. The reason is that some think that all of ethics can be explained by seeing that morality has evolved through the process of animal and human evolution, and that as such it cannot be objectively true for all.[274] But even if the morality that individuals actually have has evolved through evolution, this does not show that there cannot be ethical norms that are true for all.

Alex Rosenberg argues that evolution shows us that there are no objective moral values since people have evolved to share the same core values, which are also helpful for survival and reproduction. If these moral values are true, it cannot be a coincidence that they have been selected for. Rosenberg sees only two possible solutions: either evolution has selected the true moral norms, or the moral norms are true in virtue of having been selected. But evolution does not select for truth, nor do moral values become true by being selected for. Since neither of the solutions are possible, Rosenberg sees no other alternative than to reject that moral values are true (A. Rosenberg, 2011).

Rosenberg’s argument goes wrong in many places. Here is my explanation for the close link between shared moral assumptions and evolutionary advantage: Not all shared moral values are beneficial for survival. As Rosenberg also points out, evolution selects for prioritizing your own group over others, resulting in racism and xenophobia (A. Rosenberg, 2011, p. 111). This is how many people actually act, while most people understand that it is not morally good. We understand this by use of our reason, and even if evolution does not select for truth, it selects for survival, and the capacity to discover what is true is helpful for survival. As a side effect, we can discover much which is true without it being necessary for survival, like truths about ethical norms. Moreover, as I shall argue below, what is morally good sums up what most individuals prefer the most, and evolution has shaped our preferences (such as for food and sex and to avoid pain and death). Ethics wants more of that which is good for everyone, and evolution has selected behavior which causes more good for more individuals (in the sense of more surviving individuals preferring that which they need to survive), often through cooperation in groups. There are thus several reasons for expecting a

274 For example Ruse (1986).

non-perfect overlap between common ethics and evolutionary advantages, which is just what we see. No moral nihilism follows.

Discussing whether there can be true ethical norms or moral judgments is a main topic within metaethics, which is what this section is about. Ethical norms are sentences that can be translated into moral statements of the form "It is good/bad/right/wrong (not) to X" or "One should (not) do X", for example, "It is wrong to lie" or "One should not lie". Metaethics discusses the fundamental questions within ethics, like how to understand the status of ethical norms: Do they have truth value or not? Cognitivists say that moral judgments express beliefs, which can be true or false, and thus ethical norms have truth value.[275] Non-cognitivists say no, and argue, for example, that ethical norms just express what individuals feel.[276]

There are different ways of thinking about norms having truth value, but my interest in this chapter is true norms in the sense of true for all, and so that is what I refer to by "true norms".[277] This means I will be leaving out subjectivism, which says that norms are only true relative to the one uttering them, and error theory, which is cognitivism in saying that moral statements are propositions, but holding that they are false.[278] This leaves us with cognitivists who are also realists about moral truth, meaning that ethical norms can express true beliefs (cognitivism) because they express moral facts that are true for all (which is the kind of realism I am interested in in this chapter).

There are several reasons for holding that ethical norms have truth value in this sense. One argument is that it seems people can be wrong in their moral opinions and that they can make moral progress. Another argument is that it seems quite possible that everybody at a point of time could be wrong in their moral opinions and that the global society as a whole could make moral progress. But if ethics is relative to person or culture or does not have truth value at all, we cannot correctly say that all could actually be wrong (saying "all are wrong" would then just be a personal opinion).

If you are a cognitivist and a realist, the big question is how to understand moral facts: what gives ethical norms truth value? Three common types of answer are either supernaturalism, non-naturalism or naturalism. A *supernaturalist*

---

275 Examples of cognitivsts are Parfit (2011a) and Scanlon (1998).
276 As in the emotivism of A. J. Ayer and C. L. Stevenson: Ayer (1952, chapter 6) and C. L. Stevenson (1944).
277 This means that I am not talking about "norms" in the sense of guidelines that are actually being followed in a society, but about ethical statements and their truth value regardless of whether the norms are followed.
278 The most famous example is Mackie (1977).

will typically refer to God and say that it is the will of God which gives ethical norms their truth value. A *non-naturalist* will typically defend the view that values are ontologically irreducible entities which have their own independent ontological existence, being outside of the scope of current science, and the prime example would be values existing in a platonic world. Ethical norms are then true when they express something true about these values. A *naturalist* will refer to natural states of affairs to explain what gives ethical norms truth value.

Among naturalists we can distinguish between analytical naturalists and non-analytical naturalists. As defined by Derek Parfit, the analytical naturalist holds that normative claims can be translated into statements using natural descriptive terms, while the non-analytical naturalist holds that there are natural terms that are irreducibly normative (Parfit, 2011b, pp. 266, 295). The analytical naturalist could say that we can analyze "good" and discover that it means nothing more than "that which maximizes pleasure", while the non-analytical naturalist will say that there is an irreducibly normative component to what is good, while still being something natural. I will be defending a form of analytical naturalism, and to justify my view, I shall discuss arguments against all three main positions, starting with supernaturalism.

God can be used to explain how ethical norms get their truth value: It is true that it is good to be honest if it is true that it is the will of God that we should be honest. The classical charge against this is the Euthyphro dilemma: Is something good because God wants it, or does God want it because it is good (Plato, 1970, part two)? It would seem strange that something could become good just because the creator of the universe wants it, for it seems possible that such a creator could want something evil. Those who think not, or argue that the will of God should be obeyed, usually assume that God is good and knows what is best in the end, for it would not become good to harm someone for fun just because the creator of the universe wanted it to be so.

Those who believe that God is needed as a basis for good should note the difference between good *existing* because of God and good *being good* because of God. A believer in God should say that things that are good or sickening or funny or something else *exists* because of God, but should not need to say that good is good because God wants it or sickening because God finds it repulsive or funny because it makes God laugh, etc.

Most moral philosophers will not refer to God to explain how ethical norms get their truth value. On the other hand, it seems to be the case that normative claims must have a normative foundation. Hume famously argued that you cannot derive an ought from an is, so it seems that there must exist something normative that can give ethical claims their truth value. Non-naturalists will therefore reject the temptation to understand normative claims as having a descriptive content where normative terms have been fully defined in terms of natural states of affairs.

Another reason for being a non-naturalist as opposed to a naturalist is the open-question argument by G. E. Moore (Moore, 1903, §13). Moore argued that it was a fallacy to give a naturalistic and descriptive (as opposed to normative) definition to the term "good", since it is always an open question whether the definition given is actually what the good is. For example, if "good" is defined as that which maximizes pleasure, we can still openly and meaningfully ask whether that which maximizes pleasure is good. According to Moore, this showed that the term "good" could not be given a naturalistic descriptive definition. We shall return to this argument below.

Yet an argument for not being a naturalist is as follows: It seems that understanding values as something natural takes away the normative force of ethical claims. If good is just something that exists in nature, why should it have a normative force on us? Or if "good" is just a concept that humans suggest a definition of, this seem to make good into something random and relative. Such definitions suggested by humans are likely to be too controversial to gain any acceptance, or if they are wide enough not to be controversial, they will probably be too empty to give any content in specific cases.

Another argument against "good" being something that humans make up a definition of is the claim that even if no humans or intelligent beings ever existed on earth, something would still be right and wrong on earth, which seems to indicate that there is something more to values than humans defining something as good.

These are some reasons why non-naturalists will tend to think that values are ontologically irreducible entities with an independent existence. But there are also many arguments against non-naturalism and supernaturalism favoring naturalism. J. L. Mackie offered the argument from queerness (Mackie, 1977). It says that values are queer entities very unlike other things that exist, and that we have no idea about how it is that we discover such things or can trust our opinion about them. It seems plausible to reject that such moral values exist, according to Mackie.

Another argument against supernaturalism is the supervenience argument. All seem to agree that ethical facts supervene on descriptive facts, in the sense that there cannot be an ethical change without a descriptive change. Differently put: There cannot be two identical worlds with identical histories except that in one of them murder is wrong and in the other murder is good But this seems to contradict supernaturalism, at least a version which says that that is right which God commands. If such divine command theory is right, it seems that God could command that murder was wrong in one universe and good in another universe, which was identical in all other respects. Supervenience seems to imply that values are connected to the natural world by being identical to something natural.

We turn now to the third option, which is naturalistic realism, and I will consider an analytic version where good is a concept that humans invent. Some might disagree that this analytic version should qualify as falling under the label of moral realism, but it is moral realism in the sense that moral values can be true for all (after the moral concepts have been defined), so this is what it means here. A good example of such a strategy would be if we siad that Jeremy Bentham invented a definition of good as that which produces the most pleasure (as opposed to him discovering that that is what good really is) (Bentham, 1988, chapter 1).

This strategy avoids the objection of queerness, since the queer entities do not exist. And it explains supervenience, since the normative is defined as something descriptive. When it comes to Hume's charge that one cannot derive an *ought* from an *is*, the naturalist can respond that the *ought* has been given an *is* definition, so that ethical norms derive an *is* from an *is*. This seems to take away the normative force of ethics, which I shall discuss further below.

When it comes to Moore's open-question argument, several responses have been offered in the literature. The most common response is to say that two words can have different meanings and still refer to the same, so that it is not meaningless to ask whether one is the other. For example, it is meaningful to ask whether water is H_2O, even if water (in most forms) is indeed H_2O. Some have argued that this is a poor analogy since we know enough about water to know when we have found out what it is, but this is not the case with goodness (Horgan and Timmons, 1991).

Other arguments are that the open-question argument is question begging since it assumes for any definition of "good" that it is not right, or that the argument proves too much since it seems that we can meaningfully discuss most definitions of tricky terms (like what is knowledge, culture, religion, etc.) without concluding that they are undefinable (Lutz and Lenman, 2018, section 2.1).

I suggest a rebuttal of the open-question argument which captures some of what all these three objections say: The reason we feel that we can meaningfully ask, for any definition of "good", whether it is good is that we have a vague and/or general and/or non-conscious intuition about what goodness is. Similarly, we have a vague understanding of terms like justice, knowledge, culture and religion and can ask for any precise definition whether that is indeed justice, knowledge, culture or religion. It does not follow that there is actually more to say or something undefinable about it or a correct answer beyond how we choose to define the term. When we ask whether something controversial is, for example, religion or knowledge, we are investigating our own intuitions about the concept without that implying that our intuition can serve as a judge of what is an objectively correct definition. It is up to us to find out what the best definition is relative to the purposes that we have.

Even if the open-question argument is not a good counterargument to ethical naturalism, other arguments remain. There is a normative force in ethical claims which seems to disappear if they are given a descriptive definition invented by humans. Further, there seems to be no good definitions given on the market, for any suggested definition of good is riddled with problems (or too empty to be of interest). Third, if one holds that normative terms are just something we have decided to define a certain way, this is contradicted by the fact that it seems that there would be something morally wrong, even if no humans existed.

In the following, I will suggest a naturalistic metaethical theory which purports to answer these objections, but first a brief comment should be made on how to decide among metaethical theories. Since metaethical theories are meant to be true descriptions of how to understand ethics, we can use the same truth criteria that we use for other kinds of truth. In Chapter 2, I suggested coherence as a criterion for truth, which would then also apply to choosing between metaethical theories. Differently put, a criterion for meta-metaethics is just a normal truth criterion. With these remarks I am now ready to look at a new naturalistic theory of metaethics.

14.2 A new theory of metaethics

The theory to be defended is a form of analytical naturalist realism that argues that ethical norms can be true for all, but this presupposes that the ethical terms have first been given a descriptive definition.[279] To give an introductory example: The word "justice" can mean many things, like all getting the same amount or all ending up with the same result or all getting depending on effort or all getting depending on need – and this is just about *distributive* justice, not retributive or restorative justice. Justice is not something irreducible, it is just a concept which could have been expressed by means of other known concepts without loss.

279 If one insists that realism means that we only discover and in no way invent what is good, then one could argue that my position is not realism. However, as argued above, it is realism in the sense that it holds that normative claims express facts that are true for all, which was how realism was defined in this chapter.

When the term "justice" has been defined, ethical norms using the term can have truth value. If we define "justice" as "all getting the same amount", we can then discuss whether a specific situation is just. But we could also have defined it differently, and discussed the same situation in light of another definition of the term.

This holds for all ethical claims, and good in general. "The good" can be used as an all-encompassing term (encompassing the other ethical values), which I shall from now on refer to as The Good (with capital letters) to show this specific use. The Good is a description of a possible world, and ought-claims are claims which express means to reach the goal, and the goal is The Good. To exemplify this ought/Good relation as a means/goal relation, think of communists who claimed that The Good is the classless society and that workers ought to make a revolution in order to actualize it. Or the Nazis who defined the Aryan Empire as The Good and extermination of Jews as a means to get there. The structure is the same when having a possible world as a vision of something good which leads to specific ideas of what one ought to do to get there.

While some may think that there must be one correct meaning to the term "good", I reject this, since there is no singular "correct" or "objectively right" definition of any term, be it word or concept (Puntel, 2008, p. 107). A term requires some definition or intention before we can discuss it. If two persons disagree on the meaning of a term, they must either agree on a definition or at least explain their own use before they can have a discussion talking about the same. But they cannot appeal to the "correct" or "objectively right" definition.

I argued this in Chapter 2, but to recap, I offer some examples to support this: The term "atom" means indivisible particle and was used for what they thought was the elementary particle of the world until they discovered that it consisted of parts. They could have used the term "atom" for the newly discovered parts or keep the term "atom" for what it was already referring to, which is what happened. In Norway, "marriage" used to mean a union between one man and one woman. A couple of decades ago, there was a discussion whether a union between man and man or woman and woman should be called "marriage" or "partnership", with strong opinions on each side. "Marriage" ended up changing meaning. There is no objectively right definition of atom, marriage, good, or any other term. We just have to try to agree on what it is in the world that we refer to by using the words and concepts that we do.

What then to do if Nazis, communists and others disagree on what "good" means? They should try to agree on a common definition at a more general level, which could then be used to argue that their more specific understanding is correct. Maybe they can agree that "The Good" is what is "best for the most people", or "fulfilling the most preferences", or "what an omniscient being would know that most would prefer", or something else. Then they could argue that a classless society or an Aryan Empire or something else best fulfills that definition. If they instead discuss without knowing how the other persons use the terms, they will most likely only achieve apparent (dis)agreement.

Some will say that it is not realistic that people with very different views on what is good should find a common definition of "good" to discuss in light of. But it is not as if the discussion will be any better if they do not agree on a common definition. Trying to define the term will at least make them aware of their difference from each other. They can then discuss given either this or that definition knowing what the other person means by "good". Different definitions will then refer to different things, just as we can discuss whether something is just when given this or that definition of "justice". And they can discuss criteria for how to define terms like "good", where the usual strategy is to start with established practice and work towards as coherent a definition as possible. This is my approach as well in this text – that I try to offer a coherent ethical theory with my point of departure in the current debate within metaethics. The aim is to give the most coherent understanding of ethical goodness and ethical norms that I can.

One can object to this approach as well, and say that both humans in general and moral philosophers disagree completely on even the basic abstract principles. Ryan Muldoon and Gerald Gaus are examples of authors who have argued that since people disagree radically on both morality and how they understand the world, we should not have an ideal ethical standard, but instead negotiate step by step something that all involved will see as improvements (Muldoon, 2016; Gaus, 2016).

My response to this is that if we say that people really disagree about something – instead of just talking about completely different things – it presupposes some shared understanding of what they are talking about. If they really disagree about what is morally good or morally better, it presupposes at least some coarse-grained understanding of what "morally good" means (one cannot think that it means "green").

Note how Muldoon and Gaus also must presuppose such a moral standard in order to think that it is a morally good suggestion that we should try to have negotiations that all involved see as improvements. We cannot think that Muldoon's and Gaus' proposals are good unless we presuppose that it is good that everybody find their situation improved. They, like me, presuppose a coarse-grained abstract standard of what "morally good" means, even if they, like me, are uncertain about what concrete ethical rules are right.

Here now is my theory in more detail.[280] I suggest the following definition of The Good: The Good is that possible world which an all-knowing and all-sympathetic being would know would most probably be valuated the most by the most. Why this formulation? I will now comment upon the different parts of the definition in order to justify it. When I refer to The Good, this actually means "the best", yet there could be other possible worlds that were good without being the best. More will be said about this below.

When I refer to a possible world, I do not mean any imaginable world. I am thinking of a physically possible world, the possibility of which is rooted in our actual world. While focus is on this earth now, in the future it may extend beyond the earth, so I use the more inclusive term "the world". The term "possible world" is meant to include both the end point (the kind of world that most would valuate the most) and the way to get there (from our world today, which is *not* the kind of world that the most would valuate the most). The best possible world is thus the best possible future development of the world. Since The Good is a goal that might be reached by different means, the means to get there should also be the means valuated the most by most.

Why is "all-sympathetic" included in addition to "all-knowing"? The term "all-sympathetic" might seem like a way of smuggling in something normative and making the definition circular, if one understands sympathy as wanting good for others. But by "all-sympathethic" I only mean that this being not only knows cognitively what others want, but also emotionally feels their feelings, which gives another kind of relevant knowledge in knowing what they would valuate the most. If an all-knowing God exists, God's valuation is also relevant for the definition of good, but only if God is an all-sympathetic being. If the world had been created by an evil God who wanted all to suffer, the wants of this God should not be part of our definition of good. This is another reason to include all-sympathetic in the definition.

Instead of referring to an all-knowing and all-sympathetic being, I could have referred to what most would find best. But there are many who lack information for making the best choices, so the definition included an all-knowing being in order that all relevant information should be taken into consideration. I could have removed the all-knowing being and instead added a clause that the good is what most would valuate the most "given that all had had maximal information", but there are humans and animals who lack cognitive resources for

280 The theory to be presented contains typical structures found in an ethical theory. Together with Svein Jåvold, I have suggested an ethical model taking more into consideration other structures like mediators, typical mechanisms, power structures, unintended consequences, etc., see Jåvold and Søvik (forthcoming).

making judgments at all. If I had tried to solve this by adding "given that all have perfect cognitive resources", it would presuppose a world too different from what the actual world is like.

Deciding what to presuppose when it comes to knowledge is a tricky problem, since what we valuate, and how much, also depends on what we correctly or falsely believe. By introducing the all-knowing being, the idea is that such a being can know (as probably true as possible) both what people valuate and what they would valuate in different situations. By introducing the all-knowing being, nothing more needs to be said about the inhabitants of the world. But of course it is we who have to discuss what such a being would know when making actual judgments of good and bad, right and wrong.

The term "would know" is used to show that this all-knowing being is a hypothetical being, so that the definition does not presuppose the existence of God. It is the term "valuated" which gives the descriptive naturalistic content to ethical goodness, and I believe that the ethical goodness or value of things lie in their potential for being valuated by someone (more on that potential below).

To "valuate" something is a broad term which means that an individual experiences something as good, either individually good or ethically good and either instrumentally good or good for its own sake (again, these distinctions will be further discussed below). This implies that valuation does not have to give a sense of pleasure, but rather just be something that an individual prefers instead of something else. Valuation does not have to be an intellectual procedure – it just requires the capacity for conscious experience of something as good or preferable. It is a matter of having goals and selecting between them, not a matter of feeling pleasure.[281] I just write "the most for most" instead of for most *living beings* or most *humans*, since it is unclear which animals or individuals have the capacity for valuating something.

Note that the all-knowing and all-sympathetic being need not be understood as part of the possible world that would be valuated the most by the most. It is the valuation by the participants in the world that makes something good. In one sense, the definition could have left out this all-knowing and all-sympathetic being and just be "The Good is that possible world which would most probably

281 Daniel Hausman distinguishes between different concepts of preference: enjoyment, comparative evaluation, favoring and choice ranking (Hausman, 2012). By "valuation" I mean a comparative evaluation, which in the brain is based on enjoyment even if that enjoyment is not always consciously felt, and which comes (fallibly) to expression through choices. In order to use it as a coherent foundation for ethics, it is important to include both that it expresses appreciation and that it enables us to compare alternatives.

be valuated the most by the most". Why have I then included the all-knowing being?

The reason is that this definition – without the all-knowing being – sounds like it refers to most of those who actually exist as the basis for what is good. It sounds like we are asking for what most of those who exist now would probably prefer the most. But my point is instead to include future potential individuals as well, and this is clearer by introducing the all-knowing and all-sympathetic being. If we specify the meaning of "most", this all-knowing and all-sympathetic being can be left out of the definition. But *we* must take the role of an all-knowing and all-sympathetic being when considering what would probably be best for the most, so this part of the definition can just as well be stated explicitly in order for it to be clear.

I assume that we live in a genuinely indeterministic world, which means that not even an all-knowing being will always know what kind of actions have the best results (the best results are those valuated the most by most). At any point of time, there will thus be some uncertainty as to what is best to do in the sense of actually having the best result. This could seem to take the truth value out of ethical norms if there is no way of knowing for sure what produces the best result. For this reason, I have included "most probably", since the omniscient being can then know about the ontological uncertainty and have a true belief about what will most (ontologically) probably give the best result.

These were some comments to the parts of the definition, but a bit more should be said about its core to explain why this definition is chosen. Why should ethical goodness be defined in terms of valuation – why is valuation the central truth maker of what is ethically good? When philosophers suggest definitions of terms, the starting point should be the normal use of the term, which can then be clarified and made more coherent. What people call good are the things that they appreciate or valuate. Some things they valuate as a means to something else, some things they valuate just for their own sake, and some things can be valuated both as means and goals. Even if we just look at what people valuate for its own sake, these are very many different things. By just focusing on valuation, the definition includes as good the different things that people find good.[282]

282 "Good" should here be understood as individual good – see the next paragraphs. Valuating something as individual good includes aesthetic valuation – that one appreciates something as beautiful. I do not have a philosophy of aesthetics, I just assume that valuating something as beautiful is something that individuals do in different ways for different reasons without there being anything objectively beautiful. Rather it is a result of evolution and context only.

However, this is still a vague description of "good", and it does not explain why something valuated is good or where the goodness is located – in things or persons or both or something else? When people valuate something instrumentally, we can ask why they valuate it and they can refer to a reason. For example, they can say that they like to watch comedies because they like to laugh. But when we come to something valuated for its own sake, we seem to have reached an end to reasons. If I ask why you like to laugh, the answer is probably that you just like it. Or if I ask why you want to be happy, happiness is not a means to something else, but valuated for its own sake.

There are three different meanings of the term "good" that are important to distinguish. When something is good in the sense that it is valuated for its own sake by an individual, we could refer to this valuation as *individual good* – it is experienced as good for an individual. This is then to distinguish from *ethical good*, which is when we try to sum up all the individual goods into what would be valuated most by the most. And this again could be distinguished from the sense in which the things or events that are being valuated are good, which we could call *potential good*, since I will argue further below that the value of something lies in its potential for being valuated by someone.[283]

Derek Parfit has a long discussion of subjectivism and objectivism regarding the good, and the question is whether we make something good by desiring it (as the subjectivists say) or whether something is good and that makes us desire it (as the objectivists say) (Parfit, 2011a, p. 45). Parfit supports the objectivists and argues that something does not become good by us liking it (Parfit, 2011a, pp. 47, 55). I am closer to the subjectivists, but distinguish between potential good and actual valuation: something can be good (in the sense of potential good) even if nobody likes it, but nothing would be good if valuation was not possible at all.

These different aspects of what is good relate in the following way: The most fundamental part is that someone valuates something, which does not have a deeper explanation than just being valuation. In the chapter on qualia and evolution I explained why certain evolutionarily beneficial actions (like eating or having sex) feel good, but the basic relation that something is valuated because it feels good is just a basic conscious experience that exists without further ex-

---

283 Christine Korsgaard argues that we should distinguish between two distinctions in goodness and have a theory about how they relate to each other. The first is between instrumental goodness and goodness for its own sake, and describes how we value things. The second is between intrinsic and extrinsic goodness and describes whether the source of goodness is in the thing or outside of it (Korsgaard, 1983, pp. 169–170). I use the first distinction when describing individual and ethical good, while my term "potential good" would correspond to her "intrinsic good". I describe how I understand their relation in the following paragraphs in the main text.

planation. This valuation (individual good) is the basis for ethical good, since what is ethical good is determined (by definition) by what is most individual good for most. And this valuation (individual good) is also the basis for the value that things, people and events have (potential good).

This does not mean that something does not have value if it is not valuated, but that it would not have value if no valuation existed at all. A ring of gold or a beautiful sunset would not have value if no one ever in the past or the future could valuate it (more on that below). The reason is that the concept of value would not have meaning in such a scenario (and if you think it does, it is because you are valuating it (or imagining a possible being valuating it), but that was supposed to be ruled out).

The Good is then the possible world which gives the most ethical good, which is the most individual good for most. If I just use the term "good" without the context making it clear in what sense it is used, I mean good in the ethical good sense. There are many variables that can be altered to reach such a possible world: the world can be changed and the people who valuate can be changed, which makes it difficult to know what sort of route one should choose towards the goal. I discuss this further below, but first I will offer definitions of some other common ethical terms.

Given the suggested definition of The Good, how should standard ethical terms like value, good, bad, right, wrong, should, etc., be defined? That which gives normative claims their naturalistic descriptive content is the naturalistic descriptive definition of value: the value of something is its potential for being valuated. This understanding will be discussed below in a discussion of human dignity versus animal rights. Here I just unpack how I understand the content of the concepts.

The definition of The Good is actually a definition of what is, morally speaking, the best. A lot of possible worlds and actions could still be good without being the best, so there is a sliding scale between good, better and best. On the other side, The Bad is that possible world which an all-knowing and all-sympathetic being would know would most probably be disvaluated the most by the most. The Bad is then actually the morally worst possible world, so there is a sliding scale from bad to worse to the worst possible world.

The terms "should" and "ought" can now be descriptively defined as means to a goal. Given that The Good is the goal, one should or ought to do this and this, while one should not or ought not do that and that, given that we want to avoid the bad. This seems to take the normative force away from the terms "should" and "ought", an objection I discuss below.

How do we go from this big picture of sliding scales to what makes an individual action either good or bad? How do we define the point on this scale where we can say that you should or should not do this or that? To answer this question we must take as a starting point the actual world at the time and place where an individual exists and the resources that the person has. Everyone should, based on their resources, contribute to making the world better (including for themselves), while they should not make it worse.

Making it better is good; making it worse is bad. You should make it better and you should not make it worse. The better or worse one makes the world, the better or worse it is. If an action does not make the world better or worse, it is morally neutral. The world as it is, is then the starting point for defining an action as morally good or bad, and I use the terms "right" and "wrong" as synonyms for good and bad in this sense: making the world worse is wrong and should not be done and making the world better is right and should be done (and remember that you make the world better even when being good to yourself).

What has been said so far is still a quite coarse understanding of the term "should". The fundamental meaning of what it means that you "should" do something is just that it is the means to the goal that is presupposed when we think that someone should do something. If we then want to make a distinction between a moral and a non-moral "should (not)", we can say that morally speaking we should make the world better, we should not make it worse, but for many actions a moral "should" does not apply since they do not make the world either better or worse in any significant sense.

Saying that we should make the world better may seem like an impossible ethics to follow. Should one then always do nothing but actions that can save lives of starving children, since these are actions that will make the world better? I shall return to this objection in detail in the end of the chapter, but will briefly explain the solution here. I added the qualifier "based on your resources" above, and will return also to explain in more detail what that means. Briefly here, a very rich person giving one dollar to the poor could be said to make the world better – and still we may think that he or she should have done more. A person helping her sick mother could be said to be doing something good even if she could have saved even more starving children with the same resources. I suggest that the right way to think of these questions is to remember (as I said in the chapter on free will) that we hold people responsible by comparing people's actions with a moral standard for what we think they should have done in that situation, and this standard includes evaluating their resources. This standard of comparison should be what from here is the best way to the best world, and the standard should take people's resources and their valuation into consideration.

This means that in a standard for what is the best way to the best world, the best way to the best world entails people taking care of their closest family and very rich people giving more than a dollar to the poor, and so this is something people should do when considering their actions compared with a standard based on their resources. More on this later.

What about cases where you have to choose between two different ways of making the world better or making the world worse? I have said that we should make the world better and not make it worse, but sometimes we have to choose between different ways of making it better or worse, and is the term "should" appropriate for each alternative or just for one of them? What is it that you "should" do if you must choose between hurting or pleasing a different number of people in different ways or with different probability, etc.?

I think that it is best to let the normative force of "should" apply to making the world better and not making it worse, and then we can say that it is better to do something very good than something not so good, or that it is worse to make the world much worse than a little worse. Then we can say with normative force that everyone who is rich should give money to the poor, and also add that it is good to give much and better the more you give. But if we use the term "should" for the morally best action, and say that people should give away all their money, the term "should" loses its normative force because people will feel that the claim is too demanding and reject it as false. One may, of course, still use "should" to say that someone should do A instead of B, but it would be good if it was clear when it is just a comparative "should" (just saying that A is better than B) or when it is a normative "should" (something that a person should do to make the world better or prevent the world from getting worse based on comparison with a standard for that situation).

Note that these definitions of good and bad were relative to an individual and his or her resources. Since the world is very complex, we can always imagine special circumstances where an action generally considered to be ethically wrong would be right in that situation. If your ethics is to be able to say anything in general about actions that are good and bad, it must try to isolate actions from individual contexts, and consider whether there are actions that, in general or considered on their own, lead to a better or worse world. This general consideration is presupposed when actions in general are described as ethically good or bad, for example that lying is bad. Ethical rules are general, while an ethical judgment in a specific situation is particular.

One could argue that all we need is an ethics for individual cases instead of such coarse-grained general ethical norms or rules like "do not lie" and "do not steal". However, we often need to formulate some more general claims about what is right or wrong, for example when raising children, in education, in all kinds of rules and laws, and in discussions of ethical cases, which usually need to be simplified compared to the real world.

Some will say that an understanding of good and bad that takes all into consideration (as is typical for utilitarianism) fails to explain how an action can be wrong *against someone* or good *towards someone*. The individual seems to disappear in the whole. I think that the explanation given of what makes an action generally wrong is also the explanation of what it means that an action is wrong *against someone:* such actions are also the kinds of action that in general lead to a worse world, while doing something good *towards someone* is a kind of action that in general leads to a better (i.e., more valuated) world. Adding "against someone" or "towards someone" just adds a comment on who suffered or benefited from the action. This means that there can be a situation where doing something that is generally wrong – and which we could therefore also say was wrong towards someone – was nevertheless an overall good. I cannot see that there is more content to the phrase "wrong against person *x*" than that it was an action that is generally wrong and adding a comment on who suffered from it, namely *x*. These distinctions will be useful when we discuss problems below like the trolley dilemma or human rights.

More details and implications of these definitions will become clearer as I now relate them to the objections given to the other positions above. The objection against non-cognitivism was that it seems that people can be wrong in moral questions and make progress, and that this goes for humanity as a whole as well. If we accept that The Good is that possible world which an all-knowing and all-sympathetic being would know would most probably be valuated the most by the most, this holds: People – even *all* people – can be wrong about what is good and yet make progress toward it.

The Euthyphro dilemma is avoided, but there is an interesting twist to it given this definition, since it makes the alternatives coincide. On the one hand, God wants what is good because it is good, but on the other hand what is good is defined as that which God wants – without presupposing that God exists. As all-knowing, God knows what all would most probably valuate the most, and as all-sympathetic, God presumably wants what is good to happen,[284] which means that God's will and The Good are defined in light of each other.

The definition works equally well independently of whether or not God exists. If God exists, then God is included in the group of individuals among

284 At least this can be presupposed if we include rationality in God's omniscience, meaning that God would want what feels good for God and others to happen. An evil God is thus precluded. Something is not right just because the creator of the universe wants it – the creator must have the specified properties if the alternatives are to coincide.

whom we search what is best for most and what God values must be included in the definition.[285] If God does not exist, then there is no God to include in the term "most".

We have already seen how naturalism can respond to the deontic fallacy presented by Hume and the open-question argument presented by G. E. Moore. Ethics do not make an is-ought fallacy if it only derives *is* from is, and Moore does not offer a good argument against defining "good".

The theory presented here avoids the problem of queer entities, since values are not irreducible entities with their own independent existence but rather exist in virtue of states of affairs that can be described as parts of a possible, natural world. (Or supernatural world if God exists, in which case there would be (at least) one non-natural entity.)

The theory also avoids the problem of supervenience. We understand why ethical facts supervene on descriptive facts, because The Good is a description of a possible world. Hence, two identical universes must also be identical in ethical regard. The relationship between our world and The Good is a modal difference between our actual world now and a possible world in the future.

However, some objections against naturalism have not yet been covered, namely the problems of normative force, how to give a good definition of The Good, and why ethical truth would remain even if no persons existed. To these we now turn.

If ethical norms are given descriptive naturalistic content, it seems to take away the normative force of ethical norms. Ethical norms commend or prescribe a certain action, they motivate people for action, and they seem to say what is right or wrong regardless of whether people agree. How could this be understood if the content of the concept of The Good is something given a descriptive definition by humans?

This "normative force" can be understood in different ways. In itself, a *should claim* or an *ought claim* is here argued to be understood as a means to reach The Good. Thus, when saying that people should or ought to do a moral act, it means that if they want to reach The Good, then this or that is a means to that goal. But that alone does not seem to commend the goal in a way that normative claims seem to commend actions for people.

I believe that this commending aspect comes from saying that a possible world is not just a possible world, but calling it good. Calling it good means

285 If God exists, it raises the problem that God may have valuations that are incomprehensible to us. Since there is no rational way for us to include incomprehensible valuation in our human ethics, we must create our ethics from our human perspective with the always valid proviso that there may be relevant facts we are not aware of.

that the one who writes or utters the sentence recognizes it as true and good, and thus through words and actions commends it, i.e. thinks that all should try to reach this goal. Recognizing it as true and good means that one accepts that this is something one ought to try to reach. When others recognize it as true and good, they will act in different ways so as to make you believe the same, for example by blaming you when you do wrong. And if there is a God, most people would probably feel an extra weight from the fact that something was willed by the omnipotent creator of the universe. But if God does not exist, the recognition of norms by yourself and others is enough to explain the normative force of ethics. After all, it seems that some people do not feel the normative force.

This may still seem too weak. Is the real moral force of ethical norms just conditionally that *if* one wants to reach the goal of a good world, *then* one should act in such and such ways, and then the felt moral force comes through social conditions? Could we not say that objectively you have reason to act in such and such ways regardless of whether you or others want to or not? Derek Parfit argues that we should distinguish between instrumental reasons and substantial reasons and defines "good" in light of substantial reasons, which are impartial reasons for acting in certain ways. This is fundamental to his whole ethical project and the basis for arguing that all major ethical theories are different paths to the same goal (Parfit, 2011a, parts 1 and 3).

I find it very problematic that Parfit is not able to define what a substantial reason is. He says that reasons count in favor of something, but acknowledges that it is a circular definition and says that reasons are indefinable (Parfit, 2011a, p. 31). When he then defines the ethically good as what is implied by reasons, his concept of ethical goodness also becomes something indefinable (Parfit, 2011a, pp. 38–39). With this vague idea of goodness, Parfit can argue that all the major ethical theories are reaching for this same undefined mountaintop. In the alternative theory I present here, I want to be as coherent as possible and avoid basing everything on undefined terms. I will argue that every reason is instrumental, since what makes something a reason is that it is a means to a goal. There is not one mountaintop that is the good; rather there are many mountaintops that are possible to climb (these mountaintops represent possible futures), where we must find out how to choose which one to climb. How should we go about doing this?

To answer this, I start by answering the following question: What are reasons? Here I am not interested in causal reasons (e.g. the reason why clouds form), but in epistemic reasons, like normative or motivating reasons. Some may argue in favor of a certain ethical theory, and argue that there follow some moral facts that give us normative reasons for action, but I argue that there can be no reason without presupposing a goal. Nothing is a reason for something in itself, since what it is to be a reason is to be a means to a goal. There has to be a goal in order for anyone to have reason to do something that takes them to the goal (the goal may also be to avoid something). And so it seems that it is The Good as a goal that gives us reason for acting in such a way as to reach the goal, but that presupposes the goal first.

If reasons presuppose a goal, there are many goals, so can there also be reasons for what is the best goal? Can there be a reason for having The Good as a goal? A reason for a goal presupposes another goal, but can a goal be the best goal, the most reasonable goal, or the best justified goal? A goal could be the goal that integrates the most goals, and this is the strategy in my definition of The Good: that which is valuated the most by the most is that which makes the most individuals reach most of their goals. The Good is then the best goal in the sense of being the most goal-inclusive goal. All the individual reasons have been summed to a best reason. The morally normative reason includes most of the individually motivating reasons.

An individual may nevertheless say that he does not care about this moral goal I just defined. Against such a person one can argue that the person does not have the best goal in the sense described above. One could also give other arguments to try to persuade the person to care about the goals. One could argue that the person is not rational, since he presumably wants others to care about his goals and he should then care about other's goals in order to be consistent. And one could argue that it is in the person's own interest to care about others, since this is more likely to make you happy than being an egoistic person. All of these are reasons, but they all presuppose a goal – either The Good, being rational, or self-interest – and someone can always reply that they do not care about any of these goals.

These reasons become stronger if the person is not only not doing anything good, but also doing evil. Then the person is doing something bad, which is even more irrational and less in your self-interest, and which there are sanctions to prevent. So there are many reasons for acting morally and many kinds of normative force as described above, but there is no ontological normative force coming from ethics or the world itself (or a platonic realm of values) beyond what is described here.

If a person does not care, one can say that they are doing something they should not do – with the meaning of "should" here defined – but there is no standard or force independent of humans that they are violating or contradicting (unless God exists). If someone asks why they should do what is ethically good, my reply is that that is what "should" in the ethical sense means. You should do what is ethically good because it is ethically good; in other words, calling something ethically good is to say that it should be done. We could wish that it really was a normative force beyond what I have described, but I can see no good reason to believe that there is, since I believe I have explained why we feel that there is.

The next objection was that no good definition of The Good can be given, for it will either be too controversial or too empty to give any guidance. I have tried to offer a very general definition, but there may be a better one. I think that the best approach is to start as generally as possible in order to get most people to accept it – if people disagree about how to define "good", find some common starting point – then develop the details after that. So, I start with this general definition, and if I can convince someone about it, we should start talking about who to include in the term "most" (Which animals? Future beings? Intelligent robots?); how to balance different goods and probabilities; etc. – topics to which I will return below. Some will find this too relativistic or argue that disagreement is very likely to remain, but what would be a better alternative? It is not the case that if we do not define the good more precisely, then everybody speaks about the same and understand each other.

The last objection against this kind of metaethics was that it seems that there must exist values beyond our definitions, since presumably something could be morally wrong even if no humans and no language existed. The definition allows for this since good is defined in terms of a hypothetical all-knowing being, which does not require any humans or language to exist. According to the definition, even if no humans existed it would, for example, be morally wrong to torture animals for fun – if anyone were to come and want to do that.

But one could object that I am still referring to a definition made by a human being (namely myself), and argue instead that something would be morally wrong even if no mind had ever existed and thought about morality. However, I do not think that the thought experiment of no mind or language ever existing shows this since, when we consider this idea, we still do it from the perspective of our language and moral opinions. It thus only makes sense because mind and language exist.

These were some objections to this theory as a theory of metaethics. However, the theory does not just have an opinion on cognitivism, realism, etc., but actually suggests definitions of "good" and "should" which seem to imply a normative ethics: The Good is the best possible way to the best possible world, and we should act so as to help to actualize this world. Since it relies on valuation, the ethical theory seems quite close to the preference utilitarianism of Richard Hare (Hare, 1952). However, there are also many differences, which I will argue make the theory avoid the classical objections to utilitarianism and also helps it avoid other difficult problems.

I will explore some of these objections and questions in the next part. Some of these are huge topics and I do not pretend to give them a full treatment, but it does say quite much about the theory presented to show in basic contours how it deals with some common problems, like the trolley dilemma. Another dilemma is that if one holds – like I do – both that all humans have the same value and also that it is worse to kill a human than a rat, it is very difficult to explain both how humans can have the same value and a greater value than animals like rats. Even if such test cases are huge topics on their own, it is good to indicate how the theory presented approaches them, since it will increase the understanding of the theory. To these objections and test cases I now turn.

14.3 Objections and test cases

Before I start discussing these objections, I want to make a general comment on ethical discussions and how easily misunderstandings happen in these. Discussing ethics is quite complicated because there are so many relevant presuppositions, contexts, intentions, and interpretations that can be different, which makes it possible to make a large number of qualifications all the time, as I will soon exemplify. However, a text becomes very difficult to read if there are constant interruptions and qualifications, and so I will not do that, but make a general comment on it here.

Here are some of the things that make discussions on ethics extra complicated: When you suggest that something is ethically right and should be done, do you mean ideally or realistically? Do you suggest that this is what is best to do if everybody had done it, or do you suggest what is best to do given that many people are not going to do it? For example, you may think that it would be best for a rich country to have very open borders, given that everyone would welcome the newcomers, or you could say that since people are probably quite bad at welcoming newcomers, immigration should happen at a slow pace. Given different presuppositions about how people act, you can think that different things are the ethically best choice.

Another important complicating factor is what kind of speech act you are doing when saying what is ethically right. Are you merely describing something, or are you also trying to have an effect on those who are listening? For example, maybe somebody knows that very open borders is not a realistic suggestion in their country, yet they advocate it to influence those who listen. Or maybe you downplay the role of culture or religion when a person has done something wrong in order not to fuel prejudice, which other people may judge as a false description.

If there is an ethical debate over a question, one may choose either to present what one thinks should be the end result, or one can present what oneself would prefer as the first step in a negotiation which is supposed to end with a fair result. For example, in a discussion on who should pay how much for a common good, two different groups may both argue that the other group should pay the most, all being aware that this strategy will probably end in a compromise. But if one group starts by suggesting what they think would be the fair end result and the other groups start with an unfair suggestion to end up with a fair one, it can become extra difficult to reach a fair agreement.

Often, it is useful in a debate to distinguish a position and a metaposition (Rescher, 1985, pp. 265–266). The position is what you think about the issue being discussed, while the metaposition is what you think about your own position (and there can be different kinds of metaposition). For example, maybe your position on abortion or gay marriage is that you are against it and personally would think it was best that it was forbidden. But your metaposition is that since people have different opinions you should not force your own view onto others, so you think that it should be allowed anyway. Here it is relevant to distinguish between whether you think that your position is only meant as an ethical claim that should not have consequences for law, you think it should have consequences for law, or you think it should have consequences for law if the majority are against, for example, abortion or gay marriage.

Let us say that your position is meant as your ethical position only, not something that you think should be the law. Even then it is an important difference between whether you think of your position as a general rule or something that applies in every situation. Maybe you think that abortion is wrong as a general rule, but that there may be situations where different concerns come into play, so you do not want to say that it is never ethically acceptable.

These were some examples, and many other complicating factors could be mentioned. Since there are so many relevant distinctions in play, misunderstandings are to be expected in ethical debates. When I write about ethics in this book, unless I say otherwise, one can expect that I am stating my own ethical and general opinion given what I think is realistic to expect from people. I now move on to discussing my own ethical theory.

14.4 Test case 1: The trolley problem

The ethical theory I have suggested is a kind of teleological ethics, whereas the main competitor for this kind of ethics is deontological ethics. There are some common objections directed especially at utilitarianism, which I shall discuss here to see how the theory offered above deals with them. The most common charge is that utilitarianism leads to recommendations which seem to be unfair and morally wrong.

If what is morally right to do in a situation is that which has the best consequences, it seems that it would be good sometimes to put innocent people to jail, to take all the organs from one healthy person to save the life of six others, or to break promises. In utilitarianism, there is a distinction between act utilitarianism and rule utilitarianism. Act utilitarianism says that one should act according to what gives the best consequences considered from case to case, while rule utilitarianism says that one should act according to a rule that in general will give the best consequences if the rule is followed.

Rule utilitarianism avoids a lot of problems that act utilitarianism struggles with, such as people who think that it would be best that they do not to pay tax, etc. Rule utilitarians will also argue that they avoid problems like putting innocent people to jail, taking organs from healthy persons, or breaking promises since this would not have good consequences if followed as a general rule.

However, some argue that this response is not good enough. One critique is that rule utilitarianism is not that different from act utilitarianism, since rule utilitarianism allows for exceptions. While Immanuel Kant would say that you should not lie even to save another person's life,[286] rule utilitarians are happy to make a rule with an exception, like "do not lie unless it can save an-

---

286 While Kant interpreters discuss what Kant's position is, at least Kant seems to defend this view in a text entitled "On a Supposed Right to Lie Because of Philanthropic Concerns":

> If you have by a lie prevented someone just now bent on murder from committing the deed, then you are legally accountable for all the consequences that might arise from it. But if you have kept strictly to the truth, then public justice can hold nothing against you, whatever the unforeseen consequences might be. It is still possible that, after you have honestly answered 'yes' to the murderer's question as to whether his enemy is at home, the latter has nevertheless gone out unnoticed, so that he would not meet the murderer and the deed would not be done; but if you had lied and said that he is not at home, and he has actually gone out (though you are not aware of it), so that the murderer encounters him while going away and perpetrates his deed on him, then you can by right be prosecuted as the author of his death. For if you had told the truth to the best of your knowledge, then neighbors might have come and apprehended the murderer while he was searching the house for his enemy and the deed would have been prevented. Thus one *who tells a lie*, however well disposed he may be, must be responsible for its consequences even before a civil court and must pay the penalty for them, however unforeseen they may have been; for truthfulness is a duty that must be regarded as the basis of all duties to be grounded on contract, the laws of which is made uncertain and useless if even the least exception to it is admitted. To be *truthful* (honest) in all declarations is therefore a sacred command of reason prescribing unconditionally, one not to be restricted by any conveniences (Kant, 1976, p. 348).

other person's life". But then it seems that there is no end to such exceptions so that, in the end, all rules will have as an exception that one may do what gives the best consequence in the case in question. But then rule utilitarianism has collapsed into act utilitarianism (Lyons, 1965).

A stronger objection is that it seems that the reason it is wrong to put an innocent person in prison is that it is unfair or morally wrong *towards that person*, not just because of the total amount of well-being in the world. It is to use another person merely as a means towards a goal, which Immanuel Kant argued is always wrong. The same would apply to groups: It would be unfair for a majority to exploit or abuse a minority even if the total sum of happiness was good.

On the other hand, it seems that sometimes it can be right to kill someone to save others or to use someone just as a means towards a goal. Consider a person about to hit a button that will destroy the earth in a nuclear explosion. It does seem ethically right and good to shoot this person if that is the only way to stop him from destroying the earth. Also if it is an innocent person hypnotized to press the nuclear button. Or if by shooting an innocent person you could make him fall onto a button which in the last second saves the earth.[287]

But seeing exactly when it is okay to use someone as a means or not is tricky. In the famous trolley problem, a person has tied five persons to a rail track and a runaway trolley is approaching, about to kill them. You can make the trolley change tracks by pulling a lever, but there is another person tied to that track who will then be killed instead. Should you pull the lever? Most people say "yes". But what if a trolley is about to run over five persons, and you are standing on the bridge with a big man leaning over to see – is it then acceptable to push the big man over the bridge to stop the trolley and save five persons? This time, most people would say no, and then the challenge is to explain the morally relevant difference in the two cases. And would the ethical judgment change if the big man was the same person who had tied the five persons to the rails in the first place?

How would these problems be understood in the theory above that said that The Good is that possible world which an all-knowing and all-sympathetic being would know would most probably be valuated the most by the most? Recall that this possible world is meant to include both the end result and the way to get there. If we start by looking at the end result, it seems obvious that the possible world that would most probably be valuated the most by the most is not a world where individuals or minority groups are unfairly exploited or abused. Given a

287 The scenarios are of course unrealistic, but they are merely thought experiments to clarify ideas.

world where some are abused or exploited, it seems obvious that we can imagine a better way to a better world which did not include this.

We do not live in such an ideal world now, but if we focus on what is the best way to get to such a goal, it also seems obvious that it is not by abusing or exploiting individuals or minority groups. Given a way towards the goal which includes abuse and exploitation, we can imagine a way that more people could valuate more which does not include abuse and exploitation.

Even if it is not the best action to exploit persons or groups, could it nevertheless sometimes be right? As seen above, it could be the case that in a particular situation demanding action there and then, it could be right to use someone as a means, for example in the earth-about-to-explode scenario. Why is it sometimes right? Is there some principle which can help us also understand the trolley dilemma?

In the earth-about-to-explode scenario, it again seems quite obvious that it would lead to a better world if one person was sacrificed to save the earth instead of the whole earth exploding. The trolley dilemma, on the other hand, is trickier. In empirical research, the majority says that they would let one die to save five, but not push a big man over the bridge to save five (P. Singer, 2005), but what is the relevant moral difference between the two cases?

We saw above that exploiting someone by using that person only as a means for your own interests is in general not the best way to the best goal. In the trolley dilemma it is not exploitation in this sense, but rather using someone as a means to help someone else. Recall that an action being bad or wrong as such is a general consideration of which actions, considered on their own, lead to worse worlds, and using someone as a means is an action which, in general, contributes to worse worlds. It seems that using someone as a means also to help others is problematic. For example, if it was considered generally right to use someone merely as a means, we would all fear organ thieves or being innocently imprisoned, etc. As an action which is generally wrong, that means also that it is wrong towards someone to use them as a means.

However, sometimes it seems that doing an action which, considered on its own, is wrong to do towards someone can nevertheless overall be a morally good action given certain conditions. If we could save the whole world by pushing the big man over the bridge, that would have been the morally good thing to do, even if it is in general wrong to use someone merely as a means. Pushing him in order to save only five, on the other hand, is wrong, and would be similar to stealing organs from a healthy person in order to save five others. If that was considered morally good, we should all fear organ thieves, and it would not have been a better world. But where is the line? How many must one save before it is morally right to use a person merely as a means?

Maybe we can twist the example with organ stealing, make it happen in secret and save very many, etc. Somewhere between saving a few and saving billions, it does become right to use someone as a means, and this is of course a classical problem in ethical dilemmas connected to allowing civilians to die in war. It is generally considered that it was right to sacrifice some civilians (14 to be exact) in order to prevent Hitler from being able to get a nuclear bomb, like when the ferry *Hydro* carrying heavy water was blown up 20 February 1944. There is more disagreement on whether the Hiroshima bombing was morally right. Many would say that it was not, even given consequentialist ethics, which I think is mainly because it seems that the same effect could have been achieved without dropping the bomb. For example, it should have been tried to demonstrate the bomb to the Japanese in private and offer negotiations. Also, there are the long-term effects to consider of legitimizing nuclear weapons. But it seems that many people would accept the death of quite many thousands of civilians if it was the only way to end WWII.

What is then the morally relevant difference between pushing the big man on the bridge to save five and pulling the lever to save five? Philippa Foot has argued that there is an important difference between actively killing someone like the big man (using the person as a means) and passively letting a foreseen consequence happen, like letting the person tied to the tracks die (Foot, 1967, p. 4). But this does not seem like a right description of this scenario. For if you passively did nothing, five persons would die, but you actively killed one person instead.

If there were two trains coming to kill all six, and you could just stop one train, it is clear that the morally best thing to do would be to stop the train that is about to kill five. But in the trolley dilemma, you can save five by letting one die, who would not have died had you not done anything. On the one hand, you could then say that the trolley dilemma is then not about having to choose between one or five, but about interfering to change the situation from five dying to one dying. On the other hand, you could describe the situation as a dangerous situation where a person is standing next to a lever which can send the train towards one or five, and the person can choose between doing something or doing nothing and thereby letting one or five live. But doing nothing can be understood as a choice to do something; namely, do nothing. The whole description is thus a bit ambiguous as to whether it is a matter of saving one or five, or whether it is to actively make a choice to kill someone.

When it comes to the man on the bridge, it is clearly an active interference where the big man is used as a means for a purpose. Depending on how one describes the lever scenario, the two scenarios are then quite similar or quite different. There is also another relevant difference. In the lever-pulling scenario, you are not using someone merely as a means in the same sense as with the man on the bridge. Instead, another person has already done something morally wrong and used the five and the one merely as a means (for some bizarre purpose), then put you in the dilemma of saving one or five. While the situation is similar to the big man on the bridge, there is a difference in that something wrong has already been done to the one tied to the tracks: this person has already been put in great danger by a morally wrong deed.

Why is this an ethically relevant difference? My point with distinguishing between using someone as a means and a situation where someone has already been used as a means is the following: We draw different generalized ethical conclusions from the two examples because the one describes a scenario with much less general alternatives than the other. If it is okay to push the big man off the bridge, the general ethical conclusion to draw from this is that it is in principle okay in any situation to use a person as a means in order to reach a higher good. Then it is also okay to steal someone's organs to save two people or to steal someone's money and give to the poor.[288]

The persons on two tracks describes a narrower scenario with fewer general alternatives: Some people have already been used as a means and put in a life-threatening situation, and your only alternative is to save five or one. Even if we say it is okay to save five, it does not follow as a general ethical conclusion that it is ethically acceptable to steal organs or money. Instead it follows that if somebody had already stolen a set of organs and you could use these to save five instead of one, you should do so.

This may be clearer if we (with John Rawls) reason behind a "veil of ignorance", imaging that you could have been a person in the scenario described. We do not think that it is morally acceptable to use someone as a means "out of the blue", as it were, such as in the case with the big man on the bridge. In such a world, anyone would fear that they could be used as a means out of the blue. But given that five persons and one person have been tied to the different rail tracks, then the best world is one where five are saved. In such a world, you would be more likely to be in the pile of five than the sole person. Such rea-

288 While agreeing with the conclusion that it is acceptable to use someone merely as a means if the consequence is good enough, even if an exact line cannot be drawn when it comes to measuring how good a consequence must be before someone can be used merely as a means, Derek Parfit adds some helpful nuances. He distinguishes between *harming* someone as a means and *using* them as a means and between *using* someone as a means and *regarding* them as a means. Pushing a big man off the bridge is not an attempted killing where a person is regarded as a means (you would prefer that he survived), but it is using a body (with a foreseen killing) as a means (Parfit, 2011a, pp. 216–229).

soning behind a veil of ignorance is like asking what an all-sympathetic and all-knowing being would know was valuated most by the most.

What if the persons on the track had not been tied to the tracks by someone, but had all been trapped there by accident? Does that make a difference? Then it seems closer to a scenario where, for example, five people by accident had organ problems that could be saved by a person who happened to be nearby with all the needed organs.

I think the main point still remains: the relevant difference is which general conclusions you can draw from the examples. The lever case describes an unusual situation where pulling a lever or not is a simple choice thrown on you as to whether to save one or five. The "man on the bridge" scenario describes instead a much more general situation of using an unknown person to save someone. If you say that it is okay to push the big man off the bridge, it implies that many instances of using someone as means is okay, such as killing anyone to steal their organs, but such a world is not the best way to the best goal, and therefore the action is wrong. The lever case is a much rarer case of having to choose between using one as a means to save five where the general conclusion to draw is just that in special cases where you have to choose between saving one or five, you should save five. That is the best way to the best world, and therefore this action is right.

One could argue that it is not a rare case to have to choose between saving one or five, since if you have a certain amount of money, you could give an amount of money to one group of poor people and save five of them or give the same amount of money to another poor person and save just one. But then you are not using one as a means to save five. You are just using an amount of money to save an amount of people, and saving five is better than saving one if everything else is the same.

Another common twist to the dilemma is to ask the following question: If the big man on the bridge is the person who tied the five people to the rail tracks, is it then morally right to push him over to stop the train? My answer is yes. It would be as if he had thrown a hand grenade about to kill a lot of people but you threw the grenade back to him, and it then exploded. It would be to help someone in their right to self-defense and thus would be acceptable.

The trolley problem has returned in the debate on self-driving cars: who should a car hit if it has to make a choice? I think that one has to deal with this problem not merely as an abstract ethical question, but consider also what kind of cars people will actually buy. It is ethically good with self-driving cars because it will be much safer, but while people think that cars should protect as many lives as possible, they do not want to drive the cars unless they save the driver first (Bonnefon, Shariff, and Rahwan, 2016). It seems that at the least there must be some common principle all cars follow, and not that only some cars save the driver.

Concerning how to choose among the people not sitting in the car, I guess it would work best to consider probable deaths (e.g. 90% certain death) and choose that the fewest people should be killed. And if the car must choose between an equal number of deaths, the choice would probably have to be random. A gigantic survey throughout the world has shown that people only agree that as few as possible should be killed, but not how to prioritize when it comes to gender, age, wealth, lawfulness, etc. (Awad et al., 2018). Given this, one could believe that most people will only accept a random choice as fair. In this debate there probably are other relevant facts and dilemmas of which I am not aware. Hopefully, the problem can be avoided by making barriers that make it impossible for such cars to hit others at all.

Against all of this reasoning on the trolley problem, some may argue that it fails to consider human dignity or value or human rights, which makes many of my considerations above unacceptable. This is an important critique of teleological ethics, so I turn now to the topic of human rights. How these relate to animal rights is also a contentious issue, which will also be discussed.

14.5 Test case 2: Human value

In the following, I will discuss these questions: Do all humans have the same value, and how can such a view be justified? Do humans have a greater value than animals, and how could this view be justified? What is a human or a person if the value is connected to being a person? Like the trolley problem, I have in here selected some ethical questions I find very difficult in order to test the ethical theory being presented. These questions will now be discussed in that order before some more objections are considered.

It is clarifying to start with the concept of value. Note that in this chapter, "value" refers to value in ethics, not to the qualitative and quantitative values I have described as fundamental building blocks of the world. Above, value was defined in terms of something being valuated. Values are not something that has an irreducible independent existence, but anything that can be valued. As an ontology of value this has the advantage of not needing extra entities in the ontology, and I shall argue that it nevertheless can explain what else we think about values.

If something having value just means that something can be valuated, it raises the question of whether something just has value if it is actually being valuated by someone or whether it is enough that it has the potential for being valuated by someone. On the one hand, one could think that it should be more than actual valuation since, for example, a diamond not discovered has value even if nobody has yet actually valuated it. On the other hand, one could argue that it has value since it is a diamond, which is a kind of thing that people actually valuate. For it could seem to be too broad a concept if value is anything that can be valuated. Maybe in the future, people decide to use something of no value today as a kind of money, such as navel fluff. This would imply the conclusion that anything has potentially great value, so it seems that if value is potential value, the concept of value is too broad.

When it comes to the example that anything could potentially have value, it is clarifying to distinguish between instrumental value and value for its own sake. For it seems that anything could have instrumental value, but not that anything could have value for its own sake. Instrumental value means that it is valuated as a means to something else, while "value for its own sake" means that something is valuated as itself, not as means to anything else. This would limit the scope of what could be valuated, since there are many things that could be valuated as means to something else but not as good in a non-instrumental way.

And yet people are very different, so even if there are many things that most people would find ugly, boring, painful, etc., there are always a few who like it. Maybe someone has a hang-up on navel fluff and just loves it for its own sake. So again, the concept of value seems too broad if it just means that something can be valuated or even that it actually is valuated, since there seems always so be somebody who values just about anything.

This problem that the concept of value can seem to include anything is the same as the problem that the terms "good" and "bad" can seem to include anything if we just think of all possible individual cases. If ethics is going to say something general about value, it must consider what in general is valuated and can normally be valuated by humans, even if there can always be individual exceptions.

I believe these three distinctions then give us the concept of value that we need: between potential and actual valuation, between instrumental and "value for its own sake" valuation, and between individual and general valuation. The value of a thing or event is its general potential for being valued for its own sake. This is its objective value, while anything can have any subjective value relative to individuals.

Now I am ready to discuss human value, but first I want to draw attention to one advantage of this focus on potential value: If value was defined in terms of what is actually valuated, it would have the consequence that what we think about the value of people of different gender, race, religion, sexual orientation, etc. depended on what happened to be the preferences of the society at the time under consideration. But when value is defined in terms of potential value, then we must also consider the ones who are valuating and not just the ones who are being valuated. In other words, the people who valuate can change how and what they valuate and this possible change is relevant to what is the best way to the best world. For it seems obvious that the best way in general to the best possible world is to valuate people equally independent of gender, race, religion, etc., while we know all too well catastrophes resulting from certain groups of humans being considered of less value than others.

Now we are ready to consider the question, *Do humans have equal value?* And the answer will come as no surprise: Yes, since all have the same potential for being valuated. Even if all are not actually being valuated equally by all, the objective value is their general potential for being valuated, which is equal for all – as long as we have a wide understanding of general potential.

It is common to think – for example, in the declaration of human rights – that specific rights follow from human value and human equality, and that human value and equality can be violated. How should these claims be interpreted in light of the considerations above? I suggest the following: All humans have the same potential for being valuated. If they were actually to be equally valuated, that would very plausibly be the best possible way to the best possible world, while there is plenty of evidence from wars and genocides that not respecting human value and equality very efficiently can make the world very bad. This is what makes it true to say that respecting human value and equality is good while violating it is bad, namely because it is the best way to the best world.

What does it then mean to say that humans have *rights?* Obviously, there is the mundane sense that if a government has accepted certain duties, then the citizens have certain rights. But can it also be true to say that ethically speaking, humans have certain rights regardless of whether the government cares about its duties? Yes, and then human rights must be understood as following from the goal of ethics in the same sense as above: that humans have certain rights means that there are certain rights that are such that respecting them very plausibly is the best way to the best world, while violating them very plausibly leads to a bad/worse world. While this may be an uncommon way of formulating how to understand human rights, I believe that it is a more finely grained and coherent way of expressing things that are usually expressed in more coarsely grained frameworks. That it is true that all humans have same value and rights means more precisely that it is true that it is good and right that all humans should be valuated and treated equally.

I now move to the question of whether humans have a greater value than animals. While I think that most people would think that it is ethically preferable in all cases to save one human being instead of two pigs or rats from death, not all do, and the charge of speciesism has become more common. It may seem to follow from the considerations above that humans and animals have the same value since it seems that potentially we humans could come to valuate humans and animals equally much. However, my own view is that humans have greater value than animals. I believe that it follows from the above, which I will now show.

Recall that we are talking about objective value, which is a general value, since subjectively there can always be someone who loves a dog or a fish or a book or anything else more than they love any human. Why do humans have a greater general potential for being valuated than animals? It is because they are part of a network of potential valuation which is greater than the potential for valuation of animals. Why? The reason is the potential in human consciousness for a great variation in valuation in human interaction. There are so many different things humans can value in interaction with each other, from which it follows that humans have a wide variety of possibilities for valuation. This means that a human in general has a greater potential for being valuated than an animal.

Note the important point that this is not about the conscious capacities of individuals, for an animal may have greater conscious capacity than some human individuals. It is about the potential value in human relations overall. Even if a human individual has a non-functioning brain, it has a great and varied potential for being valuated by other fellow humans who one day may be in the same situation. Clearly, animals can take part in relations with humans in many different ways, but I find it obvious that the potential for valuation among humans is greater than between humans and animals or just among animals.

Several consequences follow from this view, which in my opinion seem correct: Animals which seem to have a richer mental life should be treated better than animals who seem not to have, therefore it is worse to kill a monkey than a mosquito. If animals were to evolve into having greater conscious capacities, they should also have greater rights. Or if we were to discover that some animals have a much richer conscious life than we thought, then they should also be treated better. And since in fact we do not know how it feels to be different animals, we should in general treat animals well.

Even if we do not know what it is like to be animals, it seems clear that they are not on the level of humans when it comes to their conscious capacities. Thus it follows that humans have greater value, and if you have to choose between saving a human or an animal, you should save the human.

In the previous paragraphs, I have focused on what gives humans the same value, which is connected to their potential for being valuated, which is how

I have defined their value. It does not necessarily follow morally that all humans should be treated equally even if they have the same value. For example, some could argue that we do not need to treat all trees or all sheep the same way even if they have the same value. Or some could argue that while all humans have the same value, we should treat them differently depending on whether they can feel suffering.

It thus seems that I should offer an understanding of equal value that implies that all should be treated equally in morally important matters. What is most important is not to show that all humans have the same value as an ontological truth, but that all humans should be considered to have the same value in the sense of having the same important rights. Let us call this having the same moral value.

I think that all humans have the same moral value (should be valuated equally) because that is the best way to the best world. I think many could agree with this even if they thought that people do not actually have the same ontological value. But one could also say that it is important to show that all people actually have the same moral value, not just the same ontological value. I think that all humans also have the same moral value because they all share the same reason for being valuated equally to all others. This same reason that they all share is that it is the best way to the best world that all are valuated equally and given the same rights, independent of their abilities, capacities, resources, etc.[289]

This line of reasoning could seem to work well when it comes to equal value among humans, but it seems to cause problems for the view that humans have a higher moral value than animals. For would it not be the best way to the best world if animals were also valuated equally with humans?

No, since the best world is the most valuated world. In a world where all humans are equally valuated, they can valuate the world more than animals and they can valuate many rights that animals cannot valuate (like freedom of speech or religion), which means that we do not have the same reason to valuate animals as much as humans – especially if we have to choose between human rights and animal rights. If we had unlimited resources, animal rights and welfare should be much better in all areas that animals could valuate, and they would have almost the same moral value as humans in the sense that the reason to valuate them was almost as good as the reason to valuate humans.

---

289 Ontological value and moral value in these senses are linked together in the way that it has higher moral value to save individuals with higher total ontological value because of the greater potential for valuation.

A related question is the following: If the best possible way to the best possible world includes that animals are also valuated as highly as possible, would it not follow that they should not be slaughtered for food? The question of eating animals is a tricky one. Comparing valuation among humans and animals is very difficult since we do not know what it is like to be different animals. Further, life on earth is fundamentally structured and functions the way that all eat all: many smaller animals get killed and eaten by bigger animals and many of the biggest animals get killed and eaten by the smallest animals (including bacteria). It seems plausible that life on earth would not be better if no animals were killed and eaten since almost all live by eating others. But life could certainly be better with much less eating of animals than today, and especially life would be better if humans were to eat less meat. It is clearly ethically good if humans eat less meat, and it may very well be the case that it was best if humans ate no meat, but here I know that I do not know enough facts about the specific question to make a well-argued decision, and so I will leave it aside.

The question of human and animal value raises an obvious question: what is a human being? And is the relevant moral category "human being" or "person"? The question is important, especially for the questions of abortion and euthanasia, such as ending the life of someone with a severely malfunctioning brain.

The concept of a human being is in the category of coarse-grained concepts with fuzzy edges. Nevertheless, it is a goal to make the concept as clear as possible and to avoid fuzziness as best as one can. Being a human is a process in time with a start and an end, and there may be fuzzy edges in the beginning and the end and in what counts as maintaining one's identity as a human over time (if one were, for example, to exchange parts of the body).

A concept needs a stable structure in order to be a concept, and in order for something to be the same over time, only small and gradual changes can happen while the internal structure remains stable. If I start building a boat, at some fuzzy time it has enough boat-structure to have become a boat, and I may make many gradual changes while it still remains a boat, but if I rebuild it into a car, it will at some fuzzy time stop being a boat.

When it comes to human beings, there is an obvious stable structure constitutive of being a human, which is being a delimited organism with a certain type of genetic code. From the time of being a fertilized egg there is a stable internal structure and only gradual changes throughout a human life. No concept of a human being can be less fuzzy or more consistent than treating the whole life process from fertilized egg to death as a human being.[290]

290 Even death is a gradual process with fuzzy borders – from when the body starts dying to

It seems that most people accept that also a fetus in the womb is a *human being*, but several ethicists will argue that what is morally relevant is who is a *person*, so that no rights follow just from being a human being.[291] Above I defended objective human value in terms of their potential for being valuated, which includes unborn babies and people who are not able to be conscious. This means that I do not think that there is an extra personhood that humans require in order to have human value and human rights.

Defining personhood is interesting for many reasons. For example, we could imagine animals, aliens or robots with personhood whose rights should be respected, and I would define personhood as having a mind and self-consciousness over time. This means that robots may become conscious persons who may even become more valuable than humans (in virtue of having greater potential for valuation), but who should nevertheless respect human rights. This also means that a human being could fail to be a person in this particular sense of "person", but that should not worry anyone, since I defend human rights based on being a human being and not on being a person in this sense.

There are some obvious objections to the view I have presented here. One argument is that it seems worse that a fetus dies late in pregnancy than early, and it seems worse to end a pregnancy late than early, so how can this be explained if a human being is a human being with the same value as any other human from the time of being a fertilized egg. Another argument is that if there was a fire at a hospital and you could either save two living children or a freezer with a hundred fertilized eggs, it seems right to save the two living children. I will now deal with these questions in this order.

Why does it seem worse that a fetus dies late in pregnancy rather than early? This seems to indicate that a greater value has been lost, but if all humans have the same value, then a greater value has not been lost. It is easy to explain why it will in most cases *feel* worse to lose a child later in pregnancy, since one has normally come to actually valuate it more through having more experiences with it. In addition, the later it is in pregnancy, the more certain it becomes that the child will be born. Losing late will then *feel* like a greater loss, and since it was more certain that the child would be born, the potential for a long life of valuation was

---

where consciousness is lost and vital functions (usually the heart and/or brain) have stopped working on their own so that it is highly probable that it is irreversible that the body will die – until all life functions have ceased. I will not here be more precise than referring to what is most normally defined as death in hospital contexts.

291 Peter Singer will, for example, say that a human being is so from the first moment of being an embryo created by sperm and egg, without this meaning that it has moral rights (P. Singer, 2000, p. 127).

greater, and thus it would also *be* a greater loss. For the same reason, it is also worse to end a pregnancy later rather than earlier. I will return below to discuss further the significance of how probable a future good or evil is.

The next counterargument was that if there is a fire at a hospital it is surely better to save two living children than a freezer with a hundred fertilized eggs. Again, it is easy to explain why it *feels* like that, since the children are actually much more valuating and valuated than the eggs. But would it objectively be morally better to save the freezer since it has the potential of becoming a hundred children instead of just two?

As in the discussion two paragraphs ago, valuation and probability are relevant. This can be shown if we add to the story that there could have been one hundred mothers and one hundred fathers waiting and longing for the eggs, to which they had a long story of finding donors, to whom they had already given names, and the best doctors were ready to make sure that everything would work well. This would increase the ethical value of saving the freezer. One could also add to the story that the two living humans were very old and either suffering intensely or with a severely destroyed brain, certain to die very soon anyway, and the choice could become less and less obvious between saving two humans or a freezer with eggs.

One can always think up individual scenarios where almost any action can be the morally right thing to do, so what we are considering is a general question of whether it is better to save two living children than one hundred fertilized eggs. In general, two living children will more probably be valuing and valuated more than one hundred fertilized eggs, so saving two children is a more probable way to a more valuated world.

This conclusion may to some seem like an argument that my reasoning is wrong and that something else must be the basis for the value of human beings than what I have suggested here. Among the most common suggestions would then be capacity for consciousness or a developed, organized brain activity or something similar. But such alternatives give the conclusion that it is morally unproblematic that a fetus early in pregnancy or a fertilized egg could just be thrown in the garbage. My view gives high value to fertilized eggs from the start, which I believe is the morally right position (the best way to the best world).

Having an abortion is a serious thing which in general is wrong, but which may be morally right in individual cases, taking everything else into account to find out what the best way to the best world is. This discussion was not meant to argue that abortion should be illegal, since there are additional reasons why it should be legal. It was merely a moral discussion, leaving out the question of laws, and the laws must take into consideration moral disagreement about abortion. One could disagree that abortion can sometimes be morally right, by saying that an abortion would be to do something wrong against the unborn child regardless of other considerations about valuation in the world. But as mentioned above, I can find no meaning to something being wrong *against someone* as opposed to just being wrong, and then "against someone" means that they are victims of the wrong that has been committed.

14.6 Further objections

Having now finished my discussion of human value and human rights, I turn in the end to other difficult questions and objections against the ethical model here presented. My goal is to show the coherence of the ethical theory through its ability to offer finely grained answers to difficult questions. The questions and objections are as follows: Can values be compared? How can values be measured? Is there a mere addition paradox – i.e. does it follow from this theory that it is better with numerous people in the world having a little joy than with fewer people having great joy? How should we choose if two values are equally good? How should one decide between a low value with high probability of occurring and a high value with low probability of occurring? Is saying that one should always act to increase value an ethic that is too demanding? How should we understand the relation and sometimes conflict between self-interest and the interests of others? In the following, these questions will be discussed in the order mentioned.

Can values be compared? How can values be measured? Some argue that everything we value and disvalue can be placed on a scale of and measured in degrees of pleasure or pain (Moen, 2012, p. 37). Others argue that values are too different to be comparable at all, such as when different people valuate and prioritize for finding truth, self-sacrificing in helping others, collecting stamps, whistling, having sex, etc. (Chang, 2002, p. xvii).

It seems to me that the things that people valuate are so different that they cannot all be considered on one scale of pleasure and pain. This is one of the reasons for me choosing the wider concept of valuation, since people have different preferences and may, for example, prefer things which are painful. But can preferences be compared and measured?

If we start by considering individuals, we all compare our own preferences all the time by choosing one thing over the other. Of course, we may sometimes choose A over B without knowing what we would actually have valuated the most. But sometimes we have experienced both A and B and know what we will choose if we have to choose again. We often know what we prefer the most. And sometimes we are unable to choose, since our preferences are equally strong.

These choices that we do all the time seem to reflect that there is some scale of desire in our brains, as argued above in the discussion of desire strength. Even if I do not know how to understand in any detail this scale of desire and its units, it seems to be of great use in ethics, and of course ethicists like Richard Hare have used preferences as the basis for their ethics (Hare, 1952).

That people make choices seems to express that they can valuate one alternative over another. But maybe they made a stupid choice. Valuating one alternative over another would most appropriately be judged by comparing how an individual would valuate her life resulting from choosing one alternative compared with how she would valuate her life had she chosen the other alternative. Peter Baumann argues that such comparisons cannot be made, and points out how different choices turn us into different people with different preferences, making it impossible for an individual to make a choice based on preferences (Baumann, 2018).

It is true that the individual cannot know which life she would have been valuated most, for example whether she would have valuated most a life with or without children, since she will only live the one life. But we can imagine the person living both lives and grading both lives even if the person would be different and having different preferences. And it is a true answer to the question of what would most probably (in the sense of epistemic probability) be the most valuated life by that person given maximal knowledge of the situation, even if nobody knows the answer. What will probably be most valuated is the best basis for making the choice, even if no certainty can be achieved. Ethical choices include uncertainty about consequences. It makes sense to compare two possible futures according to which would most probably be most valuated, which of course is something we do when making choices all the time. While Baumann says that there is no plausible metastandard for comparing possible futures, I argue that the metastandard to use is what most probably would have been valuated the most.

Ruth Chang presents nine types of arguments against comparability, but argues well that values are comparable (Chang, 2002, pp. 71, 74–76).[292] Compari-

---

292 Note that such comparison is between token and not type values. For example, while friendship and money cannot be compared in general, Chang argues that a specific friendship can be (imprecisely) compared with a specific sum of money, like giving up a dull friendship with the butcher for a million dollars (Chang, 2002, chapter 4).

son means to compare with regard to something.[293] Comparability between values implies that values can be represented with numbers, although it does not need to be precise (Chang, 2002, pp. 31–32). Using numbers this way does not mean that qualities are turned into quantities, only that qualities can be *represented* by numbers (Chang, 2002, p. 28).

Chang adds another category in addition to "better than", "worse than" and "equally good", namely "on a par" (or parity). "Equally good" means equally good measured on a common scale, such that it would be rational to flip a coin in order to choose. Parity means that there are different positive and negative consequences to each alternative, where one is not better than the other, but both are still very different. Examples could be what job to take or who to marry, and flipping a coin to choose does not seem rational in such choices (Chang, 2002, chapter 5).

I argue here that choices that are on a par are comparable on a common scale of valuation, when we consider it from the outside perspective. You can choose to become person A and live life A, and come to valuate that life as a 7 overall, or you can choose to become person B and live life B, and come to valuate that life as a 7 overall. If we disregard consequences for other people, these two choices are ethically equally good, but of course they are individually and psychologically very different, and the person choosing can live only one of the two lives.

Why is it not rational just to flip a coin with two choices that are on a par, in Chang's terminology (Chang, 2002, p. 139)? I suggest two reasons for this, different from Chang. Firstly, the fact that you choose one alternative and the reasons you have for your choice, matters for how much you valuate that life (e. g., that you made a choice to start a new business for such and such reasons matters for how much you valuate starting that new business). Making a serious life choice (as opposed to flipping a coin) is thus important for how that life becomes, even if another choice would have been important in the alternative life.

Secondly, while the choice between two different lives may be ethically and rationally equally good since the probability of a valuated outcome is the same, the actual consequences may turn out to become very different. Even if either choice is equally justified, it does not mean that the consequences will have the same value, since this is uncertain. You can improve one alternative (say, raise the wages in one career option and not the other), and still be uncertain

293 *Incomparability* would then mean that it is false that one value is either better than, worse than or equally good as another according to the standard they are measured against, while it would be *non-comparability* if the standard does not make sense (e. g., comparing the number 9 to a cup of coffee) (Chang, 2002, pp. 9, 15).

because the consequences are uncertain overall. Since the choice will matter much for your own life, it is rational to seek out as much information you can about the possible consequences, and normal to be anxious about making the choice that will in fact become the best (and there may well be loss to experience before there is balance).[294]

However, the two previous paragraphs only considered comparing preferences within one person. Can preferences be compared from person to person? If we have to choose between letting person A actualize his preference for getting a rare stamp for his collection, person B actualizing her preference for meeting a friend, and person C and a lot of other persons with different preferences – how can we decide among them?

This seems like a big problem also for those who think that all value can be placed on one scale of pleasure. For even if every individual says that they experience something as, for example, 7 on a pleasure scale of 1 to 10, we cannot compare that one person's 7 is like another person's 7. Maybe person A has a wide emotional register and person B has a very narrow register, so that a 2 for A would have felt like a 7 to B. This would then be a problem if the goal is just to achieve the biggest amount of pleasure in the world.

On the other hand, one could argue that value should be considered relative to each person. One could say that it is irrelevant that a 2 for A feels like a 7 for B, and say instead that it only matters what each one feels to be, for example, a 7. This seems to be a right move, since we cannot compare experiences between persons, that we should consider these experiences relative to each person. It is still relevant to talk about amount of value, in terms of number of people experiencing a certain amount of value over different periods of time, but the amount of value that each person experiences at a time must be considered relative to that person. To make the skeptic argument that human experiences might be completely different will make any rational ethical choice involving more than one person impossible, but we have no good reason to think that

294 This means that I disagree with Ruth Chang's understanding of parity, which she calls a primitive notion, resistant to analysis (Chang, 2002, pp. 141, 149). She says that our inability to choose is not caused by uncertainty. She gives a concrete argument and says that we may know all there is to know about how much we like a cup of coffee and a cup of tea, and still we are unable to choose which is best (Chang, 2002, p. 126). I would say that our uncertainty in that situation is what we think we will come to enjoy the most in this specific situation, since sometimes we enjoy tea the most and sometimes coffee. Then she gives an abstract argument to remove concrete uncertainty (Chang, 2002, pp. 126–128). But in real like there will always be indeterminism and uncertainty about what the consequences will in fact be, and this is not removed by the abstract argument.

they are completely different and many reasons to think they are quite similar (because of common evolution and many normal shared preferences).

Rachael Briggs offers some arguments against the idea of comparing alternatives relative to each person like this. With reference to Peter Hammond, she offers the example of a greedy person who needs a lot to experience for example a value of 7 and an undemanding person who just needs a little to experience a value of 7. Is it then the best ethical option to give a lot to the greedy person?[295] My answer to this is that as a starting point, we must measure and compare experienced value relative to each person. But note that the ethical principle I suggest is not that we should do in each situation that which satisfies the most preferences in that situation, but instead that we want to actualize the best way to the best world. In this case, that would be that the greedy person was less greedy, in which case the resources could be shared to let more people experience more valuation. I say more about this below, pointing out the importance of social equality for making everybody (poor and rich) experience the most valuation.

The conclusion in the previous paragraphs is that you cannot compare one person's 7 with another person's 7 to find which one is best, but must instead say that one person's 7 has the same value as another person's 7. What then about comparing one person's 7 with another person's 2? If you have to choose between letting one person experience a pleasure that she values to be a 7 or another person experiencing a pleasure that he values to be 2, does that mean that we should choose to let the first experience a 7 since that will bring more valuation into the world (in each case the other person will experience something of pleasure value 0)?

It seems that this might be a good choice to make in individual cases, but not on a permanent basis. If I have to choose between letting one person get 10 000 dollars or one person getting 100 000 dollars, it seems good to choose 100 000 dollars if the persons are otherwise similar. But the same reasoning does not seem to be a good choice on a permanent basis. It does not seem ethically right to prioritize person A experiencing pleasure 7 day after day instead of person B experiencing pleasure 2 day after day with the argument that prioritizing person A brings more valuation into the world.

Here it looks like teleological ethics needs a criterion of justice in order to distribute the good fairly. Does this mean that teleological ethics does not work on its own? I think the justice aspect can be brought in by seeing that when we consider human lives, there is greater value in raising somebody from a general life quality

---

295 Briggs (2015), referring to Hammond (1991, p. 216).

of 2 to a general life quality of 3 than raising somebody from a general life quality of 8 to a general life quality of 9. To exemplify, it is better that somebody who does not have food and education get food and education than that somebody rich gets even more money and holidays.

Why is that right? To explain this with the teleological approach defended here, I point to three interconnected lines of reasoning. First, it is a fact that somebody who is hungry appreciates getting food more than somebody who is full appreciates getting another ice cream. Raising someone from 2 to 3 thus involves more valuation than raising someone from 8 to 9. We can see this from a thought experiment: If we had to choose, from behind a veil of ignorance, whether to prioritize people getting from 2 to 3 or from 8 to 9, and afterwards we would ourselves either become someone at 2 or someone at 8, without knowing before where we would end, most people would prioritize that persons should be raised from 2 to 3. This is how John Rawls has argued in favor of securing basic needs (Rawls, 1971).

Secondly, it is a fact that how much people valuate something depends on what they compare themselves with. Comparing the wealth of people and their happiness shows that they do not get happier by becoming richer, because it makes people need more in order to be happy (Harari, 2017, pp. 38–40). But it is required for happiness to get above the basic requirements for survival – in other words, it is more valuated to go from 1 to 2 than from 8 to 9 in general life quality.

Rachael Briggs uses the insight that happiness depends on comparing as an objection to an ethics based on comparing how people valuate different alternatives. She argues that it implies that people can change their welfare merely based on the alternatives they consider in their mind, which seems absurd (Briggs, 2015). Instead of finding it absurd, I find it obviously true, and researchers on happiness says that the most efficient way to improve your happiness is to remind yourself of what you are grateful of, including being grateful for the problems you are not having (Emmons and McCullough, 2003). However, it clearly has a much stronger effect on your comparison what you actually see around you than merely what you choose to think about. Seeing the luxury of your neighbor influences you more than thinking about poor people in another country. For this reason, I think that it is important that societies are actually characterized by economic equality, while merely thinking about hypothetical goods and evils influence our happiness much less.

Thirdly, as seen above and closely connected to the previous point, it is not just relevant what people actually value, but what they potentially could value as well. While people who are rich today have certain needs today in order to feel happy or valuate something, it seems clear that potentially most people would valuate life the most as a total sum if most people in the world had a similar life quality since happiness depends on comparing yourself with others. Great divisions between poor and rich cause envy and conflict, while great similarities cause stability and general contentment – very broadly speaking, of course.

These things cannot be measured exactly, but if one wanted to be mathematical about it, one could make a scale with higher values at the bottom. For example, one could say that going from 1 to 2 in general life quality is worth a million points, going from 2 to 3 is worth half a million points, 3 to 4 is worth 250 000 points, from 4 to 5 is worth 125 000 points, etc. Or it could be negative points for suffering where it would be much worse to go from suffering 8 to 9 than from suffering 1 to 2. This is a way of including an element of justice in a consequentialist ethical theory by letting the theory give greater value to help those with greater need.

The logic is the same as when economists speak of marginal utility, where a typical curve shows that the utility per unit is high at first and then lower as you add more units. One could be a utilitarian, saying that all that matters is raising someone up at the scale but there is no difference between raising someone from 2 to 3 or from 8 to 9. Or one could be a prioritarian, like me, saying that it is more important to help those who are worse off. If the utilitarian agrees that going from 2 to 3 has a higher total value than going from 8 to 9, there is no disagreement between the utilitarian and the prioritarian.[296] I believe that differentiating the weight of climbing at different places of the ladder is the right way to think about this.

The main point here is that it is more important to help the poorest than the richest, but making exact values for this would, of course, be extremely difficult. In general, it is better to prioritize helping the poor than the rich, but there are many reasons why this is not always a clear rule. Often it takes less to help the poor than the rich since you can make someone who is thirsty very happy by giving them clean water, whereas a rich person might need an expensive wine to experience the same kind of happiness. But sometimes it can be very resource-demanding to help someone in need while the same amount of resources could have helped many more in less need.

For example, one person with a bad disease could be helped by very expensive medicines, but the same amount of money could have helped many more

296 Derek Parfit makes this point, saying that if we give benefits different weight, there need be no disagreement between utilitarians and prioritarians (Parfit, 2012). It will also include the point from the egalitarians that increased equality is good.

with other diseases. Or a lot of resources could be used to help people in a dictator-run country where it is very difficult to get the help through, while the same money could have helped many more in poor countries where it is easier to help. Or the amount of money spent on helping the poor could be spent by rich people investing in research that creates a better medicine, solves a big energy problem, or even learns how to remove all pain by brain manipulation. Yet another factor that complicates the discussion is whether we should discuss what is right for all to do in general or discuss what is right to do given that most people will not act in accordance with this ethical rule, but rather out of self-interest.

Maybe using a lot of money on giving poor people medicine, clean water and food is the fastest way to the best possible world or maybe letting rich people invest in research is a faster way to a better world for everyone. Prioritizing the rich could possibly be a faster and thus better way to a better world. But since we do not know, it is more probable that prioritizing helping the poor is a better way to a better world for the reason mentioned above: that more people are happier when people live under similar conditions. I will say more about the role of making probability assessments below.

Here I will end this discussion by pointing out that securing that people get their basic needs could be thought of as securing basic human rights, implying that this should be prioritized. Securing basic human rights could be thought of as first securing that everybody gets lifted to general life quality 1, and then in the next step life quality 2, etc. The list of basic human rights could be prioritized and extended as the basic rights are in place: all people should have food, clothes and shelter, but also clean water, clean air, and then it could be extended to more and more health, education, money, etc. I will return to this issue as well at the end of this chapter. This line of reasoning would imply that if superhumans or superrobots that are considered persons in the sense of having a self-conscious mind evolve or are developed – and whose mental life is far more complex than that of humans – they should also secure human rights instead of prioritizing their own pleasures.[297]

I will move on to the next question, which is: How can this kind of reasoning stand up to the mere addition paradox? Derek Parfit has argued that if what matters is the total amount of happiness, it seems better with a population of very

---

297 If robots are not conscious (but still highly intelligent), they do not have a unified conscious self, only a representation of themselves as the whole robot, similar to how our brain has a representation of our body. Then they also cannot have goals for their own sake, in the sense of something they just valuate because it consciously feels good. In practice, they could nevertheless have something very comparable to human goals, but just as dispositions for acting in certain ways. I write more about this in Chapter 15.

many people having a little happiness than a smaller population with great happiness, which he calls a repugnant conclusion (Parfit, 1984, chapter 17). This could seem even worse in the scenario I just described, since I said that low scores are worth more points. But it seems clearly wrong to say that it would be better to have 1 billion people alive and experiencing pleasure of value 1 than if there were 1 million people alive experiencing pleasure of value 10, even if the total sum of pleasure is greater in the 1 billion scenario.

I agree that it is the wrong conclusion to think it better to have many people with a low score of happiness than fewer with a higher score. When considering value of pleasure in the world, this should be divided by the number of individuals experiencing the pleasure, so that the goal is to have this kind of highest average score. Since new humans hopefully will continue to be born century after century it is best in total that they all have a higher average score. This might seem to lead to the conclusion that we should strive for having very few people alive at any given point of time, but this does not follow, since it is good for all who live that there are also many others who live at the same time and can specialize in different parts helping each other.[298]

Moving on to the next question on the list: What if two values are equally good? How does one choose if both options seem to lead to the same amount of valuation? This may happen, and in such a case both options are equally good, ethically speaking. In such a case, a random choice is the right choice, but recall the additional comments made above to what Ruth Chang called cases of parity.

The case is more difficult if there is one scenario which has a high value but a low probability of becoming actualized versus a scenario which has a low value but a high probability of becoming actualized. The problem is well known in ethics and typically turns into a question of what is most realistic. War versus pacifism, different questions in climate politics, or revolution versus revision are examples where some will argue for drastic means to reach high goals whereas others argue that walking with smaller steps will be a more realistic way to reach the goal.

Decisions must be made at a point of time by people who in most cases will not know the objective probability that what they are choosing will actually become actualized. Even in cases where there is a high probability that option X shall occur, another person may gamble and go for option Y, which may in fact

298 One could think that it follows that it was better to just have one person alive experiencing something of value 7, but this does not follow, since of course there is a much greater potential for valuation over time with many people alive, but at one point there can be too many.

turn out to be a better result. Maybe, for example, a person makes a choice to save someone which most likely will just mean that she dies and that things get worse, but as a matter of fact, she makes it and the result is much better than if she had not done so. This has happened in various heroic acts thorough history – but of course there have also been many failed attempts at heroism which have made things worse.

In successful heroic cases, we can in hindsight appreciate it as a good choice, even if it was not the choice that was best justified in the moment of choice. In general, it seems that a "better safe than sorry" strategy will produce the best results seen in total, but one can never know in advance what will actually become the best result. It is a known debate whether one should be more idealistic and revolutionary, going for big changes, or emphasize context more strongly and work for revision, and although revolution may sometimes be the best, the revisionary strategy seems to have the empirically best support as a general strategy since stability and trust are so important for economic and other factors. In other words, it is generally better to prioritize smaller goals that have a higher probability of succeeding rather than prioritizing higher goals with a smaller probability of succeeding, although the world is too complex to offer an exact definition of where to draw the line.

Two extra reasons can be given to go for the "better safe than sorry" strategy and safe bets, that is, to prioritize goals with a high probability of being achieved. The first is that the future is uncertain and it is good for a person not to risk making the world worse. The second reason is that we can be sure that there are things that we do not know that we do not know, which means that what we think is a good solution may not be good after all. In such situations it is better to make small and safe steps than to gamble with high stakes. Middle-way compromises more often reflect an average common reasoning than the more extreme measures. The uncertainty about who is right in ethical issues with much disagreement is also a reason to strive to allow much freedom for people instead of the state or others forcing their views on others in order to reach a presumed good goal.

This second point about uncertainty is worth emphasizing, and I find it to be a deep argument in favor of this ethical approach: We may all be wrong in our ethical theories. Even if I have suggested a definition of the good as the best world, which again is defined as what the most would prefer the most, this is still very abstract, and thus uncertain when it comes to concrete content: We do not know what the best world is like or what people would prefer the most, nor do we know what the best way to get there is. This uncertainty is then a good reason to move forward with small steps while learning by exploring the concrete content of what is best. Given our uncertainty (which is certain;

there may be things we do not know that we do not know), we should gradually explore what the best way to the best world is.

As Karl Popper has argued, we have no good reason to think that a quick and radical change of society will work well (Popper, 1945). Even if a very different world would be better, what people valuate depends much on their history and identity etc., which means that changes should happen gradually from where we are now. Gerald Gaus argues that a theory of how to improve society often must make a difficult choice between making an improvement relative to earlier or going in the direction of the highest goal (Gaus, 2016, p. 142). These two alternatives coincide in my proposal, since the way to reach the highest goal is to make small improvements (the way thus being part of the goal, while the content of the goal is unknown).

Can we offer any guidelines when it comes to exploring what the best way to the best world is? As mentioned before, it is not possible to give an exact recipe for how to compare alternative actions. As Derek Parfit says, ethics cannot conclude that one action is 2.36 times better than another (Parfit, 2011a, p. 132). But I will suggest that the reasoning above can help us somewhat, especially what has been said about prioritizing preferences and prioritizing safe choices. Here are some rough guidelines:

We have already seen that if the one alternative will raise one person from life quality 2 to life quality 3 and the other alternative will raise another person from 7 to 8, we should choose the one going from 2 to 3. If two alternatives will raise one person from life quality 2 to life quality 3, but one alternative has a higher probability of occurring, we should choose the one with a higher probability. Choices with high probability of success should be preferred, and choices preventing suffering or raising those with low life quality higher should be preferred. The tricky part comes when one has to balance number of people, number of life quality or suffering, probability of success, and weighing these against each other.

It becomes impossibly complex when one adds context and what resources different agents have, so it is easy to think of counterexamples to the guidelines I will now offer. Nevertheless, since it may help us clear some thinking I will make some general suggestions where we assume the same context for all the examples and the same resources for all the involved persons who are choosing what to do. Imagine just coming to a large group of people who you can help, but it has to be done in a certain order, and sometimes you have to choose one alternative over another. Helping people in this example is not about concrete instances of happiness or suffering, but considering the general life quality of people, where 1 is a life of suffering and misery and 10 is a perfectly happy life. Here is how I suggest one should prioritize:

If there are people that with a high probability can be helped, these should be helped first, starting with those suffering the most first and moving to those with less suffering up to higher and higher life quality. Mathematically put, start with those at 1 and move up the scale to 10. If we say, as above, that moving from 1 to 2 is worth many more points than moving from 7 to 8, we could use a mathematical principle to guide us as long as we are aware that it is very inexact and influenced by many other factors as well. The principle would be to take the number of points times the number of people involved times the probability of succeeding, and opt for the alternative with the highest score.

This reasoning could then be used also when deciding whether to help group A or group B. The alternative that gets the highest score is the one to choose. If there is a choice where you can help many, but a few get it worse, that counts as a minus score which it takes much to make up for. Raising a lot of people from 8 to 9 is not worth it if the price is to take a few down from 5 to 2, such as by exploiting workers. But it may well be worth raising a lot of people from 2 to 5 even if it means taking somebody down from 9 to 8, perhaps by adding taxes for the rich.

While this may seem like an absurd mix of ethics and mathematics, it does help us explain some ethical intuitions that many share. For example, it seems that the life quality of many people in North Korea is very low, meaning that helping them achieve a better life would be worth more than helping a group of people in a country suffering less. But if the probability of succeeding with a specific attempt of spending time and money in North Korea is very low while the same attempt of spending time and money somewhere else has a high probability of succeeding, it may nevertheless be ethically right to prioritize the other place.

It could seem like the suggestion here offered would imply that rich countries should give away all their money to poor countries, and while it would certainly be good that rich countries give money to poor countries, it does not follow that almost all of their money should be given. This has to do with the probability of achieving good results, which may be very poor by just giving money away while it may be very high when used in an established system. It may be that the best way to the best world is by letting countries gradually become better instead of making all equal.

Many principled questions still remain, like how to balance short-term goals versus long-term goals, and here I just presupposed short-term results. Even if answers like these are very imprecise and full of exceptions, many other ethical theories have no answers at all when it comes to what to do when you have to choose between different alternatives with different numbers of people with different needs and different probabilities of success. Unless they are clearly wrong, rough guidelines are better than no guidelines.

The next question on the list was whether this is an ethic that is too demanding. Should one always act so as to actualize a better world? How should we understand the relation and sometimes conflict between self-interest and the interests of others? This last way of articulating the question has been called "the profoundest problem" by Henry Sidgwick since it seems to be rational to do both what is best for yourself and what is impartially the best, but these two principles can often be in conflict. What then is the rational choice to make (Sidgwick, 1907, p. 386, n. 384)? This problem is discussed by Derek Parfit, looking at different cases, and a typical problem is how to decide when you have to choose between saving your own children and a larger number of strangers (Parfit, 2011a, pp. 130 – 149). At what number of strangers should you choose to save the strangers instead of your own children, and how should we be guided when making such a choice?

This is how I would reason in light of what has been said above: First of all, it is not a unique problem in ethics that we have to decide between conflicting goals. This is rather what is common to moral problems. How should be decide between a great good or evil for a few people versus more, but smaller, goods or evils for more persons? How should we balance between good or evil in the short term versus the long term; or between a small good or evil with high probability and big good or evil with lower probability? And how should we balance our own interests and the interests of others? There are many different goals, but asking the moral question is to ask how one should choose, all things considered.[299]

What then if you have to choose between saving someone you love versus a larger number of strangers? As a starting point we should consider what the people involved valuate, and we can here assume that everyone about to die has the same valuation of life. In addition, there is an extra valuation involved for a father or mother either saving or not saving their own children and for the children either being saved or not saved by their own father or mother. Seen from the outside like this, it is morally best that a person saves his or her own children versus strangers if it is the same number of people being saved, but right to save the strangers if they are a larger number.

However, there is a difference between asking what is morally best and asking what a person should do in a situation. We have seen that a person should make the world better and not worse in light of his or her own resources. Maybe

---

299 "All things considered" means "all things counting for and against considered", and not that one should consider irrelevant things like the mass of Jupiter.

it is not psychologically possible for a person to do anything other than saving his or her own children. As a clear contrast, we can imagine a case with a rich person spending money on unnecessary luxury where it would clearly be psychologically possible to use money to save lives instead of buying expensive designer furniture. Then it is morally wrong for this person to spend money on luxury, and the person can be fairly blamed for it.

It is easy to decide in extreme cases, and also quite clear that saving more lives is better than fewer, but when it comes to exactly when a person should save strangers instead of his or her beloved, it is not possible to give either an exact or a general answer. I believe that the relevant parameters are the ones here discussed: namely, the valuations of the involved and how they make the world better or worse, and the resources of the ones who act.

That the morality of an action is evaluated "in light of the resources of the person" is also quite vague, for what are the resources and how much effort based on how much resource does that imply? As seen in the examples above, "resources" has referred to different *kinds* of resources, whether economic or psychological or mental resources, and there could be physical resources as well, like how good your health is.

This qualifier, "in light of your resources", is meant to explain why it is morally wrong for someone not to help when they could easily have helped but prefer to spend money on unnecessary things instead, while it may not be morally wrong for someone else not to help because they lacked resources. I have said that we should make the world better and that we should not make the world worse, and a rich person spending money on unnecessary things could say that he or she was not making the world worse by not helping, yet it seems morally wrong to spend a lot of money on unnecessary luxury instead of relieving suffering. That is why I included the "in light of your resources" qualifier: to show that it differs from person to person how much it is reasonable that they contribute to making the world better.

But how do we determine how much people should contribute to making the world better in light of their resources? It may still seem too demanding if an ethic implies that you cannot buy new shoes since the money could have been spent to save the life of a child or that you cannot take care of your sick mother because the resources could have been spent on saving lives. The solution to this problem – which is also a solution of how to balance self-interest and the interest of others – is to think that we should compare people's actions with a standard for how to find the best way to the best world which includes taking into consideration people's resources. Given the world today and the resources of people and states, what would be the most valuated way to the most valuated world? It seems clear to me that this would be a world with room for buying new shoes and taking care of one's sick mother, since the best way to the best world would be a way where nation states, through taxation rules supported by all, take care of the ones in greatest need.[300]

Such a standard is the basis for what each person, based on their resources, should do to make the world better. If they contribute more, that is good, but they should not be blamed by saying they should have done more, since by definition the standard shows the best way to the best world. There is thus a distinction here between how much it would be *good/better/best* that you contribute, and how much you *should* contribute. It is not always the case that you *should* contribute even if it had *been good* that you contributed. Let us say that your neighbor does not buy clothes for his children and spends the money on drinking instead. He should buy clothes for them, and it is not the case that *you* should, although if he does not it may still be good that you do. Given how I defined "should" above, you should contribute to make the way towards the best world better (more likely to improve) than it is today, and you should not make it worse. And the better you make it the better it is.

How should we then understand the claim that we should not make the world worse? Is that just that we should not make it worse than it is today? No, that is too weak, since it is already bad and it is not good enough that everyone just acts so as to let the world remain as bad as it is today. The standard for what is the best way to the best world again helps us define also what we should not do, since we should not do less than what this standard says. This ethic – that you should not do less than what is the best way to the best world – may sound demanding, but actually this is not a very demanding standard since, in the best way to the best world, states should through taxation do the job of sharing goods fairly. It is good if you do more, but the moral "should" placed on your shoulders is that you (like everyone else) should contribute your fair share to making the world as good as possible for everyone (including you).

Sometimes random events will make it the case that suddenly you are the nearest person to an accident or something similar where you should do a great effort to help, since in the best way to the best world it will sometimes be the case that an individual by chance is required to do an extra effort. But in general, it is not the case that you should save all people dying from hunger in the world since, in the best way to the best world, this should be done by nation states through taxation. But you *should* vote for politicians working for glob-

---

300 Note that the best way to the best world here is how we ideally should go from where we are today towards the goal – the standard assumes that people do their part.

al justice, since the best way to the best world requires democratic support from the people. These consequences of this theory seem right to me.

This was the last question on my list, and discussing all the questions has been done with the intention of showing that one can have a coherent ethical theory even if one rejects ethical values as irreducible entities in one's ontology. As seen in this and other chapters, one can defend quite traditional views of human value, equal rights in addition to free will and responsibility also based on a naturalistic ontology. Ethical norms can be understood as a theoretical framework where we draw a map of the best way to the best world, to guide us on our way into the future.

I end with a brief comparison of this ethical model with other ethical models. Derek Parfit has made a famous comparison between Kantian ethics, consequentialist ethics and contractualism, arguing that in their best versions they are roads to the same mountain top: Kant wants to find laws that it is rational for everyone to follow, but Parfit argues that for it to be rational for everyone to follow, it must be best for all, which means that a Kantian ethics imply rule utilitarianism. Contractualism says that we should do what nobody can reasonably reject, but that is just the same as laws it is rational for everyone to follow (Parfit, 2011a, pp. 410–413).

My analysis of these models is that Kantian ethics and contractualism suggest rules we can follow in order to find the best way to the best world. We will not actualize the best way to the best world if everybody gives themselves special treatment, and thus we need some general rules for people to follow in order for the best world to be actualized.

Parfit does not think that good consequences explain why we should follow the rules, and there we disagree (Parfit, 2011a, p. 418). Parfit can say that the three ethical models are ways to the same top, and what unites them is that they give us rules to follow which we have the substantially best reason to follow, but he does not think that such substantial reasons can be defined (Parfit, 2011a, pp. 31, 39). I think that the three ethical models are ways to the same top because the top is the world which is valuated the most by the most, which is by definition The Good, and the three ethical models are suggesting general guidelines, which are in fact guidelines for how to get to that top.

15 Implications for the Future: Excursuses on the Meaning of Life, AI, and Politics for the Future

This book has mainly been about the universe as it is today (including some of its past), but in the previous chapter, the topic was also possible futures and what constitutes a good and the best future for the earth. In this chapter I want to indicate some implications of what has been said before for three different topics. The first is what to think of the meaning of life of individuals – what is the goal of an individual life and why is it worth trying to reach that goal? The second is what the theory of mind and free will in this book implies for the possibility of superintelligent machines. The third is what the ethics implies for more practical political goals – how to proceed in reaching the best way to the best world.

The contents of this chapter should be considered as brief suggestions for further research based on the previous 14 chapters, and not as anything near being thought through in detail. It should be seen as speculation – where I acknowledge that the future is very uncertain and that these topics are complex and out of my area of specialty. That may seem like a reason to exclude it, but clarifying implications, even if brief, is illuminating also for understanding what has been said before, and it is very tempting to point out some more practical implications of the many very theoretical discussions that have preceded this chapter. Even if it is uncertain, my hope is nevertheless that some of what has been said so far could also turn out to be a resource that could be used for something positive in the future, so I will try to point to some of that now.

15.1 Excursus on the meaning of life

In the previous chapter, I said that we should make the world better, and the better we make it the better it is. This sentence could be understood as a compact theory of the meaning of life, which I shall unravel a bit further in this excursus.

The term "meaning" can be understood in at least three quite different ways, which can then be further combined in various ways. First of all, "meaning" can be something understandable, which again is just to say that it is an integrated part of something coherent. For example, the string of signs "There is a cat on the mat" has meaning, whereas the string of signs "cza hafnas cjaramuish gnr" does not (as far as I know). Secondly, "meaning" can mean "intention" or "goal". For example, the meaning of an action such as the act of opening a window can be the intention of letting some fresh air in. Also events or processes can have a meaning in the sense that there is a goal or intention behind them, or

https://doi.org/10.1515/9783110770957-017

they can be meaningless in the sense that there is no goal or intention behind them. Finally, "meaning" can mean that something is valuated as something positive.[301] For example, a person can find it meaningful to work in the garden, view fine art, or something else (Puntel, 2008, pp. 342–343).

If we ask for the meaning of life, we are usually interested in the last two meanings of "meaning", namely intention or something positively valuated, and we can ask for the meaning of an individual life, all human life, or all life. In other words, we ask if there is an intention behind an individual life, all human life or all life, or we can ask for the positive value of an individual life, all human life or all life.

If one believes that the world has been created by a mind, one can answer affirmatively to the first three questions and say that there is an intention behind all life, or all human life, or even every individual life, depending on how detailed one believes the intention of the creator to be. If one does not believe that the world was created by a mind, one will answer negatively to the first three questions. However, one may still answer either affirmatively or negatively to the last three – that there is positive value to all life, all human life or all or some individual lives – or that there is not. This positive value can be further differentiated in many ways: are there one or many different kinds of values; do they differ in quality and is one the most important; are they consistent or not; and are they always actualized or could they remain unactualized possibilities?

In this excursus I shall only be discussing the meaning of life in the sense of positive valuation of life – individual life, human life and all life – and this is how the term "meaning" is now used unless otherwise specified. To valuate something is to find it meaningful, and that which is valuated is considered to have meaning. The more there is in the world that is valuated, the more meaningful the world is, and the more meaning there is in the world and in life in general.

This means that the more you do something that people valuate, the more you make life and the world meaningful. Another way of saying the same is that doing good to others and doing good to yourself is the same as making life and the world more valuable and more meaningful, since "valuable" and "meaningful" just mean something that can be valuated. Making the world better is by definition making life more meaningful.

---

301 In this sense of meaning, something negative could be meaningless (have no positive meaning, but could still be meaningful in the sense of understandable or intended).

I just described what makes the world meaningful (full of meaning) even if there is no intention behind life and the world. But if one also believes that there is a good creator behind the world, what I just described is often taken to be the intention of the world as well. To love others as you love yourself, or to treat others as you want them to treat you – the golden rule found in many religions – is said to be the will of God or the gods.

If you do not believe that there is a personal creator of the world, you could nevertheless make it the intention of your life to make the world better and find just that meaningful since that is what meaning is. Meaning is what is valuated as good, so the more good we make, the more meaningful life is. In the chapter on ethics, I also argued that we *should* make the world better, again simply because that is what the term "should" means.

You may reply that you do not care, or you may think that the goal of making the world better is not something that makes life worth living. My response is similar to the discussion on the normative force of ethics and the potential value of something: Your life and your contributions to the world (both when succeeding or when failing) have the potential for being valuated by yourself and others. If you wonder why you should try to make the world better, or whether it is worth it, I can only answer something which is true by definition, namely that you should try because it is better, and that it is worth it because it is something to be valuated (and "worth" and "valuated" mean the same).[302]

This is still just an objective description of what makes life meaningful, which means that it is possible to live a life which is meaningful (in the sense of creating a more valuated world for more people) while you still do not feel that life is meaningful for yourself. One may feel that life is meaningful (or feel happy) for different reasons (including luck), but the best way to ensure that you feel that life is meaningful (and to feel happy) is to develop a moral character where you want to do good. This is Aristotle's point: following your natural desires will not make you happy, but you can become happy if you form yourself into a person who wants to do good. If what makes you happy is to become rich and famous, you will most likely be unhappy. But if you

---

302 I guess it is possible to think of scenarios where it is impossible to improve your own life or the world, and where death seems better and more meaningful in the sense of decreasing the suffering in the world. The general rule, however, is clearly that life (as opposed to death) has the biggest potential for meaning and valuation. This topic is extra tricky since upholding the value of life makes it more valuated, while speaking of death as good can make life less valuated for some. While acknowledging the possibility of rare exceptions, I find the general rule clearly to be that life is valuable, both in the sense of having potential for being valuated and in the sense of being the cause of other things that can be valuated.

make yourself into a person who wants to do good (and that happens if you make it a habit doing good, and experiencing how good it is), then there will always be opportunities for doing good, and your chances for becoming happy are very good. Of course, happiness cannot be guaranteed, and everyone can have bad luck, but you will in any case be doing something good and meaningful, which is the best recipe for also becoming happy. Research confirms that people actually experience life as meaningful when they can do something good for the world (Schnell, 2021).

Aristotle said that forming a character through habits where you want to do good will make you excel in the art of living, but what does it mean to be excellent at the art of living? Abstractly, one could say that to be excellent at the art of living means effectively reaching well-justified life goals, which again means that it is determined both by the quality of your goals and your ability to reach them. But here one should add what I said in the ethics chapter about the "in light of your own resources"-qualifier. Given who you are and can be, you need to find out how you can best contribute to this world to the happiness of yourself and others. Only you can be excellent at living your life.

"Making the world better" in the sense of making the world more valuable and more valuated and thus more meaningful could now be seen as a description of the whole history of the universe, with meaning in all three senses of the term. It can be a goal we choose to have (or a goal that the creator has), and thus meaningful in the sense of intention or goal. It can be a description of how the world becomes more meaningful in the sense of something positively valuated. And it can be seen as a coherent description of the world, which I will now describe further. What I want to do in the following is to give a coherent description of the history of the universe which describes it as a universe becoming better and better, which we should have as our intention and meaning of life to contribute to.

The history of the universe is a history of growing complexity in the sense of individuals coming into existence with ever more complex structures. It has been an evolution from the simplest particles to more complex particles as described by physics, moving further to ever more complex molecules as described by chemistry, then further to biological cells and ever more complex individuals, then to individuals with brains and consciousness, continuing to ever more complex minds and experiences – of both bad and good – and we cannot know how it will progress or even end.

As history proceeds, individuals live longer and with less suffering, but there is still much suffering in the world. Massive problems have been greatly reduced in the last centuries, like famines, war, diseases and poverty, but still there are famines, war, diseases and poverty, and there are also new problems and risks, for example in the areas of climate change, new weapons and artificial intelligence. Is it possible to continue the good trend in making the world better (more valuated) for ever more individuals?

Through history people have realized that there are certain goods they can only achieve or at least achieve much better for everyone involved by organizing themselves. It began with families. Early in human history, they also realized that there are certain goods they can only achieve – or at least achieve much better for everyone involved – by organizing themselves into tribes.

Relatively late in human history people realized that there are certain goods they can only achieve or at least achieve much better for everyone involved by organizing themselves into national states. Through testing many different ways of organizing states, it has become clear that democracy is the best (= least bad) way of organizing the life of many people who live in the same area but do not know each other and disagree on many matters.

Now time has come to realize that there are certain problems and certain risks that we can only solve (and certain goods that we can only achieve) or at least solve and achieve much better for everyone involved by organizing international agreements. Solving the biggest problems of the world would also be the biggest improvements of the world, and hence the most meaningful project one can be part of. This will be the topic for the third excursus in this chapter, but first we shall take a closer look at an important challenge for the future which is both potentially a great risk and a great good, namely the development of artificial intelligence.

15.2 Excursus on AI: Artificial intelligence and the possibility of superintelligence

In this excursus, I will use the theories of mind, free will and metaethics presented in this book, and ask what they imply for machines. Can machines become persons? Can they be moral agents? What kind of moral agents would and should they be? Questions like these will be briefly answered in this excursus. These answers are not meant at all to be fully justified answers, but instead meant as clarifications of the main theoretical framework by showing implications.

Before answering these questions, I want to start with some words about the relevance of discussing AI and superintelligence. There are a lot of important ethical questions that I will not discuss in this excursus; rather it will remain at a very principled level and based to a large degree on Nick Bostrom's book *Superintelligence* (Bostrom, 2016).

When it comes to the development of technology, progress happens at a constantly quicker rate. If we consider the evolution of intelligence on planet earth, it seems like the universe has only explored an extremely small part of the potential of intelligence. Through billions of years, it has been a very short period from monkey brains to Einstein's brain, and there is not much of a difference in size and signal speed in a monkey brain and Einstein's brain (Bostrom, 2016, p. 85).

If intelligence depends on size and signal speed, an intelligent machine could potentially be on the size of the universe with signals at light speed (300 000 kilometers per second) compared to a 1.5 kg brain with a signal speed of maximally 120 meters per second (Bostrom, 2016, p. 72). If a machine becomes intelligent enough to create something even more intelligent, machine intelligence could very quickly become vastly more intelligent than humans can understand (Bostrom, 2016, chapter 4).

Something which is unique in history now stands before us. Before, the potential for good and evil in the future, based on what was known to be possible at the present at that time, was quite limited. Now we face the possibility of consequences both as good and as evil as it is possible to imagine: superintelligent machines might be able to create welfare and prosperity beyond comparison, but they might also be able to cause the maximally possible amount of suffering and destruction. This may be the point in history where acting wisely has (beyond comparison) the largest consequence ever.

Bostrom compares it to discovering the atomic bomb and doing research on it without knowing how it would work. What if a nuclear bomb had been something anyone could make by baking some sand in the microwave oven? We were extremely lucky that it turned out to be difficult to make, and that Hitler was not able to do it first (Bostrom, 2013, 12:37–13:00). Bostrom compares the situation now to children playing with bombs without knowing how they work, and calls upon people to gain competence, analyze the situation, and plan strategies (Bostrom, 2016, pp. 316, 319). Another illustration would be to say that the situation is like children playing with devices where we have reason to believe that some are bombs that can go off and kill us all, while some devices could give us incredible treasures, and we need to find out how to secure the good outcomes.

In the following, I shall reflect upon some principled questions connected to artificial (super-)intelligence. I will focus on three main areas that have already been covered to see what they imply for machines. The three areas are theory of mind, theory of free will and responsibility, and theory of ethics. I will ask whether machines can have different features that only humans have had so far, how they would work in machines, and what would follow if machines become superintelligent.

The first of the three areas is theory of mind. In this book I have defended a causal theory of mind in humans, which implies that it could also be something that machines could have. Human thinking is about discovering structures and organizing parts into wholes, which is something that machines can also do. With enough computing power, I believe that machines could be able to think non-consciously like humans.

Intelligence and rationality are about solving complex problems by finding means to goals, which is something machines have demonstrated well already that they can do in more and more areas. With enough computing power and learning, I see no reason why machines should not develop artificial general intelligence.

Organizing parts into wholes is what it means to understand something non-consciously. However, for a subject to be able to understand what something is about, I have argued in this book that subjectivity and intentionality are required, both of which presuppose consciousness. Since humans can do almost any thinking process non-consciously, machines should be able to do the same non-consciously, but there will be no subject or core self in them understanding what anything is about.

Could machines become conscious? Based on the theories presented in this book, there is no principled reason why not. I do not know what physical structures activate qualia structures, so possibly it can only happen with organic material, but this is unknown. In any case, it seems to matter little for what machines can do whether they are conscious or not, since I have argued that consciousness does not play any important role in concrete choices.

Would conscious machines have a consciousness like ours? Probably not, since our consciousness has been shaped through evolution, linking what we experience as objects and what we experience as good and bad very closely to our survival. We have no idea what the content of consciousness in a machine would be like.

If machines are not conscious, they seem to be very different from humans when it comes to motivation. Machines seem to operate on algorithms only, saying if *x*, then do *y*, while humans can have a conscious experience of something as good and have that as a goal in itself.

However, I have argued in this book that through evolution we have developed if-then mechanisms that are like algorithms. The causal explanation of our behavior is that we have a mechanism in our brain which in effect says if it activates a desire representing that it feels good, then do it when a certain threshold is reached. Through evolution, certain behaviors have been connected with "feeling good", such as eating, drinking and having sex. But at all levels, it seems to be the physical realizer that does the job and a physical explanation can be given for their development without need for consciousness.

It seems that motivation in machines could work the same way if they have two general algorithms saying, "If it feels good, then do it" and "If it feels bad, then avoid doing it", and some other mechanism sorting actions into categories like "*x* feels good" or "*y* feels bad".

An important difference between humans and machines is that more intelligent machines will probably be able to change their own algorithms, which is not something we humans can do (yet). We cannot choose to change what feels good and bad for us, but non-conscious, very intelligent machines probably could. Humans build a stable moral character over time, whereas a machine can turn into something else completely in no time. On the other hand, if there is an overriding algorithm guiding the behavior of the robot, the robot would in most cases not have a motive for changing it, although it could happen.

With this understanding of motivation in place, we can move on to the next of the three main areas: what the theory of free will and responsibility presented in this book would imply for machines. In this book, I have described deliberation processes and choices as causal processes, which by implication could occur in the same way in machines. But could machines also make free choices?

Machines, like humans, start with a given set of basic algorithms (in the case of machines) or dispositions and desires (in the case of humans). It is easier to think of humans developing free will compared with machines since humans start with basically the same package deal when they are born, while machines start with very different presuppositions, being more in the hands of their creators.

But something similar to the development of an increasingly more independent autobiographical self could occur in machines if they were programmed to learn and act on what they prefer to do (guided by the starting algorithms) in an undetermined world. The more intelligent machines become, the more plausible this will seem, namely that they should be selected as the causes of why they do as they do.

As long as machines do not have conscious selves, there will not be a core self. I have defined a person as a body with a mind and a core self, and a non-conscious robot will only have a body and a mind, not a core self. The robot will lack subjectivity – including a feeling of what things are like – and intentionality, but these are also pretty much the only things it will lack. Whether such non-conscious agents deserve to be called persons is up to us to choose: do we want to adjust the concept of persons to include permanently non-conscious acting minds or not? Similar reasoning would apply to concepts like free will, autonomy, etc. Do we require consciousness as part of what the concepts refer to, if everything else is similar?

Something similar about choosing definition can also be said about the concept of moral agency, but a little more should be said about it first. Could a machine have capacity for responsible behavior, such that it could be appropriate to hold it responsible and call it a morally responsible agent?[303]

I have argued in this book that holding others responsible is something we do by blaming and praising them, and that persons have capacity for responsible behavior if they can take blame and praise into consideration in a normal process of deliberation. Machine deliberation processes are not (yet) normal, but we can imagine turning a machine into a moral agent by making it follow two general algorithms: "If it feels good, then do it" and "if it feels bad, then avoid doing it", and add two more algorithms saying, "If humans praise you, it feels good" and "if humans blame you, it feels bad". More details would have to be added on what to do in conflicts and how many and who should be praising and blaming to make this work, but here I remain at an abstract and principled level. It seems that machines could learn to take blame and praise into account in their reasoning about what to do. Machines developed to do this but unable to take praise and blamed could be quarantined or given "the death penalty" – in practice, turn them off and dispose of them.

Facing the new possibility of an agent whose reasoning process can be influenced through praise and blame even if it is non-conscious at all times, we would have to decide whether or not to include this new entity in our concept of moral agents or not. What we find to be most coherent and convenient is the only guide in this question. Here I will refer to them as non-conscious moral agents. The possibility of non-conscious moral agents raises a lot of ethical questions, and I now proceed to the third main area, namely what the meta-ethics and ethics in this book imply for superintelligent machines.

Do certain motives follow from superintelligence alone? Nick Bostrom argues well that superintelligence seems compatible with almost any goal, but that any superintelligent machine with any motive would also have many shared motives of increasing their own efficiency of reaching their goals: preserving themselves, enhancing their own cognitive powers, acquiring resources and developing technology, etc. (Bostrom, 2016, pp. 130 – 137). But in accordance with what I said above, I believe that the overarching motives of a superintelligent machine will depend on what motives or algorithms it was given from the start.

303 For a long discussion of this, see Søvik (forthcoming-a).

Would a superintelligent machine know what is morally right? Would it be motivated to do what is morally right? In the chapter on metaethics I argued that there is no correct definition of what is morally good or right. It follows that a superintelligent machine would know (as well as possible) what is morally good or right given all the different possible definitions of morally good or right, but it would not select one definition as the correct one (although it could select one if given criteria for choice). It would not be motivated simply by something being moral in different senses of the terms, since motivation is given by the basic algorithms.

Bostrom uses the main part of his book *Superintelligence* to argue that it is very important to be able to control a superintelligent machine since very much can go very wrong. On the other hand, it does not seem possible for humans to control what a superintelligent machine *can* do, so our only hope is to control what it *wants* to do by giving it good overarching goals. However, one cannot specify what a superintelligent machine should do in every possible situation, so the challenge is to find a good overarching principle for the machine to follow. But what should that be?[304]

In the chapter on metaethics, I argued that the best definition of the good is the best way to the best world defined as that which is preferred the most by the most. But nobody knows what the best way to the best world is in practice – or what the most would prefer the most – so this would have to be explored gradually in a revisionary way, as argued in the chapter on metaethics.

Would a superintelligent machine know what would be preferred the most by the most? Even if you are the most superb intelligence we can imagine, it is always possible that there is something that you do not know that you do not know. Reality may have unknown deeper levels or outer areas or latent laws or hidden indeterministic possibilities that make the world today and the future tomorrow different from what we, or the most intelligent mind possible, thought. The superintelligent machine cannot be sure either that it knows what it is like for others to be like them and experience what they do. Because of this uncertainty, I think that to explore gradually what the most would prefer the most is the best principle for a superintelligent machine to follow. It should be given as a principle to find the *way* that the most would prefer the most to the *world* that the most would prefer the most in combination with some other rules like aborting the mission if many people protest because of something we have not thought about and a rule of not letting new algorithms overrule the basic moral ones.

304 For a long discussion of this problem, see Søvik (2021).

The final aspect I will discuss related to ethics are questions of when robots would require our moral concern. When do we need to care morally about how we treat robots? When would they have value and rights, and when would they be equal to each other, or even equal to – or more valuable than – humans?

As long as a robot is not conscious, there is no subject (core self) there with feelings or preferences. There is just a collection of parts and no one to harm. Because it is very similar to humans in many ways, it could be compared to animals that are never conscious. Animals deserve to be treated well, but this is important when they can or may feel pain, whereas it is not important to treat well small animals with no brains who cannot feel anything. That is not to say that you are free to destroy – in the same way as we should not destroy animals, trees and flowers, mountains, or things (including machines), all of which have potential for being valued – unless we have better reason to do so than not to do so.[305]

An interesting case would be if machines became conscious but could not feel or did not have any preferences. Would the mere presence of consciousness be a reason to let it continue existing? There would be no preference to frustrate in the machine by not letting it exist, although of course other people could value its existence. Most likely we would not know what it was like to be that machine, which would be a reason to let it exist in the same way as we should treat well animals that may be conscious without us being sure of it.

Concerning value and rights, conscious machines becoming more and more advanced would have increasing value and rights in the same way as I discussed these matters with animals and humans. Their value is determined by the potential for valuation, and this has the undeniable implication that conscious machines could surpass humans in value, which just means that they could potentially valuate and be valued more than we as a group of humans can. This seems undeniably true if we imagine minds far beyond ours in capacity and variation. As I said in the ethics chapter, such robots should respect human rights – comparable to how we should respect animal rights, even if animal rights are fewer than human rights (and human rights may be fewer than robot rights).

Writing about this third area of implication for machine ethics, I have remained at a quite principled level. There is a wide range of specific ethical questions connected to AI which cannot be treated here since they would need a lot of extra facts in order to be given a serious treatment. This concludes my discussion of implications for machines. While having some suggestions as to what to

---

305 Nolen Gertz writes well about how the way we relate to robots also form ourselves (Gertz, 2018).

do, it does not follow that people will do it, or that we can make it happen. What should we do facing the enormous potential good and risk of AI? My answer to this is similar to how we should deal with other huge problems, and this is the topic for the final excursus.

15.3 Excursus on politics for the future of the world: The best way to the best world

In this excursus, I will draw some implications[306] of what has been said earlier in the book concerning world politics. Here are some relevant points from the book that will be developed: Concerning metaphysics, the most relevant point for this excursus is that most of what we think of as things that exist are structures that are stable over time. Concerning ethics, the most relevant point for this excursus is that we should try to actualize the best way to the best world.[307] Important ethical concepts like "rights" and "justice" get their content from what is the best way to the best world, and people should contribute to actualizing this world based on their amount of resources.

What implications should we draw from this concerning goals and means for world politics? That which exists over time are stable structures, and the world needs structures that are stable over time in order for good things to happen and continue happening. The universe and we humans have managed to create a great number of stable structures, but our final goal for the world should be a robust and stable structure for lasting human coexistence in the face of any future risk, which can make us feel that the world is a good and safe place for us and for the coming generations. The world needs a stable ecosystem, stable welfare states, stable economy and stable societies. In the following paragraphs I will say a bit more about these four stable structures.

The world needs a stable ecosystem, which means an ecosystem in balance. The earth can deliver clean air, clean water and food to all if the resources are not overused. With a stable climate, the whole world is recyclable, so to speak, but it depends on us not using too much forest, oil, water, meat, etc. and that garbage is recycled. Our consumption actually is already monitored, and every year a date is set for when the resources of the year have been

306 "Implication" not in the sense of what follows deductively, but in the sense of most coherent development.
307 In this excursus, "the world" refers to our earth.

used, called the Earth Overshoot Day. Ideally, this date should be 31 December or not at all, but increasingly it comes earlier, and in 2021 it was 29 July.[308]

If the ecosystem is in balance, the earth can be a good place to live on for generations in millennia, but now we are destroying the earth for the coming generations. Imagine that you had been born into a world constantly plagued by deadly heat and forest fires, polluted air and water, lack of food, and constant extreme weather, and you learned that this was all an unnecessary consequence of selfish behavior from the previous generations. What would you have thought about them? We need to live sustainable lives now.

The world needs stable welfare states. While it may sometimes be tempting to think of a dictatorial world government fixing environmental and other problems efficiently, it is far too risky with a world government because of its potential for extreme abuse of power.[309] We need to be realistic about the human capability of doing evil. Despite all the problems of democracies, they are necessary for developing trust and stability over time in a society that is impossible with other forms of government. The lesson of history is clear that some form of democratic welfare state is best. I shall discuss objections to this below.

The world needs a stable economy. Economy does not work without stability, since people need to have trust enough in the future to dare to invest money, which again creates jobs and the welfare people need. A stable economy needs laws, agreements, and institutions to secure that stability.

The world needs stable societies. Unstable societies create social problems and extremism, which again cause violence and terror. Data show a very clear correlation between the number of social problems and the amount of economic inequality in a society (Bregman, 2017, pp. 54–55). Reducing economic differences among people through progressive taxation or more equal salaries favoring the poor is a very important means to create stable societies, and necessary to maintain a good society for all over time.

Earlier in this chapter I wrote about how establishing ever larger groups and finally nation states forcing everyone to participate has been the best way of achieving a lot of good and stable structures, like police forces, hospitals, schools, roads, etc. Humans have developed the stable structures necessary to secure basic goods first in families, then in tribes, then in nation states, and the next step needed now is to establish some international agreements to solve international problems by establishing some stable international structures.

---

308 Overshootday.org.

309 As argued by John Rawls in Rawls (1999).

Creating a good and stable world through international agreements and other means is something that many people work with already, and the most important work is that which is done by the United Nations. The UN has sustainable development goals, which are 17 goals and 169 targets, and a network of organizations working to actualize them – the United Nations Sustainable Development Solutions Network. There are also many different organizations working on actualizing the goals and raising money, such as The Giving Pledge (where about 200 of the about 2000 billionaires of the world have agreed to give at least half of their money to good causes) or movehumanity.org. However, the success in reaching the sustainable development goals is quite low, as can be seen in the annual reports that the UN makes on the status of the goals.

There is no quick fix solution to the problems of the world, and the best way to solve the problems is to take many gradual steps. My goal in this excursus is to make some suggestions on how the work already being done by the UN and others (including you as the reader) can be improved further. I acknowledge that I am very far from being an expert on world politics, and thus it would be important to have experts continually revise a process that could succeed in making the world better. But as mentioned, I will make some suggestions nevertheless, to clarify implications of what has been said earlier in the book. The rest of the excursus will be structured in three main parts: I start by saying something general about how to succeed before moving on to more specific suggestions, then end by answering some objections.

The first part, then, looks at some general considerations on what it would take to succeed in creating a good, stable and sustainable earth. Since a world government is too risky, the important decisions must be made by countries, which means that they must be made by politicians in those countries, which for democracies means that those decisions depend on the opinions of the majority of the people. In order to succeed with actualizing a good and stable world, there needs to be enough people wanting that to happen strongly enough to vote for politicians that can make it happen. Most of us are not powerful enough to make the changes that are needed, so we must make the powerful want to make those changes. To do that we must make as many as possible want to make the powerful want to make changes, and we can all make more people want that. Mass movement is thus the first requirement.

At the outset, one would think that it was in the clear interest of the majority of people to have a good, stable and sustainable earth, but there are many reasons why it nevertheless is not an interest strong enough to make them demand it from their politicians. Other short-term interests can feel more important, or one does not care, or one finds it hopeless, and there can be many other reasons why people do not demand a sustainable world from their politicians.

So what does create mass movement towards a common goal? We have seen mass movements with major consequences several times in history, with some common features. For instance, there has been the rise of religions and the rise of ideologies like Nazism or communism.

Here are some important common features. First of all, they have in common a quite simple vision about how to solve the problems of the world and make a good world instead, and for many it is motivating to work for such a meaningful cause. Secondly, it is in general very motivating to experience success in terms of partial goals being reached and feeling that you are succeeding and on your way to the goal, while it is demotivating if the goals fail and it seems hopeless to reach the final goal. Thirdly, it helps a lot to have charismatic leaders or famous people who are able to inspire and get our attention among all the other things that ask for it.

When it comes to the UN's sustainable development goals, they have not created a mass movement of people actively seeking to make them to come true. While they have a quite simple vision with some goal and sub-goals, I think it would be good with an even simpler vision and a way of organizing the goals so that one could more easily experience motivation and success by reaching the sub-goals.

One way that could have been done: First, we need a simple but grand vision, for example that we shall be the generation that finds and actualizes the best way to the best world – a stable world where everyone can be safe and free to make a happy life for themselves. Second, in order to actualize this world, the best minds of the UN should make a big list of specific goals sorted into levels of importance, where it is easy to make progress and easy to be updated on progress.

For example, there could be 20 levels from the most important to the least important goals. The top level should not have too many goals, but there can be hundreds of goals on the lower levels. Each goal should have a number, and each goal should have an indication (in percent) of how close it is to being reached. Every time a goal is reached it is transferred to a list of achieved goals, and every year there is an update on how many goals have been reached on each level and how many goals have come closer to achievement on each level.

The lower goals being reached must be such that achieving them is the way to finally reach the goals on the top level. These goals should be something like the UN sustainability goals, but maybe formulated in simple and visionary ways, like "free education for all", and maybe even some really hairy goals. The structure could be like reaching new levels in computer games.[310]

Every goal reached is a success, and every goal closer to being reached is a success. There should be many goals at the lowest level and goals reached should be counted no matter who or what it is that makes them reached. It needs to be easy to see that there is progress, and it is no problem that new goals are being added to the list. Those who want to read about details on the goals can click further and read more on obstacles, etc. As long as there is overall progress it will be motivating to be part of the project. People are motivated by success and want to be on a winning team, not a sinking ship.

In order for reaching goals to be motivating, people need to care about it in the first place. Most people already do care basically about living in a better world, but they need to care about acting in order to make it come true. When they have started acting, reaching goals is motivating, but they need to be engaged in the first place and reminded constantly.

In order to make that happen, it is extremely helpful if there are charismatic leaders and also celebrities engaged in events or sub-projects that can catch the attention of the media, so recruiting such people should be an early and important goal.[311] But the most important recruiting is to recruit ordinary people and organizations/businesses to become participants in the project and let them recruit others through their networks.

I will say more about such recruiting in the following, and now I move from the first part on overarching principles for succeeding to the second part with more concrete suggestions on how to proceed first in recruiting masses and then for getting powerful people, companies and states acting in the necessary ways.

First then, about engaging people and others to participate in the project, which I will now refer to as the best way to the best world project (for short, the BWBW project), but in content it is roughly the UN sustainable development goals. Engaging more and more people, celebrities, companies, organizations, politicians, political parties and countries as participants in the project are important goals on the list to be achieved.

---

310 Maybe even a game could be played with income going to the project, meaning that people could contribute to the project by playing computer games, and that completing game levels there would contribute to reaching levels in the project in real life.

311 World War Zero is an example of such an initiative.

Ordinary people could register as supporters and get an identity as BWBW supporter number x. Being a supporter just means that you support the BWBW project, but you can also accept to participate in various ways: receiving important information about the project, receiving requests for spreading information about something, boycotting a company, supporting a specific cause with money, etc. Famous people could register to be BWBW ambassador, agreeing to create publicity about specific causes or events/campaigns. Registering and thus explicitly being on the same team is motivating.[312]

Companies and organizations could register as BWBW partners, informing the BWBW project about relevant work they do or contributing to reaching certain goals. For example, a company could say that they can take care of goal 9 at level 4, or suggest new goals added to the list that they can take care of. Political parties could state that they will support the BWBW initiative in their political programs, and politicians could register as such and receive requests for working with specific political topics. If a political party supporting the BWBW project sits in power in a government, it could announce the whole country as supporting the BWBW project. If enough people join as BWBW supporters, one could have polls to show how many want this or that regardless of which party is in majority.

One of the most important things that people, companies and countries can do, whether or not they are participants in the BWBW project, is to live their lives now in the way that it has to be done in a good and sustainable world. Very many people and companies and countries do not act sustainably, with the excuse that my little contribution does not help in the big picture or that other people, companies and countries are not doing what they should be doing. But if you realize that everyone should already, today, live sustainable lives, those excuses are worthless and those rich people, companies and countries that do not have a sustainable conduct are the problem and have no excuse for not doing what everyone should be doing.[313] If your children ask what you did when the world was collapsing, the best answer is that you lived sustainably and voted for the politicians most concerned with sustainability.

I now move on to considering possible ways of making people, companies and countries act sustainably. There is a way to pressure people, companies and countries to have a sustainable conduct which also rewards them for doing it, and that is to use certification or labels, which could come in degrees

312 Harari makes the point that national and religious/ideological groups have usually had great success with identity markers that symbolizes for people who do not know each other that they can trust and should help each other fighting for the same cause (Harari, 2017, p. 167).
313 This point was made by Maria Berg Reinertsen in Reinertsen (2019).

also. People, companies and countries could get BWBW labels – class 1, 2, 3, etc. Companies and countries would have to be certified by a certifying institution, while persons would have to self-report and certify themselves. For example, it could simply be a "I live sustainably" graphic or picture frame to put on Facebook or something to write in your CV. A common standard would then have to be defined, such as eating less than *x* kilos of meat per year, flying less than *x* times per year, having an electric car or no car, etc.

I realize that people can have good excuses for not being able to fulfill such criteria and that it can be stigmatizing, etc. Nevertheless, the bad consequences would pale in comparison if this can make many people live much more sustainably. People and companies who do get the labels will also get benefits from them.

Often countries and companies will claim a good reason not to accept acting sustainably by arguing that they will lose something important in competition or in comparison with others if they do. A way of solving this problem is to make "if/then" agreements. Let us say that an airline company considered signing an agreement saying that airline tickets should be *x* percent more expensive as an environmental tax. All airline companies could then refuse to sign with the argument that people will instead fly with the companies that do not sign the agreement. This excuse would not be valid if it were an "if/then" agreement where the airline company is only asked to add an environmental tax if all the other airlines do the same. The pressure will then be much greater on those who do not want to sign, and making everybody (or almost everybody) sign will be easier. The same kind of reasoning would apply to international agreements on environment, weapons, etc. "If/then" agreements can also be used to make agreements between political parties in order to stop then from using the excuse that the voters will run to other parties if we put more tax on gas, etc.

It should be possible to make good international agreements by signing if-then-agreements in several rounds. Countries could start by signing that they will agree to this and this if those and those countries also sign. One could have A- and B- and C-members etc. depending on how much the countries agree too. This is how the European Union developed, and some are still B-members through the EEA-agreement. A challenge is that countries sign without following up in practice, so there should also be an agreement on accepting international control and paying into a deposit account or other ways of making sure that breaking the agreement is punished. It should be possible to put much pressure on countries not wanting to sign agreements with control mechanisms when it is obviously beneficial for all.

These ways of pressure and reward and of using "if/then" agreements could be used in dealing with two very important and specific problems: making international agreements and raising money. I will say a little about both now. Very many of the goals on the way to a good and sustainable world need money to be achieved, so raising money is an important and prioritized goal, and it is possible to use both "if/then" agreements and combinations of pressure and reward to raise the money.

There are many good suggestions on kinds of agreements that would raise much money: Removing tax havens by agreeing on an international tax level for companies and making them tax in the countries where they make money (an agreement on this was made in 2021), a progressive tax on wealth and other suggestions made by Thomas Piketty; the Tobin tax as argued by Thomas Pogge; tax on inheritance, environmental taxes, UN tax, etc. The BWBW project should have as a goal to use "if/then" agreements to make international agreements on these matters and to put pressure on those countries that block success by not signing. (I shall respond to objections against taxes below.)

Most important is probably that countries themselves put a high tax on luxury, big fortunes, and high salaries, but it would also be good with international agreements making all countries do this. A good way of making the majority of people want their politicians to make such agreements is if the taxes benefit the poorer majority. One example is how environmental taxes in Canada benefit environmentally friendly behavior directly.[314] Taxing wealth this way has the highest chance of being supported by voters when it directly benefits the majority, including the rich. Using the money for free education and health and other welfare arrangements is the most obvious strategy. Political parties should attract voters by offering them rights, making the state function as an insurance for all securing basic rights.

In 2019, Oxfam said that the richest 26 persons own more than the poorest 50 % of the world. The poorest is a much larger group than the richest, and thus it would clearly be in the interest of the poorest that the richest should have higher taxes to help the poor. For example, one could make a suggestion that when a person or company own more than some very high sum, they should give half of that sum or half of their fortune to the UN when they die. The great majority of the world should be strongly in favor of such suggestions, especially if the money is channeled as directly as possible, such as in the Canada example above. Even if it is difficult in practice to force through, maybe there

314 The Canada Climate Action Incentive makes pollution more expensive, but you also get money refunded on your tax. For example, gas gets more expensive, but you can also get money, which means that those who use much gas must pay more while those who use little gas earn money.

could be strong campaigns online encouraging named persons or companies publically to agree to such suggestions?

While waiting for more international tax agreements, funding would be possible with voluntary tax labels in the meantime, which could be there regardless of what the law demands. Very rich people and companies paying extra taxes could be labeled supertax-payers and world benefactors, and it could be made an annual list of the greatest contributors in the world.

Making international agreements is important when it comes to environmental questions and tax questions. But in order to have a stable world it is also important with international agreements to prevent global existential risks. For example, there is intense research on developing superintelligence. As seen in the previous excursus, there is also a great danger connected to such research. Those who try to develop superintelligence are putting the lives of all of us at risk without our consent, which gives us a good argument to demand something in return. It would be in the interest of every country to put an agreement into law saying that countries or companies developing superintelligence should be allowed to make a huge amount of money (say a trillion dollars, or some better- ustified amount), but the rest of the proceeds should be given to the United Nations or a project distributing the income in a fair and good way.[315]

Global existential risks can often occur because processes get their own momentum, and what seems like individual rational choices for everyone results in bad and irrational consequences for all. For example, it can be individually rational to have more arms than others or do less to save climate than others, but the result can be an arms race and a climate crisis. Something similar happened in World War I, which was a war nobody wanted to fight, yet it happened because of various mechanisms and the momentum of the process. Such risks are much easier to prevent before they happen, and the risks are bigger now than ever with the potential of new technology. There is a great need to establish some global political rules and agreements before things get out of hand, and to secure critical thinking and knowledge in schools and popular culture.

One could think that there is no hope for good international agreements on global existential risks when we have not been able to remove nuclear weapons, which would obviously have been good for all to get rid of. On the other hand, there has not been a nuclear war, and presumably the success is the NATO recipe where many countries go together and agree that an attack on one member is an attack on all (NATO article 5). I do not think that it is realistic to get rid of nuclear or autonomous weapons, so the most realistic seems to be to have as large inter-

315 This suggestion is made by Nick Bostrom, see Bostrom (2016, p. 313).

national agreement as possible of countries with these weapons agreeing that an attack on one is an attack on all.

The final overarching goal of all the sustainable development goals is to develop a robust and stable structure for peaceful coexistence facing any future risk, which can make us feel that the world is a safe place for us and the coming generations. I do not know exactly what such a structure looks like, but I think these are central elements: That parents know that their children need recognition and developing empathy trough experiencing unconditional love from their parents, and that the parents want to give this to their children; that one learns critical thinking in school and about the other points on this list; that all have basic needs covered to be safe and free in the world; and finally, that people understand well enough that these are important elements and want to actualize them by voting for the politicians who will make it happen.

To actualize these elements, it starts with people talking about it, and probably the work of teachers is the most important since they reach so many in such an important phase. There needs to be a basis on knowledge and values at the bottom. Then the majority can vote for politicians building strong institutions with highly qualified workers (police, judges, educators, those in charge of taxes and financing different rights, etc.), so that we can trust the system and experience that it works well. The system can provide security and freedom for all, which they will experience as good. This could become a structure stable enough to create a good life for all and facing any future risk.

As mentioned, it is possible to see the whole history of the universe as a gradual process of establishing more and more complex stable structures. In the end of that history, families, tribes and nations have managed to create good stable structures like houses, schools and hospitals. Now we need to establish some final international structures to make a stable and good earth for everyone. This is a grand story and a life project where everyone can participate regardless of whether they believe in a God or not. Future technology can give many goods for all if we have structures for managing it well. We need as many as possible to join this project, and it would be good with celebrations and rituals and anything that can give shared identity and support for the project.

There is extremely much more to say about specific goods and specific problems and what is the best means to reach the goals, and this is something that the best experts must work on revising all the time. This is only meant as a very small contribution to this discussion, so I will not go into more details on suggesting goals and means. In the remaining third part, I will answer objections to what has already been said.

The first objection is that this is unrealistic. Suggestions for making a better world are often called unrealistic, but it can be done in two quite different ways. Some will say that everybody has to live dramatically differently, and get the response that this is unrealistic, meaning that this is not likely to happen. They can then reply that it is not realistic to try to make win-win solutions or continue having economic growth, since that will make the earth collapse.

There are two very different ways of answering the question of what would be the best way to make the world as good as possible. One could give an answer describing what the ethically best solution would be – prioritizing the ones who need it the most, etc. – and the answer would be true, but it would presuppose that everybody acted accordingly.

Suggesting a solution that presupposes that everybody does the ethically right thing would be extremely unlikely to succeed. Many people are egoistic, and it is very common for people to fight hard against losing any benefits they may have (Kahneman, 2011, p. 305). The other way of answering the question is to ask what would be the best way to make the world as good as possible, assuming that many people will still behave egoistically. While I agree that it would be best that everybody did what was ethically best, I think that the only realistic scenario of improving the world that may actually be actualized is to make a win-win solution which makes the world better, but which includes that rich people get it better too. To my idealistic friends I will thus say that they are right and that their suggestions are ethically best, but since they are very unlikely to succeed, I suggest instead that we go for the win-win-strategy in order to actually achieve good goals.

Many will be quick to respond that it is impossible to create a sustainable world where everybody gets it better, but I think it is actually possible: For one thing, remember that better quality of life for all is not the same as increased consumption for all. I believe that a sustainable world is possible with increased quality of life for almost all (including economic growth), where everybody gets their basic needs covered and has equal basic possibilities for achieving their goals while the earth is still being sustained. Here are four important reasons why:

First, the basis and measure of welfare in a society is not the amount of money but the amount of efficient work hours available. Building infrastructure for the poor will release an enormous number of efficient work hours that can be used to create welfare.

Second, there is every reason to believe that AI will help to produce enormously more efficient work hours, allowing previously expensive goods to become much cheaper. Think of the reduced cost of online goods like movies, books, music, or games. There will also continue to be very reduced costs for robot-produced products and driverless vehicles, online education, automated health services, and many other kinds of services.

Third, the earth is clearly sustainable for fewer people living at the same time. When people get richer, they tend to have fewer children, and in rich countries couples have in average less than two children, which means that the population decreases. As a long-term goal, fewer people living on earth at the same time is a realistic goal compared to thinking that population will grow forever.

Fourth, as seen, this is physically possible with a sustainable "renewable" earth, where all energy is clean and gives clean air, clean water, and a stable climate; where all garbage is recirculated; where the consumption of food is sustainable; and where the economy contains a redistribution through tax from rich to poor which gives basic goods and opportunities to all. But it needs a mass movement to make it happen, as I have written about above.

To those who find this unrealistic, the best evidence of its realism is to point to the so-called Nordic Model, i.e. how the Nordic countries are politically and economically organized. The Nordic countries have the highest income per capita in the world, and is also on top when it comes to how evenly distributed the wealth is. They are on top in level of happiness and in employment rates. This is not due to high taxes (relevant taxes are lower than in the US) and it is not because of oil (Norway is the only Nordic country with oil) (Midttun et al., 2011). Everyone are secured housing, education, and enough money to live well, even if they cannot work. What explains the success?

The main explanation is the salary negotiation system (called the three-parts-cooperation), where most employers and employees are organized in big unions negotiating salaries and rights for all, and where the state is an active participant. This system makes sure that the gap between the highest and lowest paid jobs are not too big. While a medical doctor on average earns ten times more than a shop assistant in the US, in Norway it is three times more. This system of not paying well educated people too much allows businesses to afford employing highly qualified personnel, which is good for the whole economy (Eia, 2020a). There is no reason why a doctor should have to earn ten times as much as a shop assistant, and there are more than enough excellent people who want to become doctors in Norway, since the education is free for all.

Everybody pay taxes, and the rich pay most, but everybody (including the rich) also get very much for free or almost free: one-year maternity leave for each newborn child, financial support per child until they are 18, kindergarten, school, free education at the top universities, free health care, good roads and infrastructure, and very good financial support for those who are sick or unemployed. When the state allows companies to build apartments, one of the apartments must be an apartment the local authorities can give to poor families with income under a minimum limit, who are thus integrated in all sorts of communities and can use their money on other things than housing.

This system gives great independence and freedom to all citizens, since they have all these securities and opportunities regardless of family background or ability to work. It is like a shared insurance for all, where everybody pays according to ability, and everybody gets covered (and even a rich person can get ill or go bankrupt because of a pandemic or something else).

Many problems are solved when everybody gets the same, while the richest are taxed harder, but they also get the same benefits, e. g., a certain amount of money per child. The system is then to a large degree experienced as fair, since the rich get the same as the poor, even if the rich pay more, they also get much back. The rich can see that the system also benefits them, and the poor can see that this is a system where some getting rich also benefit the poor through taxation, and if people could realize that there is not a practically working better alternative, it would give great stability to the system.

Corruption is much more difficult when everybody gets the same, given directly to each individual. When the state finances so many kindergartens, institutions of education, health businesses and much more, they keep the economy going and create a lot of jobs. As mentioned, it is the work hours that determines the wealth of the whole society. This system creates a much bigger pie for everyone to share.

The system requires trust, since people do not want to pay taxes if they believe that the system is exploited by cheaters or corrupt workers. One could then believe that the system only works in the Nordic counties because the inhabitants are very similar to each other. But in fact, there has been a great increase of immigration in the last decades, without the level of trust going down. The reason is that people trust the institutions instead of other people. The institutions make sure that those who can work, must work, and that cheaters are caught (Eia, 2020b).

Strong institutions with highly qualified employees is what it takes to make this system work, which again requires that people vote for politicians who will have this system and ensure that there are such strong institutions. In the Nordic countries, the politicians on both the left and right side want this system.[316] And

316 The system is much easier to maintain when there are several political parties (like in Norway), and not just two. Then the political parties must make compromises in order to form a government, and it is always a moderate right government or a moderate left government who end up in power. This gives stability over time, as opposed to a two-party system, where more extreme views can end up leading the parties. The number of political parties the system includes is thus an important condition for stability in society.

the whole world could be like this: everybody could be rich, or at least sure of having a place to live, food, education, health, and opportunities in a system gaining everyone. If the majority of voters understood the value of this alternative system, they would vote for it and make it happen.

Norway was a poor country after WWII but managed to secure basic welfare rights for all in about 25 years. Different countries are different, but if good main structures are in place, details will be fixed locally. Historical experience shows that this works. People must be motivated by egoistic reasons to work, and thus it has to be less money to those who do not work, even if some are not to blame for not being able to work. There are many reasons why we need people to go to work and meet and cooperate with other people. The best way of avoiding corruption is by giving all individuals rights to the same things. Even if universities and hospitals in Norway are run by the state, they have excellent quality and efficiency since they are semi-structured as businesses with financial support partly depending on what they deliver, thus motivating them to be excellent in their niches. Private institutions have similar agreements with the state, meaning that people can study or get medical help in private institutions for very low prices.

For countries with few institutions working well, they must be built gradually with the most important first (security/police, judges, etc.). I realize that in many countries, there can be tough challenges to solve before any of the processes I have described here are realistic. If lack of democracy, conflict, corruption or mistrust has dominated for generations, it is very difficult to break that circle. In their analysis of which states succeed and which states fail, economists Acemogu and Robinson argue that the key success factor is whether there are institutions that serve everyone and not just an elite (Acemogu and Robinson, 2012). They argue that establishing such institutions is difficult to enforce by other countries and must instead be wanted and established by the country's inhabitants. It could help such internal motivation to know that there is a realistic political model that can give welfare and freedom to all that one can have as a goal. If both internal and external pressure goes in the same direction, it increases the chance of success.

These are of course extremely brief replies to objections. The BWBW project should have a commonly known good and accessible webpage dealing with objections and presenting research and uncertainties. Experts probably have better solutions than what I have presented here. To conclude the objection that it is unrealistic: It is clearly possible to reach the goal, and it can be done through solutions that are mainly win-win, but there are also several cases where the majority needs to pressure the rich minority to make it happen. The world has already seen great improvements over the decades (Harari, 2017, chapter 1), and that process needs to continue with more fervor. What is takes is mass movement, and mass movement is quite possible (even if still difficult), made much easier through internet, social media and devices such as smartphones. Even if you think BWBW is unrealistic, you are ethically obliged not to give up hope when so many people are suffering in our world: you have to opt for the best alternative and do something.

The second objection comes from those who are opposed to adding more taxes to rich people and companies, or adding a tax on wealth, inheritance, etc. To caricature the argument for simplicity: Everybody is free to create their own life, they might argue. It is not fair that hard-working rich people should pay for the welfare of lazy, poor people.

I have several responses to this. First of all, rich people also reap great benefits from a stable world with clean water, clean air, a stable climate, fewer forest fires, better roads, educated workers, fewer refugees, customers with more money, fewer hurricanes, a world society able to deal efficiently with deadly viruses or terrorists, fewer attacks from other countries or fractions, etc. – and developing these benefits further requires that some of the money accumulated at the top gets redistributed toward the bottom, which is the only way of making stable and growing economy and welfare.

I think this is the most important response, although I offer some other ethical replies below: the rich also benefits from more equality. The rich also get all the welfare services. In Norway, the best universities are free for all. Instead of paying 50 – 70 000 dollars per year (as you would on Harvard), you get the education for free when you are young, and pay tax when you earn money later in life. The same logic applies to global agreements to everybody's benefit. Rich countries have much to benefit from stable and environmentally friendly countries and highly educated majorities worldwide. A stable and secure world is to everybody's benefit, especially as things are moving faster and the dangers of instability greater.

You can argue that it is not fair or just to put a heavier taxation on the rich, but as we have seen, there is not one correct definition of justice or fairness. There are many different concepts of justice and fairness, and in this context, "fairness" defined as giving all equal opportunities is more important than ' fairness" defined as people getting to keep all the money they have worked for (at jobs which also depend on roads and education and institutions, etc.).

Luck and bad luck have a huge influence on people's lives (as seen in the discussion of luck above and shown through various experiments (Kahneman, 2011, pp. 204 – 208)) and this is in most cases more important for their wealth than whether they do much or little hard work. A random event at the genetic level before you were born could have made your IQ that much lower that you would be unable to compete for the best paid jobs, no matter how hard you work. A world policy of evening out the effects of good luck and bad luck is a good kind of fairness, which also takes into consideration millions of future earth inhabitants.

All people are not free to succeed on the same conditions. It matters greatly under what circumstances you are born. Even if you were to criticize some parents for having children in poor circumstances, the children being born cannot be blamed for the decision of their parents. We should make sure that all children born have basic needs covered and equal opportunities for a good life. Everybody is dependent on the society around them in order to be able to make money, and that societal structure should then be fair for all.

I think it is clear that we have enough world experience to have learned that a good society needs to secure basic safety and freedom for all citizens. Look at how quickly Europe rose after WWII with that model. In the US, too many people struggle to survive, and in China, too many people fear the state. Research has shown how people are unable to act rationally when their minds are occupied only with surviving (Mani, Mullainathan, Shafir, and Zhao, 2013; Mullainathan and Shafir, 2013). Securing right for all through taxing the rich is the best solution for all, the rich included, and the masses must make it happen.

"Why is that our responsibility?", some may ask. "Should it not be the responsibility of their parents or their country or themselves?" In the chapter on ethics, I argued that the content of the term "should" is that everybody should contribute to making the world a better place, and I argued that how strongly you should contribute depended on the amount of resources you have. It follows that you should help the poor, and the richer you are, the more you should help. It just follows from what the terms "should" and "responsibility" mean. It is that simple: We should increase taxes on the rich because it is the best way to the best world.

The third objection is that it seems I am pushing western ethics on the whole world. Speaking of securing basic needs and letting all be free to seek happiness is a clear reference to John Rawls' ideal society (Rawls, 1971), while people in other societies may not think of this as an ideal. A democratic welfare state may not be the best way to the best world; maybe China is a lot more efficient in finding the best way to the best world, for example by forcing people to act well with their social credit system.

Here are some responses to that objection. I have focused on the UN goals since this is a project that all countries are cooperating on. But one could reply that also the UN is dominated by western ethics. Shannon Vallor has argued that there are great similarities between western and eastern virtue ethics, and thus a great potential for common ethics in our age. She argues that for thousands of years there has been a virtue ethical approach saying that we do not know in detail what the good life is, but that the good life require that we cultivate good habits in relations to each other (Vallor, 2016). Nevertheless, there is still much disagreement on what a good life together involves.

We all disagree on ethics and we should all be self-critical and accept the undeniable truth that we may be wrong in our ethical thinking. Maybe China is able to create a better world more efficiently than the way I am describing here. But while I may be wrong, so may a Chinese person or anyone else be wrong. From the possibility of being wrong, it follows that we should try to give people quite much freedom in a good society in order to avoid that we force them to do something that is wrong. Uncertainty about ethics thus actually favors a certain kind of ethics, namely one securing much freedom and gradual exploration. The need to secure democracy and freedom are consequences of the insight that we may be wrong and others are right. When states want sovereignty, they should allow some sovereignty to their inhabitants also.

Even if uncertainty justifies freedom, we can still ask: how much freedom? The chapter on free will ended with a discussion of Berlin and Taylor on positive and negative freedom, and I said I should return to the question of how the state should secure freedom for its citizens. This is the place to again pick up that discussion. It will be brief, but I include it since it allows me to connect things I have said at different places on free will, ethics and the meaning of life.

Isaiah Berlin argues that the state must necessarily constrain people in order to secure freedom for all, otherwise one person can use her freedom to take away the freedom of others. But it is a difficult question to answer how much the state may constrain its citizens. What should be recognized, Berlin argues, is that people have different and incommensurable goals, so the state should not force one ideology or one solution on all, but rather give people freedom to choose their goals, which is a freedom that makes them human (Berlin, 1969).

What then of Taylor's point when commenting on Berlin that we may be wrong about what we really want (C. Taylor, 1979, pp. 176, 187–193)? People can be badly mistaken about what is a good life for them. It is a common experience that our long-term goals can conflict with short-term goals and that you can understand years later why it was good that your parents, your school or the state made you do and learn certain things. Should the state just let people be free to make their mistakes, or to what degree can the state try to make people have specific goals? This is the classical debate of how active the state should be, with answers ranging from extreme liberalism to authoritarianism.

In his article "The politics of recognition", Taylor discusses whether the state should be neutral on what is a good life and just try to secure for everyone a possibility of discovering it for themselves (C. Taylor, 1992, p. 57). He argues that a state can also have collective goals, in the sense of securing survival for different cultures in society, as long as diversity is respected and fundamental rights secured (C. Taylor, 1992, p. 58). Does this mean that Taylor wants the state to secure the goals of all subcultures in a society? No, he argues that not all cultures are equal, but that when cultures have survived for a long time, this gives us reason to believe that there is something objectively valuable about them. With that assumption as a point of departure, we can explore what is really good and important. Taylor is not a subjectivist, for he believes that some things really are better and important than others, but since we may all be wrong, we need to explore what is good and important (C. Taylor, 1992, pp. 62, 72).

Both Berlin and Taylor may seem overly optimistic here. It seems that people in general are not very interested in exploring what is good for all, but that various subcultures have their answers, and these answers can be quite irrational and bad, even in cultures that have survived for a long time. It could seem legitimate for the state to try to influence its citizens to act well, punishing wrong and rewarding good. It may very well be the case that it is best for us all that the state forces us to get used to getting up in the morning and doing some work and meeting other people since one's personality and preferences are shaped by habits. Life is a lot of work and we need to acquire some habits and skills in order to live well. History has shown several times how normal people can be formed into people committing atrocities (Glover, 1999, part 4), which suggests that it is important to make sure that they are formed into good citizens instead. People are being influenced anyway, so why should not the state try to give good influences instead of remaining silent, which in many cases is the same as letting bad influences happen?

The obvious reply is that we know all too well how many states have tried to force their vision on their inhabitants, and that these visions have been bad goals being realized via bad means, such as with Nazism. What is the right balance between the state trying to influence its citizens very much or not trying at all? I suggest the following answer: States should try to actualize the best way to the best world because all agents (states, parents, individuals) should do that. The reason states were created in the first place, and the goal of democratic elections, is that the elected representatives should actualize the best way to the best world for the inhabitants of that state. This justifies trying to influence people into acting good, but it needs to be balanced by the insight that states can be wrong about what is the best way to the best world. Every individual is the actual conscious core self who is going to valuate something and experience its own life, and it is highly likely that the state or others will not know what every individual would actually valuate the most, and thus what is the best way to the best world. Therefore, while we try to influence others in a good direction, this should be balanced with an attempt to secure for individuals the opportunity to themselves explore what they valuate the most, so that we can all try to find out what the best world is and what the meaning of life is for each one of us.

It could seem that these two insights (that the state should either try to influence our goals or just give us freedom to find our own goals) are in opposite directions, but to a large degree they can pull in the same direction. That happens when the state influences us to think that we should all respect each other and give each other freedom to explore what is good. It is not very realistic that people will do that when the state does nothing, but the state can actively influence us to respect others and give freedom to others. This can happen through deciding what children should learn in school (and that they have to attend school), punishing abuse of others, giving money to organizations working for respect, tolerance and freedom, etc.

This answer does not give a detailed answer to how active the state should be. I have only given a principled and abstract answer, and both liberalists and socialists could argue that their concrete answers are the best way to the best world. But I have suggested some basic principles that both liberals and socialists can use to support their views. These principles do not support any of the extreme views, but rather seem best to support social-democratic welfare states.

The fourth objection is that while much of what has been said sounds rational, there are too many people who are too irrational or even selfish to the border of being evil, such that it will not succeed. My main response to this is to point to the need for strong institutions and broad education, which are the UN sustainable development goals 4 and 16. Institutions can institutionalize good practices even when individuals are irrational, and serve to control political leaders not acting as they should. An important part of the institutions are control institutions, including systems for controlling the controllers. Securing such institutions worldwide should be a goal for international agreements and prioritized in funding.

Broad education is extremely important both for knowing what needs to be known in order to vote for the best politics and act for the best world, but also for not being misled by populism, propaganda, conspiracy theories etc. The majority of people in the world needs to have basic knowledge if we want to achieve the hairy goal of robust and eternal world peace. Even if many adults are probably lost to irrationality, the coming generations are not. High salaries for attracting high quality schoolteachers should be very high on the agenda of every country.

Highly paid teachers is not something that all countries can afford quickly. Free online education with quality certified by the UN would be a very good help. It would also be of very good help if different companies or institutions could either offer people benefits if they take the online courses or demand that people take the courses in order to be employed or get other benefits. It is in the interest of most institutions and companies that people are well educated instead of falling for populist leaders and conspiracy theories.

In order to fight extremism and false information, it would be good with an official website confirming what the majority of researchers agree on. Individual researchers could confirm on their own webpages that they agree. I think one of the most important things that everyone should learn is the self-critical insight that if the majority of experts agree on something that I disagree with, I should conclude that I am probably wrong.[317]

To sum up the whole excursus in this section, I think it is possible to create a larger mass movement pushing politicians to create a stable world. I appeal to the UN and the UNSDSN to lead that work since they have the global legitimacy needed to make it work. The dream is a world where children are born knowing that their lives will be economically secure and that they will have possibilities for them to explore when it comes to how they want to spend their lives, so that they can live without fearing the future. They should be free from threat and free to form their own life. Their life will be in a world without risk of war on our beautiful planet with clean water, clean air and a harmonious nature. You can be the generation that managed to do what the previous generations were unable to: make the ultimate dream come true!

317 Another idea which would have been fun to try, was for states to raise anti-extremism funds working this way: if some extremists launch a rally, demonstration, campaign etc. e.g., against Jews or Muslims etc., then the state anti-extremism fund would reply by giving money to the opposite goal. People from the fund could come to rallies or advertise in newspapers things like: for every Quran you burn, we will send 1000 free Qurans to people by mail; for every meter you march for Nazism, we will fund Jewish organizations with this amount of money; for every shop, mosque, synagogue etc. you attack, for every terror event, the opposite purpose will receive great funding. Extremist activities would then have the opposite effect of what they are trying to achieve and would appear to the public as failures.

16 Epilog: Why Is There Anything at All?

A suitable epilog to this book should end by saying some words about the way up and the way down compared to the level where the topics of this book have been found. By the way up, I mean to more complex levels of ontology, like social ontology. Hopefully, categories and theoretical frameworks developed in this book could be used to develop theories of how to understand the ontological status of concepts like love, evil, shame, guilt, hope, recognition, etc.

By the way down, I mean the ultimate "why" questions. To answer a question of "Why X?" is to let the answer be integrated into a larger theoretical framework. When you reach the fundamental structures of the theoretical framework, that framework cannot explain them (unless they are self-evident by being contradictory to deny, something which is very rarely the case). The choice of fundamental structures of the theoretical framework is justified by the coherence it creates in the rest of the framework (Puntel, 2008, pp. 347–349).

This book has tried to show some of the coherence that can be achieved by selecting the fundamental structures I presented in Chapter 3. Now let us try to go one step further and ask: Why are these the fundamental constituents of the world? A first answer could be that maybe they are not the deepest structures of the world, but the deepest structures that our mind can understand. But then why are these the deepest structures that our mind can understand? An answer to that could be that our mind has evolved as described in this book.

If these answers are correct, it would lead us to an explanatory circle: these are the fundamental constituents of our theory because these are the most fundamental entities we can understand, and these are the most fundamental entities we can understand because of the fundamental constituents of the world.

Have we then reached the bottom of explanatory potential, or can we go deeper? Why exactly is this explanatory circle the bottom, and not another circle? The answer seems again to be that this is the most fundamental explanatory circle our minds can understand. And since this is again because of the fundamental constituents of the world, we have reached the same explanatory circle once more.

If you try to go deeper, you may well end up in the same explanatory circle. We may get stuck now at this bottom because of the structure of our minds. However, this does not exclude the possibility that our minds can discover something smarter in the future, and this should be reason enough to continue to explore the world to see how much it is possible for us to understand.

https://doi.org/10.1515/9783110770957-018

In the following, I will try to stretch my thoughts and speculate on the question of what could be the deepest explanation of everything. Before suggesting an answer, I will offer some reflections on what such an answer would look like.

Because of how our understanding works, it is to be expected that the most fundamental explanation we can find is just a brute fact or a loose end without further explanation. But it would be nice if it were possible to find a better and ultimate explanation that did not just feel like a loose end and a brute fact. How could such a fundamental explanation be?

A fundamental answer to where the world came from or a deepest explanation of everything would also be an explanation of causation and laws of nature, so it is to be expected that it is not the usual kind of explanation where we refer to causes or laws of nature, since these are included in the things we want to explain. It would be nice if it was a kind of explanation we are used to, but it should not be a surprise if it is different from all other explanations. When something already exists, we want explanations to be simple and like the others, but explaining why anything exists at all, we have no reason to expect a simple answer similar to others, although this would be convenient.

In order to avoid a loose end which is just a brute fact, that final loose end would need to be tied to itself, like a kind of loop. As Nicholas Rescher says, an ultimate explanation has to be circular and self-explanatory if it is to be ultimate (Rescher, 2018, p. 188). While circular reasoning is often problematic, it is a virtue for an ultimate explanation since it would not be ultimate otherwise, but of course it should also be as coherent as possible. I will try to offer such an ultimate explanation below after some further reflections on the type of explanation we are looking for.

In discussing the ultimate explanation of the world, Rescher argues that it cannot be a normal fact since we are trying to explain all facts, and we cannot use a fact to explain all facts. He argues that the only possible alternative to fact is value, and that value is the ultimate explanation of the world: the world exists because it is for the best, and this is an axiological explanation (Rescher, 2018, p. 195). I think that this is not a good argument, since I think of values as being of the same kind as descriptive facts. However, Rescher also says that in order to explain the contingent world, we need to move on to the principles of the possible (Rescher, 2018, p. 184), which I think is a good point.

Previously in this book, I defined something as being possible if it is consistent with a set of presuppositions. This implies that possibility is rooted in something actual: we need presuppositions which something can then be consistent with (and thus possible) or inconsistent with (and thus impossible). But I also said that when something exists at all it had to be possible in the wide fundamental sense of being not-impossible. Fundamental possibility means just not-impossible: not prevented by anything functioning as presuppositions.

Try now to think that which it is impossible for us to think (because of how understanding works), namely to try to remove absolutely everything in order to think of a hypothetical state of absolutely nothing. Try to imagine this and try also to think another thing which is impossible to think: going back in time back to a point where absolutely nothing existed. I know that this cannot be coherently thought, but we are trying to push our thoughts to the limit here. We are trying from inside our minds to expand our understanding of the world outside of mind.

Usually we say that it is impossible for something to come from absolutely nothing. But if impossible means inconsistent with presuppositions, is something at the point of time t_1 inconsistent with absolutely nothing at a previous time, t_0? You could answer yes, since absolutely nothing at t_0 is itself inconsistent, or you could answer no since there is nothing at t_0 preventing something at t_1 from being itself consistent. The argument would then be that something is only impossible if it is made impossible by something, and something at t_1 is not made impossible by something inconsistent at t_0.

Let us explore the idea of there being absolutely nothing at t_0 and then something existing at t_1. How could the first thing or event come into being? It would have to be uncaused, which is not clearly inconsistent to believe. What kind of thing or event or structure could begin to exist uncaused? Here is a suggestion: the power of beginning to exist uncaused. This is of course circular and quite hard to believe, since it would have to be a unique power with the ability of starting to exist as a power in itself, and it would be the power to actualize anything else if it is also to be the explanation of anything else. The whole history of the cosmos would then be the unfolding of the potential of this fundamental power. Even if it sounds incredible, we have seen in this book many reasons to believe that there must be such a power.

Some will think that it is possible that at some point of time there was nothing and then something began to exist, while others will say that it is impossible; for that to happen it must have been a fundamental space of possibility where something could begin to exist and speaking of time before the first thing existed does not make sense. Here is an interesting thought: if fundamental possibility just means non-impossibility and time just begins to exist when something exists, it seems to boil down to the same thing either to say that something has always existed or that something began to exist out of nothing. Either the first structure began to exist out of nothing or it has always existed, and in any case it seems that it exists because it is the power of beginning to exist uncaused, and that "the power to begin to exist uncaused" or "the power to exist uncaused" means the same, since there is no time before the first event.

Does the power of beginning to exist uncaused presuppose a world where that was possible? The idea here is to try to imagine a world with no other structures determining what is possible and not possible, and the suggestion is that as long as the world is not one where the power of beginning to exist uncaused is impossible, the power of beginning to exist uncaused is thus possible in that deep sense. This power was then what the whole world consisted of and had its own possibility as part of.

Why has there then always been a world where such a power could begin to exist instead of there being a world where such a power could not begin to exist? Is that the ultimate brute fact? An answer could be the following: There would be no reason why there should be a world where nothing could begin to exist – there is only a reason why there should be a world where something could begin to exist: That reason is a reason which that world contains within itself, namely in the power of beginning to exist uncaused.

This means that I offer as a fundamental explanation of the world the necessarily circular hypothesis that the fundamental explanation of *everything* – where the world comes from – is the power of beginning to exist. It is the only fundamentally possible event when nothing else exists. It is the same actualizing power we have met several times as the power of actualization, the power of existence, and the power of motion. One could think of it as an alternative to God, or think of it as God, there is not much of a difference although the important difference is whether this power consciously intended the existence of our universe. This means that I suggest the following answer to the ultimate question of why the fundamental structure and thus why anything exists: Because it could.

List of References

Acemoglu, Daron and Robinson, James (2012). *Why Nations Fail: The Origins of Power, Prosperity, and Poverty.* New York: Crown Business.
Adams, Robert Merrihew (1981). Actualism and Thisness. *Synthese* 49(1), 3–41.
Albert, David Z. (1992). *Quantum Mechanics and Experience.* Cambridge, MA: Harvard University Press.
Albert, David Z. (2000). *Time and Chance.* Cambridge, MA: Harvard University Press.
Alvarez, Maria (2018). Reasons for Action, Acting for Reasons, and Rationality. *Synthese* 195(8), 3293–3310. DOI: 10.1007/s11229–015–1005–9, visited on 27 July 2021.
Armstrong, David M. (1981). *The Nature of Mind, and Other Essays.* Ithaca, NY: Cornell University Press.
Armstrong, David M. (1989). *A Combinatorial Theory of Possibility.* Cambridge: Cambridge University Press.
Armstrong, David M. (2004a). Going Through the Open Door Again: Counterfactual versus Singularist Theories of Causation. In L. A. Paul, E. J. Hall, and J. D. Collins (Eds.), *Causation and Counterfactuals* (pp. 445–454). Cambridge, MA: MIT Press.
Armstrong, David M. (2004b). *Truth and Truthmakers.* New York: Cambridge University Press.
Art of Spirit (Producer) (2014). TIME is an ILLUSION said EINSTEIN – (the space-time continuum). Retrieved from https://www.youtube.com/watch?v=VYZQxMowBsw, visited on 24 February 2019.
Ashby, Neil (2003). Relativity in the Global Positioning System. *Living Reviews in Relativity,* 6(1), 5–41. DOI: 10.12942/lrr-2003–1, visited on 2 December 2021.
Aspect, Alain, Dalibard, Jean, and Roger, Gérard (1982). Experimental Test of Bell's Inequalities Using Time-Varying Analyzers. *Physical Review Letters* 49(25), 1804–1807. DOI: 10.1103/PhysRevLett.49.1804, visited on 3 August 2019.
Aspect, Alain, Grangier, Philippe, and Roger, Gérard (1982). Experimental Realization of Einstein-Podolsky-Rosen-Bohm Gedankenexperiment: A New Violation of Bell's Inequalities. *Physical Review Letters* 49(2), 91–94. DOI: 10.1103/PhysRevLett.49.91, visited on 3 August 2019.
Awad, Edmond, Dsouza, Sohan, Kim, Richard, Schulz, Jonathan, Henrich, Joseph, Shariff, Azim, … Rahwan, Iyad (2018). The Moral Machine Experiment. *Nature* 563(7729), 59–64. DOI: 10.1038/s41586–018–0637–6, visited on 12 June 2019.
Ayer, A. J. (1952). *Language, Truth, and Logic.* New York: Dover.
Baars, Bernard J. and Gage, Nicole M. (2010). *Cognition, Brain, and Consciousness: Introduction to Cognitive Neuroscience* (2nd ed.). Burlington, MA: Academic Press/ Elsevier.
Barnes, Luke A. (2019). A Reasonable Little Question: A Formulation of the Fine-Tuning Argument. *Ergo: An Open Access Journal of Philosophy* 6, 1220–1257.
Barrow, John D. (2010). Simple Really: From Simplicity to Complexity – and Back again. In B. Bryson and J. Turney (Eds.), *Seeing Further: The Story of Science & the Royal Society* (pp. 361–384). London: Harper.
Barsalou, Lawrence W. (1999). Perceptual Symbol Systems. *Behavioral and Brain Sciences* 22, 577–609.
Barsalou, Lawrence W. (2008). Grounded Cognition. *Annual Review of Psychology* 59, 617–645.

https://doi.org/10.1515/9783110770957-019

Barsalou, Lawrence W. (2016). Can Cognition Be Reduced to Action? In A. K. Engel, K. J. Friston, and D. Kragic (Eds.), *The Pragmatic Turn: Toward Action-Oriented Views in Cognitive Science* (pp. 81–96). Boston, MA: MIT Press.

Bauer, Andrew James and Just, Marcel Adam (2015). Monitoring the Growth of the Neural Representations of New Animal Concept. *Human Brain Mapping* 36, 3213–3226.

Baumann, Peter (2018). What Will Be Best for Me? Big Decisions and the Problem of Inter-World Comparisons. *Dialectica* 72(2), 253–273. DOI: 10.1111/1746–8361.12219, visited on 11 January 2021.

Bayne, Tim (2018). On the Axiomatic Foundations of the Integrated Information Theory of Consciousness. *Neuroscience of Consciousness* 4(1), 1–8.

Bechara, Antoine, Damasio, Antonio R., Damasio, Hanna, and Anderson, Steven W. (1993). Insensitivity to Future Consequences Following Damage to Human Prefrontal Cortex. *Cognition* 50(1–3), 7–15.

Beebee, Helen, Hitchcock, Christopher, and Menzies, Peter Charles (Eds.) (2009). *The Oxford Handbook of Causation.* Oxford: Oxford University Press.

Bell, J. S. (2004). *Speakable and Unspeakable in Quantum Mechanics: Collected Papers on Quantum Philosophy* (Revised ed.). New York: Cambridge University Press.

Benacerraf, Paul (1983). Mathematical Truth. In P. Benacerraf and H. Putnam (Eds.), *Philosophy of mathematics* (pp. 403–420). Cambridge: Cambridge University Press.

Bennett, C. L., Larson, D., Weiland, J. L., Jarosik, N., Hinshaw, G., Odegard, N., … Wright, E. L. (2013). Nine-Year Wilkinson Microwave Anistropy Probe (WMAP) Observations: Final Maps and Results. *The Astrophysical Journal Supplement Series* 208(2), 20. DOI: 10.1088/0067–0049/208/2/20, visited on 5 February 2020.

Bennett, M. R. (2007). *Neuroscience and Philosophy: Brain, Mind, and Language.* New York: Columbia University Press.

Bentham, Jeremy (1988). *The Principles of Morals and Legislation.* Buffalo, NY: Prometheus.

Berg, Dann (2012). I Have a Magnet Implant in My Finger. Retrieved from http://gizmodo.com/5895555/i-have-a-magnet-implant-in-my-finger, visited on 22 August 2017.

Berlin, Isaiah (1969). *Four Essays on Liberty.* Oxford: Oxford University Press.

Bigelow, John, Ellis, Brian, and Lierse, Caroline (1992). The World as One of a Kind: Natural Necessity and Laws of Nature. *The British Journal for the Philosophy of Science* 43(3), 371–388.

Bigelow, John, Ellis, Brian, and Pargetter, Robert (1988). Forces. *Philosophy of Science* 55(4), 614–630.

Bigelow, John and Pargetter, Robert (1990). Metaphysics of Causation. *Erkenntnis* 33, 89–119.

Bird, Alexander (2007). *Nature's Metaphysics: Laws and Properties.* Oxford: Oxford University Press.

Bishop, Robert C. (2019). *The Physics of Emergence.* Bristol: Morgan & Claypool.

Blanke, Olaf, Ortigue, Stephanie, Landis, Theodor, and Seeck, Margitta (2002). Neuropsychology: Stimulating Illusory Own-Body Perceptions. *Nature* 419 (19 September), 269–270.

Bøhn, Einar Duenger (2019). *God and Abstract Objects.* Cambridge: Cambridge University Press.

Bok, Hilary (1998). *Freedom and Responsibility.* Princeton, NJ: Princeton University Press.

Bonnefon, Jean-Francois, Shariff, Azim, and Rahwan, Iyad (2016). The Social Dilemma of Autonomous Vehicles. *Science* 352(6293), 1573–1576.

Bostrom, Nick (2013). The End of Humanity: Nick Bostrom at TEDxOxford. [Online lecture]. Retrieved from https://www.youtube.com/watch?v=P0Nf3TcMiHo&t=787s, visited on 31 May 2019. .

Bostrom, Nick (2016). *Superintelligence: Paths, Dangers, Strategies*. Oxford: Oxford University Press.

Bowie, G. Lee (1979). The Similarity Approach to Counterfactuals: Some Problems. *NOÛS* 13(4), 477–498.

Bradley, Darren and Leitgeb, Hannes (2011). When Betting Odds and Credences Come Apart: More Worries for Dutch Book Arguments. In A. Eagle (Ed.), *Philosophy of Probability* (pp. 199–205). New York: Routledge.

Bregman, Rutger (2017). *Utopia for realister.* Oslo: Spartacus.

Briggs, Rachael (2015). Transformative Experience and Interpersonal Utility Comparisons. *Res Philosophica* 92(2), 189–216.

Brown, Harvey R. (2005). *Physical Relativity: Space-Time Structure from a Dynamical Perspective.* Oxford: Oxford University Press.

Budin, Itay and Szostak, Jack W. (2011). Physical Effects Underlying the Transition from Primitive to Modern Cell Membranes. *PNAS* 108(13), 5249–5254.

Callender, Craig (2011). *The Oxford Handbook of Philosophy of Time.* Oxford/New York: Oxford University Press.

Callender, Craig (2017). *What Makes Time Special?* Oxford: Oxford University Press.

Cappelen, Herman (2018). *Fixing Language.* New York: Oxford University Press.

Carnap, Rudolf (1950). *Logical Foundations of Probability.* Chicago: University of Chicago Press.

Carnap, Rudolf (1983). Empiricism, Semantics, and Ontology. In P. Benacerraf and H. Putnam (Eds.), *Philosophy of Mathematics* (pp. 241–257). Cambridge: Cambridge University Press.

Carroll, John W. (2008). Nailed to Hume's Cross. In T. Sider, J. Hawthorne, and D. W. Zimmerman (Eds.), *Contemporary Debates in Metaphysics* (pp. 67–81). Malden, MA: Blackwell.

Carroll, Sean (Producer). (2013). Particles, Fields and The Future of Physics. Retrieved from https://www.youtube.com/watch?v=gEKSpZPByD0, visited on 2 August 2016.

Carroll, Sean (2014). Why Probability in Quantum Mechanics Is Given by the Wave Function Squared. Retrieved from http://www.preposterousuniverse.com/blog/2014/07/24/why-probability-in-quantum-mechanics-is-given-by-the-wave-function-squared/, visited on 9 August 2019.

Carroll, Sean (2016). *The Big Picture: On the Origins of Life, Meaning, and the Universe Itself.* New York: Dutton.

Carruthers, Peter (2006). *The Architecture of the Mind: Massive Modularity and the Flexibility of Thought.* Oxford: Oxford University Press.

Castro, Jason B., Ramanathan, Arvind, and Chennubhotla, Chakra S. (2013). Categorical Dimensions of Human Odor Descriptor Space Revealed by Non-Negative Matrix Factorization. *PLOS ONE* 8(9), e73289. DOI: 10.1371/journal.pone.0073289, visited on 27 September 2021.

Cech, Thomas R. (2000). The Ribosome Is a Ribozyme. *Science* 289(5481), 878 – 879. DOI: 10.1126/science.289.5481.878, visited on 6 December 2019.
Cei, Angelo and French, Steven (2014). Getting Away from Governance: A Structuralist Approach to Laws and Symmetries. *Methode – Analytic Perspectives* 3(4), 25 – 48.
Chalmers, David (1995). Facing up to the Problem of Consciousness. *Journal of Consciousness Studies* 2(3), 200 – 219.
Chalmers, David (1996). *The Conscious Mind: In Search of a Fundamental Theory.* New York: Oxford University Press.
Chalmers, David (2011). A Computational Foundation for the Study of Cognition. *The Journal of Cognitive Science* 12, 323 – 357.
Chalmers, David (2012). *Constructing the World.* Oxford: Oxford University Press.
Chalmers, David (2016). Panpsychism and Panprotopsychism. In G. Brüntrup and L. Jaskolla (Eds.), *Panpsychism: Contemporary Perspectives* (pp. 19 – 47). Oxford: Oxford University Press.
Chang, Ruth (2002). *Making Comparisons Count.* New York: Routledge.
Childers, Timothy (2013). *Philosophy and Probability.* Oxford: Oxford University Press.
Churchland, Paul M. (2007). *Neurophilosophy at Work.* New York: Cambridge University Press.
Clarke, Randolph K. (2003). *Libertarian Accounts of Free Will.* New York: Oxford University Press.
Coleman, Sam (2012). Mental Chemistry: Combination for Panpsychists. *Dialectica* 66(1), 137 – 166. Retrieved from http://www.jstor.org/stable/42971283, visited on 8 September 2019.
Coleman, Sam (2015). Neuro-Cosmology. In P. Coates and S. Coleman (Eds.), *Phenomenal Qualities* (pp. 66 – 102). Oxford: Oxford Unviersity Press.
Coleman, Sam (2016). Panpsychism and Neutral Monism: How to Make Up One's Mind. In J. Brüntrup and L. Godehard (Eds.), *Panpsychism* (pp. 249 – 282). Oxford University Press.
Coleman, Sam (2018). Personhood, Consciousness, and God: How to Be a Proper Pantheist. *International Journal for Philosophy of Religion* 85. DOI: 10.1007/s11153 – 018 – 9689 – 7, visited on 23 January 2020.
Coleman, Sam (Unpublished). Unconscious Qualities as the Basis of Content. Retrieved from https://www.academia.edu/8748374/Unconscious_Qualities_as_the_Basis_of_Content, visited on 8 September 2019.
Coren, Stanley, Ward, Lawrence M., and Enns, James T. (2004). *Sensation and Perception* (6th ed.). Hoboken, NJ: Wiley.
Craig, William Lane (2001). *Time and Eternity: Exploring God's Relationship to Time.* Wheaton, IL: Crossway.
Craig, William Lane and Sinclair, James D. (2009). The Kalam Cosmological Argument. In William Lane Craig and J. P. Moreland (Eds.), *The Blackwell Companion to Natural Theology* (pp. 101 – 201). Malden, MA: Wiley-Blackwell.
Craig, William Lane and Smith, Quentin (2008). *Einstein, Relativity and Absolute Simultaneity.* London/New York: Routledge.
Crick, Francis and Koch, Christof (1998). Consciousness and Neuroscience. *Cerebral Cortex* 8(2), 97 – 107.
D'Espagnat, Bernard (2006). *On Physics and Philosophy.* Princeton: Princeton University Press.

Dainton, Barry (2017). Temporal Consciousness. Retrieved from https://plato.stanford.edu/entries/consciousness-temporal/empirical-findings.html, visited on 5 January 2018.
Damasio, Antonio R. (1994). *Descartes' Error: Emotion, Reason, and the Human Brain.* London: Penguin.
Damasio, Antonio R. (1999). *The Feeling of What Happens: Body and Emotion in the Making of Consciousness.* New York: Harcourt Brace.
Damasio, Antonio R. (2004). Wie das Gehirn Geist erzeugt. *Spektrum der Wissenschaft (Digest: Rätsel Gehirn)* 4, 6–11.
Damasio, Antonio R. (2010). *Self Comes to Mind: Constructing the Conscious Brain.* New York: Pantheon.
Davidson, Donald (1963). Actions, Reasons and Causes. *The Journal of Philosophy* 60(23), 685–700.
Davis, Tamara and Lineweaver, Charley (2003). Expanding Confusion: Common Misconceptions of Cosmological Horizons and the Superluminal Expansion of the Universe. *Publications of the Astronomical Society of Australia* 21. DOI: 10.1071/AS03040, visited on 5 February 2020.
Dawkins, Richard (1976). *The Selfish Gene.* Oxford: Oxford University Press.
De Finetti, Bruno (1974). *Theory of Probability. A Critical Introductory Treatment.* London/New York: Wiley.
Deacon, Terrence (2006). Emergence: The Hole at the Wheel's Hub. In P. Clayton and P. C. W. Davies (Eds.), *The Re-Emergence of Emergence: The Emergentist Hypothesis From Science to Religion* (pp. 111–150). New York: Oxford University Press.
Dehaene, S. and Naccache, L. (2001). Towards a Cognitive Neuroscience of Consciousness: Basic Evidence and a Workspace Framework. *Cognition* 79(1–2), 1–37.
Dennett, Daniel C. (1991). *Consciousness Explained.* Boston: Little, Brown and Co.
Dennett, Daniel C. (1995). *Darwin's Dangerous Idea: Evolution and the Meanings of Life.* New York: Simon & Schuster.
Descartes, René (2008). *Meditations on First Philosophy: With Selections from the Objections and Replies.* Oxford: Oxford University Press.
Dowe, Phil (2000). *Physical Causation.* Cambridge: Cambridge University Press.
Dretske, Fred (2010). Triggering and Structuring Causes. In T. O'Connor and C. Sandis (Eds.), *A Companion to the Philosophy of Action* (pp. 139–144). Malden, MA: Wiley-Blackwell.
Dürr, Detlef, Goldstein, Sheldon, and Zanghì, Nino (2013). *Quantum Physics without Quantum Philosophy.* Heidelberg/New York: Springer.
Eagleman, David (2009). Brain Time. *Edge.* Retrieved from https://www.edge.org/conversation/brain-time, visited on 23 November 2017.
Eagleman, David and Sejnowskij, Terrence J. (2000). Motion Integration and Postdiction in Visual Awareness. *Science* 287(5460), 2036–2038.
Earman, John (1986). *A Primer on Determinism.* Boston, MA: Reidel.
Eia, Harald (2020a). Rik og lik [Television series episode, 12 Feb.]. In Eia, Harald (program host), *Sånn er Norge:* NRK.
Eia, Harald (2020b). Tillit [Television series episode, 26 Feb.]. In Eia, Harald (program host), *Sånn er Norge:* NRK.
Eikrem, Asle (2013). *Being in Religion: A Journey in Ontology from Pragmatics through Hermeneutics to Metaphysics.* Tübingen: Mohr Siebeck.

Einstein, A., Podolsky, B., and Rosen, N. (1935). Can Quantum-Mechanical Description of Physical Reality Be Considered Complete? *Physical Review* 47(10), 777 – 780. DOI: 10.1103/PhysRev.47.777, visited on 3 August 2019.

Einstein, Albert (1905). On the Electrodynamics of Moving Bodies. *Annalen der Physik* 17, 891 – 921.

Elga, Adam (2000). Self-Locating Belief and the Sleeping Beauty Problem. *Analysis* 60(2), 143 – 147.

Emmons, R. A. and McCullough, M. E. (2003). Counting Blessings versus Burdens: An Experimental Investigation of Gratitude and Subjective Well-Being in Daily Life. *J Pers Soc Psychol* 84(2), 377 – 389. DOI: 10.1037//0022 – 3514.84.2.377, visited on 11 January 2021.

Esfeld, Michael (2017). A Proposal for a Minimalist Ontology. *Synthese* 197(5), 1 – 17. DOI: 10.1007/s11229 – 017 – 1426 – 8, visited on 9 August 2019.

Esfeld, Michael, Deckert, Dirk-André, Lazarovici, Dustin, Oldofredi, Andrea, and Vassallo, Antonio (2018). *A Minimalist Ontology of the Natural World.* New York: Routledge/Taylor & Francis Group.

Esfeld, Michael, Deckert, Dirk-André, and Oldofredi, Andrea (2015). What Is Matter? The Fundamental Ontology of Atomism and Structural Realism. Retrieved from https://arxiv.org/abs/1510.03719, visited on 9 August 2019.

Esfeld, Michael, Lazarovici, Dustin, Lam, Vincent, and Hubert, Mario (2015). The Physics and Metaphysics of Primitive Stuff. *The British Journal for the Philosophy of Science* 68(1), 133 – 161. DOI: 10.1093/bjps/axv026, visited on 9 August 2019.

Evans, Gareth (1973). The Causal Theory of Names. *Proceedings of the Aristotelian Society, Supplementary Volumes* 47, 187 – 225.

Everett, Hugh, Barrett, Jeffrey Alan, Byrne, Peter, and Everett, Hugh (2012). *The Everett Interpretation of Quantum Mechanics: Collected Works 1955 – 1980 with Commentary.* Princeton: Princeton University Press.

Fair, David (1979). Causation and the Flow of Energy. *Erkenntnis* 14(3), 219 – 250.

Fernald, Russell D. (2001). The Evolution of Eyes. *Karger Gazette* 64, 2 – 4.

Feynman, Richard (1964). Electromagnetism. Retrieved from https://www.feynmanlectures.caltech.edu/II_01.html#Ch1-S2, visited on 29 October 2021.

Fischer, John Martin and Ravizza, Mark (1998). *Responsibility and Control: A Theory of Moral Responsibility.* Cambridge: Cambridge University Press.

Floridi, Luciano (2014). Artificial Agents and Their Moral Nature. In P. Kroes and P.-P. Verbeek (Eds.), *The Moral Status of Technical Artefacts* (pp. 185 – 212). Dordrecht: Springer Netherlands.

Fodor, Jerry A. (1984). Semantics, Wisconsin Style. *Synthese* 59, 231 – 250.

Foot, Philippa (1967). The Problem of Abortion and the Doctrine of the Double Effect. *Oxford Review* 5, 5 – 15.

Foster, John (2000). *The Nature of Perception.* Oxford/New York: Oxford University Press.

Frankfurt, Harry (1971). Freedom of the Will and the Concept of a Person. *Journal of Philosophy* 68, 5 – 20.

Frege, Gottlob (1948). Sence and Reference. *Philosophical Review* 57(3), 209 – 230.

Frege, Gottlob (1956). The Thought: A Logical Inquiry. *Mind* 65(259), 289 – 311.

Fries, Pascal (2005). A Mechanism for Cognitive Dynamics: Neuronal Communication through Neuronal Coherence. *TRENDS in Cognitive Sciences* 9(10), 474 – 480.

Gallagher, Shaun (2000). Philosophical Conceptions of the Self: Implications for Cognitive Science. *TRENDS in Cognitive Sciences* 4(1), 14 – 21.
Gaus, Gerald F. (2016). *The Tyranny of the Ideal: Justice in a Diverse Society.* Princeton: Princeton University Press.
Gazzaniga, Michael S. (2005). *The Ethical Brain.* New York: Dana Press.
Gazzaniga, Michael S. (2009a). The Distributed Networks of Mind (Gifford Lecture 2). *The Gifford Lectures.* Retrieved from http://www.youtube.com/watch?v=q1fq4qTPSdg, visited on 25 April 2011.
Gazzaniga, Michael S. (2009b). The Interpreter (Gifford Lecture 3). *The Gifford Lectures.* Retrieved from http://www.youtube.com/watch?v=mJKloz2vwlc, visited on 25 April 2011.
Gazzaniga, Michael S. (Producer) (2009c). What We Are (Gifford Lecture 1). *The Gifford Lectures.* Retrieved from http://www.youtube.com/watch?v=dadT-14FkSY, visited on 25 April 2011.
Gertz, Nolen (2018). Hegel, the Struggle for Recognition, and Robots. *Techné: Research in Philosophy and Technology* 22, 138 – 157.
Geyer, Christian (2004). *Hirnforschung und Willensfreihet. Zur Deutung der neuesten Experimente.* Frankfurt am Main: Suhrkamp.
Gieler, Uwe and Walter, Bertram (2008). Scratch this! *Scientific American MIND* 19(3), 52 – 59.
Giere, Ronald N. (2011). Objective Single Case Probabilities and the Foundations of Statistics. In A. Eagle (Ed.), *Philosophy of Probability* (pp. 498 – 510). New York: Routledge.
Ginet, Carl (1990). *On Action.* Cambridge: Cambridge University Press.
Glover, Jonathan (1999). *Humanity: A Moral History of the Twentieth Century.* London: J. Cape.
Gravem, Peder (1996). Meningserfaring og livstolkning. *Prismet* 47(6), 242 – 254.
Gray, Jeffrey (2004). Mit den Ohren sehen. *Spektrum der Wissenschaft (Spezial: Bewusstsein)* 1, 62 – 69.
Hadjikhani, Nouchine, Liu, Arthur K., Dale, Anders M., Cavanagh, Patrick, and Tootell, Roger B. H. (1998). Retinotopy and Color Sensitivity in Human Visual Cortical Area V8. *Nature Neuroscience* 1(3), 235 – 241.
Hájek, Alan (2012). Intrerpretations of Probability. Retrieved from http://plato.stanford.edu/entries/probability-interpret/, visited on 28 October 2015.
Hall, Lars, Strandberg, Thomas, Pärnamets, Philip, Lind, Andreas, Tärning, Betty, and Johansson, Petter (2013). How the Polls Can Be Both Spot On and Dead Wrong: Using Choice Blindness to Shift Political Attitudes and Voter Intentions. *PLOS ONE*, April. DOI: 10.1371/journal.pone.0060554, visited on 12 October 2016.
Hammond, Peter (1991). Interpersonal Comparisons of Utility: Why and How They Are and Should Be Made. DOI: 10.1017/CBO9781139172387.008, visited on 11 Jan 2021.
Harari, Yuval N. (2017). *Homo Deus: A Brief History of Tomorrow.* New York: Harper.
Hare, Richard M. (1952). *The Language of Morals.* Oxford: Clarendon Press.
Harnik, Roni, Kribs, Graham D., and Perez, Gilad (2006). A Universe without Weak Interactions. *Physical Review D* 74(3), 035006. DOI: 10.1103/PhysRevD.74.035006, visited on 9 August 2019.
Hartshorne, Charles (1934). *The Philosophy and Psychology of Sensation.* Chicago, IL: The University of Chicago Press.
Hausman, Daniel M. (2012). *Preference, Value, Choice, and Welfare.* Cambridge: Cambridge University Press.

Haynes, John-Dylan, Sakai, Katsuyuki, Rees, Geraint, Gilbert, Sam, Frith, Chris, and Passingham, Richard E. (2007). Reading Hidden Intentions in the Human Brain. *Current Biology* 17(4), 323–328.
Heathcote, Adrian (1989). A Theory of Causality: Causality=Interaction (as Defined by a Suitable Quantum Field Theory). *Erkenntnis* 31, 77–108.
Heathcote, Adrian and Armstrong, D. M. (1991). Causes and Laws. *NOÛS* 25(1), 63–73.
Hempel, Carl G. (1983). On the Nature of Mathematical Truth. In P. Benacerraf and H. Putnam (Eds.), *Philosophy of Mathematics* (pp. 377–393). Cambridge: Cambridge University Press.
Hitchcock, Christopher (2001). Probability and Chance. In N. J. Smelser and P. B. Baltes (Eds.), *International Encyclopedia of the Social & Behavioral Sciences* (pp. 12089–12095). Amsterdam/New York: Elsevier.
Hitchcock, Christopher (2004). Beauty and the Bets. *Synthese* 139, 405–420.
Hitchcock, Christopher (2010). Probabilistic Causation. Retrieved from http://plato.stanford.edu/archives/fall2010/entries/causation-probabilistic/, visited on 3 November 2010.
Hobson, Art (2013). There Are No Particles, There Are Only Fields. *American Journal of Physics* 81(3), 211–223.
Hoffman, David (2015). Do We See Reality As It Is? Retrieved from https://www.ted.com/talks/donald_hoffman_do_we_see_reality_as_it_is, visited on 23 August 2017.
Hofstadter, Douglas R. (2007). *I Am a Strange Loop*. New York: Basic Books.
Horgan, Terence and Timmons, Mark (1991). New Wave Moral Realism Meets Moral Twin Earth. *Journal of Philosophical Research* 16, 447–465.
Howson, Colin and Uhrbach, Peter (2011). Bayesian versus Non-Bayesian Approaches to Confirmation. In A. Eagle (Ed.), *Philosophy of Probability* (pp. 222–249). New York: Routledge.
Hume, David, Selby-Bigge, L. A., and Nidditch, P. H. (1978). *A Treatise of Human Nature* (2nd ed.). Oxford: Oxford University Press.
Humphreys, Paul (1985). Why Propensities Cannot be Probabilities. *Philosophical Review* 94(4), 557–570.
Imbert, Michel (2004). Sehen ohne zu wissen. *Spektrum der Wissenschaft (Spezial: Bewusstsein)* 1, 37–43.
Inwagen, Peter van (2003). Existence, Ontological Commitment, and Fictional Entities. In M. J. Loux and D. W. Zimmerman (Eds.), *The Oxford Handbook of Metaphysics* (pp. 131–157). New York: Oxford University Press.
James, William (1890). *The Principles of Psychology*. New York: H. Holt.
Johansson, Petter, Hall, Lars, Sikström, Sverker, and Olsson, Andreas (2005). Failure to Detect Mismatches between Intention and Outcome in a Simple Decision Task. *Science* 310, 116–119.
Joseph, Rhawn (1996). *Neuropsychiatry, Neuropsychology, and Clinical Neuroscience: Emotion, Evolution, Cognition, Language, Memory, Brain Damage, and Abnormal Behavior* (2nd ed.). Baltimore: Williams & Wilkins.
Joseph, Rhawn (2006). Brain Mind Lecture 4 Parietal Lobes Body Image Phantom Limbs (sic). Retrieved from https://integral-options.blogspot.com/2009/01/brain-mind-lecture-4-parietal-lobes.html, visited on 27 April 2011.
Kahneman, Daniel (2011). *Thinking, Fast and Slow*. New York: Penguin.
Kane, Robert (1996). *The Significance of Free Will*. New York: Oxford University Press.

Kane, Robert (2007). Libertarianism. In J. M. Fischer, R. Kane, D. Pereboom, and M. Vargas (Eds.), *Four Views on Free Will* (pp. 5 – 43). Oxford: Blackwell.

Kane, Robert (2011). *The Oxford Handbook of Free Will* (2nd ed.). Oxford: Oxford University Press.

Kant, Immanuel (1976). *Critique of Practical Reason, and other Writings in Moral Philosophy* (L. W. Beck, Trans. and Ed.). New York: Garland.

Kaufman, Lloyd and Rock, Irvin (1962). The Moon Illusion. *Scientific American* 207(1), 120 – 132.

Kemmerer, David L. (2014). *Cognitive Neuroscience of Language*. New York: Psychology Press.

Kim, Jaegwon (1998). *Mind in a Physical World: An Essay on the Mind-Body Problem and Mental Causation*. Cambridge, MA: MIT Press.

Kim, Jaegwon (2006). *Philosophy of Mind* (2nd ed.). Boulder, CO: Westview.

Koch, Christof (2011). Being John Malkovich: Personal Control of Individual Brain Cells. *Scientific American MIND* 22(1), 18 – 19.

Koch, Christof and Tononi, Giulio (Producer) (2014). Christof Koch and Giulio Tononi on Consciousness at the FQXi conference 2014 in Vieques. Retrieved from https://www.youtube.com/watch?v=1cO4R_H4Kww, visited on 23 August 2017.

Kolmogorov, A. N. (1950). *Foundations of the Theory of Probability*. New York: Chelsea.

Korsgaard, Christine (1983). Two Distinctions in Goodness. *The Philosophical Review* 92(2), 169 – 195.

Kripke, Saul (1977). Speaker's Reference and Semantic Reference. *Midwest Studies in Philosophy* 2(1), 255 – 276.

Kripke, Saul (1980). *Naming and Necessity*. Cambridge, MA: Harvard University Press.

Kripke, Saul (1982). *Wittgenstein on Rules and Private Language: An Elementary Exposition*. Cambridge, MA: Harvard University Press.

Ladyman, James (2002). *Understanding Philosophy of Science*. London/New York: Routledge.

Ladyman, James, Ross, Don, Spurrett, David, and Collier, John G. (2007). *Every Thing Must Go: Metaphysics Naturalized*. Oxford: Oxford University Press.

Laitinen, Arto (2002). Interpersonal Recognition: A Response to Value or a Precondition of Personhood? *Inquiry – an Interdisciplinary Journal of Philosophy* 45, 463 – 478. DOI: 10.1080/002017402320947559, visited on 4 February 2021.

Lange, Marc (2000). *Natural Laws in Scientific Practice*. Oxford: Oxford University Press.

Lange, Marc (2002). *An Introduction to the Philosophy of Physics: Locality, Fields, Energy, and Mass*. Oxford: Blackwell.

Le Poidevin, Robin (2015). The Experience and Perception of Time. Retrieved from https://plato.stanford.edu/entries/time-experience/, visited on 23 November 2017.

Lee, Andrew Y. (2020). Does Sentience Come in Degrees? *Animal Sentience* 29(20).

Leibniz, Gottfried Wilhelm and Strickland, Lloyd (2014). *Leibniz's Monadology: A New Translation and Guide*. Edinburgh: Edinburgh University Press.

Leopold, David A. and Logothetis, Nikos K. (1996). Activity Changes in Early Visual Cortex Reflect Monkeys' Percepts during Binocular Rivalry. *Nature* 379(6565), 549 – 553. DOI: 10.1038/379549a0, visited on 25 March 2021.

Levy, Neil (2011). *Hard Luck: How Luck Undermines Free Will and Moral Responsibility*. Oxford: Oxford University Press.

Lewis, C. S. (1947). *Miracles: A Preliminary Study*. London: Bles.

Lewis, David (1966). An Argument for the Identity Theory. *Journal of Philosophy* 63(1), 17 – 25.

Lewis, David (1973a). Causation. *Journal of Philosophy* 70(17), 556–567.
Lewis, David (1973b). *Counterfactuals*. Cambridge: Harvard University Press.
Lewis, David (1979). Counterfactual Dependence and Time's Arrow. *NOÛS* 13, 455–476.
Lewis, David (1983). *Philosophical Papers* (Vol. 1). New York: Oxford University Press.
Lewis, David (1986a). *On the Plurality of Worlds*. Oxford/New York: Blackwell.
Lewis, David (1986b). *Philosophical Papers* (Vol. 2). New York: Oxford University Press.
Lewis, David (1994). Humean Supervenience Debugged. *Mind* 103(412), 473–490.
Lewis, David (2000). Causation as Influence. *Journal of Philosophy* 97(4), 182–197.
Lewis, David (2001). Sleeping Beauty: Reply to Elga. *Analysis* 61(3), 171–176.
Lewis, David (2011). A Subjectivist's Guide to Objective Chance. In A. Eagle (Ed.), *Philosophy of Probability* (pp. 458–487). New York: Routledge.
Libet, Benjamin, Freeman, Anthony, and Sutherland, Keith (1999). *The Volitional Brain: Towards a Neuroscience of Free Will*. Thorverton: Imprint Academic.
Lincoln, Tracey A. and Joyce, Gerald F. (2009). Self-Sustained Replication of an RNA Enzyme. *Science* 323(5918), 1229–1232.
Linnebo, Øystein (2017). *Philosophy of Mathematics*. Princeton, NJ: Princeton University Press.
Linnebo, Øystein (2018). *Thin Objects: An Abstractionist Account*. Oxford: Oxford University Press.
Lockwood, Michael (1989). *Mind, Brain, and the Quantum: The Compound 'I'*. New York: Blackwell.
Loria, Kevin (2015). No One Could See the Colour Blue Until Modern Times. *Business Insider*. Retrieved from https://www.businessinsider.com.au/what-is-blue-and-how-do-we-see-color-2015-2, visited on 27 September 2021.
Lowe, E. J. (2000). *An Introduction to the Philosophy of Mind*. Cambridge/New York: Cambridge University Press.
Lowe, E. J. (2002). *A Survey of Metaphysics*. Oxford: Oxford University Press.
Lutz, Matthew and Lenman, James (2018). Moral Naturalism. Retrieved from https://plato.stanford.edu/archives/fall2018/entries/naturalism-moral, visited on 3 August 2018.
Lyon, Aidan (2010). Philosophy of Probabability. In F. Allhoff (Ed.), *Philosophies of the Sciences: A Guide* (pp. 92–126). Chichester: Wiley-Blackwell.
Lyons, David (1965). *Forms and Limits of Utilitarianism*. Oxford: Clarendon Press.
Mackie, J. L. (1977). *Ethics: Inventing Right and Wrong*. New York: Penguin.
Mandik, Pete (2001). Mental Representation and the Subjectivity of Consciousness. *Philosophical Psychology* 14(2), 179–202.
Mani, Anandi, Mullainathan, Sendhil, Shafir, Eldar, and Zhao, Jiaying (2013). Poverty Impedes Cognitive Function. *Science* 341(6149), 976–980. DOI: 10.1126/science.1238041, visited on 17 January 2020.
Markowitsch, Hans J. (2004). Neuropsychologie des menschlichen Gedächtnisses. *Spektrum der Wissenschaft (Digest: Rätsel Gehirn)* 4, 52–61.
Matthews, Gary G. (2000). *Introduction to Neuroscience*. Malden, MA: Blackwell Science.
Maudlin, Tim (1995a). Three Measurement Problems. *Topoi* 14(1), 7–15.
Maudlin, Tim (1995b). Why Bohm's Theory Solves the Measurement Problem. *Philosophy of Science* 62(3), 479–483.

Maudlin, Tim (2003). Distilling Metaphysics from Quantum Physics. In M. J. Loux and D. W. Zimmerman (Eds.), *The Oxford Handbook of Metaphysics* (pp. 461–487). New York: Oxford University Press.
Maudlin, Tim (2009). *The Metaphysics within Physics*. Oxford: Oxford University Press.
Maudlin, Tim (2012). *Philosophy of Physics: Space and Time*. Princeton, NJ: Princeton University Press.
Maudlin, Tim (2019). *Philosophy of Physics: Quantum Theory*. Princeton, NJ: Princeton University Press.
McDowell, John Henry (1996). *Mind and World: With a New Introduction*. Cambridge, MA: Harvard University Press.
McKenna, Michael (2011). *Conversation and Responsibility*. New York: Oxford University Press.
McMullin, Ernan (2002). The Origins of the Field Concept in Physics. *Physics in Perspective* 4(1), 13–39.
McQueen, Kelvin J. (2015). Philosophy of Quantum Mechanics. Retrieved from https://kelvinmcqueen.files.wordpress.com/2015/08/l2.pdf, visited on 27 September 2021.
McQueen, Kelvin J. (2019). Interpretation-Neutral Integrated Information Theory. *Journal of Consciousness Studies* 26(1–2), 76–106.
McTaggart, John Ellis (1908). The Unreality of Time. *Mind* 17(4), 457–474.
Mele, Alfred R. (1987). *Irrationality: An Essay on Akrasia, Self-Deception, and Self-Control*. New York: Oxford University Press.
Mele, Alfred R. (1992). *Springs of Action: Understanding Intentional Behavior*. New York: Oxford University Press.
Mele, Alfred R. (1995). *Autonomous Agents: From Self-Control to Autonomy*. New York: Oxford University Press.
Mele, Alfred R. (2003). *Motivation and Agency*. New York: Oxford University Press.
Mele, Alfred R. (2006). *Free Will and Luck*. New York: Oxford University Press.
Midttun, Atle, Witoszek, Nina, Joly, Carlos, Karlsson-Vinkhuyzen, Sylvia, Olsen, Per, Olsson, Lennart, ... Østergård, U. (2011). The Nordic Model: Is It Sustainable and Exportable? Retrieved from https://www.academia.edu/25476815/The_Nordic_Model_is_it_sustainable_and_exportable, visited on 3. December 2021.
Minkowski, H. (1952). Space and Time. In W. Perrett and G. B. Jeffery (Eds.), *The Principle of Relativity* (pp. 73–91). New York: Dover.
Mischel, W. and Shoda, Y. (1995). A Cognitive-Affective System Theory of Personality: Reconceptualizing Situations, Dispositions, Dynamics, and Invariance in Personality Structure. *Psychological Review* 102(2), 246–268.
Moen, Ole Martin (2012). *Because It Feels Good: A Hedonistic Theory of Intrinsic Value* (PhD). University of Oslo, Oslo.
Moore, G. E. (1903). *Principia Ethica*. Cambridge: Cambridge University Press.
Mozersky, M. Joshua (2011). Presentism. In C. Callender (Ed.), *The Oxford Handbook of Philosophy of Time* (pp. 122–144). Oxford: Oxford University Press.
Muldoon, Ryan (2016). *Social Contract Theory for a Diverse World: Beyond Tolerance*. New York: Routledge.
Mullainathan, Sendhil and Shafir, Eldar (2013). *Scarcity: Why Having Too Little Means So Much*. New York: Times Books, Henry Holt.
Mumford, Stephen (2005). Laws and Lawlessness. *Synthese* 144(3), 397–413.
Nagel, Thomas (1974). What Is It Like to Be a Bat? *The Philosophical Review* 83(4), 435–450.

Narvaez, Darcia (2013). Neurobiology and Moral Mindset. In K. Heinrichs, F. Oser, & T. Lovat (Eds.), *Handbook of Moral Motivation. Theories, Models, Applications* (Vol. 1, pp. 323–340). Rotterdam: SensePublishers.

Nave, Rod (2016). More Detail on Kinetic Energy Concept. Retrieved from http://hyperphysics.phy-astr.gsu.edu/hbase/ke.html, visited on 27 September 2021.

Ney, Alyssa and Albert, David Z. (2013). *The Wave Function: Essays on the Metaphysics of Quantum Mechanics.* Oxford: Oxford University Press.

Nilsson, Nils J. (1998). *Artificial Intelligence: A New Synthesis.* San Francisco, CA: Morgan Kaufmann.

Nys, Thomas (2004). Re-Sourcing the Self? Isaiah Berlin and Charles Taylor and the Tension Between Freedom and Autonomy. *Ethical Perspectives* 11(4), 215–227.

O'Connor, Timothy (2011). Agent-Causal Theories of Freedom. In R. Kane (Ed.), *The Oxford Handbook of Free Will* (2nd ed., pp. 309–328). Oxford: Oxford University Press.

O'Connor, Timothy (2020). Emergent Properties. Retrieved from https://plato.stanford.edu/archives/fall2020/entries/properties-emergent/, visited on 2 October 2021.

Ogborn, Jon and Taylor, Edwin F. (2005). Quantum Physics Explains Newton's Laws of Motion. *Physics Education* 40(1), 26–34.

Oizumi, M., Albantakis, L., and Tononi, G. (2014). From the Phenomenology to the Mechanisms of Consciousness: Integrated Information Theory 3.0. *PLoS Comput Biol* 10(5), e103588.

Page, Don N. (2007). Susskind's Challenge to the Hartle-Hawking no-boundary proposal and possible resolutions. *Journal of Cosmology and Astroparticle Physics*, Jan. 2007. DOI: 10.1088/1475–7516/2007/01/004, visited on 5 February 2020.

Papineau, David (2009). The Causal Closure of the Physical and Naturalism. In B. P. McLaughlin, S. Walter, and A. Beckermann (Eds.), *The Oxford Handbook of Philosophy of Mind* (pp. 53–65). Oxford: Clarendon Press.

Parfit, Derek (1984). *Reasons and Persons.* Oxford: Clarendon Press.

Parfit, Derek (2011a). *On What Matters* (Vol. 1). Oxford/New York: Oxford University Press.

Parfit, Derek (2011b). *On What Matters* (Vol. 2). Oxford/New York: Oxford University Press.

Parfit, Derek (2012). Another Defence of the Priority View. *Utilitas* 24(3), 399–440.

Parrott, W. Gerrod (2001). *Emotions in Social Psychology: Essential Readings.* Philadelphia, PA: Psychology Press.

Pearl, Judea and Mackenzie, Dana (2019). *The Book of Why: The New Science of Cause and Effect.* London: Penguin.

Penrose, Roger (1989). *The Emperor's New Mind: Concerning Computers, Minds, and the Laws of Physics.* New York: Oxford University Press.

Penrose, Roger (2004). *The Road to Reality: A Complete Guide to the Laws of the Universe.* London: Jonathan Cape.

Pereboom, Derk (2007). Hard Incompatibilism. In J. M. Fischer, R. Kane, D. Pereboom, and M. Vargas (Eds.), *Four Views on Free Will* (pp. 85–125). Oxford: Blackwell.

Pereboom, Derk (2014). *Free Will, Agency, and Meaning in Life.* New York: Oxford University Press.

Peressini, Anthony F. (2018). There Is Nothing It Is Like to See Red: Holism and Subjective Experience. *Synthese* 195(10), 4637–4666. DOI: 10.1007/s11229–017–1425–9, visited on 23 January 2020.

Pettit, Philip and McDowell, John Henry (1986). *Subject, Thought, and Context.* Oxford: Oxford University Press.

Pinker, Steven (1997). *How the Mind Works.* New York: Norton.

Plantinga, Alvin (1993). *Warrant and Proper Function.* New York: Oxford University Press.

Plato (1970). *Plato's 'Euthyphro' and Earlier Theory of Forms* (R. E. Allen, Trans.). London: Routledge & K. Paul.

Popper, Karl R. (1945). *The Open Society and Its Enemies.* London: G. Routledge & Sons.

Popper, Karl R. and Eccles, John C. (1977). *The Self and Its Brain.* New York: Springer International.

Puntel, Lorenz B. (1990). *Grundlagen einer Theorie der Wahrheit.* Berlin: De Gruyter.

Puntel, Lorenz B. (2008). *Structure and Being: A Theoretical Framework for a Systematic Philosophy* (A. White, Trans.). University Park, PA: Pennsylvania State University Press.

Puntel, Lorenz B. (2014). Is the Question "Why Is There Anything Rather than Nothing?" a Meaningful Question? Retrieved from http://www.metaphysicalsociety.org/2014/Papers/puntel.pdf, visited on 19 December 2016.

Putnam, Hilary (1967). Time and Physical Geometry. *Journal of Philosophy* 64(8), 240–247.

Putnam, Hilary (1975). *Mind, Language, and Reality* (Vol. 2). Cambridge: Cambridge University Press.

Quine, W. V. (1951). Two Dogmas of Empiricism. *The Philosophical Review* 60(1), 20–43.

Ramachandran, V. S. (2007). 3 Clues to Understanding Your Brain. Retrieved from https://www.youtube.com/watch?v=Rl2LwnaUA-k, visited on 23 August 2017.

Ramachandran, V. S. and Hirstein, W. (1999). Three Laws of Qualia: What Neurology Tells Us about the Biological Functions of Consciousness, Qualia and the Self. In S. Gallagher and J. Shear (Eds.), *Models of the Self* (pp. 83–112). Thorverton, UK: Imprint Academic.

Rawls, John (1971). *A Theory of Justice.* Cambridge, MA: Belknap Press of Harvard University Press.

Rawls, John (1999). *The Law of Peoples.* Cambridge, MA: Harvard University Press.

Rebka, G.A. jr. and Pound, R. V. (1960). Variation with Temperature of the Energy of Recoil-Free Gamma Rays from Solids. *Physical Review Letters* 4, 274–275.

Reinertsen, Maria Berg (2019). Hvordan gå fra helvetsvisjon til handling? *Morgenbladet*, 17 May, pp. 4–5.

Reppert, Victor (2003). *C. S. Lewis's Dangerous Idea: A Philosophical Defense of Lewis's Argument from Reason.* Downers Grove, IL: InterVarsity Press.

Rescher, Nicholas (1973). *The Coherence Theory of Truth.* Oxford: Clarendon Press.

Rescher, Nicholas (1985). *The Strife of Systems: An Essay on the Grounds and Implications of Philosophical Diversity.* Pittsburgh, PA: University of Pittsburgh Press.

Rescher, Nicholas (2001). *Paradoxes: Their Roots, Range, and Resolution.* Chicago: Open Court.

Rescher, Nicholas (2010). *Reality and Its Appearance.* London/New York: Continuum.

Rescher, Nicholas (2018). *Understanding Reality: Metaphysics in Epistemological Perspective.* Lanham, MD: Lexington.

Roberson, Debi, Davidoff, Jules, Davies, Ian R. L., and Shapiro, Laura R. (2006). Colour Categories and Category Acquisition in Himba and English. In Nicola Pitchford and Carole P. Biggam (Eds.), *Progress in Colour Studies. Vol. 2: Psychological Aspects* (pp. 159–172). Amsterdam/Philadelphia: John Benjamins.

Roberts, John T. (2008). *The Law-Governed Universe.* Oxford/New York: Oxford University Press.

Robinson, William (2007). Evolution and Epiphenomenalism. *Journal of Consciousness Studies* 14(11), 27–42.

Rosenberg, Alexander (2011). *The Atheist's Guide to Reality: Enjoying Life without Illusions.* New York: W. W. Norton.

Rosenberg, Alexander (2018). *How History Gets Things Wrong: The Neuroscience of Our Addiction to Stories.* Cambridge, MA: The MIT Press.

Rosenberg, Gregg (2004). *A Place for Consciousness: Probing the Deep Structure of the Natural World.* New York: Oxford University Press.

Rosenthal, David M. (1991). The Independence of Consciousness and Sensory Quality. *Philosophical Issues* 1, 15–36. DOI: 10.2307/1522921, visited on 8 September 2019.

Roskies, Adina (2014). Monkey Decision-Making as a Model System for Human Decision-Making. In A. R. Mele (Ed.), *Surrounding Free Will: Philosophy, Psychology, Neuroscience* (pp. 231–254). New York: Oxford University Press.

Rowlands, Mark (2003). *Externalism: Putting Mind and World Back Together Again.* Chesham, Bucks: Acumen.

Ruse, Michael (1986). Evolutionary Ethics: A Phoenix Arisen. *Zygon* 21, 95–112.

Russell, Bertrand (1905). On Denoting. *Mind* 14(56), 479–493.

Russell, Bertrand (2010 [1903]. *Principles of Mathematics.* New York: Routledge.

Russell, Stuart J., Norvig, Peter, and Davis, Ernest (2010). *Artificial Intelligence: A Modern Approach* (3rd ed.). Upper Saddle River: Prentice Hall.

Sampaio, Eliana, Maris, Stéphane, and Bach-Y-Rita, Paul (2001). Brain Plasticity: 'Visual' Acuity of Blind Persons Via the Tongue. *Brain Research* 908, 204–207.

Scanlon, Thomas (1998). *What We Owe to Each Other.* Cambridge, MA: Belknap Press of Harvard University Press.

Schachter, S. and Singer, J. E. (1962). Cognitive, Social, and Physiological Determinants of Emotional State. *Psychological Review* 69(5), 379–399.

Schaffer, Jonathan (2000). Causation by Disconnection. *Philosophy of Science* 67(2), 285–300.

Schaffer, Jonathan (2001). Causes as Probability-Raisers of Processes. *Journal of Philosophy* 98, 75–92.

Schaffer, Jonathan (2005). Contrastive Causation. *Philosophical Review* 114(3), 327–358.

Schaffer, Jonathan (2007). The Metaphysics of Causation. Retrieved from http://plato.stanford.edu/archives/fall2007/entries/causation-metaphysics/, visited on 5 October 2007.

Schmidt, Heinz-Juergen (2019). Structuralism in Physics. Retrieved from https://plato.stanford.edu/archives/win2019/entries/physics-structuralism/, visited on 12 October 2021.

Schmidt, Richard A. (1975). A Schema Theory of Discrete Motor Skill Learning. *Psychological Review* 82(4), 225–260.

Schmidt, Richard A. (2003). Motor Schema Theory After 27 Years: Reflections and Implications for a New Theory. *Research Quarterly for Exercise and Sport* 74(4), 366–375.

Schnall, Simone, Haidt, Jonathan, Clore, Gerald L., and Jordan, Alexander H. (2008). Disgust as Embodied Moral Judgment. *Personality and Social Psychology Bulletin* 34(8), 1096–1109.

Schnell, Tatjana (2021). *The Psychology of Meaning in Life.* London: Routledge.

Schroeder, Timothy, Roskies, Adina L., and Nichols, Shaun (2010). Moral Motivation. In J. Doris (Ed.), *Moral Psychology Handbook* (pp. 72–110). Oxford University Press.
Seager, William and Allen-Hermanson, Sean (2012). Panpsychism. Retrieved from http://plato.stanford.edu/entries/panpsychism/, visited on 28 September 2012.
Searle, John R. (1980). Minds, Brains, and Programs. *Behavioral and Brain Sciences* 3, 417–457.
Searle, John R. (2002). *Consciousness and Language.* New York: Cambridge University Press.
Searle, John R. (2007). Biological Naturalism. In M. Velmans and S. Schneider (Eds.), *The Blackwell Companion to Consciousness* (pp. 325–334). Malden, MA: Blackwell.
Searle, John R. (2011a). The Mystery of Consciousness Continues. *The New York Review of Books.* Retrieved from http://www.nybooks.com/articles/archives/2011/jun/09/mystery-consciousness-continues/?pagination=false, visited on 3 April 2012.
Searle, John R. (2011b). Philosophy of Mind, Lecture 2. Retrieved from http://www.youtube.com/watch?v=c14Z180-gPo, visited on 6 November 2012.
Shermer, Michael (2012). Aunt Millie's Mind. *Scientific American* 307(1), 84.
Shoemaker, Sydney (1969). Time without Change. *The Journal of Philosophy* 66(12), 363–381.
Sider, Theodore (2011). *Writing the Book of the World.* Oxford: Oxford University Press.
Sidgwick, Henry (1907). *The Methods of Ethics* (7th ed.). London: Macmillan.
Siegel, Ethan (2018). This Is Why Physicists Think String Theory Might Be Our 'Theory of Everything'. *Forbes Magazine.* Retrieved from https://www.forbes.com/sites/startswitha bang/2018/05/31/this-is-why-physicists-think-string-theory-might-be-our-theory-of-every thing/#72e07c58c25f, visited on 5 February 2019.
Singer, Peter (2000). *Writings on an Ethical Life.* New York: Ecco Press.
Singer, Peter (2005). Ethics and Intuitions. *The Journal of Ethics* 9, 331–352.
Singer, Wolf (2004a). Ein Spiel von Spiegeln. *Spektrum der Wissenschaft (Spezial: Bewusstsein)* 1, 20–25.
Singer, Wolf (2004b). Verschaltungen legen uns fest: Wir sollten aufhören, von Freiheit zu sprechen. In C. Geyer (Ed.), *Hirnforschung und Willensfreihet. Zur Deutung der neuesten Experimente* (pp. 30–65). Frankfurt am Main: Suhrkamp.
Sklar, Lawrence (1992). *Philosophy of Physics.* Boulder: Westview.
Smolin, Lee (1997). *The Life of the Cosmos.* New York: Oxford University Press.
Soon, Chun Siong, Brass, Marcel, Heinze, Hans-Jochen, and Haynes, John-Dylan (2008). Unconcious Determinants of Free Decisions in the Human Brain. *Nature Neuroscience* 11(5), 543–545.
Søvik, Atle O. (2016). *Free Will, Causality and the Self.* Berlin: De Gruyter.
Søvik, Atle O. (2018). It Is Impossible That There Could Have Been Nothing: New Support for Cosmological Arguments for the Existence of God. *Neue Zeitschrift für Systematische Theologie und Religionsphilosophie* 60(3), 452–463.
Søvik, Atle O. (2019). A Revisionary Theoretical Framework of Responsibility: A Philosophical Exploration of Incapacity for Responsible Behaviour (utilregnelighet). *Bergen Journal of Criminal Law and Criminal Justice* 7(1), 1–26.
Søvik, Atle O. (2020). Two Objections to IIT. *Journal of Consciousness Studies* 27(9–10), 186–201.
Søvik, Atle O. (2021). What Overarching Ethical Principle Should a Superintelligent AI Follow? *AI & SOCIETY.* DOI: 10.1007/s00146-021-01229-6, visited on 29 September 2021.

Søvik, Atle O. (forthcoming-a). How a Non-Conscious Robot Could Be an Agent with Capacity for Morally Responsible Behaviour. Forthcoming in *AI & Ethics*

Søvik, Atle O. (forthcoming-b). What Kinds of Influence Reduces Freedom Negatively?

Sripada, Chandra (2016). Self-Expression: A Deep Self Theory of Moral Responsibility *Philosophical Studies* 173(5), 1203–1232. Retrieved from http://sites.lsa.umich.edu/sripada/philosophy/, visited on 26 January 2015.

Stampe, Dennis (1977). Toward a Causal Theory of Linguistic Representation. In P. French, H. K. Wettstein, and T. E. EUehling (Eds.), *Midwest Studies in Philosophy* (Vol. 2, pp. 42–63). Minneapolis, MN: University of Minnesota Press.

Steinhardt, Paul J. (2011). The Inflation Debate. *Scientific American*, April, 37–43.

Sternberg, Robert J. and Mio, Jeffery Scott (2006). *Cognitive Psychology* (4th ed.). Belmont, CA: Wadsworth.

Stevenson, Charles Leslie (1944). *Ethics and Language*. New Haven: Yale University Press.

Stevenson, Richard J. and Attuquayefio, Tuki (2013). Human Olfactory Consciousness and Cognition: Its Unusual Features May Not Result from Unusual Functions But from Limited Neocortical Processing Resources. *Frontiers in Psychology* 4, 819–819. DOI: 10.3389/fpsyg.2013.00819, visited on 27 September 2021.

Steward, Helen (2012). *A Metaphysics for Freedom*. Oxford: Oxford University Press.

Strassler, Matt (2012). How Did Einstein Do It? *Of Particular Significance*. Retrieved from https://profmattstrassler.com/articles-and-posts/particle-physics-basics/mass-energy-matter-etc/mass-and-energy/how-did-einstein-do-it/, visited on 27 September 2021.

Strawson, Galen (1994). The Impossibility of Moral Responsibility. *Philosophical Studies* 75(1/2), 5–24.

Strawson, Galen (2006). Realistic Monism: Why Physicalism Entails Panpsychism. *Journal of Consciousness Studies* 13(10–11), 3–31.

Tarski, Alfred (1944). The Semantic Conception of Truth and the Foundations of Semantics. *Philosophy and Phenomenological Research* 4(3), 341–376.

Taylor, Charles (1979). What's Wrong with Negative Liberty? In Alan Ryan (Ed.), *The Idea of Freedom. Essays in Honour of Isaiah Berlin* (pp. 175–193). Oxford: Oxford University Press.

Taylor, Charles (1992). The Politics of Recognition. In C. Taylor and A. Gutmann (Eds.), *Multiculturalism and "The Politics of Recognition": An Essay* (pp. 25–74). Princeton, NJ: Princeton University Press.

Taylor, Edwin F. and Wheeler, John Archibald (1992). *Spacetime Physics: Introduction to Special Relativity* (2nd ed.). New York: W. H. Freeman.

Taylor, Edwin F., Wheeler, John Archibald, and Bertschinger, Edmund (Unpublished). *Exploring Black Holes*. Retrieved from https://www.eftaylor.com/exploringblackholes/, visited on 4 December 2021.

Tononi, G. (2015). Integrated Information Theory. *Scholarpedia* 10(1), 4164.

Tononi, G., Boly, M., Massimini, M., and Koch, C. (2016). Integrated Information Theory: From Consciousness to Its Physical Substrate. *Nature Reviews Neuroscience* 17, 450–461.

Treisman, A. (1998). Feature Binding, Attention and Object Perception. *Philosophical Transactions of the Royal Society B: Biological Sciences* 353(1373), 1295–1306.

Unger, Peter K. (2006). *All the Power in the World*. Oxford/New York: Oxford University Press.

Unruh, W. G. (1976). Notes on Black-Hole Evaporation. *Physical Review D* 14(4), 870–892. DOI: 10.1103/PhysRevD.14.870, visited on 12 June 2019.

Vallor, Shannon (2016). *Technology and the Virtues: A Philosophical Guide to a Future Worth Wanting.* New York: Oxford University Press.
Van Fraassen, Bas C. (1989). *Laws and Symmetry.* Oxford/New York: Oxford University Press.
Van Fraassen, Bas C. (2011). Indifference: The Symmetries of Probability. In A. Eagle (Ed.), *Philosophy of Probability* (pp. 296–316). New York: Routledge.
Van Inwagen, Peter (1983). *An Essay on Free Will.* Oxford: Clarendon Press.
Vardanyan, Mihran, Trotta, Roberto, and Silk, Joseph (2011). Applications of Bayesian Model Averaging to the Curvature and Size of the Universe. *Monthly Notices of the Royal Astronomical Society: Letters* 413(1), L91–L95. DOI: 10.1111/j.1745–3933.2011.01040.x visited on 5 May 2020.
Vargas, Manuel (2013). *Building Better Beings: A Theory of Moral Responsibility.* Oxford: Oxford University Press.
Viaud-Delmon, Isabelle and Jouvent, Roland (2004). Zwischen virtuell und real. *Spektrum der Wissenschaft (Spezial: Bewusstsein)* 1, 70–75.
Wandell, Brian (2008). Colour Vision: Cortical Circuitry for Appearance. *Current Biology* 18(6), 250–251.
Weichselgartner, Erich and Sperling, George (1985). Continuous Measurement of Visible Persistence. *Journal of Experimental Psychology: Human Perception and Performance* 11(6), 711–725.
Weinberg, Steven (1989). The Cosmological Constant Problem. *Reviews of Modern Physics* 61(1), 1–23. DOI: 10.1103/RevModPhys.61.1, visited on 13 June 2021.
Weinberg, Steven (1997). What Is an Elementary Particle? *Beamline*, Spring, 17–21.
Weisberg, Michael (2006). Water Is *Not* H20. In D. Baird, E. Scerri, and L. McIntyre (Eds.), *Philosophy of Chemistry: Synthesis of a New Discipline* (pp. 337–345). Doordrecht: Springer.
Wertheimer, Max (1912). Experimentelle Studien über das Sehen von Bewegung. *Zeitschrift für Psychologie* 61(1), 161–265.
White, Alan (2014). *Toward a Philosophical Theory of Everything: Contributions to the Structural-Systematic Philosophy.* New York: Bloomsbury.
White, Benjamin W., Saunders, Frank A., Scadden, Lawrence, Bach-Y-Rita, Paul, and Collins, Carter C. (1970). Seeing with the Skin. *Perception and Psychophysics* 7(1), 23–27.
Williamson, Jon (2009). Probabilistic Theories. In H. Beebee, C. Hitchcock, and P. Menzies (Eds.), *The Oxford Handbook of Causation* (pp. 185–212). Oxford: Oxford University Press.
Wolf, Susan R. (1990). *Freedom within Reason.* New York: Oxford University Press.
Wolfe, Joe (1992). Pitch, Loudness and Timbre. *PhysClips.* Retrieved from http://www.animations.physics.unsw.edu.au/jw/sound-pitch-loudness-timbre.htm, visited on 28 September 2021.
Woodward, James (2009). Scientific Explanation. Retrieved from http://plato.stanford.edu/archives/spr2010/entries/scientific-explanation/, visited on 9 November 2010.
Wright, G. H. von (2004). *Explanation and Understanding.* Ithaca, NY: Cornell University Press.
Zihl, J., von Cramon, D., and Mai, N. (1983). Selective Disturbance of Movement Vision after Bilateral Brain Damage. *Brain* 106, 313–340.

Index of Names

https://doi.org/10.1515/9783110770957-020

Index of Subjects

https://doi.org/10.1515/9783110770957-021